ITALIAN PHYSICAL SOCIETY

PROCEEDINGS

OF THE

INTERNATIONAL SCHOOL OF PHYSICS

«ENRICO FERMI»

COURSE CXXVI

edited by V. DEGIORGIO and C. FLYTZANIS
Directors of the Course
VARENNA ON LAKE COMO
VILLA MONASTERO
20 - 30 July 1993

Nonlinear Optical Materials: Principles and Applications

1995

IOS
Press

Ohmsha

AMSTERDAM, OXFORD, TOKYO, WASHINGTON DC

ISBN 90 5199 204 1 (IOS Press)
ISBN 4 274 90029 0 C3042 (Ohmsha)
LCCC 94 - 073355

Technical Editor
P. PAPALI

Publisher
IOS Press
Van Diemenstraat 94
1013 CN Amsterdam
Netherlands

Distributor in the UK and Ireland
IOS Press/Lavis Marketing
73 Lime Walk
Headington
Oxford OX3 7AD
England

Distributor in the USA and Canada
IOS Press, Inc.
P.O. Box 10558
Burke, VA 22009-0558
USA

Distributor in Japan
Ohmsha, Ltd.
3-1 Kanda Nishiki-Cho
Chiyoda-Ku
Tokyo 101
Japan

Proprietà Letteraria Riservata
Printed in Italy

INDICE

GENERAL ASPECTS

C. Flytzanis – Nonlinear polarization.

G. P. Banfi – Nonlinear propagation in homogeneous media.

MATERIALS FOR QUADRATIC EFFECTS

C. L. Tang – Optical parametric processes and inorganic nonlinear optical crystals.

P. J. HALFPENNY, J. N. SHERWOOD and G. S. SIMPSON – The growth
and perfection of crystals of nonlinear optical materials.

G. R. MÖHLMANN and C. P. J. M. VAN DER VORST – Nonlinear optical
side-chain polymers for electro-optic switching: materials, poling,
test devices.

S. MIYATA, T. WATANABE and H. SINGH NALWA – Molecular design of nonlinear optical molecules and polymers by lambda (Λ)-type conformation.

MATERIALS FOR CUBIC EFFECTS

F. Capasso and C. Sirtori – Heterostructure engineering of nonlinear optical properties and continuum states.

C. Bubeck – Nonlinearities of conjugated polymers and dye systems.

T. Wada, M. Hosoda, H. Okawa, A. Terasaki, M. Hara and H. Sasabe – Molecular design of third-order nonlinear optically active π-conjugated compounds and preparation of thin films.

SPECIAL TOPICS

D. S. CHEMLA – Instantaneous amplitude and frequency dynamics of coherent wave mixing in semiconductor quantum wells.

D. S. CHEMLA – Nonlinear optical response of semiconductor quantum wells under high magnetic fields.

S. HUGONNARD-BRUYÈRE, C. BUSS, R. FREY and C. FLYTZANIS – Nonlinear magneto-optical materials: photo-induced gyrotropy.

Introduction.

Nonlinear optical materials play a pivotal role in the future evolution of non-linear optics in general and in its impact in technology and industrial applications in particular. The progress in nonlinear optics has been tremendous since the first demonstration of an all-optical nonlinear effect in the early sixties, but until recently the main visible emphasis was on the physical aspects of the non-linear radiation-matter interaction. In the last decade, however, this effort has also brought its fruits in applied aspects of nonlinear optics. This can be essentially traced to the improvement of the performances of the nonlinear optical materials. Our understanding of the nonlinear polarization mechanisms and their relation to the structural characteristics of the materials has been considerably improved, and, in addition, the new development of techniques for the fabrication and growth of artificial materials has dramatically contributed to this evolution.

The goal is to find and develop materials presenting large nonlinearities and satisfying at the same time all the technological requirements for applications such as wide transparency range, fast response, high damage threshold but also processability, adaptability and interfacing with other materials. These additional requirements are intrinsic to the fabrication of nonlinear integrated devices which, besides efficiently performing the expected nonlinear operation, must be miniaturized, compact, reliable and with precisely reproducible characteristics in large-scale low-cost production and long-term operation. A particularly important area is that of optical communications where devices at high speed and low optical power are needed to fully exploit the capabilities of optical-fibre transmission systems that are currently used and will be more so in the future.

For many years nonlinear optical materials have been selected mainly among inorganic crystals, especially ferroelectrics, and related crystals with oxygen polyedra. In recent years, not only new inorganic crystals have been discovered with even better performances, but also complete new families of materials have been studied and developed like the organic crystals, poled polymers, composites and artificial semiconductor microstructures. In the latter cases the electron confinement is being exploited to tailor the performances of the system in order to meet certain requirements. In the case of the organics the new molecular-engineering techniques have brought considerable improvement in their performances. All these improvements, besides rendering possible the implementation of nonlinear effects in devices, open the way to the study of new nonlinear optical effects and the introduction of new concepts.

The aim of the Course was to describe the new concepts which are emerging in the field of nonlinear optical materials, concentrating the attention on materials which seem more promising for applications in the technology of information transmission and processing. The School was attended by about 70 participants coming from all over the world. We are happy to present in this volume the final texts of nearly all the lectures and seminars presented at the Course. It should be mentioned that, during the Course, we had also several presentations of current activities made by the students and a number of informal sessions with very lively discussions.

We wish to thank the scientific secretary of the Course, Prof. G. P. BANFI, for his valuable contribution in all phases of the scientific organization, Mrs. E. MAZZI, from the Italian Physical Society, for her invaluable help before and during the Course, and Siemens S.p.A. for the generous financial support.

V. DEGIORGIO
C. FLYTZANIS

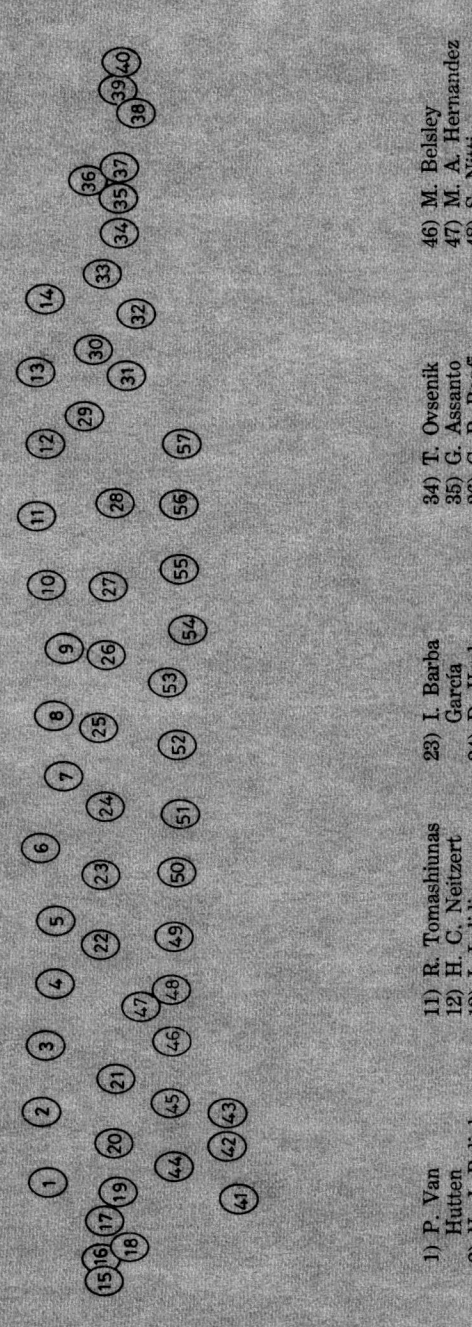

1) P. Van Hutten
2) H. J. Bohnk
3) C. Solcia
4) K. Nemoto
5) M. Murao
6) A. Cavalleri
7) M. A. Diaz García
8) D. Duca
9) V. Fiumara
10) E. Westin

11) R. Tomashiunas
12) H. C. Neitzert
13) I. Lelidis
14) C. Poulsen
15) P. Di Trapani
16) S. Burke
17) E. Mazzà
18) K. Lipinska
19) S. Mallis
20) T. Neidinger
21) L. J. Caballero

23) I. Barba García
24) D. Healy
25) D. Ricard
26) G. Meredith
27) H. J. Winkelhahn
28) U. Baier
29) C. Buss
30) G. Leo
31) C. Castiglioni
32) M. Del Zoppo
33) G. Lanzani

34) T. Ovsenik
35) G. Assanto
36) G. P. Banfi
37) N. Vaupotic
38) Huiming Tan
39) Pan Feng
40) A. Fein
41) S. Garavaglia
42) R. Brigatti
43) P. Mauri
44) C. Bubeck
45) P. Günter

46) M. Belsley
47) M. A. Hernandez
48) S. Nitti
49) S. Myata
50) C. Tang
51) V. Degiorgio
52) C. Flytzaris
53) R. Tommasi
54) C. Adant
55) L. Capogna
56) D. Dos Santos
57) L. Lapao

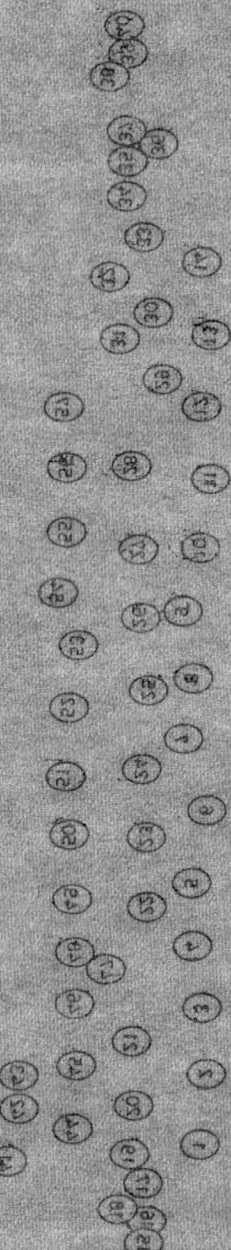

GENERAL ASPECTS

Nonlinear Polarization.

C. FLYTZANIS

Laboratoire d'Optique Quantique, Ecole Polytechnique
91128 Palaiseau Cédex, France

1. – Introduction.

Nonlinear optics is the area of optics that englobes phenomena that occur when the response of a material system to an applied electromagnetic field is nonlinear in the amplitude of the field. The relevant response here is usually the dipolar polarization induced by the field in the material. The description of this polarization and the identification as well as the evaluation of the nonlinear polarization mechanisms constitute major goals in nonlinear optics. These mechanisms mediate a coupling among the field Fourier components which can profoundly modify the spatiotemporal features of the fields and by the same token provide means to control them to an extent that can be of outmost interest for applications in optical information processing, laser technology or diagnostics, to name a few.

These modifications of the spatiotemporal characteristics of the fields result from polarization sources nonlinear in the fields that become relevant as the field amplitudes increase; thus they can be observed with intense coherent light sources, for instance lasers, and it became a tradition now to place the beginning of nonlinear optics with the first observations [1] of the second harmonic and optical rectification of the ruby laser frequency in quartz in 1961. Up to that time the optical phenomena seemed to obey the linear regime and the amplitude of the field did not matter. However, some nonlinear effects had already crept into the area of optics and were extensively studied and used long before the advent of the laser although in a different spirit and context from the present ones. They were namely treated as parametric effects and to some extent are still treated as such and here one has in mind the electro-optic effects, like the Pockels and Kerr ones, or light scattering ones like the Raman, Brillouin and Rayleigh effects, to name a few. All these effects can be actually related to the nonlinear interaction of the charges with the electromagnetic fields and constitute special cases of the area of the nonlinear optics [2,3].

3

That the optical properties of matter cannot remain insensitive to the field amplitude when the latter increases can be anticipated on very simple fundamental grounds. Indeed when this amplitude becomes comparable to the cohesive fields E_c that hold the charges together in the matter, for instance the electron to the proton in the hydrogen atom, the displacements of the charges from their equilibrium positions cannot be treated any longer within the harmonic approximation which constituted the basis of optics with incoherent light sources [4,5]; anharmonic effects must now be included which invariably lead to a nonlinear dependence of the charge displacements on the field amplitude and concomitantly of the induced dipole moments and polarizations. To fix the ideas, the light sources before the invention of the laser could only reach electric-field amplitudes of the order of $(10-10^3)$ V/m in the optical-frequency range, while E_c for the hydrogen is 10^{11} V/m, a field easily attainable nowadays with lasers. Nonlinear polarization, however, does not result only from the anharmonicity in the motion of bound charges; it can also be induced in free charges through the nonlocality of the fundamental Lorentz force [6] that drives a charge in an electromagnetic field, namely through the vector product of the charge velocity and the oscillating magnetic field.

Anharmonicity and nonlocality [7] seem to be the two features that in the last analysis mediate the coupling mechanisms for the field components and induce the nonlinear polarization in the matter. Stated in such a direct and simple picture, the description of the nonlinear polarization, however, becomes intractable and useless in practice. In order to obtain meaningful and useful results, a far more complicated and circumstantial analysis must be used that incorporates our perception and detailed description of the material and the charge organization there. It is only this way that one can gain insight into the relation between the optical nonlinearities and the material characteristics and establish guidelines and rules for investigating them and also identifying appropriate nonlinear optical materials, a *sine qua non* prerequisite for the development of nonlinear optics and its insertion in technology.

We initiate this reduction procedure by starting with the microscopic Maxwell-Lorentz equations that underlie all formal description of electromagnetic effects [6,8]

(1a)
$$\nabla \wedge e = -\frac{1}{c}\frac{\partial b}{\partial t} \,,$$

(1b)
$$\nabla \cdot e = 4\pi\rho \,,$$

(1c)
$$\nabla \wedge b = \frac{1}{c}\frac{\partial e}{\partial t} + \frac{4\pi}{c}j \,,$$

(1d)
$$\nabla \cdot b = 0 \,,$$

(2)
$$\dot{v} = \frac{q}{m}(e + v \wedge b) \,.$$

Here $e(r, t)$ and $b(r, t)$ are the complete microscopic electric and magnetic fields, $\rho(r, t)$ and $j(r, t)$ are the microscopic charge density and current distributions that satisfy the conservation law $\nabla \cdot j + \partial \rho / \partial t = 0$, while v is the velocity of a single charge. Considering that at least 10^{23} charges are crowded in a cubic centimetre, all interacting with each other and with the applied field, the above equations have only a formal utility. For the optical-frequency range we are interested in here this formal complexity can be greatly reduced by smoothing out fluctuations in the fields and distributions that occur within time and space scales, T' and λ' respectively, much shorter than the optical period T and the wavelength λ. Introducing thus the macroscopic fields $E = \langle e \rangle$ and $B = \langle b \rangle$ where $\langle \, \rangle$ stands for

$$(3) \qquad \Phi(r, t) \equiv \langle \varphi(r, t) \rangle = \frac{1}{T' \lambda'^3} \int_{T'} d\tau \int_{\lambda'^3} \varphi(r + \xi, t + \tau) \, d\xi$$

and the electric and magnetic inductions

$$(4a) \qquad D = E + 4\pi \int_{-\infty}^{t} J(t') \, dt' \, ,$$

$$(4b) \qquad H = B - 4\pi M$$

with

$$(4c) \qquad J = \frac{\partial}{\partial t}(P - \nabla \cdot Q) + c(\nabla \wedge M) ,$$

where P, Q and M are the electric-dipole, electric-quadrupole and magnetic-dipole polarization densities, respectively, one recovers the well-known macroscopic Maxwell equations

$$(5a) \qquad \nabla \wedge E = -\frac{1}{c}\frac{\partial H}{\partial t} \, ,$$

$$(5b) \qquad \nabla \cdot D = 0 \, ,$$

$$(5c) \qquad \nabla \wedge B = \frac{1}{c}\frac{\partial D}{\partial t} \, ,$$

$$(5d) \qquad \nabla \cdot B = 0 \, ,$$

where for simplicity we now assumed that there are no extraneous charges and currents and J is the inducent current density. The latter can be related to the fields E and B through the solution of the equation of motion of the charges interacting among themselves and with the applied electromagnetic field. In principle this can be done by solving the relevant Schrödinger equation that incorporates (2), but classical equations [2, 3, 7] can also be used to illustrate some key aspects.

Thus for bound charges, as in dielectrics, with their motion restricted to extensions $a \ll \lambda$ the nonlocality can be neglected in (2) and, representing the global interaction of the charges with an averaged anharmonic potential, one has the forced-damped-anharmonic-oscillator equation

$$(6a) \qquad \ddot{\boldsymbol{r}} + \Gamma\dot{\boldsymbol{r}} + \omega_0^2 \boldsymbol{r} + \beta r^2 + \gamma r^3 = \frac{e}{m} \boldsymbol{E}_l(t),$$

where $\boldsymbol{E}_l$ is the effective field that drives the charges, which besides $\boldsymbol{E}$ also includes the induced mutual interactions between the charges, and Γ is a damping that takes into account complex random interactions with other degrees of freedom (bath).

On the other hand, for free or weakly interacting charges, as in metals or plasmas, the anharmonicity can be neglected and the nonlocality is the only relevant nonlinear mechanism in (2), which then reads

$$(6b) \qquad \dot{\boldsymbol{v}} + \Gamma\boldsymbol{v} = \frac{e}{m}(\boldsymbol{E} + \boldsymbol{v} \wedge \boldsymbol{B}),$$

where now the macroscopic and effective fields coincide and Γ is again a damping. This description can also be used for all materials in the X-ray frequency range.

Even such apparently simple equations as $(6a)$, $(6b)$ cannot be solved exactly and one resorts to perturbative/iterative techniques to obtain the induced electric dipole $e\boldsymbol{r}$ or current $e\boldsymbol{v}$ per charge, respectively, in a power series of the field amplitudes when these do not exceed certain critical values. In some simple stationary regimes analytical expressions can be obtained without resorting to power series expansions. To the extent, however, that the fields may be written as superpositions of well-identified Fourier components, these series expansions also have physical relevance as they allow one to identify well-distinct physical processes according to their order, etc. Indeed writing $\boldsymbol{r} = \boldsymbol{r}^{(0)} + \boldsymbol{r}^{(1)} + \boldsymbol{r}^{(2)} + \ldots$ and $\boldsymbol{P} = \sum e\boldsymbol{r}$, where the summation runs over all charges in a unit volume (and similarly for $\boldsymbol{J} = \sum e\boldsymbol{v}$), one can write the dipolar polarization in a power series [2,3] of the electric-field amplitude $\boldsymbol{P} = \boldsymbol{P}^{(0)} + \boldsymbol{P}^{(1)} + \boldsymbol{P}^{(2)} + \boldsymbol{P}^{(3)} + \ldots$ or, assuming for simplicity no dispersion (instantaneous response),

$$(7) \qquad \boldsymbol{P} = \chi^{(1)}\boldsymbol{E} + \underset{=}{\chi^{(2)}}\boldsymbol{E}\boldsymbol{E} + \underset{\equiv}{\chi^{(3)}}\boldsymbol{E}\boldsymbol{E}\boldsymbol{E} + \ldots,$$

where the coefficient $\chi^{(n)}$ is the dipolar susceptibility of order n, a tensor of rank $n + 1$; one can also include terms in (7) induced by crossing electric and magnetic fields. One proceeds similarly for the magnetic and quadrupolar polarizations; these magnetic and quadrupolar terms are actually related to the nonlocality of the electromagnetic interactions and, except for specific cases, are weak in general compared to the electric dipolar ones and most often are neglected.

In general one can thus write $\boldsymbol{P} = \boldsymbol{P}_{\mathrm{L}} + \boldsymbol{P}_{\mathrm{NLS}}$ and similarly for $\boldsymbol{Q}$ and $\boldsymbol{M}$, where $\boldsymbol{P}_{\mathrm{L}}$ is the part of the dipole polarization linear in the amplitude $\boldsymbol{E}$, or $\boldsymbol{P}_{\mathrm{L}} = \boldsymbol{P}^{(1)}$, and $\boldsymbol{P}_{\mathrm{NLS}}$ is the nonlinear part. Combining the Maxwell equations and assuming for simplicity a neutral nonmagnetic medium, which amounts to setting $H = B$, one gets[2,3]

$$(8) \qquad \nabla \times (\nabla \times E) + \frac{1}{c^2} \frac{\partial^2 D_L}{\partial t^2} = -\frac{4\pi}{c^2} \frac{\partial^2 P_{NLS}}{\partial t^2},$$

which clearly shows that the nonlinear polarization plays the role of a radiation source. If this source is negligible, one recovers the homogeneous linear propagation equation for electromagnetic fields, which for plane monochromatic transverse waves $E = \hat{e} E_\omega \exp[ikr - i\omega t]$ with $D_L = E + 4\pi P_L \equiv \varepsilon E$ and $\varepsilon = 1 + 4\pi\chi^{(1)}$ reduces to the algebraic index equation[4,5]

$$(9) \qquad k \wedge (k \wedge \hat{e}) + \omega^2 \varepsilon \hat{e}/c^2 = 0,$$

which also defines the refractive index related to a mode $(\omega, k, \hat{e})$, namely $k^2 = \omega^2 n^2(\hat{e}, \omega)/c^2$. The intensity related to such a mode of complex amplitude E_ω is obtained from the Poynting vector $S = E \times B$ or

$$(10) \qquad I_\omega = \frac{c}{2\pi} n_\omega E_\omega E_\omega^* = N\hbar\omega c/n_\omega,$$

where N is the photon flux and n_ω is the appropriate refractive index at frequency ω; note that (10) does not contain the phase of the electric field.

As previously stated, the nonlinear term P_{NLS} acts as a polarization source generating new fields with spatiotemporal characteristics drastically different from those expected if only the linear regime applies. This is accompanied by an energy transfer with rate

$$(11) \qquad \frac{dW}{dt} = -\left\langle E \cdot \left\{ \frac{\partial}{\partial t} \{P_{NLS} - \nabla : Q_{MS}\} + c\nabla \wedge M_{NLS} \right\} \right\rangle,$$

where the brackets indicate time averaging; this transfer can be substantial[9] and the fields generated from this transfer can attain large amplitudes with specific spatiotemporal characteristics. Expression (11) can also be used as a starting point for formal thermodynamic considerations [3], but such interpretations should be made with caution since the regime is not stationary. For almost dispersionless media from (11) one can also formally set up a Hamiltonian whose equations of motion reduce to the nonlinear propagation equation for amplitude E in the slowly-varying-envelope approximation.

In the following we shall concentrate our attention in summarizing some key phenomenological features of the susceptibilities (sect. 2) and then proceed to derive the expressions and properties of the microscopic polarizabilities and their relation to the susceptibilities through the local field corrections (sect. 3) before we analyse some simple models to illustrate the main nonlinear polarization mechanisms (sect. 4) and relate them to some material characteristics (sect. 5).

2. – Phenomenological description.

Here we summarize [2, 3, 7, 10] some elementary features of the nonlinear polarization terms that are not bound to any specific microscopic description. Since the polarization induced in a medium results from the action of an external field on the charges, it must possess certain properties that are common to all responses that relate an effect to a cause. We will state and discuss them qualitatively without going to any length into the mathematical intricacies. These are obtained by invoking invariance with respect to translation of the time origin, causality, reality, symmetrization and also the symmetry of the material system. We shall also specify some conventions that are important to keep in mind. We shall concentrate our attention mainly on the dipolar polarization P and its relation to the electric field as this is related to the most important class of nonlinear optical phenomena when applications are relevant. When perturbation theory applies which in practice implies that $E < E_c$, the polarization, as previously stated, can be expressed in a power series as

$$(12) \qquad P(t) = P^{(0)}(t) + P^{(1)}(t) + P^{(2)}(t) + P^{(3)}(t) + \ldots + P^{(n)}(t) + \ldots$$

and each partial polarization term $P^{(n)}$ must possess all properties that follow from the above principles. We will in addition assume that the interaction is local or, plainly stated, the polarization at a given space point is related only to the electric-field amplitude at the same point.

2'1. Polarization response.

2'1.1. Response functions. For the linear term the most general relation between $P^{(1)}(t)$ and $E(t)$ subject to the above conditions is

$$(13a) \qquad P^{(1)}(t) = \int_{-\infty}^{\infty} d\tau_1 R^{(1)}(\tau_1) E(t - \tau_1),$$

where $R^{(1)}$ is a real second-rank tensor that vanishes for τ_1 negative. In the absence of magnetic fields it can be symmetrized and written in diagonal form with three independent components after transforming to the principal axes. The number of independent components can be further reduced by involving the symmetry of the medium; for an optically isotropic medium only a single component remains.

For the second-order term similarly one has

$$(13b) \qquad P^{(2)}(t) = \int_{-\infty}^{\infty} d\tau_1 \int_{-\infty}^{\infty} d\tau_2 R^{(2)}(\tau_1, \tau_2) E(t - \tau_1) E(t - \tau_2),$$

where $R^{(2)}$ is a real third-rank tensor that vanishes when either τ_1 or τ_2 (or both) is negative. Furthermore it can always be symmetrized, so that

$R_{ijk}^{(2)}(\tau_1, \tau_2)$ is invariant under interchange of the pairs (j, τ_1) and (k, τ_2) and this is termed intrinsic permutation symmetry. The number of independent tensor components, at most 27, is, however, greatly reduced when the symmetry of the material system is invoked: in particular $\mathbf{R}^{(2)}$ identically vanishes for materials with inversion symmetry.

Similarly for the higher-order terms one has

$$(13c) \quad \mathbf{P}^{(n)}(t) = \int_{-\infty}^{\infty} d\tau_1 \int_{-\infty}^{\infty} d\tau_2 \ldots \int_{-\infty}^{\infty} d\tau_n \times$$

$$\times \mathbf{R}^{(n)}(\tau_1, \tau_2, \ldots, \tau_n) \mathbf{E}(t - \tau_1) \mathbf{E}(t - \tau_2) \ldots \mathbf{E}(t - \tau_n),$$

where $\mathbf{R}^{(n)}$ is a real $(n + 1)$-rank tensor that vanishes for negative values of any of its time arguments. It can be symmetrized so that $R_{ij_1 \ldots j_n}(\tau_1, \ldots, \tau_n)$ is invariant under any permutation of the pairs $(j_1, \tau_1), \ldots, (j_n, \tau_n)$ and this is the intrinsic permutation symmetry. The number of independent tensor components, at most 3^{n+1}, can be substantially reduced if the material system possesses a symmetry; as a consequence the even-order polarization response functions $R^{(2n)}$ identically vanish for systems with inversion symmetry. The temporal profile of $E(\tau)$ can be arbitrary, but in the impulsive regime, with deltalike pulses, it is easily seen that each polarization term coincides with the corresponding polarization function.

2˙1.2. Susceptibilities. By introducing the Fourier representation in frequency space for the polarization and field the previous relations are reduced to algebraic ones.

Thus setting

$$(14a) \quad \mathbf{P}^{(n)}(t) = \int_{-\infty}^{\infty} \mathbf{P}^{(n)}(\omega) \exp[-i\omega t] \, d\omega,$$

$$(14b) \quad \mathbf{P}^{(n)}(\omega) = \frac{1}{2\pi} \int_{-\infty}^{\infty} \mathbf{P}^{(n)}(t) \exp[i\omega t] \, dt,$$

$$(15a) \quad \mathbf{E}(t) = \int_{-\infty}^{\infty} \mathbf{E}(\omega) \exp[-i\omega t] \, d\omega,$$

$$(15b) \quad \mathbf{E}(\omega) = \frac{1}{2\pi} \int_{-\infty}^{\infty} \mathbf{E}(t) \exp[i\omega t] \, dt,$$

one gets

$$(16) \quad \mathbf{P}^{(n)}(\omega) = \int_{-\infty}^{\infty} d\omega_1 \ldots \int_{-\infty}^{\infty} d\omega_n \underset{\equiv}{\chi}^{(n)}(\omega_1, \omega_2, \ldots, \omega_n) \mathbf{E}(\omega_1) \ldots \mathbf{E}(\omega_n) \delta(\omega - \omega_0)$$

with $\omega_0 = \sum\limits_{i=1}^{n} \omega_i$ and

(17) $\underset{\equiv}{\chi}^{(n)}(\omega_1, \omega_2, \ldots, \omega_n) =$

$$= \int_{-\infty}^{\infty} d\tau_1 \int_{-\infty}^{\infty} d\tau_2 \ldots \int_{-\infty}^{\infty} d\tau_n \mathbf{R}^{(n)}(\tau_1, \tau_2, \ldots, \tau_n) \exp\left[i \sum_{i=1}^{n} \omega_i \tau_i\right],$$

which defines the n-th–order susceptibility $\chi^{(n)}$, a $(n+1)$-rank tensor with complex components. From (17) and the properties of the response functions one shows that, because of the causality, $\chi^{(n)}$ can be analytically continued in the upper half-plane for all frequencies and

(18) $\underset{\equiv}{\chi}^{(n)}(\omega_1, \omega_2, \ldots, \omega_n)^* = \underset{\equiv}{\chi}^{(n)}(-\omega_1^*, -\omega_2^*, \ldots, -\omega_n^*)$

because of the reality of $R^{(n)}$; furthermore because of the intrinsic permutation symmetry $R_{ij_1 \ldots j_n}(\omega_1, \ldots, \omega_n)$ is invariant under all $n!$ permutations of the pairs $(j_1, \omega_1), \ldots, (j_n, \omega_n)$ and the symmetry of the material system substantially reduces the number of the independent tensor components. The latter can be obtained by imposing that the tensor components transform to themselves under any transformation of the reference space that leaves the system unchanged (symmetry operation); this leads to a number of linear relations between the tensor components from which the independent [7, 10] ones can be fixed. Use of group-theoretical techniques leads equivalently to the same result. The tensor properties of $\chi^{(n)}$ are always expressed in orthogonal coordinate systems. In some cases, however, useful insight can be gained [11] by going over to spherical coordinates.

The relations are greatly simplified when the fields are delta-functions in frequency space, namely monochromatic waves; with

$$E(t) = \mathrm{Re}\left\{\boldsymbol{E}_\omega \exp[-i\omega t]\right\},$$

$$\boldsymbol{P}^{(n)}(t) = \mathrm{Re}\left\{\boldsymbol{P}_\omega^{(n)} \exp[-i\omega t]\right\},$$

one gets

(19) $\boldsymbol{P}_{\omega_0}^{(n)} = D_n \underset{\equiv}{\chi}^{(n)}(\omega_1, \omega_2, \ldots, \omega_n)\boldsymbol{E}_{\omega_1}\boldsymbol{E}_{\omega_2}\ldots\boldsymbol{E}_{\omega_n},$

where D_n is a numerical factor that takes into account the number of distinct permutations of $\omega_1, \ldots, \omega_n$. With this convention the limiting value $\chi^{(n)}(0, 0, \ldots, 0)$ is independent of the path chosen for the frequencies to reach the values 0.

2·1.3. Parametric and photoinduced processes. From the previous definitions and conventions it is evident that the frequencies ω_i can have either sign and also be set equal to zero. Two such cases deserve special mention. The one is when all frequencies except one are zero; this is the case of the electro-optic effects, the Pockels and static (d.c.) Kerr effects which can be included in the more general class of parametric processes that will be shortly discussed. The other case is related to the odd-order effects where all frequencies are paired with opposite signs $(\omega_i, -\omega_i)$ except for the one that cannot be paired; this is the case of the optical Kerr effect and several stimulated effects and constitutes the class of photoinduced effects. In certain cases it has been customary to treat the photoinduced effects as parametric ones with $I_i \sim E_{\omega_i} E_{\omega_i}^*$, the beam intensities, playing the role of an external parameter; in essence, however, the two classes of effects, the parametric and the photoinduced, are fundamentally different.

Since the optical Kerr effect plays an important role in many applications, we wish to introduce here the relevant coefficients. This effect is related to the third-order polarization

$$P_\omega^{(3)} = 3 \, \underline{\underline{\chi}}^{(3)} (\omega, -\omega', \omega') \, | E_{\omega'} |^2 E_\omega \, ,$$

where ω and ω' can both be optical frequencies but ω' can also be zero, in which case the effect reduces to the static Kerr or quadratic Pockels effect also defined in (11). To the extent that the term (15) can be lumped together with the linear polarization term induced at frequency ω

$$P_\omega^{(1)} = \chi^{(1)} (\omega) \, E_\omega \, ,$$

one can also describe the optical Kerr effect as a photoinduced change of the refraction or absorption indices of the medium and, by writing

$$P_\omega = P_\omega^{(1)} + P_\omega^{(3)} = \widetilde{\chi}(\omega, I') \, E_\omega \, ,$$

where I' is the light intensity at frequency ω', one obtains the intensity-dependent indices

(20a) $$\widetilde{n}(I) = n_0 + n_2 I \, ,$$

(20b) $$\widetilde{\alpha}(I) = \alpha_0 + \alpha_2 I \, ,$$

for refraction and absorption, respectively, where

(21) $$n_2(\omega) = 12\pi^2 \chi^{(3)} (\omega, -\omega', \omega') / n_0^2 c$$

and similarly for α_2. We shall refer to the cases (20a) and (20b) as optical Kerr nonlinearities, dispersive and absorptive, respectively. In connection with the effective refractive index there is an additional complication compared to the linear case; in general the nonlinear one does not follow instantly the temporal intensity variations. To a good approximation[12] its temporal evolution obeys a Debye-type equation

(22) $$\tau \, \dot{\widetilde{n}}(t) + \widetilde{n}(t) = n_2 I(t)$$

or in integral form

$$\tilde{n}(t) = n_0 + \frac{n_2}{\tau} \int\limits_{-\infty}^{\infty} I(t) \exp\left[-(t-s)/\tau\right] ds \, ,$$

where τ is a phenomenological response time for the Kerr nonlinearity.

Another crucial feature of the effective indices (20a) and (20b) is that these may now become space and time dependent through their dependence on the beam intensity $I(r, t)$ and one has spatial and temporal lensing or grating effects that can lead to dramatic modifications[3, 13-16] of the spatiotemporal characteristics of a beam; these are the light self-action or photoinduced effects and include self-focusing and -defocusing, self-phase modulation, spatial or temporal soliton formation, modulational instabilities and other effects. They may lead to coupling of two beams and the appearance of optical gratings which give rise to effects like optical phase conjugation, optical bistability, instabilities and chaotic behaviour. Most frequently these effects pertain to the dispersive Kerr nonlinearity (20a), although the absorptive case can also be treated along similar lines, but all these effects cannot occur there because of the inherent energy loss.

2`2. *Kramers-Kronig relations. Energy considerations.* – The analyticity of $\chi^{(1)}(\omega)$ in the upper complex plane, that follows from the causality condition through the theorem of Titshmarsh together with Cauchy's residue theorem, leads[5, 8, 17] to the Kramers-Kronig relation between the real and imaginary parts of the linear susceptibility along the real axis, namely

(23)
$$\mathrm{Re}\,\chi^{(1)}(\omega) = \frac{1}{\pi} \mathscr{P} \int\limits_{-\infty}^{\infty} d\omega' \, \frac{\mathrm{Im}\,\chi^{(1)}(\omega)}{\omega' - \omega}$$

and its reverse relation; $\mathscr{P}$ stands for the taking of the principal value. These relations, which are also related to the dissipation-fluctuation theorem, establish an unambiguous connection between linear absorption and dispersion irrespective of the phase and amplitude of the field or

(24)
$$n(\omega) = \frac{c}{\pi} \mathscr{P} \int\limits_{0}^{\infty} d\omega' \, \frac{\alpha(\omega')}{\omega'^2 - \omega^2}$$

in terms of the absorption and refraction indices, α and n, respectively.

The causality has similar implications[2, 7, 18] on the nonlinear susceptibilities, but up to the present date these have not been cast into any useful and physically transparent form. In some isolated cases extensions of (24) in the nonlinear regime have been proposed[19] and used with scarce attention, however, as far as their applicability is concerned; this is very limited.

The difficulty in setting up K-K relations for nonlinear susceptibilities and also relating them to physically simple processes is connected [20] with the fact that nonlinear optical processes can both lead to real nonlinear absorption losses in the material system as well as energy transfer from one coherent field (Fourier) component to a large number of other field (Fourier) components through several interaction paths; as the order of interaction increases, these paths become more and more complex since cascading [7] of lower-order processes must also be included.

Actually even in the case of even-order susceptibilities, the quadratic $\chi^{(2)}$ being the lowest one, which cannot be connected with any nonlinear absorption loss but only to frequency-shifting coherent processes involving amplitude transfer between different field (Fourier) components, the attempts to derive K-K relations were not conclusive [18] with regard to their usefulness. Thus in the case of quadratic susceptibilities the derived relations mix second-harmonic, sum and difference generation as well as the electro-optic and rectification effects.

For the reasons previously evoked the situation is worse in the case of odd-order susceptibilities, the cubic $\chi^{(3)}$ being the lowest one. Here, besides all the possible coherent frequency-shifting processes, one can identify nonlinear absorption losses related to real multiphoton transitions, which are two-photon absorption, saturation and light scattering processes (Raman, Brillouin, etc.); in addition cascading of second-order processes introduces additional contributions to all these processes.

It is customary to assume that the K-K relation (23) or (24) can be set up for parametric processes when the linear susceptibility is modulated by an external «static» field ζ_0, for instance pressure or static electric or magnetic fields: one simply substitutes there the modulated susceptibility $\chi^{(1)}(\omega, \zeta_0)$, an analytic function of ζ_0. Such a procedure has been employed without any preamble [19] in the case of the photoinduced effects, namely the ones that qualitatively can be described by an intensity-dependent effective susceptibility $\widetilde{\chi}^{(1)}(\omega, I)$ disregarding the fact that the intensity I may be related to a field component of frequency $\omega' \neq \omega$. The assumed K-K relation then reads

$$(25) \qquad \Delta n(\omega, I) = \frac{c}{\pi} \mathscr{P} \int_0^\infty d\omega' \, \frac{\Delta \alpha(\omega, I)}{\omega'^2 - \omega^2} \,.$$

There is no rigorous proof for the validity of this relation, but certainly stationarity [20] must apply since expression (25) is a direct extension of the linear K-K relation which is derived under equilibrium conditions. Since in a nonlinear regime this is hardly the case, this relation is of very limited validity. In addition even this approximate relation implies that the optical Kerr coefficient is essentially related to several nonlinear absorption processes, two-photon, light scattering processes (Raman, Brillouin, Rayleigh, etc.) and not to the first one only as has been customary [19] to assume; this is a rather formidable task and effort. An indication of how such a relation can be handled could be obtained by working out a specific quantum or classical model, say the three-level model or the anharmonic oscillator.

The problem of nonlinear K-K relations is also related [21] to the efforts to set up a thermodynamic approach and formally derive the susceptibility from a postulated «free energy». This too encounters difficulties, however, when the physical interpretation is attempted, since the stationarity and equilibrium condition must be properly formulated. The starting point is the expression of the energy transfer rate which can be written

$$(26a) \qquad \frac{c}{4\pi} \boldsymbol{\nabla} \cdot (\boldsymbol{E} \wedge \boldsymbol{B}) + \frac{1}{8\pi} \frac{\partial}{\partial t} (\boldsymbol{B} \cdot \boldsymbol{B} + \boldsymbol{E} \cdot \boldsymbol{E}) + \boldsymbol{E} \cdot \boldsymbol{J} = 0$$

with $\boldsymbol{J}$ given by (4c) or, equivalently,

$$(26b) \qquad \frac{c}{4\pi} \boldsymbol{\nabla} \cdot S + \frac{1}{4\pi} (\boldsymbol{H} \cdot \dot{\boldsymbol{B}} + \boldsymbol{E} \cdot \dot{\boldsymbol{D}}) + \boldsymbol{\nabla} \boldsymbol{E} : \dot{\boldsymbol{Q}} = 0 \, ,$$

where $cS = c\boldsymbol{E} \wedge \boldsymbol{H} - 4\pi\boldsymbol{E} \cdot \dot{\boldsymbol{Q}}$ is the Poynting vector when quadrupolarization is present. Relation (26b) then simply states that the energy flow out of a volume plus the rate of work done by the fields on the material in the volume equals the decrease in the stored energy. For a lossless medium with negligible dispersion and assuming that the slowly-varying-envelope approximation applies, one can define a time-averaged energy

$$(27a) \qquad F = U - \boldsymbol{H} \cdot \boldsymbol{M} - \boldsymbol{E} \cdot \boldsymbol{P} - \boldsymbol{\nabla} \cdot (\boldsymbol{E} \cdot \boldsymbol{Q})$$

with total differential

$$(27b) \qquad \mathrm{d}F = -\boldsymbol{P} \cdot \mathrm{d}\boldsymbol{E} - \boldsymbol{M} \cdot \mathrm{d}\boldsymbol{H} - \boldsymbol{Q} : \mathrm{d}(\boldsymbol{\nabla} \boldsymbol{E}) \, ,$$

from which the polarization can be obtained by differentiation. In particular the nonlinear terms can be obtained [3,21] by a series expansion and the nonlinear propagation equation in the slowly-varying-envelope approximation in the stationary regime is simply the Euler equation related to the optimization of the functional F. This approach is the basis for deriving the Kleinman relation [22] if the nonlinear polarization is dominated by a single mechanism. The extension of this formalism to fast varying envelopes, however, is questionable, in particular when strong dispersion and absorption are present, and will not be pursued here.

2'3. *Parametric processes.* – In the previously discussed nonlinear optical effects the frequencies of the interacting fields were tacitly assumed to be comparable to each other in magnitude: all fall within the same transparency region of the medium, so that all fields couple to the same degrees of freedom of the medium. For the transparency region that extends above the vibrational frequencies and below the onset of electronic transitions, for instance, the valence electrons only contribute to the optical nonlinearities and their response can be assumed to be practically instantaneous.

If any of the frequencies, however, is lowered into a different transparency region, then the corresponding field couples [7] with additional or even different degrees of freedom. For instance, for frequencies below the vibrational frequencies the ionic motion and its coupling with the electrons set up additional mechanisms for nonlinear polarization as also does the molecular reorientation; these additional mechanisms, besides changing the magnitude and even the sign of the nonlinear coefficients, may modify some of their symmetry properties, like the breaking-down of the Kleinmann symmetry relations [22], and also introduce different time constants which can be quite crucial when considering very short light pulse interactions.

Similarly, when any of the frequencies lies much higher than the electronic transitions and falls in the X-ray region, the electrons essentially respond as free and the dipole approximation breaks down and, by the same token, the symmetry properties as well that are related to this approximation, the most conspicuous one being $\chi^{(2n)} \equiv 0$ for media with inversion symmetry. Actually the multipolar expansion must be abandoned and one must proceed from the microscopic Maxwell-Lorentz equations [8, 17] without introducing the space and time averaging that leads to the macroscopic equations (5a)-(5d); the relevant quantity now is the microscopic electron density and current distribution.

For such cases where fields of widely different frequencies are interacting it is more convenient to treat differently high- and low-frequency fields and regard the latter as external parameters that modify the medium optical properties felt by the former. This is the case in particular when one of the fields is static or, equivalently, when its frequency is zero, but other situations can be envisaged which involve static magnetic fields or other low-frequency disturbances like acoustic waves. The fields of electro-optics, magneto-optics and acousto-optics are related to some very important applications that have great impact on device technology.

2·3.1. Electro-optics. This is related to changes of the refractive index of a material brought in by a d.c. or a low-frequency electric field. If this change is linear on the static field, one has the linear electro-optic or Pockels effect, while, when this change is quadratic in the static field, one has the quadratic electro-optic effect or static Kerr effect. The former is related to a second-order nonlinear polarization

(28a) $$P_i^{(2)}(\omega) = 2\chi_{ijk}^{(2)}(\omega, 0) E_j(\omega) E_{0k}$$

and can only occur for materials without inversion symmetry, while the latter is related to a third-order polarization

$$(28b) \qquad P_i^{(3)}(\omega) = 3\chi_{ijkl}^{(3)}(\omega, 0, 0) E_j(\omega) E_{0k} E_{0l}$$

which can occur in any material. As previously stated, one usually proceeds by first deriving the changes of the refractive index induced by the static or low-frequency field and then analysing their impact on the propagation properties of an optical field in the linear regime for the latter. One important reason for adopting such a procedure is that the phase matching is irrelevant for the processes described by (28a), (28b).

More precisely one actually evaluates the modifications of the optical indicatrix of the material induced by the static electric field. For the case of the Pockels effect one thus writes[23]

$$(29) \qquad \Delta\left(\frac{1}{n^2}\right)_i = \sum_j r_{ij} E_j \,,$$

where $(1/n^2)_i$ are the optical constants that describe the optical indicatrix in an arbitrary coordinate system and can be expressed in terms of the principal-axis optical constants; r_{ij} are the electro-optic coefficients and can be related to $\chi_{ijk}^{(2)}(\omega, 0)$. An important application of the Pockels effect is the electro-optic modulation[23] of an optical beam. This effect and its variants have found several applications in optoelectronic devices.

Actually, not only the refractive index of a material is modified by a static electric field, its absorption too can be modified and many applications of the electroabsorption for device operation have been proposed, but here one must account for substantial operation losses.

2`3.2. Magneto-optics. The oscillating magnetic field associated to a light beam is very weak with respect to the electric field to induce any sizable effects, but a static magnetic field H_0 can be very intense and induce substantial changes in the optical properties as a consequence of the Zeeman splittings that the degenerate magnetic levels undergo and the concomitant changes of the energy spacings and oscillator strengths.

The lowest-order magneto-optic effect is related[3] to the polarization

$$(30a) \qquad \boldsymbol{P}_M^{(2)}(\omega) = \underline{\underline{\chi}}_M^{(2)}(\omega, 0) \boldsymbol{E}(\omega) \boldsymbol{H}_0$$

and gives rise to the Faraday effect[4] when the direction of H_0 and that of the light beam propagation coincide. As for the Pockels effect previously discussed, it is more convenient to first analyse the modifications that the refractive index undergoes in the presence of the static magnetic field. One finds that here the refractive indices for right and left circular polarization become different and these two components experience different phase shifts after propagating a distance L inside the medium; the polarization state undergoes a rotation by an angle

$$(31a) \qquad \theta_{\mathrm{F}} = \frac{\omega L}{nc}(n_- - n_+),$$

the Faraday rotation angle. In terms of susceptibilities we also have

$$(32a) \qquad n_{\mp}^2 = 1 + 4\pi(\chi^{(1)} + \Delta\chi_{\mp}^{(1)}) = n^2 + 4\pi\Delta\chi_{\mp}^{(1)},$$

where $\Delta\chi_{\mp}^{(1)}$ is the static-magnetic-field-induced change of the linear susceptibility $\chi^{(1)}$ and can be related to $\chi_M^{(2)}(\omega, 0)$ in (30a).

Very interesting is the photoinduced Faraday rotation effect[24] which is related to the nonlinear polarization

$$(30b) \qquad \boldsymbol{P}_M^{(3)}(\omega) = \chi_M^{(3)} |\boldsymbol{E}(\omega)|^2 \boldsymbol{E}(\omega)\boldsymbol{H}_0$$

and leads to a photoinduced change of the Faraday rotation angle; its implications can best be assessed as follows. If the light intensity $I \sim |E|^2$ is high, the two refractive indices $n_{\mp}$ in (31a) undergo a photoinduced modification which can be written

$$(32b) \qquad \tilde{n}_{\mp}^2 = 1 + 4\pi(\chi^{(1)} + \Delta\chi_{\mp}^{(1)} + \chi^{(3)}I + \Delta\chi_{\mp}^{(3)}I),$$

where $\chi^{(3)}$ is the optical Kerr susceptibility and $\Delta\chi_{\mp}^{(3)} \sim \chi_M^{(3)}H_0$ is its magnetic-field-induced modification or, equivalenty, the photoinduced modification of $\Delta\chi_{\mp}^{(1)}$. Accordingly the Faraday rotation angle (31a) becomes now

$$(31b) \quad \tilde{\theta}_{\mathrm{F}} = \frac{\omega L}{nc}(\tilde{n}_- - \tilde{n}_+) =$$

$$= \frac{\pi\omega L}{nc}\{(\Delta\chi_-^{(1)} - \Delta\chi_+^{(1)}) + (\Delta\chi_-^{(3)} - \Delta\chi_+^{(3)})I\} = \theta_{\mathrm{F}} + \theta_{\mathrm{NL}},$$

the second term θ_{NL} is the light-induced modification of the linear Faraday rotation angle. The photoinduced Faraday effect can have important applications[24] when fast modulation of the polarization state of a beam is required, for instance, in nonreciprocal devices.

2˙3.3. Acousto-optics. A sound wave in a medium is related to density and pressure fluctuations that modulate the refractive index of the material and concomitantly the propagation characteristics of an incident light beam[16,25]. For an anisotropic solid the changes in the refractive index are characterized by

$$(33) \qquad \Delta\left(\frac{1}{\varepsilon}\right)_{ij} = \sum_{kl} p_{ijkl} S_{lk},$$

where S_{kl} is the strain tensor and p_{ijkl} the photoelastic tensor; for an isotropic medium, liquid or gas, one can use a scalar relation. In the case of an acoustic wave of wave vector q and frequency Ω this effect induces a variation

$$\Delta\varepsilon = \Delta\varepsilon_0 \exp[i(q \cdot r - \Omega t)] + \text{c. c.}$$

in the dielectric constant. An incident beam

$$E_i = \text{Re}\{A_i \exp[ik_i \cdot r - i\omega_i t]\}$$

is scattered off this variation and produces a beam

$$E_s = \text{Re}\{A_s \exp[ik_s \cdot r - i\omega_s t]\},$$

whose characteristics depend on the interaction length. If the interaction length is sufficiently long, then phase matching is important and imposes a single diffracted beam; this is the Bragg regime. If the interaction length is very short, then the phase matching is irrelevant and several diffracted beams are obtained; this is the Raman-Nath regime. Both cases are connected with important applications[16,25].

2′4. *Nonlocality. Retardation and cascading.* – In the previous discussion only the dependence of the macroscopic nonlinear optical properties on the temporal variations of the electric field was considered; in particular it was implicitly assumed that the polarization induced in a point in the material only depends on the value of the field amplitude at this point, and its spatial variations in the neighbourhood of the point were disregarded. In certain cases the spatial variations of the field amplitude can have profound implications[7] in the optical properties in general and the nonlinear ones in particular and are related to a fundamental aspect of the interaction and propagation of electromagnetic fields in matter, namely the nonlocality. This is intrinsically contained in the Lorentz force and also in the Maxwell equations. Both lead to a wave vector dependence of the optical properties of the material which is termed spatial dispersion.

The most conspicuous manifestation of nonlocality in optics is the natural optical activity[26] which is also equivalent to assuming that the optical properties are wave vector sensitive. One writes

(34a)
$$P = \frac{\varepsilon - 1}{4\pi} E - \alpha_1 \nabla \wedge E = \chi^{(1)} E + \alpha_1 \dot{H}/c,$$

(34b)
$$M = -\alpha_2 \dot{E}/c,$$

where α_1 and α_2 are material coefficients that characterize the optical activity. One can show[17,26] that for optically isotropic media

(35)
$$D = \varepsilon E + i(G \wedge E),$$

where $G = f\hat{k}$ with $f = -4\pi(\alpha_1 + \alpha_2)\omega n/c$ and $\varepsilon = n^2 = 1 + 4\pi\chi^{(1)}$; for anisotropic media one can still write D in the form (35), but G is not along $\hat{k}$.

One can extend (34a), (34b) to the nonlinear regime by adding higher-order terms. Thus, besides the dipolar nonlinear terms previously considered in

(34a), one now will have terms like

$$(36a) \qquad\qquad \boldsymbol{P}_M^{(2)}(\omega_1 + \omega_2) = \underline{\underline{\chi}}_M^{(2)} \boldsymbol{E}_{\omega_1} \dot{\boldsymbol{H}}_{\omega_2},$$

$$(36b) \qquad\qquad \boldsymbol{P}_M^{(3)}(\omega_1 + \omega_2 + \omega_3) = \underline{\underline{\chi}}_M^{(3)} \boldsymbol{E}_{\omega_1} \boldsymbol{E}_{\omega_2} \dot{\boldsymbol{H}}_{\omega_3},$$

and similar ones in (34b). In sect. 3 we shall discuss a few aspects related to the term (36b) when $\omega_1 = -\omega_2 = \omega_3$, which is connected to processes that allow control of the polarization state of an electromagnetic field. All these effects are connected with the nonlocality in the field-matter interaction as reflected in the structure of the Lorentz force (compare eq. (2)).

As previously stated, nonlocality also enters through the Maxwell equations as reflected in the retardation effects [7]. These creep into optics in several forms. The polariton effect is such a manifestation frequently encountered in crystal optics; the phase-matching [27] problem in nonlinear optics is another such manifestation which profoundly influences the efficiency of all nonlinear processes and in particular the second-order ones. At the higher-order processes another manifestation [7] of the retardation effect is that the total polarization of a given order and a given frequency, besides the direct contribution, will contain indirect contributions arising from many-step processes involving, at the intermediate stage, nonlinear interactions with fields generated by lower-order polarization terms. Since the latter fields are strongly affected by the phase-matching conditions, at the intermediate stages the effective susceptibility for the global polarization, totalling both the direct and all indirect processes, becomes wave vector dependent. These indirect processes are also termed cascading processes [7] and must be included in an *ad hoc* way unless one uses from the beginning the true eigenstates of the matter-field system, which in the case of crystals are the polariton states.

To illustrate the problem we focus [7, 28] our attention on the case of the third-order polarization in a crystal without inversion symmetry. In such a crystal, besides the direct contribution given by (22) for $n = 3$ polarization, there will be additional contributions generated by cascading. Thus new fields at frequencies $\omega_1 + \omega_2$, $\omega_2 + \omega_3$, $\omega_3 + \omega_1$ can be generated in the crystal by two-wave mixing. These fields may beat with the fields at frequencies ω_3, ω_1, ω_2, respectively, to generate polarizations at $\omega_1 + \omega_2 + \omega_3$ by two-wave mixing; furthermore these polarizations are created with the same total wave vector as the direct one, namely $\boldsymbol{k}_1 + \boldsymbol{k}_2 + \boldsymbol{k}_3$. These additional contributions can be lumped together with the direct one (22) and one can define an effective third-order susceptibility for this total third-order polarization which, however, depends on the wave vectors of the primary and intermediate fields, in contrast to the genuine third-order susceptibility related to the direct third-order polarization contribution which does not.

Cascading processes play an important role in third- and higher-order effects and draw a growing attention in applications [29].

3. – Microscopic description.

We wish now to give a microscopic description of the induced polarization and relate it to the induced dipoles in the material. The microscopic description clearly must be compatible with the phenomenological one we outlined in the previous section; it is based on the equation of charge motion in the presence of intense coherent electromagnetic fields and of the interactions with the other charges. Because of the complexity of the problem the steps [7] that lead from this equation to the expressions of the experimentally accessible macroscopic nonlinear susceptibilities previously introduced are several and all involve certain simplifications and approximations in order to cope with the many-body interactions and obtain meaningful and useful information.

3`1. *General aspects.* – Let us assume that the material system in the absence of external fields is described by a Hamiltonian $H_0(\boldsymbol{p}_i, \boldsymbol{r}_i)$ with known eigenvalue and eigenstate spectrum, where i runs over all the charges. Under the influence of an intense electromagnetic field represented by a vector potential $\boldsymbol{A}(\boldsymbol{r}, t)$ the state of the charges making up the medium will be perturbed and oscillating currents will be induced in the system. The Hamiltonian describing the system is now $H(t) = H_0(\boldsymbol{\pi}_i, \boldsymbol{r}_i)$, where $\boldsymbol{\pi}_i = \boldsymbol{p}_i - e\boldsymbol{A}_i/c$ with $\boldsymbol{A}_i = \boldsymbol{A}(\boldsymbol{r}_i, t)$, and, assuming that H_0 is quadratic in the momenta $\boldsymbol{p}$, one has

$$(37) \qquad H(t) = H_0(\boldsymbol{p}_i, \boldsymbol{r}_i) - \sum_i \frac{e_i}{2m_i c}\{\boldsymbol{A}_i\cdot\boldsymbol{p}_i + \boldsymbol{p}_i\cdot\boldsymbol{A}_i\} + \sum_i \frac{e_i^2}{2m_i c^2}A_i^2 \,.$$

A Coulomb gauge with zero scalar potential is used and for simplicity spin effects were neglected. Beyond this stage one must resort to simplifying considerations in order to obtain tractable and useful expressions for the response of the system and also take into account the actual constitution of the medium and in particular the diversity of the charges and polarizable units involved as well as their mutual interactions. At the outset we have electrons, ions and polarizable units of molecular size, the latter in general being nonspherical. Because of their vastly different masses, their motion is characterized by eigenfrequencies that fall into well-separated spectral regions.

If we for simplicity consider monochromatic waves characterized by wavelength λ and period T, the electromagnetic field described by $\boldsymbol{A}(\boldsymbol{r}, t)$ actually samples the space and time over extensions of the order of λ and T and averages out fluctuations in the charge motion that vary faster in space and time. Otherwise stated, the response of the charges to the electromagnetic field is insensitive to all details of their motion that occur within λ and T. This qualitative argument can also be cast into a rigorous form and sets a scaling procedure.

For fast oscillating electromagnetic fields with frequencies much higher than all eigenfrequencies the charges behave[8,17] as free and the response is directly related to the electronic charge and current density distribution as is the case in X-ray scattering where the wavelength is much less than the extension of electronic wave functions.

As the frequencies become lower and the wavelength exceeds the characteristic charge extension a in the dielectric, a limited multipole expansion[30] of the interaction Hamiltonian term in (37) can be used. For most purposes it is sufficient to keep the lowest-order terms in this expansion and write

(38) $$H_I = \boldsymbol{\mu} \cdot \boldsymbol{E} - \boldsymbol{q} : \nabla \boldsymbol{E} - \boldsymbol{m} \cdot \boldsymbol{H}$$

for the interaction Hamiltonian, where $\boldsymbol{\mu} = \sum er$, $\boldsymbol{q} = \sum er(r/2)$ and $\boldsymbol{m} = (e\hbar/2mc) \sum r \wedge p$ are the electric-dipole, electric-quadrupole and magnetic-dipole operators, respectively, and the summation extends over all charges within the polarizable unit; the dipolar term is the most important one, while the quadrupolar and magnetic dipolar ones are each of order (a/λ) less than the dipolar one, but their impact in certain cases can be quite important, as will be discussed below.

As the frequencies of the electromagnetic field are lowered still further into the optical spectrum and the wavelength becomes much longer than the charge extension in a dielectric, the dipole approximation can be introduced by keeping only the first term in (38); the dipolar response to the electromagnetic field is then described in terms of the electronic polarizabilities of polarizable units. As a consequence of the adiabatic Born-Oppenheimer approximation these coefficients are functions of the ion coordinates and of the orientation of the polarizable unit; accordingly they may be modulated by the corresponding vibrational and orientational motion. In addition the polarizable units may change their relative positions and this motion introduces fluctuations in the delectric constant. The low-frequency fluctuations due to intra- and intermolecular motions result in the different light scattering effects, but they can also be directly or indirectly coupled to the electromagnetic fields and strongly modulate[7] the induced polarization.

3˙2. *Dipolar approximation. Dipoles.* – It is evident from the above qualitative discussion that for most purposes in nonlinear optics one may start by setting up the expression of the induced polarization[2,3,7] in the dipole approximation. We will restrict ourselves to the most commonly recurring case where the material is composed of identical repeat polarizable entities, atoms, molecules or clusters of dimensions much smaller than the optical wavelength; these will be either randomly but homogeneously distributed, like, for instance, in a gas, a liquid or an amorphous solid, or set in a periodic array, for instance a crystalline solid. To the extent that the charges are more or less localized in each such entity, the total induced polarization density P can be set as

$$(39) \qquad\qquad P = \sum p = N\langle p \rangle,$$

namely as the sum of the effective dipoles p induced on each entity in a volume unit by the electromagnetic field that actually acts on each entity; in (39) $\langle\;\rangle$ stands for a statistical averaging, if necessary.

The expression of the dipole moment p in a power series of this field will be obtained using perturbation theory in the dipole approximation, namely

$$(40) \qquad\qquad p = \alpha^{(1)}E + \underline{\alpha}^{(2)}EE + \underline{\underline{\alpha}}^{(3)}EEE,$$

and the coefficients $\alpha^{(n)}$ of the different powers are the microscopic polarizabilities. Their relation to the previously defined macroscopic polarizabilities and susceptibilities requires the knowledge of the relation between the effective field locally acting on the charges within each entity and the macroscopic field that enters the Maxwell equations. The latter in the optical-frequency domain is actually the average of the microscopic field over volumes of dimensions much less than the optical wavelength but still containing several polarizable entities. The two fields differ because of the mutual interactions that arise between the induced dipoles and other induced interactions. Because of the complexity of these interactions some simple models [8] have been devised to account for their effect.

The outlined procedure actually pertains to dielectrics with charges more or less well localized within each entity, for instance gases, liquids, ionic or amorphous dielectrics. For semiconductors or metals where electrons are very delocalized one must use a different procedure based [10, 31] on band theory and in addition it is more appropriate to calculate there the induced current density $J(t)$ which is related to the polarization density $P(t)$ through

$$(41) \qquad\qquad J = \frac{\partial P}{\partial t}.$$

We shall succinctly discuss aspects related to the nonlinearities of such systems as well.

3'2.1. Induced dipoles. Polarizabilities. From (38) one can show [2,3,7] that the Hamiltonian of each entity in the presence of an electric field in the dipole approximation can be written as

$$(42) \qquad\qquad H = H_0 + H_d + H_R,$$

where H_0 represents the Hamiltonian of the entity in the absence of the field, H_d is the interaction term with the electric field in the dipole approximation

$$(43) \qquad\qquad H_d = -\mu \cdot E_e,$$

where E_e is the effective field in the entity and μ is the total dipole moment operator of all charges of the entity. H_R stands for all random perturbations of the charges in the entity provoked by the environment which is represented as a bath with an infinite number of degrees of freedom. These are usually assumed to be stationary Markoffian processes uncorrelated to the coherent interaction H_d; their correlation time τ_c is assumed to be very short, much shorter than any relevant time of the problem. These random perturbations provoke relaxation processes which under certain simplifying approximations can be incorporated into the density matrix operator equation, which then reads

$$
(44) \qquad \frac{d\rho}{dt} = \frac{1}{i\hbar} [H_0 + H_d, \rho] + \frac{\partial \rho}{\partial t} \Big|_R
$$

with

$$
(45a) \qquad \frac{\partial \rho_{mn}}{\partial t} \Big|_R = -\Gamma_{mn} \rho_{mn} = -\frac{1}{T_{mn}} \rho_{mn} ,
$$

$$
(45b) \qquad \frac{\partial \rho_{mm}}{\partial t} \Big|_R = \sum_{n \neq m} [W_{mn} \rho_{nn} - W_{nm} \rho_{mm}],
$$

where Γ_{mn} and W_{mn} are positive quantities that are related to the coherence and population relaxation rates of the transition between states m and n or, equivalently, to the transverse and longitudinal relaxation times T_2 and T_1, respectively. The induced dipole moment is obtained from $p = Tr\rho\mu$. Assuming H_d a perturbation and expanding in powers of its strength, one can set $\rho = \rho^{(0)} + \rho^{(1)} + \rho^{(2)} + \dots$ in (44). Making a Fourier analysis of the field and the induced dipole and solving the resulting hierarchy of equations by an iterative procedure, one finds $p = p^{(0)} + p^{(1)} + p^{(2)} + \dots$, where $p^{(n)} = Tr\rho^{(n)}\mu$, or explicitly

$$
(46a) \qquad p^{(1)}(t) = -i \int_{-\infty}^{t} dt_1 \langle [\mu, H_d(t_1)] \rangle,
$$

$$
(46b) \qquad p^{(2)}(t) = (-i)^2 \int_{-\infty}^{t} dt_1 \int_{-\infty}^{t_1} dt_2 \langle [[\mu, H_d(t_1)], H_d(t_2)] \rangle,
$$

$$
(46c) \quad p^{(3)}(t) = (-i)^3 \int_{-\infty}^{t} dt_1 \int_{-\infty}^{t_1} dt_2 \int_{-\infty}^{t_2} dt_3 \langle [[[\mu, H_d(t_1)], H_d(t_2)], H_d(t_3)] \rangle,
$$

and averages are taken over the initial density matrix operator; these expressions are precisely of the form (13a)-(13c), which confirms the compatibility between the microscopic and macroscopic descriptions. If we apply Fourier analysis to the radiation field and to the dipoles in (46a)-(46c), we obtain

$$
(47) \qquad p^{(n)}(\omega_1 + \dots + \omega_n) = D_n \underline{\underline{\alpha}}^{(n)}(\omega_1 \dots \omega_n) E_{\omega_1} \dots E_{\omega_n} ,
$$

which defines the n-th–order microscopic polarizability. Relation (47) is tensorial and requires explicit reference to the n Fourier components of the field that are involved and the n-th–order dipole they induce together. For the three lowest-order terms, the linear, second- and third-order ones, they read[7]

(48a)
$$p_i^{(1)}(\omega) = \alpha_{ij}(\omega) E_j(\omega),$$

(48b)
$$p_i^{(2)}(\omega_1 + \omega_2) = D_2 \beta_{ijk}(\omega_1, \omega_2) E_j(\omega_1) E_k(\omega_2),$$

(48c) $$p_i^{(3)}(\omega_1 + \omega_2 + \omega_3) = D_3 \gamma_{ijkl}(\omega_1, \omega_2, \omega_3) E_j(\omega_1) E_k(\omega_2) E_l(\omega_3),$$

where

(49a) $$\alpha_{ij}^{(1)}(\omega) \equiv \alpha_{ij}(\omega) = \frac{1}{\hbar} \sum_{g, r} \rho_{gg}^{(0)} \left\{ \frac{\langle \mu_i \rangle_{gr} \langle \mu_j \rangle_{rg}}{\omega_{rg} - \omega - i\Gamma_{rg}} + \frac{\langle \mu_j \rangle_{gr} \langle \mu_i \rangle_{rg}}{\omega_{rg} + \omega + i\Gamma_{rg}} \right\},$$

(49b) $$\alpha_{ijk}^{(2)}(\omega_1, \omega_2) \equiv \beta_{ijk}(\omega_1, \omega_2) =$$

$$= \frac{1}{\hbar^2} \sum_{g, r, s} \rho_{gg}^{(0)} \left\{ \frac{\langle \mu_i \rangle_{gr} \langle \mu_j \rangle_{rs} \langle \mu_k \rangle_{st}}{(\omega_{rg} - \omega_1 - \omega_2 - i\Gamma_{rg})(\omega_{sg} - \omega_2 - i\Gamma_{sg})} + 5 \text{ terms} \right\},$$

(49c) $$\alpha_{ijkl}^{(3)}(\omega_1, \omega_2, \omega_3) \equiv \gamma_{ijkl}(\omega_1, \omega_2, \omega_3) = \frac{1}{\hbar^3} \sum_{g, r, s, t} \rho_{gg}^{(0)} \times$$

$$\times \left\{ \frac{\langle \mu_i \rangle_{gr} \langle \mu_j \rangle_{rs} \langle \mu_k \rangle_{st} \langle \mu_l \rangle_{tg}}{(\omega_{rg} - i\Gamma_{rg} - \omega_1 - \omega_2 - \omega_3)(\omega_{sg} - i\Gamma_{sg} - \omega_2 - \omega_3)(\omega_{tg} - i\Gamma_{tg} - \omega_3)} + 47 \text{ terms} \right\}$$

are the linear, second- and third-order microscopic polarizabilities and we also introduced their more conventional notations α, β and γ instead of $\alpha^{(1)}$, $\alpha^{(2)}$ and $\alpha^{(3)}$, respectively; these coefficients and the higher ones are tensors of second, third, fourth and higher order, respectively. Sometimes it is convenient to transform these expressions by introducing the momentum operator $\boldsymbol{p}$ which satisfies the identity $[H_0, \boldsymbol{r}] = -i\hbar \boldsymbol{p}/m$.

In media with very delocalized charges in periodic structures like the ones in covalent crystalline semiconductors the above approach that relies on localized wave functions is not satisfactory and one must instead use a description of the electron states in terms of band states. In addition it is more appropriate then to compute the induced current $\boldsymbol{j}$ instead of the induced dipole $\boldsymbol{p}$, the two being related by (41). The relevant expressions have been derived and discussed extensively in the literature [7, 15, 31], but will not be reproduced here. One can also proceed to directly compute the induced dipole with delocalized band states, but some care must be paid to the mathematical procedure.

3'2.2. Resonances. It is evident from the previous expressions that the nonlinear polarizabilities possess a very rich and complicated multiresonant behaviour [3,7] that is absent in the linear one, which can be only singly resonant at electric-dipole-allowed transitions. The resonant behaviour of the nonlinear susceptibilities is related to several very important nonlinear optical phenomena. As can be inferred from their expressions, the nonlinear susceptibilities can be resonantly enhanced if one or more of the complex frequency/energy denominators have their real part close or equal to zero. A simple inspection shows that these resonances fall into two categories. The first one concerns the «direct» or «primary» resonances where one or more frequencies of the fields involved in the interaction are close to electric-dipole-allowed transition frequencies of the medium; these resonances either lead to trivial resonant enhancement of the nonlinearity which, however, is counterbalanced by the concomitant linear absorption losses at the same frequency, or, in the case of very intense beams, they lead to a class of nonlinear phenomena which can be accounted for only by a nonperturbative analysis of an equivalent two-level system resonantly interacting with a monochromatic field, as will be succinctly discussed below. The other category concerns the «intermediate» resonances where an algebraic sum of primary frequencies is close to an appropriately allowed transition frequency of the medium; under certain conditions then energy can be transferred [2,3] between the field modes through the intermediary of the material mode. These indirect or secondary resonances and related energy exchange processes between the field and material modes are distinct features of the odd-order nonlinearities; as a matter of fact one can show [2,3,16] that the related processes can even grow from photon noise and give rise to stimulated effects.

In this section we shall give a brief account of these intermediate resonances in the third-order nonlinearities and the effects they give rise to; these occur whenever

(50) $$\omega_i + \omega_j = \omega_e$$

and the relevant susceptibility is $\chi^{(3)}(\omega_1, \omega_2, -\omega_1)$, where ω_e is any two-photon allowed excitation of the medium. Since condition $\omega_i + \omega_j = 0$ is satisfied, one can actually define a nonlinear energy transfer between the two electromagnetic modes ω_1 and ω_2, the difference in energy being stored in the material excitation. As was previously stated, these effects can grow even from photon noise and give rise to stimulated effects: the stimulated two-photon emission, Raman, Brillouin, Rayleigh and other light scattering effects [3,32]. We first derive the expression of the two-photon transition matrix element and relate it to the light scattering cross-section.

From (49c) one can easily show [3,7,15] that, for $\omega_i + \omega_j = \omega_{fg} = \omega_e$ and keeping only resonant terms,

(51) $$\alpha_{ijkl,\,R}^{(3)}(\omega_1, \omega_2, -\omega_1) \simeq \Delta\varrho \frac{\alpha_{ij}(\omega_1, \omega_2)\,\alpha_{kl}^*(\omega_1, \omega_2)}{\hbar(\omega_e - \omega_1 - \omega_2 - i\Gamma_{fg})},$$

where $\Delta\rho = \rho_{gg}^{(0)} - \rho_{ff}^{(0)}$ is the population difference between the ground and the excited state and

$$
(52) \qquad \alpha_{ij}(\omega_1, \omega_2) = \sum_n \left\{ \frac{\langle \mu_i \rangle_{gn} \langle \mu_j \rangle_{nf}}{\hbar(\omega_{ng} - \omega_1)} + \frac{\langle \mu_j \rangle_{gn} \langle \mu_i \rangle_{nf}}{\hbar(\omega_{ng} + \omega_2)} \right\}
$$

is a generalized polarizability which for $f \equiv g$ and $\omega_1 \equiv \omega_2$ reduces to the linear polarizability (49a); it can also be regarded [7] as the matrix element between states f and g of an effective two-photon operator $H^{(2)}$ whose diagonal elements for $\omega_1 \equiv \omega_2$ are the linear polarizabilities. Note that $\alpha_{ij}(\omega_1, \omega_2)$ can be enhanced if either ω_1 or ω_2 is close to an electric-dipole-allowed transition.

The previous discussion actually only concerned two-photon resonant behaviour related to a transition between the ground state which is initially populated and an excited state which can be unpopulated. A careful inspection of expression (49c) shows [33] that under certain conditions one can also have such a resonant behaviour even between two unpopulated excited states n and n' when $\omega_1 - \omega_2 = \omega_{nn'}$. This can occur when the double-resonant condition $\omega_1 = \omega_{n'g}$ and $\omega_2 = \omega_{ng}$ is satisfied and in addition $\Gamma_{n'g} + \Gamma_{ng} \neq \Gamma_{n'n}$. The latter condition implies that the relaxation is not exclusively radiative but is also due to «collisions» and other nonradiative processes. If this is the case, one can show that the resonant term is

$$
(53) \qquad \alpha_R^{(3)}(\omega_1, -\omega_2, \omega_1) \sim \frac{1}{\omega_{n'g} - \omega_2 - i\Gamma_{n'g}} \frac{1}{-\omega_{ng} - \omega_1 - i\Gamma_{ng}} \times
$$

$$
\times \left\{ 1 - \frac{\Gamma_{ng} + \Gamma_{n'g} - \Gamma_{n'n}}{\omega_{n'n} - (\omega_2 - \omega_1) - i\Gamma_{n'n}} \right\}.
$$

Such a resonant behaviour has been observed [34]; note that an analogous resonant behaviour is also present in the second-order polarizability $\alpha^{(2)}(\omega_1, -\omega_2)$.

Before proceeding further, we wish to classify the resonances by relating them to the different charge motions that take place in a polarizable unit. The degrees of freedom of the polarizable unit can be either internal, like the bound-electron motion and the vibrational motion, or external, like the orientational and translational motion of the whole unit. If one assumes that the internal and external degrees of freedom are decoupled and that the adiabatic approximation applies for the internal motion, one may write

$$
\Psi(r, Q; \widehat{n}; R) = \phi(r, Q) \chi(Q) \theta(\widehat{n}) \psi(R)
$$

for the wave function, where r is the electronic coordinate, Q the vibrational coordinate, $\widehat{n}$ the direction coordinate of the polarizable unit and R the centre-of-mass coordinate. At room temperature the orientational and translational motion can be treated as classical with appropriate statistical averaging. On the other hand, the electronic and vibrational motion can be treated quantum-mechanically [7] within the Born-Oppenheimer approximation. This allows one to obtain the different contributions to the optical nonlinearities. These are as follows [7]:

purely electronic with the ions being fixed in an equilibrium position,

hybrid with the electronic distribution being modulated by the driven vibrational motion of the ions,

purely ionic due to the electric and mechanical anharmonicity of the ions,

orientational,

translational.

The order of magnitude of the different contributions varies substantially as also does their response time. As far as the choice of nonlinear optical materials is concerned, the relevant contribution is the electronic one. Far from resonances this is essentially instantaneous but usually weak; close to an electronic resonance it can be enhanced but becomes slow.

For systems with very delocalized electrons one can substantially modify the electronic spectrum by introducing an artificial confinement of the delocalized electrons. This introduces a new class of resonances which can be termed morphological resonances [34]; their characteristics can be modified by the artificial fabrication technique. One has two main classes of confinement, the dielectric and the quantum confinement, which are extensively used to enhance the optical nonlinearities in artificial materials.

3·2.3. Susceptibilities and local field corrections. As previously stated, the connection between macroscopic susceptibilities and the microscopic polarizabilites requires [7] the relation between the macroscopic and the effective fields. This is a central problem in the whole theory of the dielectrics and touches its very fundamental aspects. To gain an insight into the problem, let us assume that the charge clouds of the different entities do not overlap and we subdivide the space into identical boxes each containing one polarizable entity. The field E_e that acts on a given entity is the one that results [35] from all sources outside its box; it is the microscopic field E_{mic} which is the same in every box with the exclusion of the field from the entity itself E_s or $\boldsymbol{E}_e = \boldsymbol{E}_{mic} - \boldsymbol{E}_s$. The macroscopic field that enters Maxwell equations is the space average of E_{mic} over many boxes which in our case reduces to its average within any box or $E = \langle E_{mic} \rangle_b$. The field E_e is actually not constant throughout its box and the effective field will be defined as a weighted average of E_e over the box, the weight being roughly the charge density distribution and similarly for E_s, so that

$$E_f \equiv \langle E_e \rangle = \langle E_{mic} \rangle - \langle E_s \rangle.$$

Without going into a more lengthy analysis, it is quite evident from the above considerations that the effective-field corrections are a measure of the inhomogeneity [36,37] of the charge density distribution within a box; the smoother this distribution is, the less important are the local field corrections. In the extreme case of a metal in which the charge density is uniform throughout each box, the local field corrections disappear, while in the extreme case of pointlike induced dipoles the Lorentz model may apply [8] and one finds an isotropic dielectric

$$(54) \qquad\qquad E_f = E + \frac{4\pi}{3} P .$$

For the cases intermediate between these two extremes the calculation of the local field corrections become exceedingly difficult because of the overlap of charge distributions in different boxes. Let us assume [7] for simplicity that

$$(55) \qquad\qquad E_f = E + LP ,$$

where L is a frequency-independent second-rank tensor whose components vanish for a metal and reduce to (54) in an isotropic dielectric with pointlike nonoverlaping induced dipoles. Thus the effective field contains a contribution from the induced polarization whose dynamics in particular are related to those of eqs. (45a), (45b). A more rigorous theory in crystals relates [37] the local field corrections to the Fourier series expansion of the microscopic field in the inverse lattice space.

Assuming for simplicity isotropic medium and using expression (55) with L a scalar, inserting in (40) and substituting in (39), one obtains

$$(56) \quad P = N\alpha(E + LP) + N\beta(E + LP)(E + LP) +$$

$$+ N\gamma(E + LP)(E + LP)(E + LP),$$

where for simplicity no explicit reference to the frequencies is made. Setting

$$(57) \qquad\qquad f(\omega) = 1/(1 - N\alpha(\omega)L) ,$$

rearranging and keeping track of terms of the same order in the electric field, one can cast this expression into the macroscopic form

$$(58) \quad P = N\widetilde{\alpha}E + N\widetilde{\underset{=}{\beta}}EE + N\widetilde{\underset{\equiv}{\gamma}}EEE \equiv \chi^{(1)}E + \underset{=}{\chi^{(2)}}EE + \underset{\equiv}{\chi^{(3)}}EEE ,$$

where the macroscopic polarizabilities $\widetilde{\alpha}, \widetilde{\beta} \dots$ are

$$(59a) \qquad\qquad \widetilde{\alpha}(\omega) = f(\omega)\,\alpha ,$$

$$(59b) \qquad\qquad \widetilde{\underset{=}{\beta}}(\omega_1, \omega_2) = f(\omega_1 + \omega_2)f(\omega_1)f(\omega_2)\,\underset{=}{\beta}(\omega_1, \omega_2),$$

and the susceptibilities are given by

$$(60) \qquad\qquad \chi^{(1)} = N\widetilde{\pmb{\alpha}} = \widetilde{\pmb{\alpha}}/v\,,$$

$$(60b) \qquad\qquad \underline{\underline{\chi}}^{(2)} = N\,\underline{\underline{\widetilde{\pmb{\beta}}}} = \underline{\underline{\widetilde{\pmb{\beta}}}}/v\,,$$

$$(60c) \qquad\qquad \underline{\underline{\chi}}^{(3)} = N\widetilde{\gamma} = \widetilde{\gamma}/v\,,$$

where $v = 1/N$ is the effective volume of the polarizable unit.

The expression of $\widetilde{\gamma}$ is somewhat involved[7] and is not reproduced here. It suffices to state that it contains additional terms to γ which correspond to microscopic cascading processes. For optically anisotropic media f is a second-rank tensor, the inverse of $1 - N\,\pmb{L\alpha}$, and the previous relations must be accordingly transcribed. Local field corrections may be substantial[7] and must be included; in the following we shall assume that they most often lead to a renormalization of the oscillator strengths and will be lumped together with them in the expressions of the susceptibilities. These oscillators strengths in the following will be understood to be the renormalized ones.

At this level we wish to remind that for third or higher polarizations additional contributions must be included[7] which result from the so-called cascading processes. These processes result from the interaction of the primary beams with intermediate ones that are generated by lower-order induced nonlinear polarizations compatible with the symmetry of the medium. At the macroscopic level these cascading processes can be lumped together with the direct contribution since they have the same total wave vector, frequency and electric-field dependence; the resulting effective susceptibility has the same symmetry properties as the direct one, but also depends on the wave vector because of the phase matching at the intermediate interactions.

3'3. *Nonlocality. Spatial dispersion.* – In certain cases the quadrupole and magnetic-dipole terms in (38) cannot the altogether disregarded even in the optical spectrum range. This is the case, for instance, in media whose optical properties are not invariant under reflection on a plane like in the chiral molecular systems[17, 26]. These media show optical activity in the linear regime which results precisely from the second and third term in (38); in the nonlinear regime they lead to more complex effects and in particular to photoinduced optical activity. All these effects and the corresponding coefficients can be derived by perturbation approach along the same lines as the one outlined in the previous case, namely the dipole (or local) approximation.

The two simplest cases are the dipoles induced by the magnetic interaction term at a frequency ω, namely

(61a)
$$\boldsymbol{p}_M^{(1)}(\omega) = \boldsymbol{\alpha}_M^{(1)}(\omega)\, \boldsymbol{H}_\omega\,,$$

(61b)
$$\boldsymbol{p}_M^{(3)}(\omega) = \alpha_M^{(3)}(\omega,\, -\omega,\, \omega)\, \boldsymbol{E}_\omega \boldsymbol{E}_\omega^* \boldsymbol{H}_\omega\,.$$

Applying perturbation theory as in paragraph 3'2.1 and deriving the expression of the induced dipole to first order in the magnetic-dipole interaction in (38), one finds [38] that

(62)
$$\boldsymbol{\alpha}_M^{(1)}(\omega) = \sum_l \left\{ \frac{\langle \boldsymbol{\mu} \rangle_{gl} \langle \boldsymbol{m} \rangle_{lg}}{\hbar(\omega_{lg} - \omega)} + \frac{\langle \boldsymbol{m} \rangle_{gl} \langle \boldsymbol{\mu} \rangle_{lg}}{\hbar(\omega_{lg} + \omega)} \right\},$$

which has the same form as (49a) but with one of the electric-dipole matrix elements replaced by a magnetic-dipole one; similarly $\alpha_M^{(3)}(\omega,\, -\omega,\, \omega)$ can be obtained [39] from (49c) by performing the same exchange.

One can show that α_1 and α_2 introduced in (34a), (34b), apart from some evident factors, are both equal to (62) and for an isotropic medium can be written [40]

(63)
$$\alpha_1 = \alpha_2 = \frac{i\omega}{\hbar} \sum_l \left\{ \frac{R_{gl}}{\omega_{lg} - \omega} + \frac{R_{gl}}{\omega_{lg} + \omega} \right\},$$

where $R_{gl} = \text{Im}\,(\boldsymbol{\mu}_{gl} \cdot \boldsymbol{m}_{gl})$ are the Rosenfeld factors [38] which play the same role in the optical activity as the optical oscillator strengths do in the linear refraction. One can also show that they satisfy [40] a sum rule which has a simple physical interpretation. Similarly one can show that in $\alpha_M^{(3)}$ the quantity

(64)
$$\boldsymbol{\alpha}_M(\omega_1,\, \omega_2) = \frac{1}{\hbar} \sum_R \left\{ \frac{\langle \boldsymbol{\mu} \rangle_{gl} \langle \boldsymbol{m} \rangle_{lf}}{\omega_{lg} - \omega_1} + \frac{\langle \boldsymbol{m} \rangle_{gl} \langle \boldsymbol{\mu} \rangle_{lf}}{\omega_{lg} + \omega_2} \right\}$$

plays a role analogous to that of the two-photon transition strength (52) in $\alpha^{(3)}$. One can also show that the above microscopic description is compatible with the macroscopic one outlined in subsect. 3'4. Indeed, since from Maxwell equations $\omega \boldsymbol{H}_\omega = \boldsymbol{k} \wedge \boldsymbol{E}_\omega$, introducing this relation in (61a) and (61b) one sees that these can be regarded as corrections to (48a), (48c), respectively, and, when lumped together, they can be cast in the form (35). The calculation of the coefficients $\alpha_M^{(1)}(\omega)$ and $\alpha_M^{(3)}(\omega)$ introduces very subtle considerations concerning the wave functions; the phase of the latter in particular cannot be disregarded as is in practice the case when calculating the dipolar polarizabilities (49a)-(49c). Clearly the latter are real quantities when $\omega_i \to 0$, while, when optical activity is present, the relevant coefficients are complex numbers, as can be inferred from (35).

4. – Nonlinear polarization mechanisms.

The mechanisms that underlie the nonlinear polarization terms in (7) can be qualitatively discussed without making appeal to the detailed quantum-mechanical expressions of $\chi^{(2)}$, $\chi^{(3)}$ .... A classification can be obtained by setting up the expression of the linear susceptibility

$$(65) \qquad \chi_{ij}^{(1)}(\omega) = N \sum_{g,e} \Delta\varrho_{ge} \left\{ \frac{\langle\mu_i\rangle_{ge}\langle\mu_j\rangle_{eg}}{\hbar(\omega_{eg} - \omega - iT_{2,eg}^{-1})} \right\},$$

where $\Delta\varrho_{ge}$ is a population difference for states g and e and $\hbar\omega_{eg} = E_e - E_g$, $\langle\mu\rangle_{eg}$ and $T_{2,eg}$ are the transition energies, dipole moments and coherence times, respectively. In the presence of an intense electric field E all these quantities are perturbed and become E-field dependent and the higher-order terms in (7) can be thought to arise from the modification of (65) by the electric field.

Thus the population difference $\Delta\varrho_{ge}$ can be perturbed by real transitions with energy (photons) supplied by the field and hence depends on powers of the beam intensity $I \sim EE^*$ only and the coherence here is irrelevant; consequently the lowest-order effect related to this mechanism is cubic in the electric-field intensity and is sometimes termed photoinduced. In contrast the quantities $\langle\mu\rangle_{eg}$ and E_{eg} in general depend on both even- and odd-order powers of the field E and the coherence here is in general relevant; these mechanisms give rise to both quadratic and cubic terms in the electric-field intensity in (7). This distinction is very important to keep in mind because the decay time of the photoinduced effects that exploit population changes consecutive to real transitions is T_1, the longitudinal or population relaxation time, which is much longer than T_2, the transverse or coherence relaxation time, and consequently the exploitation of incoherent processes strongly reduces the operation speed of the material.

The above approach is of very limited use and one must resort to a more detailed description of the mechanisms in order to quantitatively predict the optical nonlinearities. Such an approach was briefly outlined in paragraph 3˙2.3, where it was also pointed out that the most relevant nonlinearities are the ones due to the valence electrons. We shall succinctly discuss below certain models that will also single out some salient features of these nonlinearities and relate them to the characteristics of the electronic-charge density distribution.

4˙1. *Nonresonant regime.* – Of particular interest in most applications are the nonlinear optical properties of materials in their transparency region that extends above the vibrational frequencies and below the onset of electronic transitions. These are due to the polarization of the valence electrons and the relevant nonlinear susceptibilities can be related to some important structural characteristics of their density distribution. Besides its intrinsic fundamental interest this relationship plays a key role in the ongoing nonlinear-optical-material research. Several computational techniques have been devised[7] to account for the magnitude and behaviour of these susceptibilities, but of equal value are also some simple models that have been worked out to illustrate the nonlinear polarization mechanism, in particular the one underlying the quadratic polarization. Before discussing some of the most recurring models, we wish to succinctly point out a couple of important points regarding the nonresonant electronic nonlinearities.

One such point is the Kleinman relation which was initially formulated[22] for the quadratic susceptibility. This relation simply states that $\chi^{(2)}_{ijk}$ is invariant under all permutations of the indices i, j and k irrespective of the frequencies. This relation was initially derived by assuming the existence of an energy term in (27b) which is cubic in the field amplitudes or $U^{(3)} = -(1/3)\chi^{(2)}_{ijk}E_iE_jE_k$; as was pointed out in sect. 2, this can be done only if the dispersion and absorption are negligible. A more careful examination[7] of the expression of $\chi^{(2)}$ within the framework of the Born-Oppenheimer approximation indicates that this relation is approximate and only applies when a single polarization mechanism is present that can be modelled by a single anharmonic oscillator; stated in this form, one can easily see that the extension of the Kleinman relation to nonlinearities higher than the quadratic breaks down in most cases.

The other point we wish to bring up here is the phase of the nonresonant susceptibilities. In the dipolar approximation one can easily see that $\chi^{(2)}(0, 0)$ and $\chi^{(3)}(0, 0, 0)$ are real. When spatial dispersion, however, becomes important and terms like (61a), (61b) must be included, then the effective nonresonant nonlinearities can be complex as is also the case of the effective linear susceptibility; the latter can be directly inferred from expression (35) for the electric induction when optical activity (spatial dispersion) is present.

The final point we wish to stress here is that the nonresonant dipolar susceptibilities $\chi^{(2)}(0, 0)$ and $\chi^{(3)}(0, 0, 0)$ which are real can have either sign. The sign of $\chi^{(2)}(0, 0)$, however, cannot be specified without stating[7] the axis convention; the sign of $\chi^{(3)}(0, 0, 0)$ is independent of the axis convention. The signs of the nonresonant susceptibilities are important material characteristics and convey crucial information regarding the electronic distribution.

The expressions of the nonresonant polarizabilities can be easily derived[7] from (49a)-(49c) and read

$$\alpha = 2\sum_{sg}' f_g \frac{\mu_{gr}\mu_{rg}}{E_{sg}},$$

$$\beta = 3\sum_{g, r, t}' f_g \left[\frac{\mu_{gr}\mu_{rs}\mu_{sg}}{E_{rg}E_{sg}} - \mu_{gg}\sum_{\sigma}' \frac{\mu_{gs}\mu_{sg}}{E_{sg}^2} \right],$$

$$\gamma = 4\sum_{g, r, s, t}' f_g \left[\frac{\mu_{gr}\mu_{rs}\mu_{st}\mu_{tg}}{E_{rg}E_{sg}E_{tg}} - \sum_{t}' \frac{\mu_{gt}\mu_{tg}}{E_{tg}} \sum_{s}' \frac{\mu_{gs}\mu_{sg}}{E_{sg}^2} \right],$$

where $f_g \equiv \rho_{gg}^{(0)}$ and $\sum_{s,\,\dots}'$ means that the term with $E_{sg} = 0, \dots$ will be excluded from the summation. These coefficients are also related to the electric-field-induced shifts in the energy of the ground state g. Throughout the previous discussion, for simplicity, we assumed that kT, where T is the temperature, is much smaller than any electronic transition energy $\hbar\omega_{sg}$, so that thermodynamic considerations will not enter the picture; we remind that we neglect all contributions to the response from other degrees of freedom than electronic.

For systems with infinitely extended electronic states like the Bloch band states in bulk crystals certain care must be paid in the use of the dipolar interaction Hamiltonian (43) as a perturbation, since the matrix elements of the dipole moment operator $\boldsymbol{\mu} = e\boldsymbol{r}$ between states within the same band, the so-called intraband elements, are singular and can only be defined in terms of distributions.

One way to circumvent this problem is to go over to the momentum operator $\boldsymbol{p}$, whose intraband terms are zero, but this causes difficulties in the comparison of the nonlinearities of a bulk crystal with those of a microstructure of the same material, the latter being expressed in terms of $\boldsymbol{\mu}$; evidently one may revert to the $\boldsymbol{r}$-operator elements, but this is exceedingly cumbersome. An alternative and very elegant approach is the one devised by GENKIN and MEDNIS[7,31], which by a proper transformation in state phase space allows one to directly derive the expressions in terms of the effective transition dipole moment matrix elements.

4'1.1. Quadratic polarizabilities. These mainly result from heteropolar bonds whose polarizabilities, first and second order, can be assumed to be additive and transferable. We may then focus our attention to such a bond which for simplicity will be assumed to be unidimensional.

a) *Unsöld approximation.* With $\omega_i = 0$ in (49a), (49b) we set[41] all energy denominators $E_{rg} = \hbar\omega_{rg}$ equal to an average energy E_a that is determined by using the same approximation in the Thomas-Kuhn rule

$$(66) \qquad \sum_r |\langle r|x|g\rangle|^2 E_{rg} = \hbar^2/2m$$

per electron. One obtains $E_a = \hbar^2/2m\langle \Delta x^2\rangle$ and

$$(67a) \qquad \alpha_{xx} = 4\langle \Delta x^2\rangle^2/a_0 ,$$

$$(67b) \qquad \beta_{xxx} = 6\langle \Delta x^3\rangle\langle \Delta x^2\rangle^2/e\,a_0^2 ,$$

where $\Delta x = x - \langle x\rangle$ and $\langle\ \rangle$ stands for $\langle\ \rangle_{gg}$; expressions (67a), (67b) clearly show that the bonds must be very polarizable and possess a large octupole moment in order to have large second-order polarizability; the energy E_a is also a measure of the electron delocalization along the bond. This simple approximation reproduces quite well the essential trends. Note that only the ground electronic distribution matters.

b) *Charge transfer model.* To gain more insight we represent[42] the ground electronic distribution of an heteropolar bond AB of length R with two pointlike charges $(1 - \varepsilon)\, e$ and $(1 + \varepsilon)\, e$ at its two ends. When an electric field E is applied in the direction from A to B, a charge amount $e\Delta\varepsilon$ flows from B to A such that $\alpha E = e\Delta\varepsilon R$, where $\alpha = (1 - \varepsilon^2)^2 R^4/4a_0$ is the bond polarizability (67a) in the Unsöld approximation with the assumed charge distribution. The intra-bond charge flow results in a modification of this polarizability by replacing ε with $\varepsilon - \Delta\varepsilon$, which now becomes electric-field dependent. Developing into powers of the electric field one can identify

$$(68) \qquad\qquad \beta = (1 - \varepsilon)^3\, \varepsilon R^7/4ea_0^2 \,,$$

which can also be obtained directly from (67b) with the assumed charge distribution. This expression clearly shows that β vanishes for $\varepsilon = 0$ (homopolar bond) and $\varepsilon = 1$ (ionic bond), the ions being spherical. It acquires a maximum value for $\varepsilon = 1\sqrt{7}$, a result which can used as a guide for the optimal choice of heteropolar bonds; ε can be related to the bond polarity and electronegativity difference of atoms A and B. One also sees that β increases with the polarizability and the bond length.

In organic systems, because of the strong overlap of the π-electrons, the relevant units where additivity and transferability can apply are larger than a bond; this is the case, for instance, in monosubstituted benzenes $X\text{-}\diamondsuit$ or charge transfer complexes. We briefly present these two cases.

c) *Equivalent field model.* In the first case one can relate[43] the magnitude of β to the distortion of the π-electrons of the symmetric benzene ring caused by the substituent radical which is measured by the mesomeric dipole moment μ_M. This is done by assuming that this distortion is induced by an equivalent electric field E_R such that

$$\mu_M = \alpha_R E_R \,,$$

where α_R is the linear polarizability of the π-electrons of the symmetric ring. In the presence of a field E, then, the induced dipole moment results from the compound effect of $E + E_R$ on the symmetric ring and one obtains

$$(69) \qquad\qquad \beta = 3\,\frac{\gamma_r}{\alpha_r}\,\mu_M \,,$$

where γ_r is the third-order polarizability.

d) *Two-level model.* For the charge complexes an estimate of β can be obtained[44] by simulating the system as a quantum two-level system, the ground (g) and the excited (e), with wave functions $\psi_g = a\psi_0 + b\psi_1$ and $\psi_e = a^*\psi_1 - b^*\psi_0$, the Mulliken states, where ψ_0 and ψ_1 are the «no bond» and «dative» states, respectively. For a two-level system one has from (49b) with $\omega_i = 0$

(70)
$$\beta = 3\mu_{ge}^2 \Delta\mu_{eg}/E_{eg}^2 .$$

μ_{eg} is the transition dipole moment, $\Delta\mu_{eg} = \mu_e - \mu_g$ the dipole difference in the ground and excited states, and E_{eg} the energy difference between them. The important point to notice in (70) is that for such a system β can be large, although μ_g might be small or even vanish as long as μ_e is large. This remark is essential[45] in designing molecular crystals.

4`1.2. Cubic susceptibilities. This case is more complex to analyse with microscopic models both because of the increased computational complexity and also because of more fundamental grounds. If we assume for simplicity centrosymmetric media, it is evident by inspection of the expression of γ with $\omega_i = 0$ (see (49c)) that large values can be obtained if the electrons are highly delocalized not only within a molecular unit but also across several such identical units. This delocalization over several identical repeat units in a periodic array is *a sine qua non* prerequisite for large cubic susceptibilities. Bond additivity and transferability then do not apply. This is in particular the case in semiconductors and conjugated polymers.

a) *Finite chains.* Free-electron model. An extreme case then is to consider the electrons as moving along a chain of N identical bonds. Treating this model either as a free electron in a box[46] or in the Hückel approximation[47], one gets

(71a)
$$\alpha \sim N^3$$

and

(71b)
$$\gamma \sim N^5 ,$$

which strikingly shows the breakdown of additivity which would lead to $\alpha \sim N$ and $\gamma \sim N$.

b) *One-dimensional delocalization.* If the bonds are not all identical but alternately so, with resonance energies β_1 and β_2, respectively, for the «long» and «short» bonds, and one assumes an infinite chain, one can model[48] the system as a one-dimensional semiconductor and describe the electrons with Bloch band states; within the two-band approximation the expressions of $\chi^{(1)}$ and $\chi^{(3)}$ are

(72a)
$$\chi^{(1)} = \frac{4e^2}{\hbar V} \int_{BZ} \Omega_{cv} S_{cv} \, dk ,$$

(72b)
$$\chi^{(3)} = \frac{8e^4}{\hbar V} \int_{BZ} \left[\frac{1}{\omega_{cv}} \frac{\partial S_{cv}}{\partial k} \frac{\partial S_{vc}}{\partial k} - \Omega_{vc} S_{cv} S_{vc} S_{cv} \right] dk ,$$

where $\hbar\omega_{\mathrm{cv}} = \varepsilon_{\mathrm{c}} - \varepsilon_{\mathrm{v}}$, Ω_{vc} is the transition dipole moment matrix element between the two bands, a valence (v) and a conduction (c) band of energies ε_{v} and ε_{c}, respectively, and $S_{\mathrm{vc}} = \Omega_{\mathrm{vc}}/\omega_{\mathrm{cv}}$. An important point to notice here is that $\chi^{(3)}$ in (72b) is determined by the competition of an intraband term and an interband term; in contrast the linear susceptibility (72a) only involves interband terms. For highly delocalized systems (strong overlap between wave functions of unit cells) the quantities ω_{cv} and Ω_{cv} vary strongly over the Brillouin zone (BZ) and, therefore, the intraband term in $\chi^{(3)}$ becomes the dominant one. Another important point that reflects (72b) is that the main contribution to $\chi^{(3)}$ comes from a few nonoverlapping critical points in the joint density of states, the ones where $\nabla_k \omega_{\mathrm{cv}} = 0$, which greatly simplifies the behaviour of $\chi^{(3)}$ and allows one to establish[49] scaling laws and study the impact of the dimensionality. So in contrast to $\chi^{(2)}$, where bond additivity, a property of real space, can be assumed, and is used to account for its trends, in $\chi^{(3)}$ one can assume critical-point contribution additivity, a k-space property. Thus for a one-dimensional semiconductor or an infinite conjugated chain one finds[49]

$$(73) \qquad\qquad \chi^{(3)} \sim (E_{\mathrm{F}}/E_{\mathrm{g}})^6 \,,$$

where E_{F} is the valence electron Fermi level, essentially the band width, and E_{g} is the optical gap; this behaviour of $\chi^{(3)}$ is characteristic of one-dimensional semiconductors.

For large dye molecules where the electron states can be described by the free electron[46,48] in a box model one finds that $\gamma \sim L^5$, where L is an effective molecular dimension.

It is quite plausible from these simple considerations that β and γ are related to some important features of the electron charge density distribution in molecules and solids and justifies the effort that is being concentrated on calculating and predicting their trends. We remind that previous and ongoing studies on the linear polarizability provided a wealth of information concerning the structure of molecules and solids.

More recently a particular effort is being concentrated on evaluating the impact of quantum confinement on the optical nonlinearities: this occurs when very delocalized electrons are confined within regions smaller than their natural delocalization length. This is particularly relevant for third-order nonlinearities which are very sensitive to electron delocalization. This is the case of several artificial semiconductor structures like quantum wells, wires and dots. Here, however, the resonant behaviour of the nonlinearities close to the quantum confined transitions is more interesting and more directly related to the nonlinear-optical-material research which will be briefly discussed in sect. 5.

c) *White self-transparency.* As was previously pointed out, $\chi^{(3)}$ can have either sign irrespective of the axis convention. This is quite evident from the quantum-mechanical expression of $\chi^{(3)}(0, 0, 0)$ which contains two contributions of opposite sign (see, for instance, $(72b)$) in contrast to the linear one (see, for instance, $(72a)$) which has only one positive contribution when the system is in thermodynamic equilibrium. The fact that $\chi^{(3)}$ or, equivalently, the optical Kerr coefficient can have either sign can have profound implications in nonlinear composite or other inhomogeneous materials. We will illustrate this point with the white self-transparency[50,51].

In the previous cases the medium is assumed homogeneous with an index of refraction n_0 that has the same value throughout the medium. If n_0 fluctuates randomly in space but not in time, the spatial fluctuations of the refractive index may be compensated with the optical Kerr nonlinearity and the losses due to elastic scattering are then suppressed leading to a new class of self-action effects and in particular to self-transparency. To simplify let us assume that the inhomegenous medium is a composite consisting of small spherical dielectric particles of refractive index n_B and number density N randomly dispersed into a homogeneous dielectric of refractive index n_A, both without absorption losses. A beam propagating in the z-direction experiences losses because of elastic scattering which in the Rayleigh and Rayleigh-Debye-Gans regimes, neglecting multiple scattering[52], is expressed with an elastic diffusion constant $\alpha_s \sim (\Delta n)^2$, where $\Delta n_0 = n_A - n_B$. If the incident-beam intensity is high, then each refractive index is modified by the optical Kerr nonlinearity or $\widetilde{n}_A = n_A + n_{2A}I$ and $\widetilde{n}_B = n_B + n_{2B}I$. If we assume that the elastic diffusion constant can be written[50]

$$(74) \qquad \widetilde{\alpha}_s \sim (\Delta\widetilde{n})^2 \sim (\Delta n_0 + \Delta n_2 I)^2 \,,$$

where $\Delta n_2 = n_{2A} - n_{2B}$, one easily sees that for sign $\Delta n_0 \Delta n_2 = -1$ there is a light intensity $I_c = |\Delta n_0/\Delta n_2|$ for which $\widetilde{\alpha}_s$ in (74) vanishes: below this intensity the beam is totally attenuated and above it is transmitted but with intensity I_c; in a certain way the photons in the scattered modes are recycled into the incident one. In particular that part of the spatiotemporal intensity profile $I(r, t)$ of a beam that has intensity below I_c is attenuated, while the one that has higher intensities propagates without further loss once its intensity has been attenuated to I_c; one can obtain square-form spatiotemporal profiles. One also can achieve bistable operation with a single mirror and several other interesting effects[50,51].

4˙2. *Resonant regime.* – It is quite clear that the nonlinear susceptibilities have a very rich and complex resonant behaviour that extends from the X-ray frequency region down to frequencies related to very slow translational motions of the polarizable units. This was succinctly discussed in paragraph 3˙3.3. Among the different resonances the most interesting ones for applications are the ones related to two-photon transitions, namely the ones where the condition $\omega_i + \omega_j = \omega_e$ is satisfied. The expression of the relevant resonant susceptibility was derived in (51) and we will now discuss some important physical aspects.

Before doing this, we will succinctly present two models that are extensively used [2, 3, 7] to illustrate different aspects of the optical nonlinearities in general and their frequency dependence in particular. These are the anharmonic-oscillator model for bound electrons and the free-electron model which is relevant for nonlinearities of electrons in metals or plasmas and also in all materials in the X-ray frequency region.

4'2.1. Anharmonic-oscillator model. In many instances one can model the bound-charge motion in the presence of an external oscillating field in terms of a driven anharmonic oscillator, namely

$$(75) \qquad \ddot{x} + \Gamma\dot{x} + \omega_0^2 x + \beta x^2 + \gamma x^3 = \frac{e}{m} E_\omega \cos \omega t \,,$$

where for simplicity we assume one-dimensional motion and a single oscillating electric field. Equation (75) cannot be solved analytically. One usually resorts to a perturbative/iterative technique by setting $x = x^{(0)} + x^{(1)} + x^{(2)} + \ldots$ in (75) making a Fourier series expansion and identifying terms of the same power in the electric-field amplitude E_ω. This approach has been extensively discussed in the literature [2, 3, 7] and will not be reproduced here.

When the field amplitude E_ω is large and exceeds certain critical values but still remains below the value of the cohesive field amplitude E_c, the above perturbative/iterative approach breaks down and one must resort to a nonperturbative approach. This was discussed [53] in the case of the driven Duffing oscillator

$$(76) \qquad \ddot{x} + \Gamma\dot{x} + \omega_0^2 x + \gamma x^3 = \frac{e}{m} E_\omega \cos \omega t \,,$$

where it was shown that the response shows an intrinsic bistable behaviour. As the field intensity increases, one has transition to a cascade of instabilities and finally to a chaotic behaviour.

4'2.2. Free-electron model. X-ray nonlinearities. For free electrons, for instance, in plasmas or in the X-ray frequency region where the anharmonicity can be neglected the only mechanism that induces nonlinearities is the nonlocality of the Lorentz equation

$$(77) \qquad \dot{v} + \Gamma v = \frac{e}{m}(E + v \wedge B) \,,$$

where E and B are interrelated through Maxwell equation. Equation (77) too cannot be solved analytically and one must resort to a perturbative/iterative procedure by setting $v = v^{(1)} + v^{(2)} + \ldots$ making a Fourier series expansion and identifying terms of the same power in the electric-field amplitude. We shall not reproduce here these expressions [3, 7], but we wish to stress a couple of important points since this model is of relevance when discussing nonlinearities in plasmas and in the X-ray frequency region.

If is easy to infer from (77) that because of the nonlocal character of this equation one can have quadratic nonlinearities which in principle would be absent if the dipolar approximation was applied. This in particular implies that in the X-ray frequency region one can have second-harmonic generation even in a material that is centrosymmetric, for instance diamond.

The other point is that the term in the induced current that is quadratic in the field amplitude results from three mechanisms [54], namely the nonlinear Lorentz, the displacement and the Doppler mechanisms.

4·2.3. Two-photon resonances. Stimulated effects. The expression of the two-photon resonant cubic susceptibility was derived in 3·2.2. As can be seen, there the two-photon resonant susceptibility is complex because of the combined effect of the energy denominator and the generalized polarizabilities. If $i = k$ and $j = l$ in (51), the product $\alpha_{ij}(\omega_1, \omega_2)\,\alpha_{ij}^*(\omega_1, \omega_2)$ is a real number and the separation of (51) into real and imaginary parts is related only to that of the energy denominator. The associated energy transfer

$$(78) \qquad \frac{\mathrm{d}W_i^{(4)}}{\mathrm{d}t} = \left\langle \boldsymbol{E}_i \cdot \frac{\mathrm{d}\boldsymbol{P}_i^{(3)}}{\mathrm{d}t} \right\rangle$$

is proportional to the energy density flows $I_i \sim \boldsymbol{E}_i \boldsymbol{E}_i^*$, $i = 1, 2$, only and unaffected by the phases of the electric fields. Hence one can unambiguously define a nonlinear energy gain or loss in the involved modes 1 or 2 of the electromagnetic field, the difference being taken or given up by the material ω_e mode so that energy conservation is satisfied.

The energy density flow can be expressed as photon density flow or $I_i = = N_i \hbar \omega_i c / n_i$, where N_i is the photon number density in mode i and n_i is the refractive index at frequency ω_i; if one introduces the compound light-matter states $|N_1 N_2 a\rangle$ of energy $N_1 \hbar \omega_1 + N_2 \hbar \omega_2 + E_a$, then one can interpret the whole energy transfer process as transitions between such isoenergetic states. As can be inferred from the previous comments, one can describe and analyse these effects within a unified frame. This, however, is quite cumbersome to set up and we only present a brief analysis [2, 3, 16].

The relation between the spontaneous and stimulated light scattering processes can be easily established starting [2, 3, 16] with the expression

$$(79) \qquad \frac{\mathrm{d}P}{\mathrm{d}t} = BN_1(N_2 + 1)$$

for the probability per unit time for mode 1 losing one photon, mode 2 gaining one and the material system transiting from state g to state f such that $\omega_f - - \omega_g \equiv \omega_e = \omega_1 - \omega_2$ or, equivalently, the transition from state $|N_1 N_2 g\rangle$ to state $|N_1 - 1, N_2 + 1, f\rangle$; the coefficient B can be derived from Fermi's golden rule using the appropriate two-photon transition operator $H^{(2)}$. If the propagation direction of mode i coincides with the z-axis and we assume a stationary regime, then the above expression can be cast into the form

(80)
$$\frac{dN_2}{dz} = B' N_1 (N_2 + 1).$$

For $N_2 \ll 1$ the spontaneous scattering is dominant and $N_2(L) \sim B' N_1 L$, where L is the interaction length, while for $N_j \gg 1$ the stimulated effect becomes dominant and grows as

(81)
$$N_2(L) = N_2(0) \exp[GL],$$

where $G = B' N_1$. This growth can be initiated by injected photons at frequency ω_2 or by the photon noise provided by the spontaneous process.

Before proceeding to analyse an example of a stimulated process, we wish to make here their connection with the light scattering processes more explicit [33]. The latter are due to the temporal and spatial inhomogeneities of the dielectric constant because of the intrinsic fluctuations of the different degrees of freedom of the polarizable units. The fluctuations introduce additional terms in the induced polarization that oscillate at frequencies shifted below (Stokes side) and above (anti-Stokes side) the incident frequency by amounts equal to the eigenfrequencies of the degrees of freedom. Because the dielectric properties of a medium at a given frequency ω corresponding to a wavelength λ are essentially obtained by averaging the charge motion over space and time domains of extensions less than λ and $T = 2\pi/\omega$, respectively, only fluctuations with eigenfrequencies much less than ω are relevant in light scattering. With this is mind we may write

(82)
$$\varepsilon = 1 + 4\pi N\alpha$$

for the dielectric constant of a dielectric medium formed by identical polarizable units of effective polarizability α and number density N; both α and N are fluctuating quantities because they are modulated by internal and external degrees of freedom, respectively. The internal degrees of freedom, for instance vibration or rotation, the one available to a single polarizable unit being confined in space, are local and essentially possess a discrete spectrum; in contrast the external ones like translation of the polarizable unit involve extended motion of a large number of units and can be related to thermodynamic fluctuations like density, pressure and temperature whose evolution is described by the hydrodynamic equations of continuity, momentum transfer and heat transport.

The incident light then is scattered off the fluctuations of either the polarizability $\Delta\alpha$ or the density ΔN to produce Stokes and anti-Stokes components in addition to the incident one. For an intense incident light these additional components can become intense because each added photon in the scattered-light modes enhances the probability for getting an additional one in the same mode provoking an avalanche of photons in a single mode which leads to the build-up of a large-amplitude coherent electric field. The generation of this additional field on the Stokes side is accompanied by the build-up of a large-amplitude fluctuation at the relevant material eigenfrequency, which, because of its resonant coupling, enhances the energy transfer. This coupling can be best understood by reminding that the energy stored in a polarizable-unit interaction with an electric field is

$$(83) \qquad W = -\frac{1}{2}\alpha E^2(z, t),$$

where $E = E_i + E_s$, E_i and E_s being the electric fields in the incident and scattered mode. Since this energy is a function of the internal and external degrees of freedom of the polarizable unit, it can be regarded as potential energy and one can define a force conjugated to each degree of freedom Q by

$$(84) \qquad F_Q = \nabla_Q\left(\frac{1}{2}\alpha E^2\right)$$

or

$$(85) \qquad F_Q = \frac{1}{2}\frac{\partial\alpha}{\partial Q}E^2,$$

$$(86) \qquad F_E = \frac{1}{2}\alpha\nabla E^2$$

for an internal and external (position) degrees of freedom, respectively. This force may drive the corresponding degree of freedom conjugated to a large amplitude. This is the origin of the stimulated process.

a) *Stimulated Raman effect.* We illustrate this effective-force approach[55] with the stimulated Raman scattering. We consider the case of a dielectric in the presence of an electromagnetic field

$$(87) \qquad E(z, t) = \widehat{e}_i A_i \exp[ik_i z - i\omega_i t] + \widehat{e}A_j \exp[ik_j z - i\omega_j t] + \text{c. c.}$$

We assume for simplicity that the dielectric is isotropic, possesses inversion symmetry and its scalar dielectric constant is given by (82), while the induced dipole is $p = \alpha E$ and the polarization density

$$(88) \qquad P = N\alpha E.$$

Now let us assume that the polarizability α can be modulated by a single internal degree of freedom of coordinate q and can be written $\alpha = \alpha_0 + q\partial\alpha/\partial q$ to first order in this coordinate; this can be a vibration or rotation coordinate. The force (86) drives its conjugated coordinate according to the equation

$$(89) \qquad \ddot{q} + \Gamma_q\dot{q} + \omega_q^2 q = F_q/m_q = \frac{1}{2m_q}\frac{\partial\alpha}{\partial q}E^2,$$

where we also introduce a phenomenological damping for the material-mode coordinate q and ω_q and m_q are the eigenfrequency and effective mass associated to q.

From (87) one sees that E^2 contains components at several frequencies and in particular at frequency $\omega_1 - \omega_2$; if $\omega_1 - \omega_2 \approx \omega_q$, only this component is relevant for driving the coordinate q to an appreciable amplitude which can be written as

$$q(z, t) = \text{Re}\{q \exp[i(k_i - k_j)z - i(\omega_i - \omega_j)t]\}$$

with

$$q = \frac{1}{2m_q} \frac{\partial \alpha}{\partial q} \frac{A_i A_j^*}{\omega_q^2 - (\omega_i - \omega_j)^2 - i(\omega_i - \omega_j)\Gamma_q}.$$

This modulation of the polarization (88) leads to a nonlinear term

(90) $$P_{NL}^{(3)} = N \frac{\partial \alpha}{\partial q} qE$$

cubic in the field intensities. With a collinear geometry for the wave vectors k_i and k_j the Stokes component in this polarization,

$$P_j^{(3)} = \text{Re}\{\sigma\chi_R^{(3)} A_i A_i^* A_j \exp[ik_j z - i\omega_j t]\}$$

with

$$\chi_R^{(3)}(\omega_2) = -\frac{iN}{12m_q\omega_q\Gamma_q}\left(\frac{\partial \alpha}{\partial q}\right)^2$$

for the exact resonant case $\omega_1 - \omega_2 = \omega_q$, is phase matched; note that $\chi^{(3)}$ is purely imaginary and indeed can be related to a nonlinear absorption loss that contributes to α_2 in (34). If we disregard the other components in (90), then from nonlinear propagation equation[2,9] one has

$$\frac{dA_2}{dz} = gI_1 A_2$$

with

(91) $$g = \frac{2\pi^2\omega_2}{n_1 n_2 c^2} \frac{N}{m_q\omega_q\Gamma_q}\left(\frac{\partial \alpha}{\partial q}\right)^2$$

for the spatial evolution of the field in the stationary regime. Assuming undepleted pump one gets

$$A_2(z) = A_2(0)\exp[gz],$$

which implies an exponential growth the same as the one derived by (82) if we notice that $G = 2g$. It is more convenient to relate[2,3,16] the gain coefficient to the Raman cross-section related to mode q; one has

(92)
$$G = \frac{4\pi^3 Nc^2}{\hbar\omega_1^2 \omega_2 n_2^2} \frac{\partial^2\sigma}{\partial\omega \, \partial\Omega},$$

where $\partial^2\sigma/\partial\omega \, \partial\Omega$ is the differential spectral cross-section.

The above approach is valid whenever the coordinate q is related to a motion with well-defined frequency ω_q, vibrational or rotational, of the polarizable unit; in quantum mechanics this implies discrete level spacings and their dynamics in the presence of the driving fields can still be described by eq. (89). If the thermal energy kT, however, becomes comparable to or larger than the level spacing as is commonly the case for rotation where $kT \geq \hbar\omega_q$, then the motion becomes diffusive and appropriate provisions must be made in the equation of motion (89). In the case of rotation this leads to the classical reorientational motion of the polarizable units if these are nonspherical, which can be described by the Debye model. One can similarly proceed with the stimulated Brillouin and Rayleigh scattering, which result from driven fluctuations of the dielectric constant as the result of fluctuations in the external degrees of freedom of the polarizable units, namely changes of their position which can be related to density fluctuations $\Delta\varepsilon = (\partial\varepsilon/\partial\rho)\Delta\rho$; since from thermodynamics $\Delta p = (\partial\rho/\partial p)_s \Delta p +$ $+ (\partial\rho/\partial s)_p \Delta s$ and s and p are independent, to a good approximation one can treat [3, 16] independently the two effects.

b) Optical balance. Very frequently in multiphoton excitation processes one sees suppression of a process instead of its enhancement when the exciting frequencies are tuned to intermediate resonances. Such effects have been reported in numerous cases. This suppression occurs [56] because another concurrent process which shares the same intermediate resonance gets enhanced and drains the energy available to the compound process. The order of the competing processes may be very different as are also the resulting spectral and spatial characteristics. Thus the system may switch from the one regime to the other and *vice versa* by changing certain parameters. This optical balance behaviour has been reported in numerous cases, for instance in the competition of high-harmonic generation and multiphoton ionization or fluorescence or in the Stokes–anti-Stokes competition in stimulated Raman scattering [57].

4'2.4. Resonant nonlinear effects in two-level systems. A very important class of phenomena in nonlinear optics is related with the behaviour of the matter when the frequency of an intense optical field is in close resonance with an electric-dipole-allowed transition frequency of the medium. The nonlinear susceptibilities then are multiresonantly enhanced, increasingly so as their order increases, and the first few terms of the perturbation expansion in powers of the electric field may not adequately describe the nonlinear response of the system even for moderately intense optical fields. However, because all other transitions of the medium contribute very little in comparison with the resonant one, one may altogether ignore them and model[58] the behaviour with that of an assembly of two-level systems resonantly interacting with the optical field. The latter to a large extent can be treated analytically without recourse to perturbation theory. Because this approach constitutes the basis of numerous investigations, we quickly review this model and present a sample of effects that one may expect if such is the case.

a) *Bloch equations.* Let N be the number density of the two-level systems; the two levels are labelled a and b with energies E_a and E_b respectively $(E_b > E_a)$ and $\hbar\omega_0 = E_b - E_a$. The density matrix operator is a 2×2 matrix and its four elements satisfy the equations

$$(93a) \qquad \dot\rho_{ba} = - \left(i\omega_{ba} + \frac{1}{T_2} \right) \rho_{ba} + \frac{i}{\hbar} H_{ba}(t) \Delta ,$$

$$(93b) \qquad \dot\Delta = - \frac{\Delta - \Delta^{(0)}}{T_1} + \frac{2i}{\hbar} (H_{ba}\rho_{ab} - \rho_{ba}H_{ab}) ,$$

where $\Delta = \rho_{bb} - \rho_{aa}$, $Tr\rho = $ const, T_1 and T_2 are the population and coherence relaxation times, and $H_{ba} = -\mu_{ba}(A\exp[-i\omega t] + A^*\exp[i\omega t])$ is the dipole interaction term in the Hamiltonian with $\mu_{ba} \neq 0$ and $\mu_{aa} = \mu_{bb} = 0$. Since $\omega \approx \omega_0$, we can use[58] the rotating-wave approximation by suppressing antiresonant terms. With a slight rearrangement of the terms and introducing the expectation value of the induced transition dipole moment $p = Tr\rho\mu = \mu_{ba}(\rho_{ab} + \rho_{ba})$ one gets

$$(94a) \qquad \ddot p + \frac{2}{T_2}\dot p + \left(\omega_0^2 + \frac{1}{T_2^2} \right) p = - \frac{2\omega_0}{\hbar} |\mu_{ba}|^2 A(t)\Delta ,$$

$$(94b) \qquad \dot\Delta + \frac{\Delta - \Delta^{(0)}}{T_1} = - \frac{2}{\hbar\omega_0} E\left(\dot p + \frac{p}{T_2} \right) \approx - \frac{2}{\hbar\omega_0} E\dot p .$$

Equations $(94a)$, $(94b)$ are the essential starting point for deriving the behaviour of the quantal two-level system interacting with an electric field of frequency $\omega \approx \omega_0$. In particular equations $(94a)$, $(94b)$ show that the problem is reduced to that of an harmonic oscillator forced by a quantal force which is connected with an energy rate equation. With the adjunction of the field envelope eq. (11) one can also treat their impact on the propagation of the field.

Actually Bloch equations can be transcribed in vector form[59] by use of the fact that any 2×2 matrix M can be cast into the form $M = M_0\sigma_0 + \boldsymbol{M}\cdot\boldsymbol{\sigma}$, where $\sigma = \{\sigma_x, \sigma_y, \sigma_z\}$ and $M = \{M_x, M_y, M_z\}$ with

$$\sigma_0 = \begin{vmatrix} 1 & 0 \\ 0 & 1 \end{vmatrix}, \qquad \sigma_x = \begin{vmatrix} 0 & 1 \\ 1 & 0 \end{vmatrix}, \qquad \sigma_y = \begin{vmatrix} 0 & -i \\ i & 0 \end{vmatrix}, \qquad \sigma_z = \begin{vmatrix} 1 & 0 \\ 0 & -1 \end{vmatrix}$$

and $M_0 = (M_{aa} + M_{bb})/2$, $M_x = (M_{ab} + M_{ba})/2$, $M_y = (M_{ba} - M_{ab})/2i$, $M_z = (M_{aa} - M_{bb})/2$. Using this property and setting $H = \hbar(\Omega_0\sigma_0 + \boldsymbol{\Omega}\cdot\boldsymbol{\sigma})$, $\rho = (R_0\sigma_0 + \boldsymbol{R}\cdot\boldsymbol{\sigma})$, one obtains the equation

$$(95) \qquad \frac{d\boldsymbol{R}}{dt} = (\boldsymbol{\Omega} \wedge \boldsymbol{R}) + \frac{d\boldsymbol{R}}{dt}\bigg|_{\mathrm{R}}$$

in the vector space of the Pauli matrices σ_i. Equation (95) is the same as that for the precession of a magnetic moment in a magnetic field in real space; in deriving (95) the rotating-wave approximation was used, which now has a very plausible geometrical justification. The vector model has been also extended to the two-photon resonant case [60].

In terms of this equation several effects can be given a simple geometrical interpretation by noticing that $\boldsymbol{\Omega} = \{\Omega_{\mathrm{R}}(t), 0, \Delta\omega\}$, where $\Omega_{\mathrm{R}}(t) = 2|\mu_{ba}|A(t)/\hbar$ is the on-resonance Rabi frequency and $\Delta\omega = \omega_0 - \omega$ is the frequency detuning, so that $\Omega = \sqrt{\Omega_{\mathrm{R}}^2(t) + \Delta\omega^2}$ is the precession frequency of the state. Effects like superradiance, photon echos, Rabi precession, self-induced transparency and others can be given [58] a simple geometrical interpretation in terms of the vector model; we shall give a brief account [61] of the self-induced transparency because it strikingly shows the impact of resonant interaction on the nonlinear propagation.

b) Nonlinear response. Saturation. The induced polarization in the steady-state regime is $P = \tilde{\chi}A$ with

$$(96) \qquad \tilde{\chi} = \frac{N|\mu_{ba}|^2\Delta}{\hbar(\omega - \omega_0 + i/T_2)} = -\frac{N|\mu_{ba}|^2\Delta_0 T_2(\Delta\omega T_2 + i)}{\hbar(1 + (\Delta\omega T_2)^2) + \Omega_{\mathrm{R}}^2 T_1 T_2}$$

and $\Omega_p = 2|\mu_{ba}|A/\hbar$. This expression clearly exhibits [58] the saturation behaviour with a saturation intensity $|A_{\mathrm{s}}|^2 = \hbar^2/2|\mu_{ba}|^2 T_1 T_2$ at zero detuning. Besides obtaining the saturation behaviour, which is widely used as nonlinear spectroscopy technique, with the adjunction of the field envelope equation one can analyse [16] several other effects like absorptive optical bistability, optical phase conjugation, etc. Note that the saturation is a purely quantum-mechanical effect.

Simple results can also be obtained in the limit of the adiabatic following approximation where the relaxation is assumed to be negligible and the dipole exactly follows the optical pulse. Setting now $\Omega_{\mathrm{R}} = 2|\mu_{ba}|A(t)/\hbar$, one gets

$$(97) \qquad P = -\frac{1}{2}\frac{N\mu_{ab}\Omega_{\mathrm{R}}(t)}{\sqrt{\Delta\omega^2 + \Omega_{\mathrm{R}}^2(t)}}\,\mathrm{sign}\,\Delta\omega.$$

It is easy to verify that for low field intensities by expanding in powers of the intensity one recovers the odd-order susceptibilities of a two-level system in the appropriate limits.

Quite instructive for the understanding of the phenomena that can take place in an assembly of two-level systems resonantly driven by a monochromatic optical field is the expression of the induced dipole moment when T_1 and T_2 are infinite, in which case the equations have the simple solutions $\rho_{aa} = \cos^2(\Omega t/2) + (\Delta\omega/\Omega)^2 \sin^2(\Omega t/2)$, $\rho_{bb} = (\Omega_R/\Omega)^2 \sin^2(\Omega t/2)$ for the initial condition $\rho_{aa}(0) = 1$ and $\rho_{bb}(0) = 0$ and

$$\rho_{ab} = \mu_{ab} \frac{\Omega_R}{\Omega} \left[-\frac{\Delta\omega}{2\Omega} \exp[-i\omega t] + \frac{1}{4}\left(\frac{\Delta\omega}{\Omega} - 1\right)\exp[-i(\omega + \Omega)t] + \right.$$

$$\left. + \frac{1}{4}\left(\frac{\Delta\omega}{\Omega} + 1\right)\exp[-i(\omega - \Omega)t] \right],$$

which clearly shows that the induced transition dipole radiates not only at frequency ω but also at the sidebands $\omega \pm \Omega$ and accordingly we expect[16] enhanced interactions with signal beams at these frequencies.

  c) *Self-induced transparency*. In terms of the vector model one can anticipate[61] that, when $T_1 \gg \tau_\pi$, where τ_π is the optical-pulse duration, and in addition

$$\int_{-\infty}^{\infty} \Omega_R(t)\, dt = 2\pi$$

is satisfied for $\tau_\pi < T_2$, then the state vector coherently undergoes a complete Rabi precession where in the first half gains «energy» from the field which it restores back in the second half of the Rabi precession period; this implies a threshold condition

$$|A| \geqslant \pi\hbar/T_2\mu_{ba}$$

and the optical pulse propagates through the medium without substantial loss, but suffers a delay as a consequence of the Rabi precession that is needed to recover the energy. This is the self-induced transparency effect[61] that only occurs close to a resonance in contrast to the white self-induced transparency [50,51] in a composite dielectric that can occur at any frequency of the transparency region (paragraph 4'1.2c).

5. – Nonlinear optical materials.

  5'1. *General aspects*. – Despite the remarkable progress that nonlinear optics has witnessed over the three decades of its existence, the implementation of related devices in large-scale technology is still not satisfactory. This can be traced to several causes, both structural and conceptual, but the most severe ones can be traced to the present-day performances of nonlinear optical materials. Nature has not been generous with the optical nonlinearities of bulk optical materials; in the case of integrated nonlinear optics the situation is aggravated by additional requirements [62] on the materials as regards their processability, adaptability and interfacing with other materials. These additional requirements are intrinsically related to the fabrication of nonlinear integrated devices, which, besides efficiently performing the expected nonlinear operation, must be miniaturized, compact, reliable and with precisely reproducible characteristics in large-scale production and long-term operation. We succinctly present the main representatives of different classes of nonlinear optical materials that hold promise in this field and discuss the relation between optical nonlinearities and structural and dynamic characteristics of the valence electron distribution.

In order to introduce some systematics in the presentation, we remind that the nonlinear optical effects irrespective of their order can be divided into two major classes:

the frequency-preserving effects like the optical Kerr effect that can be used for bistable operation, phase conjugation, soliton formation and other operations on the spatiotemporal profile of coherent light pulses; here one can also include the hybrid effects like electro-optic, magneto-optic and acousto-optic effects that can be used for parametric devices;

the frequency-shifting effects like the sum or difference frequency generation processes, which allow one to up/down convert the frequencies of the existing light sources.

There is a major reason for this division. Indeed the efficiency [3, 9, 16] of the latter effects is conditioned by the ability to match optical characteristics in the material, like the phase and group velocities, at widely different frequencies. This is a formidable task and drastically restricts the possible choices and the situation is further complicated by the fact that the material characteristics cannot be optimized over the wide frequency range often required in frequency up and down conversion processes.

In contrast the efficiency of the frequency-preserving effects is not affected by such problems, since only frequencies within a very narrow spectral range are involved and the matching conditions can be automatically satisfied; as a matter of fact to a large extent the interactions here can be assumed to be local and are in general much easier to analyse.

As a consequence the nonlinear frequency-preserving effects are the ones that are most seriously considered [62] in integrated optical devices. They can be either all-optical or hybrid (parametric) effects. The all-optical nonlinearities [7] essentially involve valence electron motion and are in general weaker than the hybrid ones where the ionic motion, vibrational, orientational or translational, can set up very large nonlinearities; the situation, however, is reversed as regards the speed of establishing and erasing these nonlinearities, the electronic one being much faster than the ionic ones. The magnitude and speed of the nonlinearities are essential characteristics in any assessment of the materials for nonlinear optical devices and must be properly introduced [63] in any figure of merit.

The magnitude and speed of the nonlinearity, however, are strongly dependent on how close to a material resonance is the operating frequency, or a multiple thereof, as this introduces resonant enhancement of the nonlinearity and absorption losses; furthermore the speed is limited by the relevant relaxation processes close to a resonance. One may have linear (one-photon) or nonlinear (multiphoton) absorption losses which can be related to imaginary parts of odd-order susceptibilities; we stress the fact that there are no absorption losses related to even-order nonlinear processes. Thus the amount of linear and/or nonlinear absorption losses is another important criterium for assessing nonlinear optical materials along with the magnitude and speed of the nonlinearity at the frequency of interest.

Clearly the above criteria are relevant to the extent that the material is not irreversibly modified by the optical radiation. Such modifications can result from photochemical or thermal bond breaking, state-selective photochemistry, optical breakdown and many others directly related to the interaction of light and matter. In addition material growth, doping and aging considerations, processability, interfacing and packaging and many other aspects will heavily weigh on the final choice, since for a wide class of materials the nonlinear optical figures of merit are of comparable magnitude.

On the basis of the types of cohesive forces that bind the charges and polarizable units together, the nonlinear optical materials can be assigned to one of the following classes:

ionic crystals, essentially oxygen-polyedra-based solids;

covalent crystals, essentially semiconductors;

molecular crystals, in particular organic and polymeric;

disordered and amorphous solids, in particular glasses and polymers;

composites and inhomogeneous artificial solids.

We shall succinctly review the main features of some materials representative of these classes. At this stage, however, it is appropriate to make certain remarks concerning the evaluation criteria for the corresponding nonlinearities.

A rough order-of-magnitude estimate of $\chi^{(n)}$ can be obtained [2] through

$$\chi^{(n+1)} \approx 1/E_c^{-n} \, ,$$

where E_c is the effective cohesive field that keeps attached the polarizable charges or units to each other. All arguments concerning the improvement of the nonlinear coefficients in the last analysis amount in appropriately modifying

TABLE I. – *Indicative (*) table of second-order nonlinear coefficients.*

Crystal	Symmetry	n_0	d_{21}	d_{14}	d_{33} $(10^{-12}$ m/V)	d_{eff} (a)	C	I_{ob} (GW/cm²) (b)	Transmission range (μm)
LiNbO$_3$	$3m$	2.232	−2.1	0	−27	5.1	70	10	0.35–5
BaBO$_4$	$3m$	1.655	−2.3	0	0	1.9	16	14	0.2–2.6
LiIO$_3$	$6m$	1.857	0	0	4.5	1.8	13	2	0.34–4
KDP	$\overline{4}2m$	1.493	0	0.37	0	0.35	1	5	0.18–1.8
ADP	$\overline{4}2m$	1.509	0	0.47	0	0.39	1.2	6	0.18–1.5
AgGaGe$_2$	$\overline{4}2m$	2.594	0	33	0	28	81	0.3	0.78–18
ZnGeP$_2$	$\overline{4}2m$	3.073	0	69	0	70	292	0.05	0.74–12
KTP	$mm2$	1.737	0	0	8.3	3.2	47	15	0.35–4.5
KNbO$_3$	$m2$	2.119	0	0	−19.5	−11	312	7	0.4–5.5
					d_{11}				
urea	—	—	—	—	12	—	—	5	0.2–1.4
MAP	—	1.508	16.8	—	—	—	—	3	0.5–2.5
POM	—	1.663	—	9.2	—	—	—	2	0.5–1.7
MNA	—	2.0	—	—	168	—	1000	—	0.48–2.0
NPP	—	—	—	—	85	—	—	0.05	0.48–2.0

(*) Values selected from ref. [64].
(a) Divide each value by $4.2 \cdot 10^{-4}$ to convert to e.s.u.
(b) Optical breakdown.

this field. In tables I and II we have collected the values for some well-known materials.

To the extent that in device applications one essentially exploits [63, 65] the phase shifts provoked by the nonlinear terms, the relevant parameters in the assessment of the nonlinear materials are not the bare nonlinear susceptibilities but appropriately renormalized expressions thereof that measure the induced relative phase shift; these are called figures of merit and are also directly related to the efficiency of the nonlinear interaction. Their precise definition is actually closely related to the operation or the device one has in mind.

Thus, if we concentrate [65] our attention on all optical guided wave-switching devices like nonlinear directional coupler, nonlinear Bragg reflector, nonlinear Mach-Zehnder interferometer, nonlinear mode sorter or nonlinear X-junc-

TABLE II. – *Indicative table of third-order nonlinear coefficient.*

Material	$\chi^{(3)} (10^{12} \text{e.s.u.})$
glass	10^{-4}
NaCl	10^{-2}
CS_2	1
Si	1
GaAs	10
Ge	10^2
PTS	10^2
SDG $(^a)$ (resonant)	$> 10^4$

$(^a)$ Semiconductor doped glasses.

tion, the relevant figure of merit can be derived from the requirement that the nonlinear phase shift per unit absorption length is larger than 2π. The total phase shift over a length L of the nonlinear material being

$$(98) \qquad \Delta\phi = (n_0 + n_2 I) k_0 L = \Delta\phi_0 + \Delta\phi_{NL} ,$$

where n_0 and k_0 are the linear refractive index and wave vector, while n_2 is the optical Kerr effect coefficient (21). The absorption coefficient being $\alpha_0 + \alpha_2 I$, where α_0 is the one-photon (linear) loss and α_2 the «two-photon» (nonlinear) loss, the above criterium implies that

$$(99a) \qquad \Delta\phi_{NL}/2\pi = n_2 I/\lambda_0 \alpha_0 \geqslant 1$$

if one-photon absorption is dominant and

$$(99b) \qquad n_2/\lambda_0 \alpha_2 > 1$$

if two-photon absorption is dominant. Thus possible figures of merit are $F_0 = n_2/\alpha_0$ and $F_2 = n_2/\alpha_2$, respectively; these considerations pertain to steady-state operation regime. In actual devices using a high-repetition pulse rate regime one rather uses [66] the figure of merit

$$(100) \qquad F_d = \frac{n_2 I}{\alpha_0 c\tau} ,$$

where τ is the recovery time of the nonlinearity, which in practice is the population relaxation time T_1. One can similarly introduce figures of merit for quadratic nonlinearities. One such figure of merit can be derived as

$$(101) \qquad C = d_{\text{eff}}^2/n^3 ,$$

where $2d_{\mathrm{eff}} \equiv \chi^{(2)}$ and n is the refractive index. Arguments based on figures of merit should be made with some caution as they may lead to irrelevant controversies.

5˙2. *Second-order nonlinearities.* – There is only one second-order frequency-preserving effect, namely the electro-optic or Pockels effect which was discussed in subsect. 2˙3. It is related to the nonlinear second-order polarization

$$\boldsymbol{P}_\omega^{(2)} = 2\,\underline{\underline{\chi}}_{\mathrm{EO}}^{(2)}(\omega,\,0)\,\boldsymbol{E}_\omega\boldsymbol{E}_0\,,$$

where ω is an optical frequency within the transparency region of the crystal and «0» is a frequency well below any vibrational or orientational frequency. The second-order susceptibility $\chi_{\mathrm{EO}}^{(2)}(\omega,\,0)$ contains[7] two contributions: one purely electronic, the same as that for the optical second-order harmonic generation $\chi_{\mathrm{E}}^{(2)}(0,\,0)$, and another $\chi_Q^{(2)}$ which arises from a rearrangement of the valence electron distribution consecutive to nuclear displacements induced by the static field. The first is directly related to the characteristics of the valence electronic-charge distribution, while the second is related to the modulation of the latter by the phonons and can be evaluated from infrared and Raman spectroscopic data. Except for the ferroelectrics[39], where the ionic displacements related to the driven soft mode can be substantial, the latter contribution is in general smaller although comparable to the purely electronic one and of either relative sign; the overall signs of $\chi_{\mathrm{EO}}^{(2)}$ and $\chi_{\mathrm{E}}^{(2)}$ (or $\chi_Q^{(2)}$) are of certain relevance[7] and must be explicitly stated together with the adopted axis orientation. In the following we will concern ourselves only with the electronic contribution $\chi_{\mathrm{E}}^{(2)}(0,\,0)$ to the electro-optic coefficient which is also relevant for second-harmonic generation.

In the early days of nonlinear optics it was suggested[67] that an estimate of $\chi_{ijk}^{(2)}$ could be obtained through the conjecture that the coefficient

(102) $\qquad \delta_{ijk}(\omega_1 + \omega_2) = \chi_{ijk}^{(2)}(\omega_1,\,\omega_2)/\chi_{ii}^{(1)}(\omega_1 + \omega_2)\,\chi_{jj}^{(1)}(\omega_1)\,\chi_{kk}^{(1)}(\omega_2)$

is a constant for all crystals. A more careful examination[41], however, of the nonlinear polarization mechanisms and values of $\chi^{(2)}$ for different crystal classes showed that this conjecture does not apply and δ may differ by orders of magnitude among the different crystal classes and does not constitute a useful figure of merit for identifying crystals with large nonlinearity. However, the fact that the spread of δ_{ijk}-values is narrower than that of $\chi_{ijk}^{(2)}$ indicates that among the noncentrosymmetric crystals those with large $\chi^{(1)}$, which implies crystals with covalent bonding, would also have large $\chi^{(2)}$ in general. The question is then to what extent the bond configuration can be optimized for a crystal to show a large second-order nonlinearity; this is the key problem in noncentrosymmetric-material research.

One may also assume[7] that the second-order polarizability coefficients of these complexes, after certain provisions regarding local field corrections are made, satisfy the additivity and transferability property as was found to be the case for the linear polarizabilities. This implies that the second-order suscepti- bility is entirely determined by the asymmetric charge distribution within the constituent repeat polarizable unit and not by charge delocalization across such units. This greatly simplifies the search of materials with large second-order nonlinearities, on the one hand, and, on the other, points the road for fabricat- ing new ones starting with molecules that possess large second-order polariz- abilities. These remarks have been essential in improving the second-order nonlinear coefficients of existing materials by judicious modification of the con- stituent unit and in conceiving new materials by using appropriate molecular- engineering techniques.

5·2.1. Inorganic covalent crystals.

a) Heteropolar semiconductors. This is the simplest class of noncen- trosymmetric crystals[68]. The building-block unit here is one heteropolar bond that serves to form the whole crystal through tetrahedral connection; the bonding is sp^3-type and the structure is cubic zincblende (for example, GaAs) or hexagonal wurtzite (for example, CdS), the former being optically isotropic, while the latter is uniaxial and differs from the former by compression or ex- pansion of the bonds along the 111-direction which then becomes the c-axis. Their valence electronic structure has been extensively studied by different ap- proaches. Using the bond additivity conjecture and making provisions for local field corrections, one may express[7,41] their linear and second-order optical coefficients in terms of those of the heteropolar bonds, namely the linear and second-order polarizabilities α and β, respectively; one actually may assume the bonds to be unidimensional. The values of $\chi^{(2)}$ can be thus predicted to high ac- curacy and related to the charge asymmetry along the heteropolar bond. Using (68) one can show[7,42] that, for an isoelectronic series of IV-IV, III-V, II-VI and I-VII compounds, the III-V compound has the optimal second-order nonlin- earity; furthermore the d-hybridization of the sp^3-orbitals (for example, CuCl) may lead to a reversal of sign of $\chi_E^{(2)}$.

As a general rule $\chi^{(2)}$ is large in these materials, but so is also $\varepsilon_\infty = n^2 = 1 + + 4\pi\chi^{(1)}$, so that their figures of merit are not impressively larger than those of other materials we will discuss below. Their most severe drawbacks, however, are their optical isotropy, which prevents phase matching, and their narrow transparency region that hardly extends beyond $2\,\text{eV}$ for any of them. Al- though the first one can now be compensated by fabricating periodic stacked structures with the polar-axis direction periodically reversed from layer to layer, the second one remains a major drawback because real transitions and photocarrier generation can easily take place in these materials which greatly reduce their electro-optic modulation efficiency. With new fabrication techniques[69] that have been introduced in semiconductor technology, however, these materials have surfaced again in the competition for the development of nonlinear integrated optical devices and compounds like InP, which possess a large electro-optic coefficient and a wide energy gap, are now seriously considered there.

b) *Ionic crystals. Crystals with oxygen polyedra*. This is the class of materials that presently sets the standards in designing prototype integrated nonlinear devices. These can be viewed as ionic crystals[23,68] where one of the ions is replaced by an oxygen polyedron AO_n, in a ionized state, and stabilized in a noncentrosymmetric configuration by the surrounding ions (usually of metal elements). The second-order optical nonlinearity here results from the unbalanced system of heteropolar AO bonds which has a rather complex configuration around the element A and in contrast to the heteropolar semiconductors the bonds here cannot be assumed unidimensional. Indeed in all models, in order to account for the magnitude and sign of $\chi^{(2)}$, one must introduce[23] for each bond a substantial transverse contribution $\beta_\perp$ for the second-order polarizability; this is because the AO bonds are usually counterdirected in pairs and their longitudinal components $\beta_\parallel$ cancel to a large extent. Actually the additivity of bond polarizabilities has never been convincingly established for these compounds. Rather one should treat the whole AO_n polyedron as a single unit, but this makes the calculation and prediction of the nonlinearities quite cumbersome and subtle, since the modelization of the electronic distribution must be compatible with the one expected from the phase transitions that almost all these materials undergo, *e.g.* para- to ferroelectric.

As a general rule the values of $\chi_E^{(2)}$ for these materials are lower than those of the heteropolar semiconductors, but most of their other bulk properties are superior to those of the semiconductors. In particular they are optically anisotropic allowing for phase matching of nonlinear interactions, their transparency region can extend up to the near ultraviolet, they are robust against photochemical degradation or optical breakdown and can be produced in high optical quality. Because of their complex chemical structure, a superposition of ionic and covalent ones, however, they cannot be processed by the same sophisticated fabrication techniques as the semiconductors; furthermore their surfaces, because of their high polarity and the establishment of charged layers, are easily attacked by moisture and other degradation and must be properly treated, encapsuled or packaged.

By weighting all these factors at present only $LiNbO_3$ seems to gain the favour of the potential manufacturers and actually constitutes the reference for comparing and evaluating all other potentially useful materials. The progress, however, in growing other oxygen-polyedron-based crystals with larger second-harmonic efficiency than $LiNbO_3$ has been spectacular[64] over the last years and has led to many applications in ultrashort-laser-pulse technology and may well lead to similar improvements in nonlinear integrated optical devices that exploit the electro-optic effect. Certainly here the introduction of new crystal growth and fabrication techniques will be needed in order to reach the standard required for integrated optical devices.

5`2.2. Molecular crystals. Organics. Much expectations were laid [70-73] on noncentrosymmetric molecular crystals, in particular the organic ones, for second-order nonlinear effects and a considerable effort has been concentrated in the last two decades on identifying such crystals or growing new ones with second-order nonlinearities higher than those of $LiNbO_3$. The results have been mitigated; such molecular crystals with large values of $\chi^{(2)}$ formed by organic molecules have now been grown, but their other performances fall short of all expectations and their potential implementation in nonlinear integrated optical devices seems problematic and still open to debate. The major limitation is their photochemical instability and aging against long-term exposure to intense light pulses at high repetition rate as required in most applications.

All advantages but also disadvantages of the molecular organic crystals with respect to the covalent and ionic ones for applications in nonlinear integrated optics can be traced to the difference between inter- and intramolecular forces that prevail in these systems. As a consequence the constituent molecules preserve their main characteristics even when assembled to form the molecular crystal. Indeed, in contrast to the covalent and ionic crystals where inter- and intracell forces are similar and of equal strength, in the molecular ones the intramolecular forces are covalent directive and already saturate with the formation of the molecule, while the intermolecular ones are essentially unsaturated van der Waals or dipole-dipole interactions which are more than one order of magnitude weaker than the intramolecular ones and do not significantly alter the intramolecular electron distribution. In several cases other forces like the hydrogen bonding [70], intermediate in strength between the previous two, are operative.

Thus the process of search and growth of noncentrosymmetric molecular crystals with large second-order nonlinearities at first sight is greatly facilitated by these facts, since it is reduced to that of the identification and synthesis of stable asymmetric molecules with large second-order polarizabilities β and wide transparency range between the highest vibrational and the lowest electronic transitions. The latter requirement is very restrictive regarding organic molecules and excludes the majority of them for further consideration in nonlinear optics. Among the remaining ones a large proportion is also excluded because they cannot form stable noncentrosymmetric crystals. This is mainly because asymmetric molecules most frequently carry a dipole moment in their ground electronic state and, in order to reduce the dipole-dipole interaction, which is dominant over the van der Waals in the lattice, a head-to-tail antiparallel configuration will be favoured when forming the crystal most frequently resulting in centrosymmetric crystalline structures and consequently vanishing $\chi_E^{(2)}$ unless certain precautions[70] are taken to prevent this from happening. There are several features that can be exploited in this respect:

poling in an externally applied field,

synthesis of chiral molecules,

synthesis of asymmetric molecules with vanishing dipole moment in the ground electronic state,

hydrogen bonding.

The first one is not very favourable for crystalline structures, but has been successfully applied and exploited in polymers[74-76], as will be discussed below. The other three, on the other hand, have been successfully used for growth of good optical-quality asymmetric crystals with large second-order susceptibilities. We cursively present the most representative crystal cases, namely MAP (methyl-(2,4-dinitrophenyl)-aminopropanoate), POM (3-methyl-4-nytropyridine-1-oxide), NPP (N-(4-nitrophenyl)-(L)-prolinol) and urea ($CO(NH_2)_2$).

Because of the wide difference between inter- and intramolecular forces in molecular crystals one may safely assume that the molecular-polarizability coefficients α and β satisfy the additivity property. For most organic molecules the dominant contribution to β comes from an asymmetric part of the electronic distribution which can be modelled with a quantum two-level system; the expression of β was derived in (70). The coefficient β is a third-rank tensor and clearly its maximum value is obtained for directions where $\Delta\mu$ and μ_{eg} are maximal but not necessarily collinear. For most organic molecules the most favourable case occurs when a charge transfer complex is formed[70] and this can be modelled with a two-level system involving a nonbonding π-donor orbital and a vacant π-antibonding acceptor. The important point to notice in (70) is that β only depends on the difference $\Delta\mu$ and not on the individual dipole moments μ_{gg} and μ_{ee}; in particular μ_{gg} can be zero and yet have $\beta \neq 0$ as long as $\mu_{ee} \neq 0$. Accordingly one can have molecules with large β but vanishing μ_{gg}, in which case dipole-dipole interactions in the lattice are weak and cannot prevent the noncentrosymmetric lattice configuration from being formed. Such is the case of POM[77] and presently good-quality large crystals for this compound have been grown and used in second-harmonic generation.

This situation, however, is not common and one must cope with molecules having large β-values but also appreciable μ_{gg}-values; the remedy then is either to render the molecule chiral by the adjunction of an appropriate molecular unit, either left- or right-handed, in which case the lattice structure can never be centrosymmetric, this is the case of MAP[78], or MNA[79], where one can judiciously introduce hydrogen bonding between the molecules to overcome the dipole-dipole forces and combine with chirality to favour noncentrosymmetric structures as in case of NPP[80]. Presumably this situation is prevailing also in urea[81]. In all three cases very-good-quality crystals with large $\chi_E^{(2)}$ have been grown and used in several applications; urea seems to be the most stable of them. But clearly these approaches can be used for a wide range of organic materials.

Despite a formidable pluridisciplinary effort[70-72] the bulk organic and molecular crystals, with the possible exception of urea[81], have not reached yet the required standards for applications in nonlinear optics and much less so in integrated optics. The reasons are several and can be traced back, as was previously pointed out, to the weakness of the intermolecular forces with respect to the intramolecular ones and to the sensitivity of organic molecules to light. As a consequence the organic crystals have much larger defect density than the covalent or ionic crystals; also their mechanical and optical resistance is weaker than in the inorganic crystals. The second issue is even more serious as all organic molecules easily undergo bond breaking, bond arrangement or other photochemical reactions once the photon energy or a multiple thereof is close to a photosensitive transition. In addition organic crystals are apt to aging even without exposure to light. Another serious drawback regarding the use of organic crystals in nonlinear integrated optical devices is the difficulties encountered in doping, processing, polishing or interfacing them with techniques currently used in semiconductors, ionic and other inorganic crystal.

5˙2.3. Disordered oriented media. Crystallinity is not an imperative requirement for second-order effects to occur in a material. It is sufficient for the constituent asymmetric molecules to be oriented on the average so that the medium macroscopically lacks inversion symmetry. This implies that the requirement of translational periodicity is irrelevant as long as the molecules are uniformly distributed but with their axis pointing on the average in the same direction. This can also occur in an otherwise centrosymmetric medium, if one uniformly induces oriented noncentrosymmetric molecular complexes. These two approaches have been successfully put into practice in the case of poled polymers[71,74-76] and in certain glasses[82,83]. While the first class[74-76] was the outcome of a well-thought strategy that led to materials and prototype devices that compete or even surpass those based on $LiNbO_3$, the second class was an unexpected[82,83] finding and the origin of the second-harmonic generation there is still not well understood and accounted for by the different models that have been proposed. We remind that solid polymers[84] and glasses can be prepared in excellent optical quality and all show a phase transition around a temperature T_g to a plastic phase where collective flexibility and molecular motion at large distances can take place; below this temperature the medium is rigid and only slight local motion is allowed like in a crystal. T_g can be very high, well above the room temperature.

a) Poled polymers. Here organic molecules with large charge transfer type polarizability β are introduced and uniformly dispersed[76,84] into an excellent optical-quality amorphous organic polymer. The solution of the polymer and guest molecules is then evaporated by spin coating, film casting or other techniques; at this stage the guest molecules are randomly oriented. They are oriented by poling, namely by applying a strong static electric field (electrode poling) or corona discharge (corona poling) at a temperature T close to T_g; by a proper choice of the polymer its T_g can be very high, much higher than the extreme temperatures that a transparent nonlinear material will be expected to operate in practical applications. Several other precautions allow one to obtain excellent optical-quality poled polymers, photoresistant and long aging. Concerning the guest-molecule relation to the polymer several situations are possible[76,84]:

guest-host system: here the nonlinear molecule is not attached to the polymer chain and thus can freely rotate close to T_g;

side-chain (co)polymers: here one end of the nonlinear molecule chemically reacts and is covalently attached or grafted on the polymer chain but its other end is free;

cross-linked polymers: here both ends of the nonlinear molecule chemically react and are covalently attached with the polymer chains;

main-chain systems: here the guest molecule is inserted in the chain.

The poling efficiency depends on several factors, but the most important one is the ratio $\mu_{gg} E/kT$, where E is the poling electric field; corona and electrode poling have their own advantages and disadvantages. The techniques are well understood and controlled and allow one to obtain poled polymers with large orientational order for the guest molecules, namely large $\langle \cos^3 \theta \rangle$, where θ is the angle between the guest-molecule axis and the poling-field E-direction.

Poled polymers exhibit[76,84] substantial advantages over all crystalline materials regarding most requirements, nonlinearity, optical quality, processability, interfacing, phase matching, etc., except for one: their stability against aging is not yet satisfactorily solved. Clearly this is connected to the excess of T_g over the highest temperature that the material reaches under practical operation. The best results were obtained with the cross-linked polymers where aging can be retarded to any desirable length of time, but more recently equally good results were also obtained with side-chain polymers. The aging manifests itself with the lowering of $\langle \cos^3 \theta \rangle$, namely the guest molecules lose preferential orientation and disorientation and $\chi^{(2)} \sim N\beta\langle \cos^3 \theta \rangle$ decreases. Such a decrease of $\chi^{(2)}$ always occurs immediately after the poling is interrupted and the material is cooled to the room or the operating temperature; however, after a few hours or days the average orientation parameter $\langle \cos^3 \theta \rangle$ and $\chi^{(2)}$ stabilize to values that are well above those of most crystalline noncentrosymmetric materials inclusive $LiNbO_3$ and can remain so for years with a judicious choice of the guest molecule and the polymer. Another problem that recently surfaced in connection with their implementation in electro-optic waveguided devices is the occurrence[85] of a photorefractive effect, which is not well understood and clearly sets limitations in the long-term operation since it implies the existence of charged molecular defects. Notwithstanding these and some other drawbacks, poled polymers presently constitute the most serious and advantageous contenders in the competition and development of nonlinear integrated devices.

b) Glasses. Glasses and in particular fused silica have now reached outstanding optical quality, more than any other optical material, as their introduction in modern technology becomes increasingly sophisticated and irreversible. These materials were always expected to be macroscopically centrosymmetric and the observation[82] presently of second-harmonic generation in Si-Ge glass fibres, but also in other glassy media, came as a surprise and aroused immense interest and expectations[86]. Despite the substantial and understandable effort in delineating the mechanisms that provoke this effect, its understanding is still not satisfactory and falls short of our capability to control and exploit it. The proposed models like electric-field-induced nonlinearities, the static electric field being provided by a third-order rectification process, or the photovoltaic effect based on the interference between the fundamental and the harmonic field, or other mechanisms, do not simultaneously account for all observations related to this effect.

The fact is that a glass fibre of particular constitution, for instance Si-Ge, which initially does not show second-order generation, after irradiation with a laser beam within a certain time lapse and over a certain distance that never exceeds a few tens of centimetres, evolves to a fibre that subsequently allows instantaneous and highly efficient second-harmonic generation at the same wavelength as at its preparation stage; this is because of a built-in photoinduced spatial modulation along with noncentrosymmetricity that allows phase matching at a given wavelength. The SH conversion efficiency can be very large by more than 20%. It has not been proved yet that the same fibre can also be used for efficient electro-optic modulation or other applications related to the electro-optic effect.

More recently it was shown[83] that a large second-order nonlinearity less than but still comparable to that of $LiNbO_3$ can be induced within a layer near the surface region of commercial fused-silica optical plates at a temperature close to the Orbach temperature and with a poling process in the presence of a static field; the layer can be of the order of a few micrometres, namely of the order of or larger than typical optical wavelengths.

5‘3. *Third-order nonlinearities.* – The main frequency-preserving third-order effect is the optical Kerr effect which is related to the third-order polarization

$$(103) \qquad \boldsymbol{P}_{\omega}^{(3)} = 3 \, \underset{\equiv}{\chi}^{(3)} (\omega, \, -\omega', \, \omega') |\boldsymbol{E}_{\omega'}|^2 \boldsymbol{E}_{\omega} \,,$$

where ω and ω' can both be optical frequencies, but ω' can also be zero, in which case the effect reduces to the static Kerr or quadratic Pockels effect also defined in (44b). It is also convenient to introduce the related intensity-dependent indices $\widetilde{n}(I) = n_0 + n_2 I$ and $\widetilde{\alpha}(I) = \alpha_0 + \alpha_2 I$ by (20a), (20b).

There are several mechanisms [7] that can lead to such photoinduced changes of the refraction or absorption: electronic, vibrational, orientational, translational. With the exception of the first one all the others involve ionic or molecular motion and can be substantial in magnitude, but, when all factors are taken into consideration, these are not exploitable in applications. Hence the following discussion will only concern the electronic cubic nonlinearities.

In line with the introductory discussion of sect. 4 the electron delocalization over several identical units has been identified [70] as the principal material feature that markedly influences the magnitude of the cubic nonlinearities; this is in contrast with the case of quadratic nonlinearities where the charge asymmetry within a single unit is the material feature that most matters. The main implications are that the structure must be periodic at least over the electron delocalization length and the polarizability additivity which is a real space property is irrelevant for materials with large cubic nonlinearities like the inorganic semiconductors and the linear conjugated polymers. Instead in such materials where the electron states are delocalized band states one can establish [87] an additivity in k-space which involves the contributions to $\chi^{(3)}$ from a small number of critical points in the joint density of states. This allows one to derive scaling laws that strongly depend on the electron charge dimensionality. In such materials with very delocalized electrons the cubic nonlinearities [88,89] can be sensitive to many-body effects like the Fermi exclusion principle or change screening.

It has been suggested [7,28,29] that third-order nonlinear processes can also take place through two cascading second-order processes if the material is noncentrosymmetric. Here the electric field generated by a second-order polarization, induced by any pair of incident beams inside the material, interacts with one of the incident beams to create a second-order polarization which, however, is effectively equivalent to a third-order one and can be cast in the form (103) and lumped together with this term; the corresponding effective third-order nonlinearity which is now nonlocal and depends on the wavelengths can be comparable to or larger than the intrinsic cubic nonlinearity in particular if phase matching is achieved at the intermediate second-order process.

In general the values of $\chi^{(3)} (\omega, \, -\omega', \, \omega')$ due to the electrons, when ω and

ω' are in the transparency region of the material, are very weak to be exploited for applications involving interactions that occur over a short distance as is the case in miniaturized single-pass integrated devices. The only alternative is to resonantly enhance the nonlinearities. The use of resonances, however, introduces two drawbacks which in general can be very severe: linear and/or nonlinear absorption losses, on the one hand, and, on the other, a long nonlinearity recovery time. One can resonantly enhance $\chi^{(3)}(\omega, -\omega', \omega')$ by having ω or ω' (or both) close to dipole-allowed resonances of the material; here the absorption losses are linear at low intensity and nonlinear at high intensities (saturation effect). One can also enhance $\chi^{(3)}(\omega, -\omega', \omega')$ by having $\omega + \omega'$ (or $\omega - \omega'$) close to a two-photon (or Raman) allowed resonance without any of the frequencies ω or ω' being close to a resonance; here the losses are nonlinear at low intensity.

These drawbacks related to the resonant enhancement of the cubic nonlinearities can be brought under control to some extent by exploiting [66, 89, 90] morphological resonances whose position, width, oscillator strength can be externally controlled and tailored to meet certain requirements related to the desired performances of the device. Such morphological resonances can be introduced through confinement of the delocalized electrons, for instance the quantum and dielectric confinements. The new artificial-material fabrication techniques allow [64, 69] one to control the interfacing and confinement to better than an atomic layer. Several classes of such artificial semiconductor-based microstructures are now studied for this purpose and in the case of quantum wells these studies have reached a high degree of sophistication. For applications, however, the artificial materials that emerge as the most promising [66, 90] are the composites obtained by uniformly dispersing semiconductor or even metal nanocrystals in a glass or in a transparent polymer matrix of high optical quality.

Based on the above general remarks the materials that presently show promises in devices that exploit cubic nonlinearities are

covalent semiconductors,

composite materials,

linear conjugated polymers,

noncentrosymmetric media with large $\chi^{(2)}$.

We shall briefly discuss the main nonlinear mechanisms that are responsible for large cubic nonlinearities and how they relate to the electron distribution and dynamics.

5˙3.1. Semiconductors. Covalent semiconductors like Ge or GaAs possess [91] the largest nonresonant cubic nonlinearities among all known crystals, but these are unexploitable for several reasons, the most severe being the low absorption threshold in these compounds. On the other hand, large and exploitable cubic optical nonlinearities can be produced [88, 92, 93] by generating a finite concentration of electron-hole pairs by resonant excitation above the band gap.

These large resonant nonlinearities have their origin in the band-filling mechanisms whereby the photocreated electrons and holes quickly thermalize and fill all states at the bottom of the conduction and at the top of the valence bands, respectively, up to levels that depend on the light intensity and pulse or recombination times, thus excluding these states from further occupation because of the Fermi principle. This blocking mechanism [88, 93] appears as a repulsion of the states on either side of the forbidden energy gap or, equivalently, as a blue shift of the absorption threshold; at very high intensities this mechanism leads to a saturation of the nonlinearity.

This band-filling mechanism is not the sole cause of nonlinearities; it is accompanied [88, 89] by additional many-electron effects, like exciton formation and band renormalization, that have a certain impact on the global nonlinearity. These additional effects have received a lot of attention, but their impact on device applications is overestimated and to some extent unreliable; in particular their relative contribution to the total nonlinearity is greatly reduced at room temperature with respect to that of the band filling.

As previously stated, the major problem with resonant nonlinearities and in particular the ones in semiconductors is the concomitant absorption losses and the long times required for the photoinduced nonlinearities to decay. The latter are related to the evolution of the electron and hole populations. These can be modified to some extent with appropriate doping or through the quantum confinement by using sophisticated but well-established fabrication techniques of artificial materials.

5˙3.2. Composites. Semiconductor and metal nanocrystals in glasses. A way to enhance the cubic nonlinearities of materials with very delocalized electrons, like metals, semiconductors, or conjugated polymers, is to artificially confine the valence electrons in regions much shorter than their natural delocalization length in the bulk, which extends over many unit cells or even to infinity; its more conspicuous feature is the appearance [66, 89, 90] of broad but discrete optical resonances whose position, oscillator strength and dynamics depend on the extension of the artificial confinement and hence can be modified to meet certain requirements. These morphology-related [90] resonances result from two types of confinement: quantum and dielectric. The first one prevails in semiconductor nanocrystals and the second one in metal nanocrystals.

The dielectric confinement resides [66, 94, 95] in the difference of the dielectric constants between the crystallites and their surrounding transparent medium. Because of this dielectric inhomogeneity the electric field $E_{\omega l}$ that effectively polarizes the charges in these crystallites can be substantially different from the macroscopic Maxwell field E_ω in the composite. Under certain simplifying conditions one shows that the relation between these two fields is

$$(104) \qquad E_{\omega l} = \frac{3\varepsilon_0}{\varepsilon_m(\omega) + 2\varepsilon_0} E_\omega \equiv f_l(\omega) E_\omega$$

and the effective cubic susceptibility of the composite is [94]

$$(105) \qquad \widetilde{\chi}^{(3)}(\omega) = p|f_l(\omega)|^2 f_l(\omega)^2 \chi_m^{(3)}(\omega)$$

for spherical particles of volume concentration $p \ll 1$, dielectric constant $\varepsilon_m(\omega)$ and cubic susceptibility $\chi_m^{(3)}$ embedded in a transparent dielectric of dielectric constant ε_0. To the extent that only $\varepsilon_m(\omega)$ is complex and frequency dependent in the optical range, the field $E_{\omega l}$ in (104) is enhanced close to the resonance ω_s, such that

$$(106) \qquad \mathrm{Re}\,\varepsilon_m(\omega_s) + 2\varepsilon_0 = 0\,,$$

which is the condition for the surface plasmon frequency; the cubic nonlinearity (105) is enhanced by the fourth power of the same resonance. The position of the resonance is controlled by modifying ε_0, while its width depends on the size of the crystallites.

Careful studies [66, 95] of the nonlinearities of composites formed by uniform dispersion of gold particles of different average sizes in glass matrices revealed that the quantum confinement is irrelevant in these compounds and that the nonlinearity $\chi_m^{(3)}$ of the crystallites results from essentially the saturation of the interband transition and the rearrangement of the hot conduction electron population.

The quantum confinement occurs [90, 96] when the electron and hole envelopes are restricted within a region of extension L equal to or smaller than the electron and hole Bohr radii, a_e and a_h, respectively; the latter are defined in the bulk by the condition that the average value of the electron or hole kinetic energy roughly equals that of the potential. The confinement perturbs this balance, since these energies now vary as $(a_c/L)^2$ and a_c/L, respectively, where $a_c = a_e, a_h$. The characteristic energy of the confinement is the kinetic energy

$$(107) \qquad E_c = \frac{1}{2}\frac{e^2}{a_c}\left(\frac{a_c}{L}\right)^2,$$

and, as L decreases, E_c increases and gradually suppresses the effect of the other interactions. Otherwise stated, the charges behave as free within the confinement region, which to a good approximation can be considered as a spherical quantum well of infinite potential height. The main consequence is that the initially continuous energy spectrum is replaced[96] by a discrete one with a spacing that varies as (107); the widths of these resonances as well as their oscillator strengths can dependent on L. Because of the selection rules that prevail for the transitions between these quantum confined states each crystallite essentially behaves as a quantum two-level system and the main nonlinearity results[97] from the saturation of these transitions. The interface and impurity states, however, can drastically modify this behaviour. In this respect the technique employed to fabricate these composites plays a crucial role since it allows one to control these states.

These materials are formed[66] by uniformly dispersing semiconductor nanocrystals, such as CdS_xSe_{1-x}, in a glass where x varies from 1 to 0, but also other II-VI and I-VIII compounds can be used as well. The technique is a more or less thermally or chemically controlled precipitation in transparent solid matrices like glasses or polymers. It evolves in two steps: the nucleation and striking stages. In the nucleation stage one first forms widely supersaturated glass melts with uniformly dispersed semiconductor clusters; at this stage the resulting batch material is transparent like the initial undoped material, the glass or the polymer. In the ensuing striking stage, that occurs close to the Orbach temperature, these clusters grow to crystallites by coalescence, namely larger clusters grow further at the expense of the smaller ones, and reach sizes where volume properties overtake surface properties and solid-state-like features are acquired. Several structural techniques have revealed that these crystallites beyond the nucleation stage have the same structural features as the bulk ones. The size distribution can to some extent be controlled by the temperature and duration of the striking process, and this also fixes the colour of the doped glass. It has been found that the optical properties of these materials are strongly affected by a photodarkening process that takes place when they are exposed to high fluences[98]. More sophisticated growth techniques like growth in zeolites, on polymers or colloids together with controlled doping allow one to produce composites with narrower size distribution and improved characterization of the interface. Similar techniques are used to obtain metal nanocrystals uniformly dispersed in glasses or polymers.

Although the nonlinearities of these artificial composites and their response times can be tailored to arbitrary values by varying several parameters during the fabrication process, their figures of merit are not significantly improved over those that prevail in the bulk semiconductors. However, many of their other properties like robustness, interfacing, processability... are superior to those of the bulk semiconductors and this has led to intensive studies to implement them in device applications in integrated nonlinear optics; with narrower size distribution of the crystallites, improvement in the doping process and interface characterization these materials will become the most serious candidates for applications there.

5'3.3. Linear conjugated polymers. There is a large class of conjugated polymers that can be obtained by several stereochemical processes. The most studied one is the polydiacetylene class [71, 72, 99]. Here one initially forms good-quality crystals with diacetylene monomers $R_1 - C \equiv C - C \equiv C - R_2$, where R_1 and R_2 are appropriately chosen radicals, and then polymerizes them by heat, mechanical shear force or UV radiation and obtains the conjugated-polymer crystals which, however, contain defects of several types irrespective of the employed fabrication technique. The obtained crystals are photochemically rather stable and of good optical quality but occur only in small sizes and still with poor mechanical resistance.

The nonresonant cubic nonlinearity along the chain direction in these compounds is among the largest known [100] and comparable to that of bulk semiconductors like Ge or GaAs; in the direction transverse to the chain direction $\chi^{(3)}$ is two or three orders of magnitude lower. The magnitude and anisotropy of $\chi^{(3)}$ is due to the extensive delocalization that the valence π-electrons undergo along the chain direction. Visualizing these chains as unidimensional semiconductors and using a simple tight-binding model to describe their electron states one obtains the expressions (see (97a), (97b))

$$(108a) \qquad\qquad \chi^{(3)} \sim \left(\frac{E_{\mathrm{F}}}{E_{\mathrm{g}}} \right)^6 ,$$

while

$$(108b) \qquad\qquad \chi^{(1)} \sim \left(\frac{E_{\mathrm{F}}}{E_{\mathrm{g}}} \right)^2 ,$$

where E_{F} is the Fermi level and E_{g} the energy gap; the parameter $E_{\mathrm{F}}/E_{\mathrm{g}}$ plays a very important role as it measures the optical delocalization length and also the minimum chain length beyond which the nonlinearities are insensitive to the chain length. This is certainly important since these chains are never infinite even in the best-quality crystals. The power behaviour of the nonlinearity (108a) is intrinsically related [87] to the unidimensional character of the electron distribution and is also reproduced with more sophisticated descriptions of the electron states. On the other hand, the conjugation defects, like the Pople-Wamsley ones also termed solitons, seem irrelevant for the magnitude of the nonresonant cubic nonlinearities contrary to certain claims which were never substantiated. As in the case of bulk semiconductors the nonresonant cubic nonlinearities of the conjugated polymers are not exploitable mainly because of the inconvenient transparency range of these compounds.

The origin of the resonant cubic nonlinearity in these one-dimensional organic semiconductors is still a question of debate [101, 102] as is also the origin and width of the main absorption peak. This controversy is related to the uncertainties that still persist regarding the assignment of the levels in these conjugated chains and the inclusion of the delocalized or π-electron correlation effects; another is the strong electron-phonon or vibronic coupling in these chains which introduces a polaronlike character in the electron states. Both these features lead to complicated state configurations and nonlinear mechanisms that are only partially supported by experimental findings. Thus state-state interaction mediated by phonons similar to the inverse Raman scattering has been claimed[102] as being the origin of the resonant cubic nonlinearity. Whatever the mechanism may be, these π-electron-conjugated polymers are photochemically unstable[101] and their use in device applications becomes rather questionable.

5˙3.4. Cascading of second-order nonlinearities.

In noncentrosymmetric media the cubic nonlinearity besides the intrinsic contribution also contains a contribution that results by cascading of second-order nonlinearities. This is actually a retardation effect[7] that leads to a nonlocal or wave-vector-dependent nonlinearity. The origin was briefly described at the beginning of this section. Thus in the case of the optical Kerr effect in a noncentrosymmetric material the third-order polarization at frequency ω besides the direct contribution (102) also contains two additional contributions: a contribution that results from a second-order polarization induced by a field of frequency $-\omega$ and another at 2ω, the latter being the field generated by a second-order polarization at frequency 2ω induced by the field at frequency ω; and a contribution that results from a Pockels effect induced by a static electric field generated by an optical rectification process. One can easily show that all these contributions lumped together lead to[7] an effective third-order polarization of the same form as (102) but with a nonlinear susceptibility that now depends on the wave vector mismatch at the intermediate second-order process.

These cubic nonlinearities can be equal to or larger[7, 29] than the intrinsic ones in the nonresonant regime and are related to coherent processes. All noncentrosymmetric materials discussed in relation to the second-order nonlinearities can be used here and the advantage with respect to the centrosymmetric ones is evident; they benefit by the more developed material technology related to noncentrosymmetric materials with large second-order nonlinearities and in addition the same class of materials will be used for quadratic- and cubic-nonlinearity-based devices.

5˙4. Hybrid nonlinearities.

– Besides the applications that exploit the photoinduced or all optical modifications of the optical characteristics of a material, a whole class of other or similar applications in integrated nonlinear optics can be envisaged that exploit optical parametric effects, where the modifications of the optical characteristics of the material are provoked by an external agent. Here we have in mind modifications and modulations of the characteristics mediated through electro-optic, acousto-optic or magneto-optic coupling. The first two cases and the related devices and materials are well documented in the literature, but much less so the third one, and we briefly discuss here certain aspects of interest for integrated devices and the related materials.

Nonlinear magneto-optical effects can be treated within the framework of nonlinear optics in general, but in line with the criteria stated in this section we shall only consider the main frequency-preserving effects, namely the photoinduced Faraday rotation[24, 103] and gyrotropy. These effects allow a very efficient photoinduced control of the polarization state of an optical field and the development of nonreciprocal optical devices like optical valves and others.

The photoinduced Faraday effect originates from the combined effect of the Faraday rotation and the optical Kerr effect. In an isotropic medium in the presence of a static magnetic field H_0 the two eigenmodes of frequency ω in the direction of H_0 are the left and right circularly polarized waves with indices n_- and n_+, respectively; through the optical Kerr effect the latter become intensity dependent for high light intensities. Accordingly the polarization direction of a linearly polarized input wave E of frequency ω after propagation through a length L in such a medium collinear with H_0 rotates by an angle

$$(109) \qquad \qquad \tilde{\theta}_F = \theta_F + \Delta\theta_{NL}(I),$$

where θ_F is the usual linear Faraday rotation angle $\theta_F = \omega L(n_- - n_+)/2nc$ and $\Delta\theta_{NL}(I)$ is the photoinduced change of the latter and is proportional to the difference of the optical Kerr coefficients for left and right circular polarizations.

These photoinduced changes are in general very weak even in materials with large optical-Kerr-effect coefficients, like the semiconductors, unless certain provisions are made such that the difference of the optical Kerr coefficients for left and right polarizations is large. This can be done in II-VI semiconductors like CdSe or CdTe doped with magnetic impurities, the so-called[104] semi-magnetic semiconductors. Through the spin exchange interaction of the band and impurity electrons the Landé factor of the band electrons is enhanced by almost two orders of magnitude and similarly the magneto-optical coupling and the Zeeman splitting. Otherwise stated, the magnetic impurities act as «local amplifiers» of the static magnetic field. Without optimizing the interaction configuration, giant photoinduced Faraday rotations, for instance, exceeding 90° have been measured in these compounds under resonant conditions around the liquid-nitrogen temperature and for magnetic fields of the order of half tesla; these performances can be substantially improved to meet the device requirements.

Similar effects are expected without the presence of a magnetic field, in media with rotatory (gyratory) power and here one has a large choice of organic materials but also their severe problems regarding photochemical stability etc. At present the photoinduced natural gyrotropy has been studied in a couple of isolated cases[105], but the measured rotations were not sufficiently large.

5˙5. *General remarks.* – It would be an euphemism to state that there is no consensus yet regarding the class of nonlinear optical materials that will be used in integrated nonlinear optics. It is a fact that the choice of such materials cannot reside only on the figures of merit for the nonlinear operation but must also take into account many other criteria crucial for the design, production and maintenance of the nonlinear devices. Actually, in view of the rather limited nonlinear optical performances of the available materials, these other criteria will be the most crucial ones in making a choice. And this choice must be made with less ambitious goals than one would wish or had wished as our understanding of nonlinear integrated optics is progressing. There are now certain materials that can be used not only to conceive prototype nonlinear optical devices but also to proceed to their massive development and production and their insertion in large-scale technology.

REFERENCES

[1]   P. A. FRANKEN, A. E. HILL, C. W. PETERS and G. WEINREICH: *Phys. Rev. Lett.*, **7**, 118 (1961); M. BASS, P. A. FRANKEN, J. P. WARD and G. WEINREICH: *Phys. Rev. Lett.*, **9**, 446 (1962).

[2]   N. BLOEMBERGEN: *Nonlinear Optics* (Benjamin, New York, N.Y., 1964).

[3]   See, for instance, Y. R. SHEN: *The Principles of Nonlinear Optics* (Wiley, New York, N.Y., 1984).

[4]   G. FOWLES: *Introduction to Modern Optics* (Holt, Rinehart and Winston, New York, N.Y., 1975).

[5]   M. BORN and E. WOLF: *Principles of Optics* (Pergamon Press, Oxford, 1975).

[6]   W. K. H. PANOFSKY and M. PHILLIPS: *Classical Electricity and Magnetism* (Addison-Wesley, New York, N.Y., 1964).

[7]   C. FLYTZANIS: in *Quantum Electronics: A Treatise*, 1 A, edited by H. RABIN and C. L. TANG (Academic Press, New York, N.Y., 1975), p. 9.

[8]   J. D. JACKSON: *Classical Electrodynamics* (J. Wiley, New York, N.Y., 1962).

[9]   J. A. ARMSTRONG, N. BLOEMBERGEN, J. DUCUING and P. S. PERSHAN: *Phys. Rev.*, **127**, 1918 (1962).

[10] P. N. BUTCHER and D. COTTER: *The Elements of Nonlinear Optics* (Cambridge University Press, Cambridge, 1991).

[11] J. JERPHAGNON: *Phys. Rev. B*, **2**, 1091 (1970).

[12] S. A. AKHMANOV, R. V. KHOKHLOV and I. SUKHUROV: in *Laser Handbook*, Vol. E, edited by T. ARECCHI and E. O. SCHULZ-DUBOIS (North-Holland, Amsterdam, 1975), p. 1151.

[13] See, for instance, *Optical Phase Conjugation*, edited by R. A. FISHER (Academic Press, Orlando, Fla., 1983); B. YA. ZELDOVICH, N. F. PILIPETSKY and V. V. SHKUNOV: *Principles of Phase Conjugation* (Springer-Verlag, Berlin, 1985); see also the contribution of G. P. Banfi in this volume, p. 73.

[14] F. S. FELBER and J. M. MARBURGER: *Appl. Phys. Lett.*, **28**, 231 (1976); *Phys. Rev. A*, **17**, 335 (1978).

[15] H. EICHLER: *Optical Gratings* (Springer-Verlag, Berlin, 1980).

[16] See also R. W. BOYD: *Nonlinear Optics* (Academic Press, New York, N.Y., 1992).

[17] L. D. LANDAU and E. M. LIFSHITZ: *Electrodynamics of Continuous Media* (Pergamon Press, New York, N.Y., 1960).

[18] S. M. KOGAN: *Sov. Phys. JETP*, **43**, 217 (1963); P. C. PRICE: *Phys. Rev.*, **130**, 1792 (1963).

[19] D. A. B. MILLER, C. T. SEATON, M. E. PRICE and S. D. SMITH: *Phys. Rev. Lett.*, **47**, 197 (1981).

[20] C. FLYTZANIS and D. RICARD: to be published.

[21] P. S. PERSHAN: *Phys. Rev.*, **128**, 2903 (1962); **130**, 919 (1963).

[22] D. A. KLEINMAN: *Phys. Rev.*, **126**, 1977 (1962).

[23] See, for instance, M. E. LINES and A. M. GLASS: *Principles of Ferroelectrics and Related Materials* (Clarendon Press, Oxford, 1979).

[24] R. FREY, J. FREY, C. MERIAUX and C. FLYTZANIS: in *Guided Wave Nonlinear Optics*, edited by D. B. OSTROWKY and R. REINISCH, *NATO ASI Series* (Kluwer, Dordrecht, 1992), p. 75; see also contribution of J. Frey *et al.* in this volume, p. 437.

[25] See, for instance, A. YARIV: *Introduction to Optical Electronics* (Holt, Rinehart and Winston, New York, N.Y., 1971).

[26] See, for instance, M. BORN: *Optik* (Springer-Verlag, Berlin, 1954); also W. PAULI: *Optics and the Theory of Electrons*, edited by P. ENZ (MIT Press, Cambridge, Mass., 1973).

[27] J. GIORDEMAINE: *Phys. Rev. Lett.*, **8**, 407 (1962); R. W. TERHUNE, P. D. MAKER and C. M. SAVAGE: *Phys. Rev. Lett.*, **8**, 404 (1962).

[28] E. YABLONOVITCH, C. FLYTZANIS and N. BLOEMBERGEN: *Phys. Rev. Lett.*, **29**, 865 (1972).

[29] G. I. STEGEMAN, M. SHEIK-BAHAE, E. VAN STRYLAND and G. ASSANTO: *Opt. Lett.*, **18**, 13 (1993).

[30] J. FIUTAK: *Can. J. Phys.*, **41**, 12 (1968).

[31] V. N. GENKIN and P. M. MEDNIS: *Sov. Phys. JETP*, **27**, 609 (1968).

[32] N. BLOEMBERGEN, H. LOTEM and R. T. LYNCH: *Ind. J. Pure Appl. Phys.*, **16**, 151 (1978); Y. PRIOR, A. R. BOGDAN, M. DAGENAIS and N. BLOEMBERGEN: *Phys. Rev. Lett.*, **46**, 111 (1981).

[33] I. L. FABELINSKII: *Molecular Scattering of Light* (Plenum, New York, N.Y., 1968).

[34] C. FLYTZANIS and J. HUTTER: in *Contemporary Nonlinear Optics*, edited by G. P. AGRAWAL and R. W. BOYD (Academic Press, Orlando, Fla., 1992), p. 297.

[35] E. M. PURCELL: *Electricity and Magnetism*, Berkeley Physics Course, Vol. 2 (1965).

[36] G. DARWIN: *Proc. R. Soc. London, Ser. A*, **146**, 17 (1934).

[37] S. L. ADLER: *Phys. Rev.*, **126**, 413 (1962); N. WISER: *Phys. Rev.*, **179**, 62 (1963).

[38] L. ROSENFELD: *Z. Phys.*, **52**, 161 (1928).

[39] G. WAGNIERE: *J. Chem. Phys.*, **77**, 2786 (1982).

[40] W. KRAUZMAN: *Quantum Chemistry* (Academic Press, New York, N.Y., 1957).

[41] C. FLYTZANIS and J. DUCUING: *Phys. Rev.*, **178**, 1218 (1969).

[42] C. FLYTZANIS and C. L. TANG: *Phys. Rev. B*, **4**, 2520 (1971).

[43] J. L. OUDAR and D. S. CHEMLA: *Opt. Commun.*, **13**, 164 (1975).

[44] B. F. LEVINE and C. G. BETHEA: *J. Chem. Phys.*, **66**, 1070 (1977).

[45] B. DAVYDOV, L. DERKACHEVA, L. D. DUBINA, V. V. ZHABOTINSKII, M. E. ZOLIN, L. G. KORENEVA and M. A. SAMOKHINA: *JETP Lett.*, **12**, 16 (1970).

[46] K. RUSTAGI and J. DUCUING: *Opt. Commun.*, **10**, 258 (1974).

[47] G. P. AGRAWAL and C. FLYTZANIS: *Chem. Phys. Lett.*, **44**, 366 (1976).

[48] G. P. AGRAWAL, C. COJAN and C. FLYTZANIS: *Phys. Rev. B*, **17**, 776 (1978).

[49] C. FLYTZANIS: in *Nonlinear Optical Properties of Organic Molecules and Crystals*, edited by D. CHEMLA and J. ZYSS, Vol. II (Academic Press, New York, N.Y., 1987), p. 121.

[50] G. B. ALTSHULER, V. S. ERMOLAEV, K. I. KRYLOV, A. A. MANENKOV and A. M. PROKHOROV: *J. Opt. Soc. Am. B*, **3**, 660 (1986); C. FLYTZANIS: *Ann. Phys.*, **15**, Supp. 2, 49 (1990).

[51] N. KOTHARI and C. FLYTZANIS: *Opt. Lett.*, **11**, 806 (1986); **12**, 492 (1987).

[52] A. ISHIMARU: *Wave Propagation and Scattering in Random Media* (Academic Press, New York, N.Y., 1978).

[53] C. FLYTZANIS and C. L. TANG: *Phys. Rev. Lett.*, **45**, 441 (1980).

[54] P. EISENBERGER and S. L. MCCALL: *Phys. Rev. Lett.*, **26**, 684 (1971).

[55] E. GARMIRE, F. PANDARESE and C. H. TOWNES: *Phys. Rev. Lett.*, **11**, 160 (1963).

[56] J. J. WYNNE: *Phys. Rev. Lett.*, **52**, 751 (1984).

[57] F. DE ROUGEMONT and R. FREY: *Phys. Rev. Lett.*, **60**, 2010 (1988).

[58] M. SARGENT III, M. O. SCULLY and W. E. LAMB jr.: *Laser Physics* (Addison-Wesley, Reading, Mass., 1974).

[59] R. P. FEYNMAN, F. L. VERNON and R. W. HELLWARTH: *J. Appl. Phys.*, **28**, 49 (1957).

[60] D. GRISCHKOWSKY, M. T. LOY and P. F. LIAO: *Phys. Rev. Lett. A*, **12**, 2514 (1975).

[61] S. L. MC CALL and E. L. HAHN: *Phys. Rev. Lett.*, **18**, 908 (1967); *Phys. Rev.*, **183**, 457 (1969); R. E. SLUSHER and H. M. GIBBS: *Phys. Rev. A*, 5, 1634 (1972).

[62] See, for instance, *Nonlinear Optical Materials and Devices and their Applications in Information Technology*, edited by A. MILLER, B. DAINO and K. WELFORD, *NATO ASI Series* (Kluwer Publishers, Dordrecht, to appear) and also contribution of G. Assanto in this volume, p. 457.

[63] See contribution of G. Stegeman in ref. [62].

[64] F. C. ZUMSTEG, J. D. BIERLEIN and T. E. DIER: *J. Appl. Phys.*, **47**, 4986 (1976); for a recent review see, for instance, F. BORDUI and M. M. FEJER, *Annu. Rev. Mat. Sci.*, **23**, 321 (1993); G. I. CHEN and G. Z. LIU: *Annu. Rev. Mat. Sci.*, **16**, 203 (1986).

[65] See, for instance, *Nonlinear Optics in Signal Processing*, edited by R. W. EASON and A. MILLER (Chapman and Hall, London, 1993).

[66] See D. RICARD in this volume, p. 289; and C. FLYTZANIS, F. HACHE, M. C. KLEIN, D. RICARD and PH. ROUSSIGNOL: in *Progress in Optics*, Vol. XXIX, edited by E. WOLF (Elsevier, Amsterdam, 1991), p. 321.

[67] D. MILLER: *Appl. Phys. Lett.*, 5, 17 (1964).

[68] See, for instance, C. KITTEL: *Introduction to Solid State Physics* (J. Wiley, New York, N.Y., 1984).

[69] See, for instance, *The Physics and Fabrication of Microstructures and Microdevices*, edited by M. J. KELLY and C. WEISBUCH (Springer-Verlag, Berlin, 1988); *Quantum Semiconductor Structures; Fundamentals and Applications* (Academic Press, New York, N.Y., 1991); see also contributions of V. Degiorgio, and G. P. Banfi, F. Capasso and C. Sirtori in this volume, p. 267, 335.

[70] See, for instance, *Nonlinear Optical Properties of Organic Molecules and Crystals*, Vol. **1** and 2, edited by D. S. CHEMLA and J. ZYSS (Academic Press, New York, N.Y., 1987).

[71] See, for instance, *Organic Molecules for Nonlinear Optics and Photonics*, edited by J. MESSIER, F. KAJZAR and P. PRASAD, *NATO ASI Series* (Kluwer Publishers, Dordrecht, 1991); P. N. PRASAD and D. J. WILLIAMS: *Introduction to Nonlinear Optical Effects in Molecules and Polymers* (J. Wiley Interscience, New York, N.Y., 1990).

[72] See, for instance, *Principles and Applications of Nonlinear Optical Materials*, edited by R. W. MUNN and C. N. IRONSIDE (Blackie Academic, London, 1993).

[73] B. L. DAVYDOV, S. G. KOTOVCHIKOV and V. A. NEFERLOV: *Sov. J. Quant. Electron.*, **7**, 129 (1977).

[74] G. R. MEREDITH, J. G. VAN DUSEN and D. J. WILLIAMS: *Macromolecules*, **15**, 1385 (1982).

[75] K. D. SINGER, J. E. SOHN and J. LALAMA: *Appl. Phys. Lett.*, **49**, 248 (1986).

[76] See G. R. MÖHLMANN and C. P. J. M. VAN DER VORST in this volume, p. 175.

[77] J. ZYSS, J. F. NICOUD and D. S. CHEMLA: *J. Chem. Phys.*, **74**, 4800 (1981).

[78] J. L. OUDAR and R. HIERLE: *J. Appl. Phys.*, **48**, 2699 (1977).

[79] B. F. LEVINE, C. G. BETHEA, C. D. THURMOND, R. T. LYNCH and J. L. BERNSTEIN: *J. Appl. Phys.*, **50**, 2523 (1979).

[80] J. ZYSS, J. F. NICOUD and M. COQUILLAY: *J. Chem. Phys.*, **81**, 4160 (1984); R. MASE and J. ZYSS: *Mol. Eng.*, **1**, 141 (1991).

[81] J. M. HALBOUT, S. BLIT, W. DONALDSON and C. L. TANG: *IEEE J. Quantum Elctron.*, **15**, 1176 (1979).

[82] V. OSTERBERG and W. MARGULIS: *Opt. Lett.*, **11**, 516 (1986).

[83] R. A. MYERS, N. MURKHERJEE and S. R. J. BRUECK: *Opt. Lett.*, **16**, 1734 (1991).

[84] See contribution of G. Zerbi in ref.[62].

[85] M. C. J. DONCKERS, S. M. SILENCE, C. A. WALSH, F. HACHE, D. M. BURLAUD, W. E. MOERNER and R. J. TWIEG: *Opt. Lett.*, **18**, 1044 (1993).

[86] R. H. STOLEN and H. TOM: *Opt. Lett.*, **12**, 587 (1987); D. Z. ANDERSSON, V. MIZRAHI and J. E. SIPE: *Opt. Lett.*, **16**, 796 (1991).

[87] See C. FLYTZANIS in ref.[70], Vol. **2**, p. 121; also M. CARDONA and F. H. POLLACK: in *Optoelectronic Materials*, edited by G. A. ALBERS (Plenum, New York, N.Y., 1971), p. 81.

[88] See, for instance, *Optical Nonlinearities and Instabilities in Semiconductors*, edited by H. HAUG (Academic Press, New York, N.Y., 1988).

[89] S. SCHMITT-RINK, D. S. CHEMLA and D. A. B. MILLER: *Adv. Phys.*, **38**, 89 (1989).

[90] See, for instance, C. FLYTZANIS and J. HUTTER: in *Contemporary Nonlinear Optics*, edited by G. P. AGRAWAL and R. W. BOYD (Academic Press, New York, N.Y., 1992), p. 297.

[91] J. J. WYNNE: *Phys. Rev.*, **178**, 1295 (1969).

[92] D. WEAIRE, B. S. WHERETT, D. A. B. MILLER and S. D. SMITH: *Opt. Lett.*, **4**, 331 (1979); D. A. B. MILLER, C. T. SEATON and S. D. SMITH: *Phys. Rev. Lett.*, **47**, 197 (1981).

[93] B. S. WHERETT: *Proc. R. Soc. London, Ser. A*, **390**, 373 (1983).

[94] D. RICARD, PH. ROUSSIGNOL and C. FLYTZANIS: *Opt. Lett.*, **10**, 511 (1985); K. C. RUSTAGI and C. FLYTZANIS: *Appl. Phys. Lett.*, **9**, 344 (1984).

[95] F. HACHE, D. RICARD, C. FLYTZANIS and U. KREIBIG: *Appl. Phys. Lett. A*, **47**, 347 (1988).

[96] AL. L. EFROS and A. L. EFROS: *Sov. Phys. Semicond.*, **16**, 772 (1982).

[97] D. A. B. MILLER, D. S. CHEMLA and S. SCHMITT-RINK: *Phys. Rev. B*, **35**, 8113 (1987).

[98] PH. ROUSSIGNOL, D. RICARD, J. LUKASIK and C. FLYTZANIS: *J. Opt. Soc. Am. B*, **4**, 5 (1987).

[99] See, for instance, G. WEGNER: *Makromol. Chem.*, **35**, 154 (1971).

[100] C. SAUTERET, J. P. HERMANN, R. FREY, F. PRADÈRE, J. DUCUING, R. R. CHANCE and R. H. BAUGHMAN: *Phys. Rev. Lett.*, **36**, 956 (1976).

[101] B. I. GREENE, J. ORENSTEIN and S. SCHMITT-RINK: *Science*, **247**, 679 (1990).

[102] B. I. GREENE, J. F. MUELLER, J. ORENSTEIN, D. H. RAPKINE, S. SCHMITT-RINK and M. THAKUR: *Phys. Rev. Lett.*, **61**, 325 (1988); G. J. BLANCHARD, J. P. HERITAGE, A. C. VON LEHMEN, M. K. KELLY, G. L. BAKER and S. ETERNAD: *Phys. Rev. Lett.*, **63**, 887 (1989).

[103] J. FREY, R. FREY and C. FLYTZANIS: *Phys. Rev. B*, **45**, 4056 (1992).

[104] See, for instance, J. K. FURDYNA: *Appl. Phys.*, **64**, R29 (1988).

[105] H. ASHITAKA, Y. YOKOH, R. SHIMIZU, T. YOKOZAWA, K. MORITA, T. SNEHIRO and Y. MATSUMOTO: *Nonlinear Opt.*, **4**, 281 (1993).

Nonlinear Propagation in Homogeneous Media.

G. P. BANFI

Dipartimento di Elettronica dell'Università - 27100 Pavia, Italia

1. – Propagation equations in a nonlinear medium.

The propagation equations are derived from the macroscopic Maxwell equations. In the dipole approximation and neglecting the magnetization of the medium, the current J is given by $J = \partial P/\partial t$, with P the polarization induced by the electric field. The polarization can be divided into two parts, the first linear with the field and the second nonlinear:

$$(1) \qquad\qquad P = P^{(L)} + P^{(NL)} .$$

The Maxwell equations $\vec{\nabla} \times E = -\partial B/\partial t$ and $\vec{\nabla} \times B = \mu_0(J + \varepsilon_0 \partial B/\partial t)$ are then combined to give the wave equation:

$$(2) \qquad \nabla^2 E - \frac{1}{c^2} \frac{\partial^2}{\partial t^2} (E + e_0^{-1} P^{(L)}) = -\mu_0 \frac{\partial^2}{\partial t^2} P^{(NL)} ,$$

where we neglected for simplicity the term $\nabla(\nabla \cdot E)$.

The linear polarization is related to the field E through

$$(3) \qquad P^{(L)}(r, t) = \varepsilon_0 \int_{-\infty}^{t} dt' \chi^{(1)}(t - t') E(r, t'),$$

where the tensor $\chi^{(1)}$ is the linear susceptibility.

1'1. *Frequency domain.* – E and P are represented through the frequency decomposition

$$(4) \qquad E(r, t) = \int_{-\infty}^{\infty} d\omega\, E_\omega(r) \exp[-i\omega t].$$

In the frequency domain, eq. (3) simplifies to $P_\omega^{(L)}(r) = \varepsilon_0 \chi_\omega^{(1)} E_\omega(r)$, and eq. (4) becomes

(5) $$\nabla^2 E_\omega(r) + \frac{\omega^2}{c^2}(1 + \chi_\omega^{(1)}) E_\omega(r) = -\mu_0 \omega^2 P_\omega^{(NL)}(r).$$

We shall always assume that $\chi_\omega^{(1)}$ is real and does not depend on the position r (this means that there is no linear absorption or scattering). When not specified otherwise, we shall also neglect birefringence and drop the tensor notation; in this simplified picture the dielectric constant, defined as $\varepsilon = I + \chi_\omega^{(1)}$, becomes a scalar and is connected to the refractive index n_ω by the scalar relation $\varepsilon = n_\omega^2$.

When $P^{(NL)} = 0$, the homogeneous form of eq. (5) describes the free propagation of the field at frequency ω in a medium with linear dielectric constant. The power of the wave at the entrance of the medium remains constant and does not change during propagation. In this case a solution of eq. (5) is given by plane waves, with $E_\omega(r) \approx \exp[i k \cdot r]$ and the wave vector k satisfying the dispersion relation

(6) $$k^2 = \frac{\omega^2}{c^2} n_\omega^2.$$

When $P^{(NL)} \neq 0$, the term which appears at the right-hand side of eq. (5) acts as a source for the field E_ω. In most cases $|P^{(NL)}| \ll |P^{(L)}|$, and the effect of the nonlinear polarization is modest for propagation distances of some wavelengths. One is then tempted to retain the idea of a propagating free wave, but with its amplitude and phase modified by the nonlinearity.

1'2. *Slowly-varying-envelope approximation* (in space). – For a monochromatic field propagating with a small divergence along the direction z, one writes

(7) $$E(r, t) = \mathrm{Re}\, E_\omega(r) \exp[-i\omega t] = \mathrm{Re}\, A(z, \rho) \exp[i(kz - \omega t)]$$

with $r \equiv (z, \rho)$, and ρ the coordinates perpendicular to z. Making use of eq. (7), the quantity $\nabla^2 E_\omega$ can be rewritten as $\nabla^2 E_\omega = (-k^2 A + \nabla_\perp^2 A + 2ik\, \partial A/\partial z + \partial^2 A/\partial z^2) \exp[ikz]$, and, under the assumption that the envelope A changes on a scale length much larger than the wavelength λ, $\lambda = 2\pi/k$, the last term in parenthesis can be neglected in comparison with the previous one. With this approximation (called «slowly-varying-envelope approximation» and abbreviated with SVEA), one derives from eqs. (5)-(7) the following equation for A:

(8) $$\nabla_\perp^2 A + 2ik \frac{\partial A}{\partial z} = -\frac{\omega^2}{e_0 c^2} P_\omega^{(NL)}(z, \rho) \exp[-ikz].$$

The homogeneous form of eq. (8) is the usual parabolic, or paraxial, wave equation. It accounts for the changes of the transverse pattern of the field during free propagation (diffraction effect). When the effect of diffraction is not important, the dependence of A upon the transverse coordinate can be neglected (infinite-plane-wave approximation) so that eq. (8) further simplifies to

(9)
$$\frac{\mathrm{d}A(z)}{\mathrm{d}z} = i\frac{\omega}{2c\varepsilon_0 n_\omega} P_\omega^{(\mathrm{NL})}(z)\exp[-ikz].$$

One should notice that the nonlinear polarization must have a spatial component close to $\exp[ikz]$ in order to appreciably affect A. This requirement will lead to the phase-matching conditions.

1'3. *Time-dependent propagation.* – Equations (8), (9) describe the behaviour of a monochromatic wave. One can approximate the field with a single monochromatic wave only in the case of stationary processes. With pulses of short time duration (picosecond or femtosecond regime) the steady-state solution can become an inadequate picture. We remind that, for a pulse of time duration $\Delta\tau$, the width of the frequency spectrum $\Delta\omega$ is such that $\Delta\omega \geqslant 1/\Delta\tau$.

With short pulses some effects not accounted for by eqs. (8)-(9) become important. For example, consider a field E, generated by a nonlinear polarization $P^{(\mathrm{NL})}$ induced by a field E_2. The envelope of E propagates with a group velocity v which, due to the frequency dispersion of the medium, can be different from that of E_2. If $E(r, t)$ and its source $P^{(\mathrm{NL})}(r, t)$ travel with different velocities, it is likely, with short pulses, that after some propagation distance they fail to overlap. With short pulses, always due to the frequency dispersion of the medium, another effect, which becomes relevant even for short propagation distances, is the broadening of the pulse duration.

In principle, one could work also in the case of time-dependent propagation in the frequency domain, solving eq. (5), or eqs. (8)-(9), for every frequency component and then integrating eq. (4) to recover $E(r, t)$, but the approach would turn out to be quite complex. Whenever possible, the usual procedure is to generalize to the time domain the slowly-varying-amplitude approximation introduced for the space domain in the previous subsection. For an optical pulse with a frequency spectrum centred around ω_c and which propagates in the direction z, one writes $E(z, t)$ as

(10)
$$E(z, t) = \mathrm{Re}\, A(z, t)\exp[i(k_c z - \omega_c t)]$$

and tries to derive an equation for the envelope $A(z, t)$. For simplicity, in the procedure we neglect the dependence of A on the transverse coordinates. From the dispersion relation $k = k(\omega)$, expanding to second order in $\omega - \omega_c$, one has

(11)
$$k = k_c + K'(\omega - \omega_c) + K''(\omega - \omega_c)^2$$

with

(11')
$$K' = \frac{\partial k}{\partial \omega}\bigg|_{c} = v_c^{-1},$$

(11'')
$$K'' = \frac{\partial^2 k}{\partial \omega^2}\bigg|_{c} = -\frac{1}{v_c^2}\frac{\partial v}{\partial \omega}\bigg|_{c} = -\frac{1}{v_c^2}V_c'.$$

The quantity v_c is the group velocity at ω_c, while $V' = \partial v/\partial \omega$ is the group velocity dispersion. One assumes that also $\boldsymbol{P}^{(NL)}$ has a frequency spectrum centred around ω_c, and writes it as $\boldsymbol{P}^{(NL)}(z, t) = \mathrm{Re}\,\boldsymbol{P}^{(NL)}(z, t)\exp[-i\omega_c t]$. Substituting eq. (10) into eq. (2), taking into account the definition (4), making use of the truncated expansion (11), and performing some mathematical manipulations, one arrives at

(12)
$$\left(\frac{\partial}{\partial z} + \frac{1}{v_c}\frac{\partial}{\partial t}\right)A(z, t) + i\frac{K''}{2}\frac{\partial^2 A}{\partial^2 t} = i\frac{\omega_c}{2\varepsilon_0 cn}\boldsymbol{P}^{(NL)}(z, t)\exp[-ik_c z].$$

A few approximations are necessary in order to write eq. (12), the first one being that A, $\boldsymbol{P}^{(NL)}$ and their time derivatives change on a temporal scale much longer than the optical period $1/\omega_c$. There is also a limitation on the maximum spectral width (and hence on the minimum pulse duration) which are implicit in the truncation of eq. (11).

When $\boldsymbol{P}^{(NL)} = 0$, eq. (12) describes the linear propagation of a pulse in dispersive media. It is instructive to see in this linear case the meaning of the terms in the left-hand side. When on neglects the second-order dispersion (which implies $K'' = 0$), the relation $(\partial/\partial z + (1/v_c)\partial/\partial t)A(z, t) = 0$ is verified by $A = A(z - v_c t)$. Such an envelope represents a pulse propagating with no distortion at velocity v_c. The term in K'' gives rise to a broadening of the pulse during propagation, as shown in fig. 1. Assume that the pulse is initially transform limited, with $\Delta\tau_0\Delta\omega \approx 1$. Its width in z is given by $\Delta z_0 = v_c\Delta\tau_0$ at $t = 0$, and by $\Delta z_t \approx \sqrt{\Delta z_0^2 + (\Delta vt)^2}$ at the later time t. The quantity Δv denotes the spread of

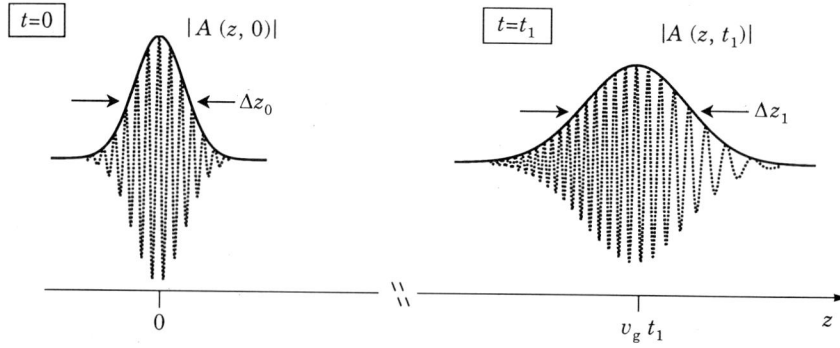

Fig. 1.

group velocities; since $v(\omega) = v_c + V_c'(\omega - \omega_c)$, Δv is given by $|V_c'|\Delta \omega \approx$ $\approx |V_c'|/\Delta \tau_0$. One can notice that the effect of pulse broadening becomes more relevant for shorter pulses. The contribution of Δvt to the spread of the pulse is due to the fact that the different frequencies of the pulse travel with a different group velocity. After some propagation, in the case $K'' > 0$ considered in fig. 1, the lower frequencies will move at the front end of the pulse and the higher ones at the trailing edge. The opposite situation is found for $K'' < 0$.

2. – Second-order effects.

The second-order nonlinear polarization, denoted $P^{(2)}$, is given by

(13) $$P^{(2)}(t) = \int dt' \, dt'' \, \chi^{(2)}(t - t', t - t'') E(t') E(t''),$$

where the third-order tensor $\chi^{(2)}$ denotes the second-order susceptibility. In eq. (13), the relation between $P^{(2)}$ and the field E has been assumed local in space: $P^{(2)}$ at a position r depends on the value of the field at the same point r.

Assume now that E is constituted by two monochromatic fields at frequencies ω_1 and ω_2, linearly polarized along e_1 and e_2:

(14) $$E(r, t) = \mathrm{Re}\,(e_1 E_1(r) \exp[-i\omega_1 t]) + \mathrm{Re}\,(e_2 E_2(r) \exp[-i\omega_2 t]).$$

Consider $P^{(2)}(t)$, the component of $P^{(2)}$ along a direction defined by e_p. Substituting eq. (14) into eq. (13), $P^{(2)}(t)$ is given by the sum of the terms listed below:

	$P^{(2)}(t)$	amplitude of $P^{(2)}$
(15a)	$\mathrm{Re}\,P^{(2)}(\omega_1 + \omega_2)\exp[-i(\omega_1 + \omega_2)t]$,	$\varepsilon_0 \chi^{(2)}(-(\omega_1 + \omega_2); \omega_1, \omega_2)E_1 E_2$,
(15b)	$\mathrm{Re}\,P^{(2)}(2\omega_1)\exp[-i2\omega_1 t] +$	$(1/2)\varepsilon_0 \chi^{(2)}(-2\omega_1; \omega_1, \omega_1)E_1 E_1$,
	$\quad$ + similarly in $2\omega_2$,	
(15c)	$\mathrm{Re}\,P^{(2)}(\omega_1 - \omega_2)\exp[-i(\omega_1 - \omega_2)t]$,	$\varepsilon_0 \chi^{(2)}(-(\omega_1 - \omega_2); \omega_1, \omega_2)E_1 E_2^*$,
(15d)	$P^{(2)}(0)$,	$(1/2)\varepsilon_0 \chi^{(2)}(0; \omega_1, -\omega_1)E_1 E_1^* +$
		$\quad$ + term in ω_2.

One defines $\chi^{(2)}(\omega_f; \omega_i, \omega_j)$ to be the Fourier transform of $\chi^{(2)}(t - t', t - t'')$ at the frequencies ω_i and ω_j. The quantity $\chi^{(2)}$ which appears in eqs. (15) is the effective second-order susceptibility: it is a scalar quantity given by $\chi^{(2)} =$ $= e_p \chi^{(2)} e_1 e_2 = \sum_{j, m, n = 1}^{3} e_{p,j} \chi^{(2)}_{j, m, n} e_{1, m} e_{2, n}$ for the terms (15a) and (15c), and similarly for the other terms ($e_1 e_2$ is replaced by $e_1 e_1$ or $e_2 e_2$).

The various terms in eqs. (15) are responsible for the generation of (15a), a field at the sum frequency; (15b) is the degenerate case of (15a) and leads to second-harmonic generation; (15c) a field at the difference frequency; (15d) a static polarization, which is associated to a static, and hence nonpropagating, electric field.

As far as the notation in eqs. (15) is concerned:

It is convenient to consider the ω's positive. The first argument of $\chi^{(2)}$ refers to the frequency of the generated nonlinear polarization, the others to those of the generating fields. A positive sign in front of ω is associated with a quantum taken from the field at the same frequency, *vice versa* a negative sign is associated with a quantum given to the field.

The coefficients $1/2, 1, \ldots$, which appear in the amplitude of $P^{(2)}$ depend on the definition chosen for the fields and also on the number of terms which give rise to the same nonlinear polarization once eq. (14) is substituted in eq. (13).

The Pockels effect can be cast under the same formalism. Consider E to be made by an optical pulse plus a static field:

(16) $$E(t) = \mathrm{Re}\,(e_1 E_1 \exp[-i\omega_1 t]) + e_0 E(0).$$

In this case, the nonlinear polarization includes the term $P^{(2)}(t) = \mathrm{Re}\,(P^{(2)}(\omega_1)\exp[-i\omega_1 t])$ with $P^{(2)}(\omega_1) = 2\varepsilon_0\chi^{(2)}(-\omega_1;\omega_1,0)E_1 E(0)$. Adding the linear and the nonlinear contribution, the total polarization at frequency ω_1 can be written as

$$P^{(\mathrm{tot})}(\omega_1) = \varepsilon_0\chi^{(\mathrm{tot})}E_1(\omega_1),$$

where $\chi^{(\mathrm{tot})}$ depends on the static field. More precisely the tensor components of $\chi^{(\mathrm{tot})}$ are given by

(17) $$\chi_{lm}^{(\mathrm{tot})} = \chi_{lm}^{(1)} + 2\chi_{lnm}^{(2)}E_n(0).$$

The optical field E_1 experiences during its propagation the dielectric constant $\varepsilon = I + \chi^{(\mathrm{tot})}$, and its phase (and its polarization) can be modified by varying the static field.

The magnitude of $\chi^{(2)}$ of the sum-frequency processes can be quite different from that of the Pockels effect. In fact, the slow ionic displacement contributes only to the $\chi^{(2)}(-\omega_1;\omega_1,0)$ of the Pockels effect. Typical is the case of ferroelectric materials which exhibit also a large static linear polarizability due to the ions.

2'1. *Second-harmonic generation.* – Second-harmonic generation denotes the process in which energy is transferred from the fundamental field E_1 to the field E_2 at the double frequency. A simple equation for the generated SH can be derived under the following simplifying assumptions: both fields are monochromatic, linearly polarized plane waves propagating in the $+z$ direction; the fundamental field $E_1(z, t) = \mathrm{Re}\,(e_1 A_1(z)\exp[i(k_1 z - \omega_1 t)])$ enters in the nonlinear medium at $z = 0$, the SH field, $E_2(z, t) = \mathrm{Re}\,(e_2 A_2(z)\exp[i(k_2 z - \omega_2 t)])$ with

$\omega_2 = 2\omega_1$, is absent at the input, so that $A_2(0) = 0$; the frequency dispersion of $\chi^{(2)}$ is negligible (one can then avoid specifying the frequencies in the argument of $\chi^{(2)}$).

Substituting $E_1 = A_1(z)\exp[ik_1 z]$ into eq. (15b), one obtains the amplitude of the polarization at ω_2:

(18) $$P^{(2)}(\omega_2, z) = (1/2)\varepsilon_0\chi^{(2)}A_1(z)A_1(z)\exp[i2k_1 z]$$

which, substituted into eq. (9), gives

(19) $$\frac{d}{dz}A_2(z) = i\frac{\omega_1}{2cn_2}\chi^{(2)}A_1^2(z)\exp[i\Delta kz]$$

with

(20) $$\Delta k = 2k_1 - k_2 .$$

The quantity Δk is the wave vector mismatch of the process $(\omega_1, k_1) + (\omega_1, k_1) \to (\omega_2, k_2)$. The conversion efficiency η is defined through $\eta = \Phi_2(L)/\Phi_1(0)$, with $\Phi_2(L)$ the SH intensity at the output and $\Phi_1(0)$ the intensity of the fundamental beam at the input.

We here assume the medium to be homogeneous with $\chi^{(2)}$ constant in space. When η is small, the pump beam suffers a negligible depletion and one can set $A_1(z) \approx A_1(0)$ in eq. (19). Integrating for $A_2(z)$, making use of the relation $\Phi = (1/2)\varepsilon_0 n|A|^2$, the SH intensity is given by

(21) $$\Phi_2(L) = \frac{\omega_1^2}{2\varepsilon_0 c^3 n_1^2 n_2}\left|\chi^{(2)}\right|^2\Phi_1^2 L^2 \,\text{sinc}^2\,\frac{\Delta kL}{2} .$$

When η is larger than a few percent, the depletion of the fundamental field must be considered. The depletion of A_1 during SH generation is accounted for by solving eq. (19) together with the equation for dA_1/dz. Substituting the nonlinear polarization $P^{(2)}(\omega_1, z) = \varepsilon_0\chi^{(2)}A_2(z)A_1^*(z)\exp[i(k_2 - k_1)z]$ into eq. (9), one obtains for dA_1/dz

(22) $$\frac{d}{dz}A_1(z) = i\frac{\omega_1}{2cn_1}\chi^{(2)}A_2(z)A_1^*(z)\exp[-i\Delta kz] .$$

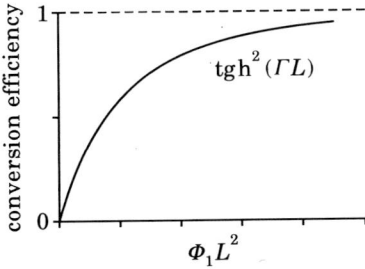

Fig. 2.

For $\Delta k = 0$, the solution of the two coupled equations (19) and (22), with the condition $A_2(0) = 0$, is given by

$$(23) \qquad\qquad A_2(L) = A_1(0) \sinh(\Gamma L)$$

with $\Gamma = (\omega_1/2cn_2)|\chi^{(2)}| A_1(0)$.

The behaviour of η vs. the input intensity is plotted in fig. 2.

2'2. *Phase matching.* – The role of the wave vector mismatch Δk can be appreciated from eq. (19). The plot of η vs. Δk for a fixed L (fig. 3a)) shows that a large SH generation requires the condition $\Delta kL < \pi$ to be satisfied. When this is not the case, $\Phi_2(z)$ oscillates with z (fig. 3b)): it achieves a maximum after a distance equal to the coherence length $l_c = \pi/\Delta k$ and vanishes at $2l_c$. What happens is that the nonlinear polarization has a phase dictated by the fundamental field, and then it changes in space according to the wave vector $2k_1$, while the SH field propagates with a wave vector k_2. Phase matching is required in order to add in phase the SH fields generated at the different positions.

2'3. *How to achieve phase matching.*

Birefringence. In an homogeneous medium, the phase-matching condition is exactly fulfilled whenever $2k_1 = k_2$, and, since $k = \omega/cn$, it requires $n_1 = n_2$. Due to the frequency dispersion, usually it is not possible to have the same refractive index at the two frequencies in a material with isotropic ε; for example, in a region of normal dispersion, n grows with ω and one can never obtain $n_1 = n_2$. On the contrary, this possibility is given in a birefringent crystal. Consider the case of a uniaxial negative crystal in the optical or near-i.r. frequency

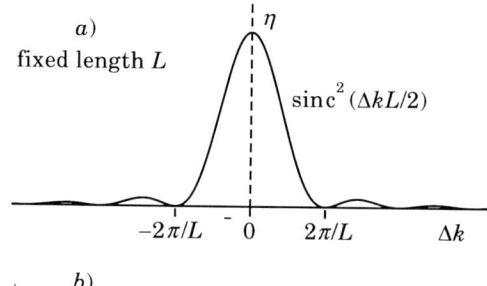

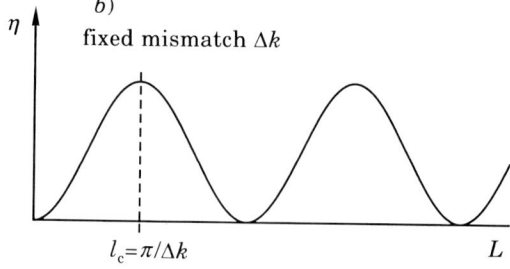

Fig. 3.

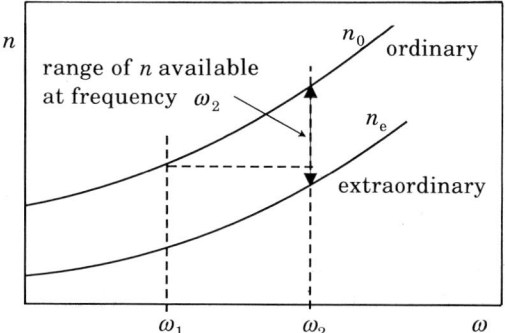

Fig. 4.

range, and denote with X, Y, Z the principal dielectric axes. The behaviour of $n_x = n_y = n_0$ and of $n_z = n_e$ vs. ω is depicted in fig. 4. Suppose that $\boldsymbol{e}_1 \| Y$ (fundamental field polarized along Y), while $\boldsymbol{e}_2$ belongs to the (X, Z)-plane and makes an angle α with the Z-axis. $\boldsymbol{E}_1$ propagates with the ordinary refractive index $n_0(\omega_1)$, while the refractive index of $\boldsymbol{E}_2$ can be adjusted to any value in the range between $n_0(\omega_2)$ and $n_e(\omega_2)$ by choosing the proper α. Crystals with high birefringence are more likely to fulfil the condition $n_1 = n_2$ in a wide range of frequencies. A change of the crystal temperature slightly modifies the relative positions and the ω dependence of the two curves in fig. 4, and it can be used for fine adjustments. When $n_e(\omega_2)$ is close to $n_0(\omega_1)$, a temperature adjustment can allow one to satisfy the phase-matching condition with $\boldsymbol{E}_2 \| Z$ (noncritical phase matching). This is a favourable configuration since efficient SH generation can take place with large angular divergences of the fundamental beam.

Inverted layers. Consider a nonlinear medium with a $\chi^{(2)}$ dependent on z. By solving eq. (19), one notices that A_2 can grow in propagation provided that $\chi^{(2)}$ has a spatial modulation in z with wave vector Δk. In principle, alternating signs of $\chi^{(2)}$ could be obtained by staking inverted layers. Phase matching is obtained with layers of thickness l_c. A practical application of this concept is being pursued in waveguides designed for SH generation by alternate poling of the active material.

Propagation in a waveguide offers the additional possibility to rely on mode dispersion. In fact, in a waveguide, different propagation modes with the same ω have different wave vectors, and in principle one could satisfy the conditions of phase matching by involving modes of higher orders.

2˙4. *Coupling of three different waves.* – Equations (19)-(22) can be readily extended to the case of three waves whose frequencies satisfy the relation $\omega_1 + \omega_2 = \omega_3$. We assume that all the waves propagate along $+z$.

With the total field written as $E(z, t) = \sum\limits_{j=1,2,3} E_j(z, t)$, with $\boldsymbol{E}_j(z, t) =$

$= \mathrm{Re}(e_j A_j(z) \exp[i(k_j z - \omega_j t)])$, the nonlinear polarizations that couple the E_j's are

(24)
$$\begin{cases} P^{(2)}(\omega_1) = \varepsilon_0 \chi^{(2)} A_3 A_2^* \exp[i(k_3 - k_2)z], \\ P^{(2)}(\omega_2) = \varepsilon_0 \chi^{(2)} A_3 A_1^* \exp[i(k_3 - k_1)z], \\ P^{(2)}(\omega_3) = \varepsilon_0 \chi^{(2)} A_1 A_2 \exp[i(k_1 + k_2)z]. \end{cases}$$

When these $P^{(2)}$ are substituted into eq. (9), one has the following set of equations:

(25)
$$\begin{cases} dA_1/dz = c_1 A_3 A_2^* \exp[-i\Delta kz], \\ dA_2/dz = c_2 A_3 A_1^* \exp[-i\Delta kz], \\ dA_3/dz = c_3 A_1 A_2 \exp[-i\Delta kz] \end{cases}$$

with $c_j = i(\omega_j^2/2c^2 k_j)\chi^{(2)}$, $\Delta k = k_1 + k_2 - k_3$.

The effects described by eqs. (25) depend on the conditions at $z = 0$. We assume that $\Delta k = 0$. If $A_2(0) \neq 0$, $A_1(0) \neq 0$ and $A_3(0) = 0$, the field E_3 is generated at the expenses of the other two.

On the contrary, if $A_3(0) \neq 0$, $A_2(0) \neq (0)$, $A_1(0) = 0$, energy will flow from the field at high frequency to the other two; in terms of photons: $\hbar\omega_3 \to \hbar\omega_1 + \hbar\omega_2$. This is the mechanism of parametric amplification, an effect which has relevant applications and which will be discussed in the lecture by TANG. When the gain of the field E_2 is modest, the process is usually called difference-frequency generation.

3. – Third-order effects.

The procedure adopted for eqs. (13)-(15) can be readily extended to the case of the third-order nonlinear polarization:

(26) $$P^{(3)}(t) = \int dt'\, dt''\, dt''' \chi^{(3)}(t - t', t - t'', t - t''') E(t') E(t'') E(t''').$$

Assume that $E(t)$ is made by three monochromatic fields E_j, with frequency ω_j, linearly polarized with versor e_j. Consider the component $P^{(3)} = e_p \cdot P^{(3)}$ and denote with $\chi^{(3)}$ the effective third-order susceptibility: $\chi^{(3)} = e_p \chi^{(3)} e_j e_l e_m$, with $j, l, m = 1, 2, 3$ according to the fields involved. The terms of $P^{(3)}$ at the various frequencies are:

sum-frequency: $\omega_s = \omega_j + \omega_j + \omega_3$,

(27a) $$P^{(3)}(\omega_s) = (3/2)\varepsilon_0 \chi^{(3)}(-\omega_s; \omega_1, \omega_2, \omega_3) E_1 E_2 E_3 \qquad (\omega_1 \neq \omega_2 \neq \omega_3);$$

the degenerate case $\omega_1 = \omega_2 = \omega_3$ leads to third-harmonic generation with

$$P^{(3)}(3\omega_j) = (1/4)\varepsilon_0 \chi^{(3)}(-3\omega_j; \omega_j, \omega_j, \omega_j) E_j E_j E_j;$$

frequency mixing: $\omega_d = \omega_1 + \omega_2 - \omega_3$

(27b) $\qquad P^{(3)}(\omega_d) = (3/2)\,\varepsilon_0\,\chi^{(3)}(-\omega_d;\omega_1,\omega_2,-\omega_3)\,E_1 E_1 E_2^* \qquad (\omega_1 \neq \omega_2);$

electric-field-induced second harmonic

$$P^{(3)}(2\omega_1) = (3/2)\,\varepsilon_0\,\chi^{(3)}(-2\omega_1;\omega_1,\omega_1,0)\,E_1 E_1 E_0\,;$$

d.c. Kerr effect

$$P^{(3)}(\omega_1) = 3\varepsilon_0\,\chi^{(3)}(-\omega_1;0,0,\omega_1)\,E_0 E_0 E_1\,;$$

optical Kerr effect

(27c) $\qquad \begin{cases} P^{(3)}(\omega_1) = (3/2)\,\varepsilon_0\,\chi^{(3)}(-\omega_1;\omega_2,-\omega_2,\omega_1)\,E_2 E_2^* E_1 \qquad (\omega_1 \neq \omega_2), \\ P^{(3)}(\omega_1) = (3/4)\,\varepsilon_0\,\chi^{(3)}(-\omega_1;\omega_1,-\omega_1,\omega_1)\,E_1 E_1^* E_1\,. \end{cases}$

In the literature dealing with the degenerate optical Kerr effect, one often finds $\chi^{(3)}$ defined through

(28) $$P^{(3)} = \frac{1}{2}\varepsilon_0\,\chi^{(3)}\,|E|^2 E\,,$$

which differs from eq. (27c) in the numerical factor, since 1/2 replaces 3/4. We shall adhere to this convention in the following, and to keep consistency we adopt the same change of numerical factors also in the case of wave mixing.

The optical Kerr effect gives rise to a change of the refractive index experienced by E_1 which depends on its intensity (or on the intensity of another field E_2). The effect is particularly important for applications since it allows an optical control of the behaviour of propagation. Neglecting possible birefringence and adopting a scalar notation, for a beam of frequency ω, the total polarization at the same frequency is given by $P^{(\text{tot})} = \varepsilon_0\,\chi^{(\text{tot})} E_1$, with $\chi^{(\text{tot})} = \chi^{(1)} + \chi^{(3)}|E|^2/2$. The refractive index n is given by $n = \varepsilon^{1/2} = (1 + \chi^{(\text{tot})})^{1/2}$, and, since $|P^{(\text{linear})}| \gg |P^{(3)}|$, one can expand n around $n_0 = (1 + \chi^{(1)})^{1/2}$ to obtain

$$n = n_0 + n_2\,|E|^2/2 \qquad\qquad \text{with } n_2 = \chi^{(3)}/2n_0\,,$$

or

(29) $\qquad\qquad\qquad n = n_0 + \gamma\Phi \qquad\qquad \text{with } \gamma = \chi^{(3)}/2\varepsilon_0\,c n_0^2\,.$

The imaginary part of $\chi^{(3)}$ is related to the two-photon absorption. In fact, from eqs. (9), (28) one has $dA/dz = i\omega/(4cn)\chi^{(3)}A^2 A^*$. Multiplying by A^*, adding the complex conjugate, substituting $\Phi = (1/2)\varepsilon_0\,nc\,|A|^2$, one obtains

(30) $\qquad\qquad\qquad\qquad d\Phi/dz = -\beta\Phi^2\,.$

The absorption is quadratic in the intensity, and depicts a process where two photons are absorbed simultaneously. In eq. (30), β is the two-photon absorption coefficient:

$$\beta = \omega c^{-2} \varepsilon_0^{-1} n^{-2} \operatorname{Im} \chi^{(3)} . \tag{30'}$$

In the following we outline some of the third-order effects which are relevant in applications and are also employed in the characterization of the nonlinear material. We shall assume that $\chi^{(3)}$ is real (which implies neglecting any nonlinear absorption) and constant thoughout the medium.

3'1. Effects with a single beam.

3'1.1. Self-phase modulation. Consider a monochromatic plane wave, with $E \approx \exp[i(kz - \omega t)]$, which propagates in a nonlinear medium of length L (fig. 5). The phase of the field at $z = L$ is $\phi(L) = kL = (\omega/c)nL = (\omega/c)(n_0 + \gamma \Phi)L$. There is then an intensity-dependent phase shift given by

$$\phi^{(\mathrm{NL})} = (\omega/c)\gamma \Phi L . \tag{31}$$

Figure 5 shows how this phase shift can be employed in a nonlinear interferometer to obtain a transmission which depends on the intensity. Beam a propagates through the nonlinear material before recombining with b. Assume that at low intensity (when $\phi^{(\mathrm{NL})} = 0$) the relative phases of a and b are such that they interfere destructively at the output O. By increasing the intensity, the phase of a changes, and a constructive maximum, with a and b adding in phase, occurs whenever $\phi^{(\mathrm{NL})} = \pm \pi (+\text{multiple of } 2\pi)$.

3'1.2. Self-focusing. Consider now a monochromatic beam of finite size, with a smooth transverse intensity profile, as depicted in fig. 6. The beam propagates toward $+z$, and ρ denotes the transverse coordinates. An intuitive picture of self-focusing is given by the following argument: when the beam enters the nonlinear material, it experiences a refractive index given by eq. (29). If $\gamma > 0$, the refractive index will be larger at the centre of the beam where the intensity is higher. The effect is similar to that of a positive lens, and the beam will be focused. Defocusing occurs for $\gamma < 0$. For a quantitative evaluation, one

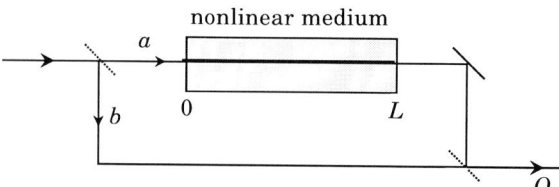

Fig. 5.

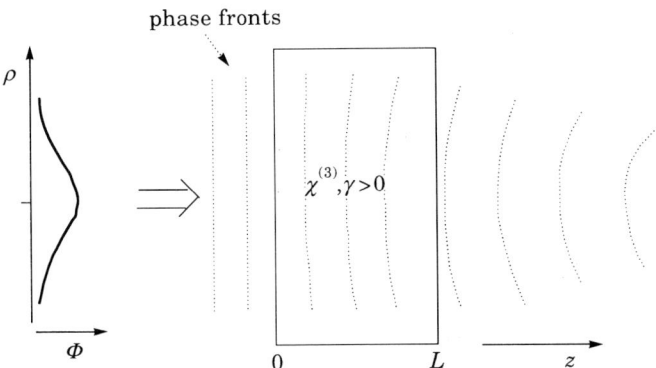

Fig. 6.

can write the fields as in eq. (7), so that the nonlinear polarization is given by

$$(32) \qquad P^{(3)} = \frac{1}{2} \varepsilon_0 \chi^{(3)} |A|^2 A \exp[ikz].$$

Substituting eq. (32) into eq. (8), one derives for the envelope $A(z, \rho)$

$$(33) \qquad \nabla_\perp^2 A + 2ik \frac{\partial A}{\partial z} = -\frac{\omega^2}{c^2} \frac{\chi^{(3)}}{2} |A|^2 A.$$

If the medium is sufficiently thin, the role of the diffraction term can be neglected (this is equivalent to assuming $\Phi(\rho)$ constant in z, and hence no changes
of the beam size in the nonlinear medium). In this simple case, omitting
the inessential constant phase change $(\omega/c)n_0 L$, one obtains $A(z = L, \rho) =$
$= A(z = 0, \rho)\exp[i\phi^{(\mathrm{NL})}(\rho)]$ with

$$(34) \qquad \phi^{(\mathrm{NL})}(\rho) = (\omega/c)\gamma\Phi(\rho)L.$$

The curvature of the phase front is schematically shown in fig. 6. In the case
$\gamma > 0$, a beam which is collimated at the entrance will be convergent at the exit.

In a thick medium, the beam size can decrease appreciably within the nonlinear medium itself. The increased intensity will make self-focusing even
stronger, and the minimum size that the beam can reach is limited by the
diffraction (the term $\nabla_\perp^2 A$ in eq. (33)). The pattern is quite complex, and cru-
cially dependent on the beam profile and intensity at the entrance. In many
cases, self-focusing can lead to extremely high intensities which cause the
breakdown of the medium. The two-dimensional version of eq. (33), with x as
the only transverse coordinate, coincides with the nonlinear Schrödinger equa-
tion $i\partial u/\partial z + (1/2)\partial^2 u/\partial x^2 + |u|^2 u = 0$. In this case, there is a stable solution
which depicts a beam propagating in z with no change in x. The mechanism
leading to this spatial soliton relies on a proper balance of the opposite effects of
diffraction and of self-focusing.

The nonlinear interferometer and focusing/defocusing effects can be exploited in some devices. Waveguide devices based on nonlinear interferometers will be discussed in the lecture by ASSANTO. Focusing effects play an an important role in the so-called «Kerr lens mode looking» employed for the generation of ultrashort pulses: due to the higher intensity, the peak of a pulse is more focused (for $\gamma > 0$) than the front and trailing edges, so that a laser resonator, when designed to give smaller losses when the beam is more focused, will favour the shortening of the pulse.

3'2. *Wave mixing.* – In the more general case, the third-order nonlinearity couples four fields. The names «three-wave mixing», «four-wave mixing» refer to the number of different beams which enter in the nonlinear process. Even in the stationary case, the problem becomes quite complicated if one includes finite sizes of the beams, self-actions and so on. In the following, we adopt the infinite-plane-wave approximation and consider only those terms of $P^{(3)}$ which are essential for the process under consideration.

3'2.1. Phase conjugation. Consider the case of four-wave mixing depicted in fig. 7a). The three fields a, b, c overlap in the nonlinear medium: the strong beams a and b are counterpropagating, with $\mathbf{k}_a = -\mathbf{k}_b$, and act as pumps. We shall find an equation for d, the new beam generated in the interaction, and show that d is the phase-conjugated field of c. We assume that all beams have the same frequency (degenerate four-wave mixing, shortened to DFWM), and we adopt the infinite-plane-wave approximation. From eq. (28) one notices that the nonlinear polarization contains the term

(35) $$P^{(3)} = \varepsilon_0 \chi^{(3)} A_a A_b A_c^* \exp[i(\mathbf{k}_a + \mathbf{k}_b - \mathbf{k}_c)\mathbf{r}].$$

Since $\mathbf{k}_a = -\mathbf{k}_b$, one has $P^{(3)} \approx \exp[-i\mathbf{k}_c\mathbf{r}]$ which becomes the source of a field

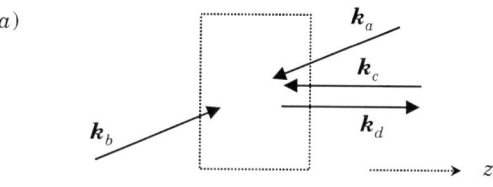

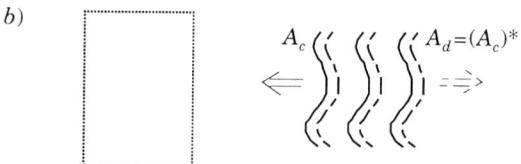

Fig. 7.

d with a wave vector $\mathbf{k}_d$ given by

$$(36) \qquad\qquad \mathbf{k}_d = -\mathbf{k}_c .$$

Taking the z-axis parallel to $\mathbf{k}_c$, and substituting $P^{(3)}$ into eq. (9), the equation for the amplitude of d is given by

$$(37) \qquad\qquad \frac{\mathrm{d}A_d}{\mathrm{d}z} = i\frac{\omega}{2cn}\chi^{(3)}A_aA_bA_c^* \exp[-i(k_d + k_c)z].$$

When the intensity of d is much smaller than that of the other beams, the amplitudes A_a, A_b, A_c can be assumed to remain constant. With no input at $z = 0$, one sets $A_d(z = 0) = 0$, and the integration of eq. (34) gives

$$(38) \qquad\qquad A_d(L) = [i\omega/(2cn)L\chi^{(3)}A_aA_b]A_c^* .$$

Denoting with Φ_d the intensity of d, one has

$$\Phi_d(L) = \omega^2/(2\varepsilon_0 c^3 n^3)L^2|\chi^{(3)}|^2\Phi_a\Phi_b\Phi_c .$$

Equations (36), (38) show that field d is the phase conjugated of c. The plane-wave approximation we relied upon may not evidence the features of wave front reversal. However, a signal beam as depicted in fig. 7b) can be decomposed in plane waves, each giving rise to its phase-conjugated reflection. Since the quantity in square brackets in eq. (38) is the same for all spatial components, the total reflected field will be phase conjugated of the impinging one. Applications of phase conjugation are discussed in the lecture by GUNTER.

A solution which accounts for the modification of A_c during the interaction can be obtained by coupling eq. (37) to the equation for $\mathrm{d}A_c/\mathrm{d}z$. The analytic solution (available under the assumption that the pumps a and b are not depleted) shows that also the intensity of c increases. In the FWM process, quanta are exchanged among the fields according to

$$(39) \qquad\qquad h\omega_a + h\omega_b \rightarrow h\omega_c + h\omega_d .$$

One photon is absorbed from each pump beam a and b, one photon is created in c and one in d: the generation of the reflected beam is always accompanied by a simultaneous growth of the signal pulse.

3$^{\cdot}$2.2. Self-diffraction. When two waves intersect with an angle a in a Kerr medium, they can give rise to diffracted fields as shown in fig. 8. Assume that the two waves, of wave vectors $\mathbf{k}_a$ and $\mathbf{k}_c$, have the same frequency ω. The source term for the field d is

$$P^{(3)} = (1/2)\varepsilon_0\chi^{(3)}A_aA_aA_c^* \exp[i(2\mathbf{k}_a - \mathbf{k}_b)\mathbf{r}].$$

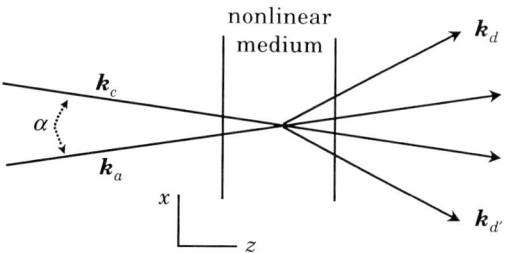

Fig. 8.

Writing the spatial part of the diffracted field as $A_d(z)\exp[i\boldsymbol{k}_d\boldsymbol{r}]$, applying the slowly-varying-amplitude approximation, for small α's one derives the following equation for $A_d(z)$:

$$(40) \qquad \frac{\mathrm{d}A_d(z)}{\mathrm{d}z} = +i(\omega/4cn)\exp[i\Delta hz]\chi^{(3)}A_a^2A_c^*$$

with the condition $\Delta k_x = (2\boldsymbol{k}_a - \boldsymbol{k}_c - \boldsymbol{k}_d)_x = 0$. The fact that the x components of the wave vector must be matched stems from the assumption of infinite plane waves. In eq. (40) the wave vector mismatch involves only the z components, and is given by

$$\Delta h = (2\boldsymbol{k}_a - \boldsymbol{k}_c - \boldsymbol{k}_d)_z \ .$$

Equation (40) is similar to eq. (37), and it can be integrated under the same assumptions to give $A_d(L)$. The physical process is here accounted by $h\boldsymbol{k}_a + h\boldsymbol{k}_a \to h\boldsymbol{k}_c + h\boldsymbol{k}_d$. The difference with the previous case of DFWM is that phase matching is not exactly satisfied in degenerate self-diffraction. Since the intensity of the diffracted field behaves as $\mathrm{sinc}^2(\Delta hL/2)$, the phase matching is practically satisfied whenever $\Delta hL < \pi$, which is often called the thin-grating condition. Denoting with Λ the period of the intensity modulation along x, which arises from the interference of a and c, the condition $\Delta hL < \pi$ implies $\lambda/\Lambda < \Lambda/L$. This last inequality can be interpreted by saying that, within a «channel» of width Λ, the diffraction angle is smaller than the geometrical one. Since $\Lambda = 2k\sin(\alpha/2)$, the angle α must be small whenever $L \gg \lambda$.

3`2.3. Frequency mixing. We consider now the process of frequency mixing shown in fig. 9. Two collinear beams a and c have frequencies ω_a and ω_c and generate in the nonlinear medium the field d with frequency $\omega_d = 2\omega_a - \omega_c$. This process is the analogue in the frequency domain of self-diffraction in the space domain, and it can be described through similar equations. The source of the field d is $P^{(3)}(\omega_d) = (1/2)\varepsilon_0\chi^{(3)}A_aA_aA_c^*\exp[i(2k_a - k_c)z]$, and, when the equation for $\mathrm{d}A_d(z)/\mathrm{d}z$ is solved under the assumption that the driving fields are

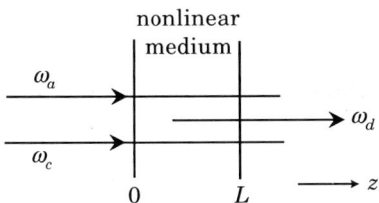

Fig. 9.

negligibly affected by the interaction, one has

(41)
$$A_d(L) = i\omega_s L(4cn_s)^{-1}(S_0 - iC_0)\chi_0^{(3)}A_a^2 A_c^*$$

with $S_0 = \sin \Delta hL/\Delta hL$, $C_0 = (\cos \Delta hL - 1)/\Delta hL$. With all the beams propagating along $+z$, the mismatch Δh is given by $\Delta h = 2k_a - k_c - k_d$, and it is now due to the frequency dispersion of the refractive indices.

3·3. *Measurements of* $\chi^{(3)}$. – The configurations discussed above are also the most common arrangements employed to measure the magnitude of the nonlinear susceptibility. Each arrangement has its own advantages and disadvantages. In phase conjugation, self-diffraction and frequency mixing, Φ_d, the intensity of the generated field, has the dependence $\Phi_d \propto |\chi^{(3)}|^2 \Phi_a \Phi_b \Phi_c$. Measuring Φ_d allows one to determine $|\chi^{(3)}|^2$, once the intensities of the other beams are known (or, more simply, by performing a relative measurement with a material of known $\chi^{(3)}$). Some tricks are necessary to determine, beside a probable contribution of $\mathrm{Im}\,\chi^{(3)}$, the sign of $\mathrm{Re}\,\chi^{(3)}$. On the contrary, measurements based on the effect of focusing/defocusing (such as the recent technique of Z-scan) give a direct access to the sign of $\mathrm{Re}\,\chi^{(3)}$.

3·4. *Time-dependent optical Kerr effect.* – In the previous examples, we have considered stationary cases. Suppose now that a pulse of time duration $\Delta\tau$, central frequency ω_0 propagates in a medium with a «fast» $\chi^{(3)}$. By «fast» we intend that the memory time in eq. (26) is much shorter than $\Delta\tau$, so that the nonlinear polarization can be considered to follow instantaneously the changes in the field amplitude. Equation (31) remains valid also in this case with $n(t) = = n_0 + \gamma\Phi(t)$. Neglecting pulse broadening, one can imagine that every slice of the pulse, with intensity Φ, after propagating a distance L in the Kerr medium, has cumulated a total phase φ given by $\varphi = kL - \omega_0 t$, with $k = (\omega_0/c)(n_0 + \gamma\Phi)$. The instantaneous frequency is then

(42)
$$\omega(t) = -\partial\varphi/\partial t = \omega_0 - \gamma L\partial\Phi/\partial t .$$

As shown in fig. 10, the frequency sweeps in time, according to the rate of change of the intensity. The pulse with a time-dependent frequency is said to be frequency chirped.

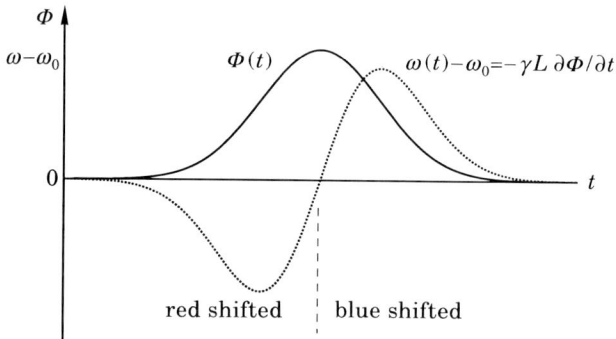

Fig. 10.

Frequency modulation plays an important role in the generation of ultra-short pulses with a large frequency spectrum. Pulses with a «continuum» spectrum are employed in time-resolved spectroscopy, and, being not directly available from a laser source, they are usually obtained by focusing an ultrashort pump pulse to high intensity in water or fused silica.

The time duration of a chirped pulse can be shortened through propagation in a dispersive medium. If $\gamma > 0$ as in fig. 10, the nonlinearity gives rise to a pulse with the front end red shifted, while the tail is blue shifted. If this chirped pulse enters in a dispersive medium with $V' = \partial v / \partial \omega < 0$ (a case opposite to that shown in fig. 1), the lower frequencies travel at a lower speed than the higher frequencies, and one can arrange for a situation where the pulse is compressed in time. A pulse can be chirped in a controlled manner in a fibre, where it is possible to maintain a uniform intensity for long propagation lengths.

The effects of the material dispersion and of the Kerr nonlinearity can be considered together by substituting the nonlinear polarization given by eq. (32) into eq. (12). The equation for $A(z, \tau)$ becomes more transparent when written in the frame of reference which moves with the velocity v, the group velocity of the pulse. Performing the change of variables $z = z$, $\tau = t - z/v$, one obtains

$$(43) \qquad i\frac{\partial}{\partial z}A - \frac{K''}{2}\frac{\partial^2 A}{\partial \tau^2} + \frac{\omega_0}{2cn}\chi^{(3)}|A|^2 A = 0 \,.$$

When $\chi^{(3)} > 0$ and $K'' < 0$, eq. (43) is equivalent to the nonlinear Schrödinger equation. One can then find a pulse with a particular temporal profile and amplitude (soliton) which propagates with no change of its envelope. Notice that eq. (43) is not suitable to describing the free propagation over significant lengths in a bulk material, since in this case the transverse profile of A and diffraction effects cannot be neglected. The proper context for eq. (41) and soliton propagation occurs in monomode optical fibres.

4. – Cascade effects.

In a material with a second-order susceptibility, two second-order processes in cascade give rise to many of the effects described for a third-order medium. Figure 11 gives a schematic representation of what happens in cascading; the two fields a and b give rise, through a sum frequency, to the intermediate field s, which can in turn interact with field c to generate field d. Wiping out the intermediate field, the overall process can be seen as a four-beam interaction among fields a, b, c, d, and, under some restrictions, it can be described by an induced «third-order» susceptibility $\chi_{\text{cas}}^{(3)}$.

The magnitude of $\chi_{\text{cas}}^{(3)}$, the good transparency of the materials that can be used and the fast response (purely electronic) make cascading quite attractive for many applications for which pure third-order materials have been so far considered. This is the main reason of the present revival of interest in cascading, an effect which has been left unexplored after early investigations [1-3].

The cascade effect manifests itself through the enhancement of self-diffraction [4], frequency mixing [3,5], self-phase modulation [1,2,6]. In the practical arrangements one chooses $a \equiv b$, so that the intermediate field s is given by the second harmonic of a. In the arrangement for self-phase modulation one makes use of a single beam; for this degenerate case also c and d are taken to coincide with a.

We consider the case of frequency mixing, with $a \equiv b$. The equations for the amplitudes of the intermediate SH field and of d are given by

$$(44a) \qquad \frac{\mathrm{d}A_s(z)}{\mathrm{d}z} = i\frac{\omega_a}{2cn_s}\chi^{(2)}A_a^2\exp[i\Delta kz],$$

$$(44b) \qquad \frac{\mathrm{d}A_d(z)}{\mathrm{d}z} = i\frac{\omega_d}{2cn_d}\exp[i\Delta hz][\chi_0^{(3)}/2A_a^2A_c^* + \chi^{(2)}A_s(z)A_c^*\exp[-i\Delta kz]]$$

with

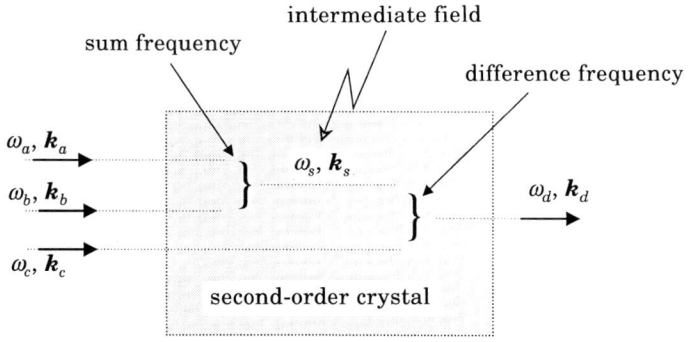

Fig. 11.

$\chi^{(2)}$ the effective second-order susceptibility,

$\Delta k = 2k_a - k_s$ the wave vector mismatch for the SH generation,

$\Delta h = 2k_a - k_c - k_d$ the wave vector mismatch for the process $2k_a \to k_c + k_d$,

$\chi_0^{(3)}$ the intrinsic third-order susceptibility.

Integrating eq. (44a) for $A_s(z)$, substituting it into eq. (44b), and then solving for $A_d(z)$, always assuming negligible depletion of the generating fields, one finds

$$A_d(L) = i\omega_s L(4cn_s)^{-1}(S_0 - iC_0)\chi_{\text{eff}}^{(3)}A_a^2 A_c^*$$

with

(45) $$\chi_{\text{eff}}^{(3)} = \chi_0^{(3)} + \chi_{\text{cas}}^{(3)} = \chi_0^{(3)} + \frac{\omega_a}{2cn_s}L(\chi^{(2)})^2 F(\Delta kL, \Delta hL).$$

The expression for $A_d(L)$ coincides with that given in eq. (41) for frequency mixing in a pure third-order material. The only difference is that here $\chi^{(3)}$ is replaced by $\chi_{\text{eff}}^{(3)}$.

In eq. (45), the function F is given by $F = (2/\Delta kL)((S - iC)/(S_0 - iC_0))$, with $S = S_0 - \sin(\Delta h - \Delta k)L/(\Delta h - \Delta k)L$ and $C = C_0 - (\cos(\Delta h - \Delta k)L - 1)/(\Delta h - \Delta k)L$. The real and the imaginary parts of F are plotted vs. ΔkL in fig. 12. The main features of the plots do not change as far as $\Delta hL < \pi$. For $|\Delta kL| > 2\pi$ (far from SH phase matching): $\chi_{\text{cas}}^{(3)} \approx (\omega_a/cn_s)(\chi^{(2)})^2/\Delta k$. Approaching SH phase matching, $\chi_{\text{cas}}^{(3)}$ becomes complex and dependent on L. At $\Delta k = 0$, $\chi_{\text{cas}}^{(3)}$ is imaginary and given by $\chi_{\text{cas}}^{(3)} = -i\omega_a(2cn_s)^{-1}(\chi^{(2)})^2 L$.

The formal equivalence of $\chi_{\text{cas}}^{(3)}$ with the intrinsic $\chi^{(3)}$ includes the imaginary part. In fact, one can show that $\omega_0 c^{-2}\varepsilon_0^{-1}n_0^{-2}\,\text{Im}\,\chi_{\text{cas}}^{(3)}$ represents the coefficient of two-photon absorption experienced by the pump beams as a consequence of SH generation, and it is then equivalent to β defined in eq. (30′). $\text{Im}\,\chi_{\text{cas}}^{(3)}$ vanish-

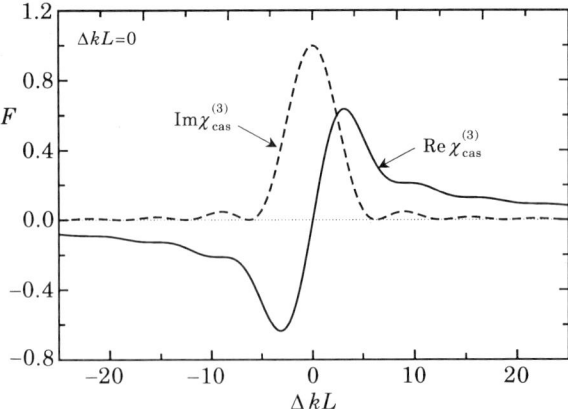

Fig. 12.

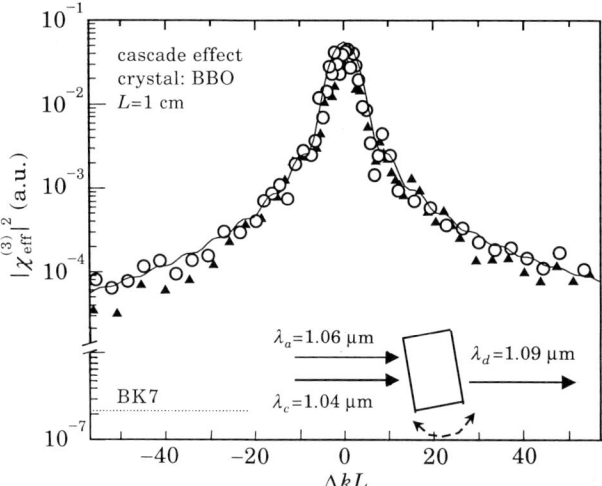

Fig. 13.

es for $|\Delta kL| = 2\pi \times$ integer, when there is no energy left in the intermediate field at the output of the crystal. The effective nonlinear susceptibility given by eq. (45) is not peculiar of wave mixing, but holds as well for self-diffraction and phase modulation (one should set $\Delta h = 0$ in this last case). We remark that eq. (43) has been derived under the assumption that $|A_s| \ll |A_a|$, so that A_a could be taken constant and independent of z. This is not the case when the intermediate field becomes comparable to the pump, a fact which leads to some saturation of the nonlinearity.

Experimental results on the cascade-enhanced frequency mixing are shown in fig. 13. The quantity on the vertical scale is proportional to Φ_d, the intensity of the beam d at the new frequency ω_d, and it is then proportional to $|\chi_{\text{eff}}^{(3)}|^2$. The nonlinear material is a 1 cm long BBO crystal, and the change of Δk is obtained by rotating the crystal through the SH phase-matching angle. One notices a large increase of $|\chi^{(3)}|^2$ at small ΔkL, consistent with the prediction (solid line) of eq. (45). The horizontal line gives $|\chi^{(3)}|^2$ of the BK7 glass (used as a reference) where cascading is absent since $\chi^{(2)} = 0$.

The effect of cascading can be made quite large by operating in the proximity of the phase matching for SH generation. For example, at $|\Delta kL| \approx 2\pi$ one calculates $F \approx 0.3$. For $\chi^{(2)}/2 = d = 2$ pm/V, $n = 1.6$, $L = 1$ cm, $\lambda = 1$ µm, one derives $|\chi_{\text{cas}}^{(3)}| = \pi L/\lambda(\chi^{(2)})^2 F \approx 10^{-19}$ m^2/V^2. These were the parameters and the value of $|\chi_{\text{cas}}^{(3)}|$ which was found in the experiment of fig. 12. As presented in other lectures, there are second-order materials with a much higher second-order nonlinearity. A material with $d \approx 50$ pm/V should allow one to achieve, for the same L and ΔkL of the previous case, effective $\chi^{(3)}$ as large as 10^{-16} m^2/V^2.

BIBLIOGRAPHY for sect. 1-3

N. BLOEMBERGEN: *Nonlinear Optics* (Benjamin, New York, N.Y., 1965).
F. ZERNIKE and J. E. MIDWINTER: *Applied Nonlinear Optics* (Wiley, New York, N.Y., 1973).
Quantum Electronics, Vol. 1A and B, edited by H. RABIN and C. H. TANG (Academic Press, New York, N.Y., 1975).
Y. R. SHEN: *The Principles of Nonlinear Optics* (John Wiley, New York, N.Y., 1984).
P. N. BUTCHER and D. COTTER: *The Elements of Nonlinear Optics* (Cambridge University Press, New York, N.Y., 1990).
A. C. NEWELL and J. V. MOLONEY: *Nonlinear Optics* (Addison-Wesley, Reading, Mass., 1992).
Optical Phase Conjugation and Instabilities, special issue of *Journal de Physique*, 44, C2, March 1983.
Optical Phase Conjugation, edited by R. A. FISHER (Academic Press, New York, N.Y., 1983).

REFERENCES for sect. 4

[1] L. A. OSTROVSKIJ: *JEPT Lett.*, **10**, 281 (1967).
[2] E. YABLONOVITCH, C. FLYTZANIS and N. BLOEMBERGEN: *Phys. Rev. Lett.*, **29**, 865 (1972).
[3] J. M. JARBOROUGH and O. AMMANN: *Appl. Phys. Lett.*, **18**, 145 (1976).
[4] R. DANIELIUS, P. DI TRAPANI, A. PISKARSKAS, D. PODENAS, A. VARANAVICIUS and G. P. BANFI: *Opt. Lett.*, **18**, 574 (1993).
[5] H. TAN, G. P. BANFI and A. TOMASELLI: *Appl. Phys. Lett.*, **63**, 2472 (1993); S. NITTI, H. M. TAN, G. P. BANFI and V. DEGIORGIO: *Opt. Commun.*, **106**, 263 (1994).
[6] R. DE SALVO, D. J. HAGAN, M. SHEIK-BAHAE, G. STEGEMAN, E. W. VAN STRYLAND and H. VANHERZELE: *Opt. Lett.*, **17**, 28 (1992); G. I. STEGEMAN, M. SHEIK-BAHAE, E. W. VAN STRYLAND and G. ASSANTO: *Opt. Lett.*, **18**, 13 (1993).

MATERIALS FOR QUADRATIC EFFECTS

Optical Parametric Processes and Inorganic Nonlinear Optical Crystals.

C. L. TANG

Cornell University - Ithaca, NY 14853

1. – Introduction.

One of the most important recent developments in the applications of nonlinear optics involves optical parametric oscillators and amplifiers that can operate in the nanosecond to the femtosecond time domain. These are powerful solid-state sources of continuosly tunable coherent radiation with potentially broad applications in research and industry. The basic concept of the parametric process is, of course, not new[1-3], but the practical development of the devices had been very slow in coming due to the lack of suitable nonlinear optical materials. Progress in the development of a number of new nonlinear optical crystals in the past decade has finally led to the recent rapid advances in optical parametric devices.

A brief review will first be given of the basic concepts in optical parametric processes and the key considerations in the search for useful nonlinear optical crystals. This will be followed by a more detailed discussion in sect. 2 of some of the recently discovered inorganic crystals such as β-barium borate (BBO), lithium triborate (LBO) and potassium titanyl phosphate (KTP) crystals. The basic properties of these crystals as examples of important practical inorganic nonlinear optical crystals for applications in optical parametric devices will be emphasized. Finally, in sect. 3, examples of nanosecond and femtosecond types of optical parametric oscillators coupled with second-harmonic and sum-frequency generation that can be tuned from the near UV to the mid i.r. will be discussed.

The parametric process is one of the most elementary nonlinear optical processes involving three photons and can be respresented schematically by the simplest kind of Feynman diagram (fig. 1.1). It describes the process in which one high-frequency photon is annihilated and two lower-frequency photons are created. This process has its origin in the second-order nonlinear polarization in

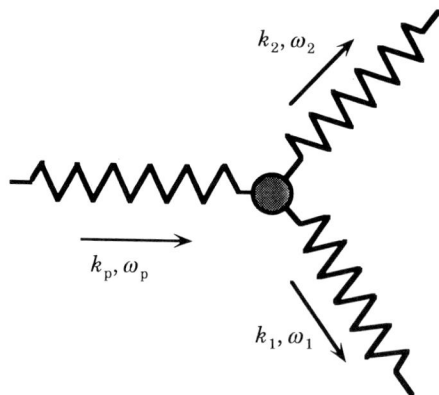

Fig. 1.1. – Breakdown of a pump photon into a signal and an idler photon by the spontaneous parametric process.

the expansion of the induced macroscopic polarization P in the nonlinear medium in powers of the electric field E (for units and definitions, see appendix):

$$(1.1) \quad P = \varepsilon_0 \chi^{(1)} E + \varepsilon_0 \chi^{(2)} EE + \varepsilon_0 \chi^{(3)} EEE + \ldots = P^{(1)} + P^{(2)} + P^{(3)} + \ldots .$$

This second-order polarization term, $P^{(2)}$, leads to a term proportional to E^3 in the Hamiltonian of the field in the medium:

$$(1.2) \qquad\qquad\qquad H = H_0 + H_1 \, ,$$

where H_0 is the field Hamiltonian corresponding to the medium in the absence of the nonlinearity:

$$(1.3) \qquad\qquad\qquad H_0 = \frac{1}{8\pi} \int [D \cdot E + B \cdot H] \, dr \, .$$

For optical parametric processes, the transparency region of the optical medium is of interest. The response of the medium to the fields can then be assumed instantaneous and local. For moderately strong fields, the total Hamiltonian can also be expanded in a Taylor series, as in the case of the induced polarization in the medium. The corresponding interaction Hamiltonian H_1 describing the nonlinear response of the medium is, therefore, of the form

$$(1.4) \qquad\qquad H_1 = \frac{1}{3} \sum_{ijk} \int \chi^{(2)}_{ijk} E_i(r, t) \, E_j(r, t) \, E_k(r, t) \, dr \, .$$

Quantizing the fields leads in turn to a term of the form $\chi^{(2)} a_1^- a_2^+ a_3^+$ in the interaction part of the Hamiltonian, where the a^+'s and a^-'s are the creation and annihilation operators, respectively, of the appropriate photons. This term corresponds to the parametric process in which a photon at ω_1 is annihilated to create two photons at ω_2 and ω_3.

It is obvious that the parametric process can be generalized to include four or more photons corresponding to the higher-order terms in the induced polarization in the nonlinear medium. Four-photon parametric processes, also known as four-wave mixing processes, are used extensively for various fundamental measurements of atomic, molecular, or macroscopic properties of materials. Practical device applications based upon these higher-order nonlinear optical processes are, however, not yet well developed. In this lecture, the focus will be on the basic three-photon parametric process. It will be used as a specific example to illustrate the more important requirements in nonlinear optical materials.

In a three-photon parametric process, if the initial state, $|i\rangle$, of the field contains only N_1 pump photons in mode 1 at frequency ω_1 but no photon in mode 2 and mode 3 at frequencies ω_2 or ω_3, respectively, the corresponding parametric process is the spontaneous parametric emission process through which one pump photon at ω_1 spontaneously breaks down into two lower-frequency photons at ω_2 and ω_3 in the final state $|f\rangle$, or

$$(1.5) \qquad |i\rangle = |N_1 00\rangle \to |f\rangle \propto \sqrt{N_1}|(N_1 - 1)\,11\rangle.$$

In this case, the corresponding transition probability is proportional to N_1 or to the intensity of the pump beam.

If there are already photons at ω_2 and ω_3 (which are also referred to as signal and idler frequencies ω_s and ω_i using the nomenclature borrowed from earlier microwave parametric-oscillator work), then stimulated emission takes place, or

$$(1.6) \quad |i\rangle = |N_1 N_2 N_3\rangle \to |f\rangle \propto$$

$$\propto \sqrt{N_1 (N_2 + 1)(N_3 + 1)}|(N_1 - 1)(N_2 + 1)(N_3 + 1)\rangle,$$

which shows that the transition probability to the final state is proportional to not only the intensity of the pump beam, but also those of the signal and idler beams. This means that the more the signal or idler photons are present in the medium already, the more signal and idler photons will be emitted through this process. This is just like the stimulated-emission process in ordinary lasers or masers. Here it corresponds to amplification of the signal and idler beams through the parametric amplification process.

Given the parametric amplification process, an optical parametric oscillator can be constructed with the addition of suitable optical feedback, such as the use of a Fabry-Perot cavity. The basic oscillator configuration is, therefore, extremely simple, as shown schematically in fig. 1.2. It is completely analogous to that of conventional lasers, except that the active medium is a nonlinear optical crystal and no unique set of discrete energy levels of the medium

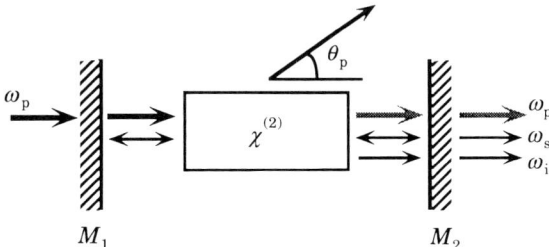

Fig. 1.2. – Schematic of singly resonant optical parametric oscillator.

is directly involved in converting the pump photons into the signal and idler photons (fig. 1.3).

The tuning of the optical parametric oscillator comes about as follows. As in any multiphoton process, the energy in the optical parametric process is always conserved: $\omega_1 = \omega_2 + \omega_3$, of course. However, in going from one high frequency to two lower frequencies, the splitting of ω_1 into ω_2 and ω_3 is not unique. As long as the the sum is conserved, ω_2 or ω_3 can each have basically any values from zero to ω_1. This flexibility is, in fact, the origin of the basic tunability of the optical parametric process. The specific pair of frequencies resulting from each ω_1 and crystal orientation is ultimately determined by the conservation of photon momentum $k_1 = k_2 + k_3$, or the phase-matching condition, taking into account the material dispersion and the birefringence of the crystal. By rotating the crystal relative to the direction of propagation of the waves, the corresponding birefringence can be tuned, which in turn tunes ω_2 and ω_3 (fig. 1.4).

The OPO is in many respects analogous to the optically pumped 3-level laser (fig. 1.3), but with some important differences. Instead of the discrete energy levels in the laser, tunable photon energy levels are involved in the OPO. The pump source for a laser can be a coherent or an incoherent source. The OPO only

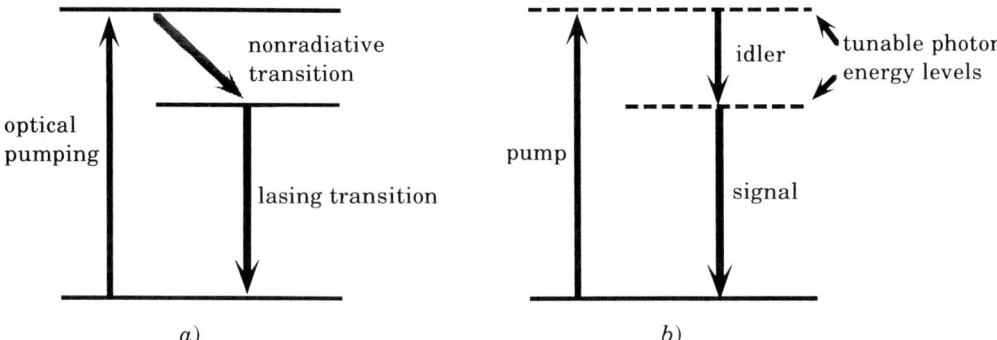

Fig. 1.3. – a) Optically pumped three-level laser and b) three-photon optical parametric process.

converts coherent light at one frequency into coherent radiation in a range of tunable frequencies. The pump source of the OPO must, therefore, be a laser. Without the laser, there would not have been any OPO. In the laser, nonradiative transitions are often involved leading to losses in converting the absorbed pump energy into useful laser output. And in the laser, because the radiative processes involved are single-photon transitions, and because of the possible involvement of nonradiation transitions, the temporal and the coherence properties of the laser output may not be related to those of the pump radiation. In the OPO, there are no nonradiative transitions involved and the whole process is a single three-photon process. The temporal and coherence properties of the OPO are, therefore, directly related to those of the pump. In the OPO, there is in principle no loss; the quantum efficiency of the OPO can potentially be 100% with all the absorbed pump energy converted into useful outputs.

The OPO is basically a wavelength conversion device which converts the laser output at one wavelength into a broad range of wavelengths where coherent radiation is needed for applications. Nonetheless, because of the similarities between the OPO and the optically pumped three-level laser, many of the device research issues and techniques used in connection with the lasers are also relevant to the OPO's, such as the effects of phase conjugation mirrors, squeezing and microcavity on the emission process, soliton formation, etc. Thus the OPO is not only an important practical source of tunable coherent radiation, it is also a fertile ground for studying various physics and materials problems.

The development of practical OPO's is, however, not a simple matter. It took many years primarily because of the the lack of suitable nonlinear optical

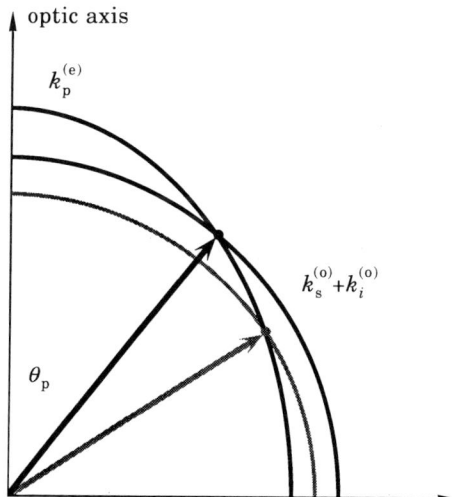

Fig. 1.4. – Use of birefringence to compensate for material dispersion. Rotating the crystal relative to the direction of propagation of the waves leads to tuning of the frequencies of the signal and idler waves.

crystals. It is not just a question of finding a material with a large enough non-linearity. In fact, of all the relevant material properties, optical nonlinearity is often one of the easier problems to deal with. There are a host of other issues. To give a simple example, the optical parametric gain coefficient, g, squared is of the form

$$g^2 = \frac{(2\pi)^4 \, |\chi^{(2)}|^2 \, |E_1|^2}{n_2^2 \, n_3^2 \, \lambda_2 \, \lambda_3} \, ,$$

where the n's and λ's are the indices of refraction and wavelengths, respectively. The total single-pass gain depends on the product of g and the effective crystal length L_{crystal}. A figure of merit Q can be defined in terms of the ratio of oscillation threshold to the optical-damage threshold of the crystal, Γ_{max}, or $Q = = |\chi^{(2)}|^2 L_{\text{crystal}}^2 \Gamma_{\text{max}}/n^3$. A comparison of different materials based upon this figure of merit using the information given in tables I-III, sect. 2, of various well-known nonlinear optical crystals will show immediately that the materials with the larger Q's are not always those with the larger $\chi^{(2)}$'s. Beyond the considerations of the gain of the material and the available crystal size, there are many other critical properties that a good OPO material must possess, such as suitable mechanical, thermal, chemical and phase-matching properties. The OPO is now a practical reality, because new materials such as BBO[4], LBO[5] and KTP[6] are finally meeting the basic requirements in all the critical areas for the first time. Since the development of such OPO materials, the progress has been rapid.

2. – Material considerations.

The basic material requirements for all three-photon processes are similar. The important ones are: suitable transparency, high nonlinearity, sufficient birefringence for phase matching in the wavelength range of interest, high optical-damage threshold, good thermal, mechanical and chemical properties, and, above all, possibility to grow sufficiently large crystals of good optical quality. For femtosecond applications, the crystal should also have low group velocity dispersion to reduce pulse broadening. For optical-parametric-device application, transparency and the possibility of phase matching over a large spectral range are particularly important.

In the past two decades, a great deal of effort has been made in the development of organic molecular crystals for nonlinear optics[7,8]. Systematic investigations have led to considerable understanding of the molecular nonlinearity, and crystals with phenomenally large optical nonlinearity have been reported. Unfortunately, the application of these crystals in optical parametric processes has been limited due to their relatively narrow transparency range. Their short-wavelength absorption cut-offs are generally limited to $\sim (0.4–0.45) \, \mu m$

by the π–π^* transitions of extended conjugation, while overtone C-H absorptions become a problem in the near $\sim (1.0$–$1.2)\,\mu m$ region. In addition, typical molecular crystals are generally mechanically soft and hard to polish. Though index-matching fluid has been used to reduce this difficulty, it makes the optical-parametric-device design more cumbersome and often introduces problems such as thermal defocusing of the pump lasers at high peak intensity. These problems make organic molecular crystals generally unattractive for optical-parametric-device applications. For other nonlinear optical applications limited to specific narrow wavelength regions, organic crystals with their large nonlinearity could be useful. Significant improvement in mechanical property can be obtained by using the ionic bonding available in crystals of molecular salts. Also, as illustrated by crystalline urea, good UV transparency can be obtained by using organic groups with shorter conjugation. The new nonlinear L-arginine phosphate and its analogs [9] are good examples of this approach. A similar approach using metal-organic complexes is also being tried. Though promising, these approaches have not yet yielded any crystals with properties that can compete with inorganic oxide crystals such as BBO, LBO and KTiOPO$_4$. In the following brief review, emphasis will be on more recently developed inorganic crystals primarily for applications in optical parametric processes.

The more commonly used general second-order nonlinear optical crystals [10] in the bulk form tend to be inorganic crystals such as the ADP-isomorphs NH$_4$H$_2$PO$_4$(ADP), KH$_2$PO$_4$(KDP), NH$_4$H$_2$AsO$_4$(ADA), CsH$_2$AsO$_4$ (CDA), etc. and the corresponding deuterated version; the ABO$_3$ type of ferroelectrics such as LiIO$_3$, LiNbO$_3$, KNbO$_3$, etc.; the borates such as β–BaB$_2$O$_4$, LiB$_3$O$_5$, etc.; and the isomorphic family of nonlinear optical crystals with the generic chemical formula MTiOXO$_4$, where $X = \{$P and As$\}$ [5], $M = \{$NH$_4$, K, Rb, Tl and Cs (for $X =$ As only)$\}$. Although the binary III-V and II-VI compounds, such as GaAs, InSb, GaP, ZeTe, etc., generally have large d-coefficients, because their structures are cubic, there is no birefringence that can be used to compensate for material dispersion. Therefore, they cannot be phasematched in the bulk and are useful only in waveguide forms. The ternary chalcopyrites (such as AgGaSe$_2$, AgGaS$_2$, ZnGeP$_2$ and CdGaAs$_2$) have electronic structures and optical nonlinearities comparable to those of the III-V and II-VI compounds, but are of tetragonal symmetry and have birefringence that can be used to compensate for material dispersion. The growth of these ternary compounds has been difficult, however.

For one reason or another, optical-parametric-device technology did not really take off until the growth technologies of β–BaB$_2$O$_4$, LiB$_3$O$_5$ and KTiOPO$_4$ were sufficiently well developed. In terms of the materials that are reasonably well developed today, the materials of choice for parametric-device applications in the mid infrared from about 2 to 20 μm are the chalcopyrites (AgGaSe$_2$, AgGaS$_2$, ZnGeP$_2$ and CdGaAs$_2$), Tl$_3$AsSe$_3$ and GaSe; for the visible and infrared, the newer materials are KNbO$_3$[11], KTiOPO$_4$ and its isomorphs

such as KTiOAsO$_4$[12] and CsTiOAsO$_4$[13]; for the ultraviolet and visible, mainly the borates such as β–BaB$_2$O$_4$ and LiB$_3$O$_5$.

In tables I-III, key properties of several selected nonlinear optical crystals
are tabulated.

TABLE I. – *Properties of some nonlinear optical crystals for OPO applications.* Data
shown are at 1.064 μm unless otherwise indicated. Γ_{max} surface damage threshold; $l\Delta T$
temperature-tuning bandwidth; $l\Delta\Theta$, CPM critical phase-matching acceptance angle;
$l^{1/2}\Delta\Theta$ noncritical phase-matching acceptance angle; $l\Delta\lambda$ SHG bandwidth; Δv_g^{-1} group velocity dispersion for SHG at 630 nm.

Crystal	LiB$_3$O$_5$	β-BaB$_2$O$_4$ (^{f})
point group	$mm2$ (^{a})	$3m$
birefringence	$n_{x=a} = 1.5656$ (^{b})	$n_e = 1.54254$
	$n_{y=c} = 1.5905$	$n_o = 1.65510$
	$n_{z=b} = 1.6055$	—
nonlinearity (pm/V)	—	$d_{22} = 1.6$
	$d_{32} = 1.16$ (^{b})	$d_{31} = 0.08$
transparency (μm)	0.16–2.6 (^{c})	0.19–2.5
Γ_{max} (GW/cm^2)	~ 25 (^{b})	~ 5 (^{g})
SHG cut-off (nm)	555 (^{d})	411
$l\Delta T$ (°C · cm)	3.9 (^{e})	55
$l\Delta\Theta$ (mrad · cm), CPM	31.3 (^{e})	0.52
$l^{1/2}\Delta\Theta$	71.9 (^{e})	not available
(mrad (cm)$^{1/2}$)	NCPM at 148.0 °C	—
$l\Delta\lambda$ (Å·cm)	not available	21.1
Δv_g^{-1} at 630 nm (fs/mm)	240 (^{d})	360
OPO tuning range (nm)	~ 415–2500 (^{d}) ($\lambda_p = 355$)	~410–2500 ($\lambda_p = 355$)
boule size	$(20 \times 20 \times 15)$ mm^3 (^{e})	Ø 84 mm × 18 mm
growth	TSSF (^{e}) at ~ 810 °C	TSSG from Na$_2$O at ~ 900 °C
predominant growth defects	flux (^{e}) inclusions	flux and bubble inclusions
chemical properties	nonhygroscopic (^{e}) (m.p. ~ 834°C)	slightly hygroscopic ($\beta \to \alpha$ ~ 925 °C)

Source references:

(a) H. VON KONIG and A. HOPPE: Z. Anorg. Allg. Chem., **439**, 71 (1978); M. IHARA, M. YUGE and J. KROG-MOE: Yogyo Kyokai Shi, **88**, 179 (1980); Z. SHUQUING, H. CHAOEN and Z. HONGWU: J. Cryst. Growth, **99**, 805 (1990).
(b) C. CHEN, Y. WU, A. JIANG, B. WU, G. YOU, R. LI and S. LIN: J. Opt. Soc. Am. B, **6**, 616 (1989); S. LIU, Z. SUN, B. WU and C. CHEN: J. Appl. Phys., **67**, 634 (1989). On the basis of $d_{32} = 2.69 \cdot d_{36}$ (KDP) and using the value d_{36}(KDP) = 0.39 pm/V according to R. C. EKART et al.: J. Quantum Electron., **26**, 922 (1990).
(c) (0.16–2.6) μm: C. CHEN, Y. WU, A. JIANG, B. WU, G. YOU, R. LI and S. LIN: J. Opt. Soc. Am. B, **6**, 616 (1989). (0.165–3.2) μm: A. ZHAO, C. HUANG and H. ZHANG: J. Cryst. Growth, **99**, 805 (1990).
(d) Calculated by using Sellmeier equations reported in B. WU, N. CHEN, C. CHEN, D. DENG and Z. XU: Opt. Lett., **14**, 1080 (1989).
(e) T. UKACHI and R. J. LANE, measurements carried out on Cornell LBO crystals grown by the self-flux method.
(f) Reference souces given in Growth and Characterization of Nonlinear Optical Crystals Suitable for Frequency Conversion, edited by L. K. CHENG, W. R. BOSENBERG and C. L. TANG, review article in Progress in Crystal Growth and Characterization, **20**, 9 (1990), unless indicated otherwise.
(g) Estimated surface damage threshold scaled from detailed bulk damage results reported by H. NAKATANI et al.: Appl. Phys. Lett., **53**, 2587 (1988).

TABLE II. – Properties of several visible-near i.r. nonlinear optical crystals. Unless otherwise specifies, data are for $\lambda = 1.064$ μm. (Data taken from $(a, e\text{-}i), (a, b, c)$ and (a, d) respectively.)

Characteristics	KNbO₃ (*)	LiNbO₃ (**)	Ba₂ NaNb₅ O₁₅
point group	$mm2$	$3m$	$mm2$
transparency (μm)	0.4–5.5	0.4–5.0	0.37–5.0
birefringence	negative biaxial	negative uniaxial	negative biaxial
	$n_{x=c} = 2.2574$	—	$n_{x=b} = 2.2580$
	$n_{y=a} = 2.2200$	$n^\circ = 2.2325$	$n_{y=a} = 2.2567$
	$n_{z=b} = 2.1196$	$n^e = 2.1560$	$n_{z=c} = 2.1700$
second-order nonlinearity (pm/V)	$d_{32} = 12.9,$ $d_{31} = -11.3$	$d_{33} = -29.7$	$d_{32} = -12.8,$ $d_{31} = -12.8$
	$d_{24} = 11.9,$ $d_{15} = -12.4$	$d_{31} = -4.8$	$d_{24} = 12.8,$ $d_{15} = -12.8$
	$d_{33} = -19.6$	$d_{22} = 2.3$	$d_{33} = -17.6$
$\partial(n^\omega - n^{2\omega})/\partial T(°C^{-1})$	$1.6 \cdot 10^{-4}$	$-5.9 \cdot 10^{-5}$	$1.05 \cdot 10^{-4}$
$T_{pm}(°C)$	181, d_{32}	-8, d_{31}	89, d_{39}
	—	—	101, d_{31}
$l\Delta T$ (°C-cm)	0.3	0.8	0.5
λ_{SHG} (cut-off)(μm) at 25 °C	0.860	~ 1.08	1.01
Γ_{max} (MW/cm²)	not available	~ 120	40
phase transition temperature (°C)	225 and 435	~ 1000	300

TABLE II (*continued*).

Characteristics	$KNbO_3$ (*)	$LiNbO_3$ (**)	$Ba_2 NaNb_5 O_{15}$
growth technique	TSSG from K_2O at ~ 1050 °C	Czochralski at ~ 1200 °C	Czochralski at ~ 1440 °C
predominant growth problems	cracks, blue coloration, multidomains	temperature-induced compositional striations	striations microtwinning, multidomains
post-growth processing	poling	poling	poling and detwinning
crystal size	$(20 \times 20 \times 20)\,mm^3$ (single domain)	$\varnothing\,100\,mm \times 200\,mm$ (as grown boule)	$\varnothing\,20\,mm \times 50\,mm$ (with striations)

(*) There is disagreement on the sign of the nonlinear coefficients of $KNbO_3$ in the literature. Data used here are taken from reference (^{e}) with the appropriate correction for the IRE convention (^{a}).
(**) Data are for congruent melting $LiNbO_3$ (^{b}). 5% MgO doped crystals give photorefractive damage threshold about (10–100) times higher (^{k,l}). The phase-matching properties for these crystals may differ due to the resulting changes in the lattice constants (^{j}).

Source references:

(*a*) S. SINGH: in *CRC Handbook of Laser Science and Technology*, Vol. 4, *Optical Materials*, Part I, edited by M. J. WEBER (CRC Press, Boca Raton, Fla., 1986), p. 3.
(*b*) R. L. BYER, J. F. YOUNG and R. S. FEIGELSON: *J. Appl. Phys.*, **41**, 2320 (1970).
(*c*) R. L. BYER: in *Quantum Electronics: A Treatise*, edited by H. RABIN and C. L. TANG, Vol. 1, Part A (Academic Press, New York, N.Y., 1975), p. 588.
(*d*) S. SINGH, D. A. DRAEGERT and J. E. GEUSIC: *Phys. Rev. B*, **2**, 2709 (1970).
(*e*) Y. UEMATSU: *Jpn. J. Appl. Phys.*, **13**, 1362 (1974).
(*f*) P. GUNTER: *Appl. Phys. Lett.*, **34**, 650 (1979).
(*g*) W. XING, H. LOOSER, H. WUEST and H. AREND: *J. Cryst. Growth*, **78**, 431 (1986).
(*h*) D. SHEN: *Mater. Res. Bull.*, **21**, 1375 (1986).
(*i*) T. FUKUDA and Y. UEMATSU: *Jpn. J. Appl. Phys.*, **11**, 163 (1972).
(*j*) B. C. GRABMAIER and F. OTTO: *J. Cryst. Growth*, **79**, 682 (1986).
(*k*) D. A. BRYAN, R. GERSON and H. E. TOMASCHKE: *Appl. Phys. Lett.*, **44**, 847 (1984).
(*l*) G. ZHONG, J. JIAN and Z. WU: in *11th International Quantum Electronics Conference*, IEEE Cat. No. 80 CH 1561-0, June 1980, p. 631.

TABLE III. – *Properties of several* UV, *visible and near*-i.r. *crystals.* Unless otherwise stated, all data for 1064 nm. (Data taken from $(^{c, e})$, $(^{a, b, f, m})$ and $(^{d, g^{-i}})$, respectively.)

Crystal	KDP	KTP (II)(*)
point group	$42\,m$	$mm\,2$
birefringence	$n_e = 1.4599$	$n_{x = a} = 1.7367$
	$n_o = 1.4938$	$n_{y = b} = 1.7395$
	—	$n_{z = c} = 1.8305$
nonlinearity (pm/V)	$d_{36} = 0.39$	$d_{32} = 5.0, d_{31} = 6.5$
	—	$d_{24} = 7.6, d_{15} = 6.1$
	—	$d_{33} = 13.7$

TABLE III (*continued*).

Crystal	KDP	KTP (II)(*)
transparency (μm)	0.2–1.4	0.35–4.4
Γ_{max} (GW/cm^2)	~ 3.5	~ 15.0
SHG cut-off (nm)	487	~ 990
$l\Delta T$ (°C · cm)	7	22
$l\Delta\theta$ (mrad · cm),	1.2	15.7
$l\Delta\lambda$ (Å · cm)	208 (**)	4.5
Δv_g^{-1} at 630 nm (fs/mm)	185	not applicable
OPO tuning range (nm)	~ 430–700 ($\lambda_p = 266$)	~ 610–4200 ($\lambda_p = 532$)
ΔT_F (°C)	12	not available
boule size	$(40 \times 40 \times 100)$ cm^3	$\sim (20 \times 20 \times 20)$ mm^3
growth technique	solution growth from H_2O	TSSG from $2KPO_3$-$K_4P_2O_7$ at ~ 1000 °C
predominant growth defects	organic impurities	flux inclusions
chemical properties	hygroscopic (m.p. ~ 253 °C)	nonhygroscopic (m.p. ~ 1172 °C)

(*) KTP type-I interaction gives $d_{eff} \sim d_{36}$(KDP) or less for most processes (^{m}). The d_{ij} values (^{d}) are for crystals grown by the hydrothermal technique ($^{j-l}$). Significantly lower damage thresholds were reported for hydrothermally grown crystals (^{d}).

(**) The anomalously large spectral bandwidth is a manifestation of the λ-noncritical phase-matching (^{n}). This is equivalent to a very good group velocity matching ($\Delta v_g^{-1} \sim 8$ fs/mm) for this interaction in KDP.

Source references:

(*a*) D. EIMERL: *IEEE J. Quantum Electron*, QE-23, 575 (1987).
(*b*) D. EIMERL, L. DAVIS, S. VELSKO, E. K. GRAHAM and A. ZALKIN: *J. Appl. Phys.*, **62**, 1968 (1987).
(*c*) D. EIMERL: *Ferroelectrics*, **72**, 95 (1987).
(*d*) Y. S. LIU, L. DRAFALL, D. DENTZ and R. BELT: G.E. Technical Information Series Report, 82CRD016 (February, 1982).
(*e*) Y. NISHIDA, A. YOKOTANI, T. SASAKI, K. YOSHIDA, T. YAMANAKA and C. YAMANAKA: *Appl. Phys. Lett.*, **52**, 420 (1988).
(*f*) A. JIANG, F. CHENG, Q. LIN, Z. CHENG and Y. ZHENG: *J. Cryst. Growth*, **79**, 963 (1986).
(*g*) P. BORDUI: in *Crystal Growth of KTiOPO4 from High Temperature Solution*, Ph. D. thesis, Massachusetts Institute of Technology (1987).
(*h*) Information sheet on $KTiOPO_4$, Ferroxcube, Division of Amperex Electronic Corp., Saugerties, New York (1987).
(*i*) P. BORDUI, J. C. JACCO, G. M. LOIACONO, R. A. STOLZENBERGER and J. J. ZOLA: *J. Cryst. Growth*, **84**, 403 (1987).
(*j*) F. C. ZUMSTEG, J. D. BIERLEIN and T. E. GIER: *J. Appl. Phys.*, **47**, 4980 (1976).
(*k*) R. A. LAUDIS, R. J. CAVA and A. J. CAPORASO: *J. Cryst. Growth*, **74**, 275 (1986).
(*l*) S. JIA, P. JIANG, H. NIU, D. LI and X. FAN: *J. Cryst. Growth*, **79**, 970 (1986).
(*m*) L. K. CHENG: unpublished.
(*n*) J. ZYSS and D. S. CHEMLA: in *Nonlinear Optical Properties of Organic Molecules and Crystals*, Vol. 1, edited by D. S. CHEMLA and J. ZYSS (Academic Press, New York, N.Y., 1987), p. 146.

3. – Optical parametric sources from nanoseconds to femtoseconds.

3`1. *Basic concepts.* – Although the elementary spontaneous parametric process can be understood only on the basis of quantum field theory, the stimulated parametric process, or the parametric amplification process, can be treated easily using classical wave equation[14]:

$$(3.1) \qquad \nabla^2 E_2(r, t) - \mu_0 \varepsilon_0 \frac{\partial^2}{\partial t^2} E_2(r, t) = \frac{4\pi}{c^2} \frac{\partial^2}{\partial t^2} P(r, t).$$

In the case of the basic three-photon parametric process, the E-field in the medium of interest contains three spectral components:

$$(3.2) \quad E(r, t) = \frac{1}{2} \{E_1(r) \exp[ik_1 \cdot r - i\omega_1 t] + E_2(r) \exp[ik_2 \cdot r - i\omega_2 t] +$$

$$+ E_3(r) \exp[ik_3 \cdot r - i\omega_3 t] + \text{complex conjugates}\}.$$

The corresponding induced macroscopic polarization in the medium contains a linear term and a nonlinear term:

$$P(r, t) = \frac{1}{2} \{P^L(r, t) + P^{NL}(r, t) + \text{complex conjugates}\}$$

which are in turn made up of terms of different physical origins:

$$P^L(r, t) = \chi^{(1)}(\omega_1) \cdot E_1 \cdot \exp[ik_1 \cdot r - i\omega_1 t] + \chi^{(1)}(\omega_2) \cdot E_2(z) \cdot \exp[ik_2 \cdot r - i\omega_2 t] +$$

$$+ \chi^{(1)}(\omega_3) \cdot E_3(z) \cdot \exp[ik_3 \cdot r - i\omega_3 t]$$

and

$$(3.3) \quad P^{NL}(r, t) = \chi^{(2)}(\omega_3 = \omega_1 - \omega_2): E_1 E_2^* \exp[i(k_1 - k_2) \cdot r - i\omega_3 t] +$$

$$+ \chi^{(2)}(\omega_2 = \omega_1 - \omega_3): E_1 E_3^* \exp[i(k_1 - k_3) \cdot r - i\omega_2 t] +$$

$$+ \chi^{(2)}(\omega_1 = \omega_2 + \omega_3): E_2 E_3 \exp[i(k_2 + k_3) \cdot r - i\omega_1 t] +$$

$$+ \text{other nonlinear terms}.$$

The key results of the parametric amplification process can be derived by using a one-dimensional scalar model of the wave equation. Furthermore, for the linear amplification regime, the depletion, or the spatial variation, of the «pump wave» at ω_1 can be neglected; thus $E_1(r)$ in this case can be assumed to be a fixed parameter E_p equal to the amplitude, $E_1(0)$, of the incident pump wave. Effects associated with the depletion of the pump wave due to parametric conversion will be discussed in the following section. Making all these approximations and assuming that the energy and the momentum conservation conditions

$$(3.4) \qquad\qquad\qquad \omega_1 = \omega_2 + \omega_3$$

and

(3.5) $$k_1 = k_2 + k_3$$

are satisfied, one obtains from eqs. (3.1)-(3.3) the coupled-wave equations for the complex amplitudes of the signal and idler waves:

(3.6) $$\left[\frac{\partial^2}{\partial z^2} + \frac{n_2^2 \omega_2^2}{c^2} \right] E_2(z) \exp[ik_2 z] = - \frac{4\pi\omega_2^2}{c^2} d_{\mathrm{eff}} E_{\mathrm{p}} E_3^* \exp[ik_2 z]$$

and

(3.7) $$\left[\frac{\partial^2}{\partial z^2} + \frac{n_3^2 \omega_3^2}{c^2} \right] E_3(z) \exp[ik_3 z] = - \frac{4\pi\omega_3^2}{c^2} d_{\mathrm{eff}} E_{\mathrm{p}} E_2^* \exp[ik_3 z].$$

The effective Kleinman d-coefficient is defined as

$$d_{\mathrm{eff}} = \sum_{i,j,k=1}^{3} \chi_{ijk}^{(2)} \varepsilon_{\mathrm{p}i} \varepsilon_{2j} \varepsilon_{3k} ,$$

where ε_{ij} is the direction cosine of the E_i-field relative to the $\hat{j}$ axis. In eqs. (3.6) and (3.7), the indices of refraction are defined in terms of the linear susceptibilities:

(3.8) $$n_2^2 = \varepsilon_0 + 4\pi\chi^{(1)}(\omega_2), \qquad n_3^2 = \varepsilon_0 + 4\pi\chi^{(1)}(\omega_3).$$

The dispersion in the nonlinear coefficients in eqs. (3.6) and (3.7) can usually be neglected if all the relevant frequencies are well within the transparency region of the nonlinear material. The effective nonlinear optical coefficient d_{eff} takes into account the tensor nature of $\chi^{(2)}$ and the three E-fields and must be carefully calculated for each specific case from tabulated values of the relevant components of $\chi^{(2)}$, or the corresponding Kleinman d-tensor.

Equations (3.6) and (3.7) show that the propagation of the signal wave at ω_2 is coupled to the idler wave at ω_3 through the pump wave and the nonlinear susceptibility d_{eff}, and *vice versa*. The algebra involved in solving the coupled second-order differential equations can be complicated. Fortunately, for optical problems, it is often possible to simplify these equations further by using the so-called «slowly varying amplitude approximation», in which it is assumed that the change in the complex amplitude of the waves of interest in a distance on the order of one wavelength is much smaller than the complex amplitude itself, or, equivalently,

(3.9) $$\left| \lambda \frac{\partial}{\partial z} E \right| \ll |E| \qquad \text{or} \qquad \left| \frac{\partial^2}{\partial z^2} E \right| \ll \left| k \frac{\partial}{\partial z} E \right| ,$$

which is almost always a good approximation. Making use of this approximation, the second-order equations (3.6) and (3.7) can be approximated by the coupled-amplitude equations:

$$(3.10) \qquad \frac{\partial}{\partial z} E_2(z) = i \frac{2\pi k_2}{n_2^2} d_{\text{eff}} E_p E_3^*(z),$$

$$(3.11) \qquad \frac{\partial}{\partial z} E_3(z) = i \frac{2\pi k_3}{n_3^2} d_{\text{eff}} E_p E_2^*(z).$$

Combining eqs. (3.10) and (3.11) leads to

$$(3.12) \qquad \frac{\partial^2}{\partial z^2} E_2(z) = \frac{(2\pi)^2 k_2 k_3 d_{\text{eff}}^2 |E_p|^2}{n_2^2 n_3^2} E_2(z),$$

and the spatial gain coefficient

$$(3.13) \qquad g = \frac{2\pi d_{\text{eff}} |E_p| \sqrt{k_2 k_3}}{n_2 n_3},$$

which is real and positive. As a numerical example, for a pump intensity of, for example, 10^7 W/cm^2 which is easily achievable, a typical d_{eff} value of $5 \cdot 10^{-9}$ e. s. u., $n_2 \sim n_3 \sim 1.5$, g is approximately 0.4 cm^{-1} at, for example, $\lambda_2 \lambda_3 = (700 \text{ nm})^2$. Such a spatial-gain-coefficient value makes a practical optical parametric oscillator quite feasible.

Solution of eq. (3.12) subject to the boundary condition that there is an input in channel 2 equal to $E_s(0)$ but no input in channel 3 leads to the spatial variations of the complex amplitude of the signal and idler waves:

$$(3.14) \qquad E_2(z) = E_s(0) \cosh gz$$

and

$$(3.15) \qquad E_3(z) = i \frac{n_2}{n_3} \sqrt{\frac{k_3}{k_2}} E_s^*(0) \sinh gz.$$

Equation (3.14) shows explicitly the amplitude- and phase-sensitive nature of the amplification of the signal wave. For $gz \gg 1$, $E_2(z) \sim E_2(0) \exp[gz]$, just like in a laser amplifying medium.

While this simple linearized classical theory is quite adequate for a description of the amplification process, it cannot describe the spontaneous-emission process nor the saturation effects as the signal becomes nonnegligible compared to the pump.

For the spontaneous-emission process, the field must be quantized and the transition probability corresponding to the first-order Feynman diagram shown in fig. 1.1 calculated and integrated over the appropriate phase space. The algebra involved is somewhat tedious. The details can be found in the literature[15, 16]. The final result in the single-mode case is relatively simple, however. In terms of the photon flux densities, Π (defined as $n\varepsilon_0 E^2 c/8\pi h\nu$), and generalizing to include inputs in both the signal and idler channels, the signal and idler flux densities including spontaneous parametric emission are

(3.16) $$\Pi_s(z) = \Pi_s(0)\cosh^2 gz + [\Pi_i(0) + \Pi_{zp}]\sinh^2 gz,$$

(3.17) $$\Pi_i(z) = [\Pi_s(0) + \Pi_{zp}]\sinh^2 gz + \Pi_i(0)\cosh^2 gz,$$

where Π_{zp} represents the equivalent noise input due to the zero-point fluctuations in the signal and idler channels that leads to the spontaneous parametric emission in the idler and signal channels, respectively. Quantum theory shows that the zero-point fluctuation is equivalent to one photon per volume per mode in free space or $\Pi_{zp} = c$. The application of the single-mode results is complicated by the fact that, in the optical domain, open medium and, hence, continuous-mode spectrum are usually involved. These results must be carefully integrated over all the relevant modes in the phase space. Detailed development of such a theory can be found in ref. [15] and [16].

Spontaneous parametric emission is the signal that initiates the oscillation in optical parametric oscillators. For c.w. optical parametric oscillators, the oscillator output level is determined by the condition where the saturated gain is equal to the total cavity loss, as in a laser. For pulsed optical parametric oscillators, the oscillator output level is equal to the amplified spontaneous emission in the pump pulse duration taking into account the saturation effect.

To account for the saturation effect, the spatial depletion of the pump wave due to parametric conversion must be included that, in turn, leads to three coupled nonlinear wave equations rather than eqs. (3.6) and (3.7). The solution is, of course, much more complicated than (3.14) and (3.15). It involves Jacobian elliptic functions and is discussed in the ref. [14] and [17].

Apart from these complications, eqs. (3.14) and (3.15) and the boundary conditions

(3.18) $$E_s(0) = r_1 r_2 E_s(L) \quad \text{and} \quad E_i(0) = 0$$

lead to the threshold condition for singly resonant parametric oscillators (SRO) consisting of a nonlinear crystal of length L and a Fabry-Perot cavity resonating one wave (signal wave):

(3.19) $$E_s(L) = E_s(0)\cosh gL = r_1 r_2 E_s(L)\cosh gL, \quad \text{or} \quad 1 = r_1 r_2 \cosh gL,$$

where r_1 and r_2 are the amplitude reflectivities of the mirrors at the signal wavelength. In the limit of small gL and assuming the two mirror reflectivities are the same, $r_1 = r_2 \equiv R^{1/2}$, eq. (3.19) gives the oscillation threshold condition of a singly resonant OPO:

(3.20) $$(I_p L^2)_{\text{threshold}}^{\text{(SRO)}} \simeq \frac{n_p n_s n_i \lambda_s \lambda_i c}{128\pi^5 d^2} \frac{1-R}{R},$$

where the λ's are the free-space wavelengths.

As a numerical example, for nonlinear optical crystals such as β-barium borate (BBO) pumped by the 3rd harmonic of a Nd:YAG laser, an $(I_p L^2)_{\text{threshold}}$ value of $3.5 \cdot 10^6$ W is obtained for an R value of 90%. A BBO crystal length on the order of 1 cm is easily achievable giving a threshold pump intensity of 3.5 MW/cm^2 or on the order of 30 mJ/cm^2 for a typical 8 ns pulse from a wavelength-tripled pulsed Nd:YAG laser, which is not difficult to achieve. For crystals with larger nonlinearities, such as KTP, the threshold intensity will be even lower. For c.w. mode-locked sources, this kind of peak intensity is also now achievable. With the large variety of pump sources and nonlinear crystals now available, OPO's are becoming a practical source for a wide range of applications. We consider some specific examples.

3'2. *Optical parametric sources.* – Optical parametric oscillators and amplifiers are powerful practical sources of coherent radiation tunable over a large spectral range. Because of the threshold pump power requirement, stable purely c.w. optical parametric oscillation is still difficult to achieve. Most require pumping by high-energy Q-switched laser pulses, generally at low repetition rate but relatively high energy per pulse, or c.w. beams of high-repetition-rate femtosecond pulse trains with a very low duty cycle and only the signal wave is resonated. Resonating both the signal and idler waves generally leads to highly unstable parametric oscillation.

In the nanosecond time domain, the pulse length of typical pump lasers is on the order of (5–10) ns at a pulse repetition rate of 10 to 50 Hz. The number of round-trip passes (typically 15 to 20 passes) in the OPO cavity, which is generally a few centimeters long, during a single quasi-stationary pump pulse is large enough for the oscillator to build up from the spontaneous-parametric-emission noise level to a sufficiently high output level.

In the picosecond time domain, the pump sources generally available typically generate low-repetition-rate (less than 100 Hz) bursts of picosecond pulses (on the order of 10 ps) or continuous trains of pulses at (1–100) kHz rates depending on the energy per pulse. The temporal characteristics of such pump sources are usually more suited for low-repetition rate pulsed oscillators or single- or double-pass optical-parametric-amplifier purposes. Broadly tunable multimegawatt peak power in the picosecond time domain can be achieved using such inorganic crystals as BBO and LBO. These sources are particularly useful for nonlinear optical studies where very high intensities are needed. The trade-off is that the time resolution is poorer compared to the femtosecond sources with lower peak powers.

The challenges related to femtosecond sources are different. For ultra-high time resolution spectroscopy, the desired characteristics of femtosecond sources are broad tunability, high repetition rate, high average power and the shortest pulse possible. Broad tunability allows a greater variety of materials and processes to be studied. The broad-tunability requirement means that some sort of parametric device is needed. To take advantage of the higher time resolution possible with the shorter pulses, higher-quality data become even more important, because, with higher time resolution, more complicated relaxation dynamics will inevitably be revealed. The corresponding problem of analyzing such data will be more complicated because of the possibility of, for example, multiple exponentials or even non-Markovian processess. Higher repetition rate and higher average power lead to higher signal-to-noise ratio and, hence, data of better quality. Finally, for the shortest pulse width possible, the parametric-conversion crystal must be very thin to avoid group velocity dispersion. On the other hand, for very thin crystals, the parametric-conversion efficiency would be low. Therefore, the parametric device cannot simply be an amplifier; it must be an oscillator. These basic requirement—broad tunability, high average power, high repetition rate and ultrashort pulse—are generally difficult to satisfy simultaneously in a femtosecond source. The recent excitement about the development of the femtosecond optical parametric oscillator stems from the fact that, for the first time, these requirements are being met as a result of recent advances in the development of new nonlinear optical materials and pump sources for OPO's, thus opening the way for a great variety of new ultrafast processes and materials to be studied.

The pumping scheme of the optical parametric oscillator depends on that of the pump pulse. For 100 fs or less, the spatial extent of the pulse is 30 μm or less. The pumping cannot be quasi-c.w. It must be synchronous in the sense that each pulse in the pump pulse train must arrive at the nonlinear crystal in synchronism with the OPO signal pulse as it travels back and forth in the oscilator cavity. The thickness of the nonlinear crystal in the OPO is in general extremely thin (a millimeter or less, depending on the crystal and wavelength range). Maintaining synchronism for extremely short pulses in very thin crystals is one of the difficulties in achieving oscillation in fs OPO. On the other hand, because of the low duty cycle and high peak power of the individual femtosecond pulses in a typical pump laser beam, the threshold power for oscillation in synchronous pumped fs OPO can be reached even with c.w. mode-locked fs lasers such as Rh6G dye or Ti:sapphire lasers. This has led to broadly tunable high-repetition-rate c.w. femtosecond sources of exceptionally high quality for the first time, which should in turn make it possible to study a large variety ultrafast processes with great precision.

3˙2.1. Nanosecond BBO OPO. The first nonlinear crystal to meet all the criteria for broad practical OPO applications is β-barium borate (BBO). Although growth of large BBO crystals of good optical quality was initially problematic, recent development of the immersion-seeded growth method of large barium borate crystals [18] from sodium chloride solution seems to have solved the problem making commercialization of BBO OPO possible. The UV absorption edge of BBO is near 200 nm. For OPO interactions, it can be pumped at the

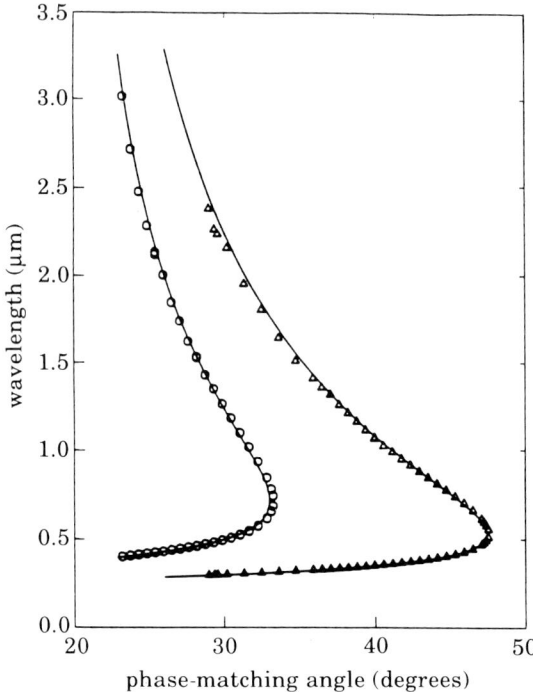

Fig. 3.1. – Type-I 355 nm pumped (○) and 266 nm pumped (△) BBO OPO tuning curves determined via the parametric fluorescence technique. The solid lines are predictions based known Sellmeier equations.

third harmonic of the Nd:YAG laser at 355 nm. The corresponding tuning ranges of the signal and idler outputs extend from 400 nm to 2.5 μm (fig. 3.1). In fact, this entire tuning range can be covered[19] with a single set of mirrors resonating the signal branch in the visible. With the addition of an efficient BBO frequency-doubler, continuously tunable radiation from 200 nm to 2.5 μm is now available from a single BBO OPO system. Some basic properties of BBO OPO are reviewed in this section to illustrate the basic design considerations of practical OPO's.

BBO OPO's pumped by the second[20], third[19], or fourth[21] harmonic output of the Nd:YAG laser have been reported extensively in the literature. The more optimal choice is pumping by the third harmonic at 355 nm using either type-I or type-II interaction. (In the type-I phase-matching condition, the signal and idler waves are both polarized orthogonal to the pump wave. In type-II interaction, only one of the two waves is polarized orthogonal to that of the pump wave.) It gives a broad tuning range (from approximately 400 nm to 2.5 μm) and yet the mirror UV damage problem is still manageable at this pump wavelength. With a fourth-harmonic pump at 266 nm, a tuning range from 300 nm to 2.5 μm (fig. 3.1) becomes possible using multiple sets of mir-

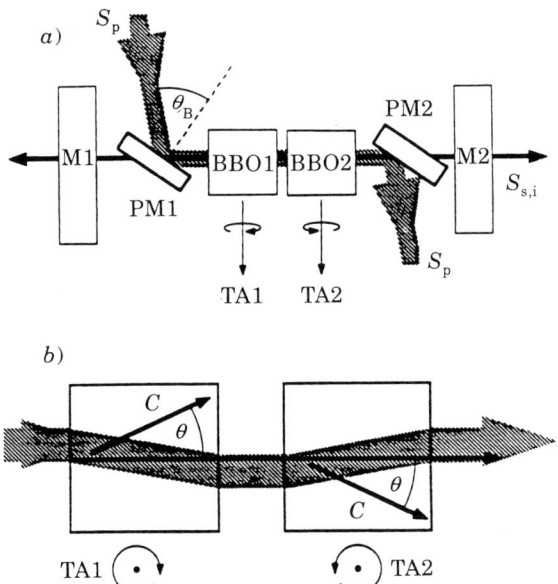

Fig. 3.2. – *a*) Top view of a 355 nm, type-I walk-off compensated BBO OPO. PM1 and 2 = 355 nm high reflectors placed at Brewster's angle (θ_B) relative to the cavity axis. M1 = broad-band visible high reflector, M2 is a 50% output coupler. TA1 and 2 = crystal tuning axes. S_p, S_s, S_i = Poynting vectors of the pump (striped) signal and idler (black). The pump beam is polarized out of the page; the signal and idler beams are polarized in the plane of the page. *b*) Close-up side view of the crystal arrangement. Walk-off compensation is achieved by placing the crystals with their optic axes at an angle of twice the phase-matching angle (θ) relative to one another. The extraordinary pump beam (striped) walks off the ordinary signal/idler beam (black in the first crystal, and back on in the second).

rors, but the UV mirror coating damage problem becomes severe. The tuning range with second-harmonic pumping is more limited.

Figure 3.2 shows the schematic of a third-harmonic YAG-laser-pumped BBO OPO operating in the nanosecond range [21]. This OPO embodies two design features that might be useful for other types of OPO's as well. The first feature is the two-crystal walk-off compensation arrangement. In type-I phase matching, because BBO is a negative uniaxial crystal, the pump wave is an extraordinary wave and the signal is an ordinary wave. Because the Poynting vector and the wave vector of the extraordinary pump beam are not in the same direction, while those for the ordinary signal beam are in the phase-matched direction, the pump beam walks off the signal beam. In the case of BBO pumped at the third harmonic of YAG, the walk-off length is typically about 1 cm at the wavelengths of interest, which is the effective limit of the interaction length. To achieve a longer interaction length, for example twice as long, one way to compensate for this walk-off effect is to use two crystals with the *c*-axes of the two oriented in such a way that the pump beam walks away from the signal beam in one crystal and walks back on in the following crystal as shown in fig. 3.2. A significantly lower threshold and higher efficiency were achieved using the two-crystal walk-off compensation scheme.

The second useful design feature is the use of the beam-steering mirrors to couple the pump beam into and out of the OPO cavity. The purpose of these mirrors is to avoid UV damage of the OPO cavity mirror coatings at the pump wavelength. Because the damage threshold of UV-transmitting and visible-reflecting mirror coatings is generally much lower than that of UV-reflecting and visible-transmitting mirror coatings, the use of beam-steering mirrors substantially increases the pump power level that can be tolerated in practical UV-pumped OPO's. As the energy level of third-harmonic YAG-laser-pumped BBO OPO reaches the hundreds of mJ level, mirror damage is an increasingly important design consideration. The beam-steering scheme allows pumping at 266 nm leading to parametric oscillation in BBO down to 300 nm.

For applications of nanosecond type of OPO's, the linewidth of the oscillator output is an issue [22]. For type-I phase-matched BBO OPO pumped at 355 nm, the linewidth can vary from a few ångström far from degeneracy to nearly 100 Å near degeneracy without the use of any special line-narrowing scheme. Although oscillator linewidth of this order of magnitude may be adequate for some applications, it is important to reduce the linewidth for many other applications. The dominant line-broadening mechanisms are due to pump beam linewidth and finite pump beam divergence. At the degenerate point where the signal and idler wavelengths are the same and the wavelength *vs.* phase-matching angle curve is vertical, a small pump beam divergence can lead to a very large parametric linewidth. To reduce the linewidth, additional line-narrowing elements must be introduced into the oscillator cavity. In the case of BBO OPO, the use of a grating in Littrow configuration replacing one of the OPO mirrors typically can reduce the linewidth down to 2 or 3 Å throughout the tuning range. Using the grating in the Littman configuration can reduce the tuning range to approximately 0.3 Å. In both cases, the threshold for oscillation is typically raised by 50 to 200% with a corresponding reduction in the output power and efficiency.

The use of type-II interaction can also lead to narrower linewidth because the signal and idler waves have different polarizations and, hence, the tuning curves of these waves cross at an angle (fig. 3.3) where the signal and idler wavelengths are the same. There is, therefore, no true degenerate point where the signal and idler wavelengths and dispersion are the same. With type-II phase-matched BBO OPO pumped at 355 nm, it is possible to maintain an oscillator linewidth less than 1 Å over most of the tuning range (fig. 3.4). The trade-off is that the threshold for oscillation is higher for type-II interaction because of a lower effective nonlinear coefficient.

To obtain narrow linewidth at high output power levels, a commonly used

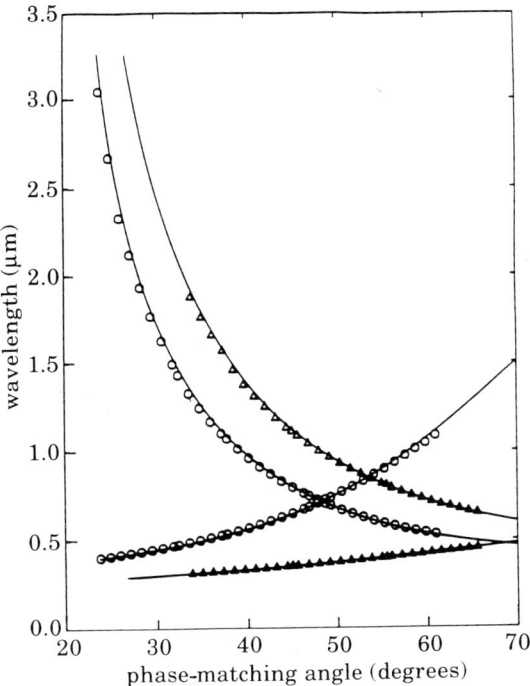

Fig. 3.3. – Type-II 355 nm pumped (○) and 266 nm pumped (△) BBO OPO tuning curves determined via parametric fluorescence. The solid curves are predictions based on known Sellmeier equations.

approach is to injection-lock a high-power oscillator with a low-power narrow-linewidth oscillator. Injection locking the pump laser to achieve narrow pump linewidth and better collimation will further improve the OPO linewidth. The use of a combination of these schemes has led to single longitudinal-mode operation of practical BBO OPO's at power levels on the order of 100 mJ or more. This approach has been successfully demonstrated in commercial BBO and KTP OPO's.

After nearly thirty years of development, the first practical commercial OPO's have finally appeared on the scene. Many more choices of different pump sources and nonlinear crystals are still to be explored and evaluated. As the range of applications expands, one of the principal applications of high-power lasers in the future will be to pump the OPO's. Also, it is entirely possible that, eventually, every laser of sufficiently high power will have an OPO as an attachment for wavelength extension.

3˙2.2. Femtosecond OPO. Although Q-switched laser-pumped nanosecond type of OPO's dominated the early development, recent efforts have

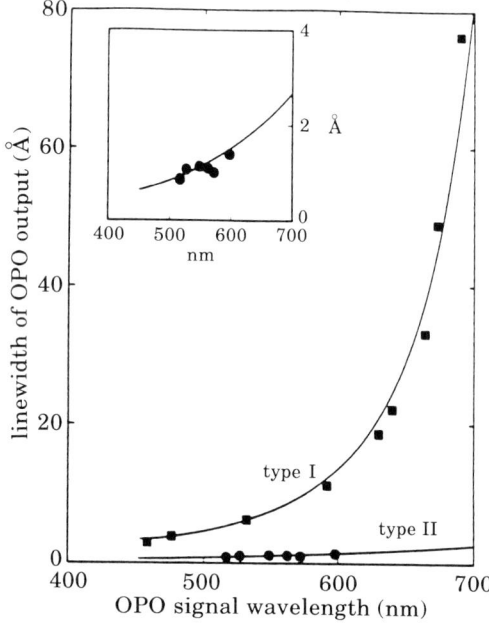

Fig. 3.4. – The linewidth *vs.* tuning for the type-I (solid squares) and type-II (solid circles) 355 nm pumped BBO OPO's. The solid lines represent calculated linewidths. The type-II OPO has a significantly narrower linewidth throughout the tuning range. The inset shows the type-II data with an expanded ordinate scale.

broadened to include the development of picosecond and femtosecond optical-parametric-amplifier *oscillators.*

High repetition rate and broad tunability are particularly important for the study of ultrafast processes. Higher repetition rate leads to higher signal-to-noise ratio in such experiments and broad tunability allows a greater variety of materials and processes to be studied. There have been dramatic advances in the development of such sources in the last few years since the first demonstration of the broadly tunable femtosecond optical parametric oscillator [23] (fs OPO). Initially, only the Rh6G fs dye laser was available as the pump source and, because of the power requirement, intracavity pumping had to be employed. This made the operation of the fs OPO very difficult and the output power relatively low. With the development of the relatively high-power mode-locked Ti:sapphire lasers [24], the situation has changed fundamentally. Using the fs Ti:sapphire laser as the pump, externally pumped high-repetition-rate fs OPO's producing hundreds of milliwatt broadly tunable from 900 nm to over 3.5 µm have been shown to be possible [25] and this development was quickly followed by the introduction of commercial fs OPO's. Recent developments have involved the extension of such sources to new wavelength regions through the use of new nonlinear optical crystals and techniques. Recent advances include the highly efficient conversion of various femtosecond sources [26-28], including the dye laser and Ti:sapphire laser pumped fs OPO outputs, into the ultraviolet and the visible through intracavity doubling, and the use of new nonlinear crystals, such as KTA ($KTiOPO_4$), in femtosecond OPO's which can potentially operate in the important (3–5) μm region [29] and beyond. The basics of the KTP OPO and some recent advances are discussed in this section to illustrate the special considerations underlying the design considerations and the characteristics of the high-repetition-rate broadly tunable femtosecond sources.

In c.w. mode-locked femtosecond lasers or synchronously pumped OPO's, a single femtosecond signal pulse propagates back and forth within the laser or OPO cavity. Each time the pulse strikes the output coupling mirror, a femtosecond pulse is transmitted through the mirror leading to the emission of a train of femtosecond signal pulses separated by the round-trip propagation time in the cavity. For intracavity pumped OPO, the nonlinear crystal for the OPO is in the cavity of the pump laser. The pump wave as a train of femtosecond pulses is also generated by a single femtosecond pulse circulating in the pump laser cavity. This pump pulse must be in synchronism with the intracavity signal pulse in the OPO. They must always meet at the nonlinear crystal so that, when the signal pulse arrives at the crystal, it is pumped and experiences parametric gain on each passage. Because the typical gain in the OPO is relatively low requiring hundreds of round-trips in the OPO cavity to build the signal up to the steady-state level from the intracavity noise, the tolerance in the synchronization of the pump pulses and the signal pulses is very tight. For pulses on the order of 100 fs or a spatial extent of 30 μm, the OPO cavity length must be controlled to the order of 100 nm.

When the first experiment [23] on c.w. femtosecond OPO was carried out, the only c.w. femtosecond pump source available was the Rh6G-DODCI mode-locked dye laser with an average power of a few tens mW and peak powers on the order of a few kW. The primary design concern was to optimize gain. Considering the effective nonlinear coefficient and group velocity dispersion, the best choice of nonlinear optical crystal in the sub-100 fs time domain was KTP using type-II interaction (o → e + o). For this type of interaction in biaxial KTP, at a pump wavelength of 630 nm and signal wavelength of 850 nm, for example, the corresponding group velocity dispersion Δv_g^{-1} is 53 fs/mm. This implies that the length of the nonlinear crystal should be on the order of a millimeter. Even though KTP is a material with a relatively large effective nonlinear coefficient, the effective gain coefficient at a few kW focused down to a spot of a diameter on the order of 100 μm is still too small to reach the threshold for optical parametric oscillation external to the dye-laser cavity. The only possibility is to place the nonlinear crystal inside the Rh6G dye-laser cavity to take advantage of its higher intracavity power to pump the OPO. Peak intensities

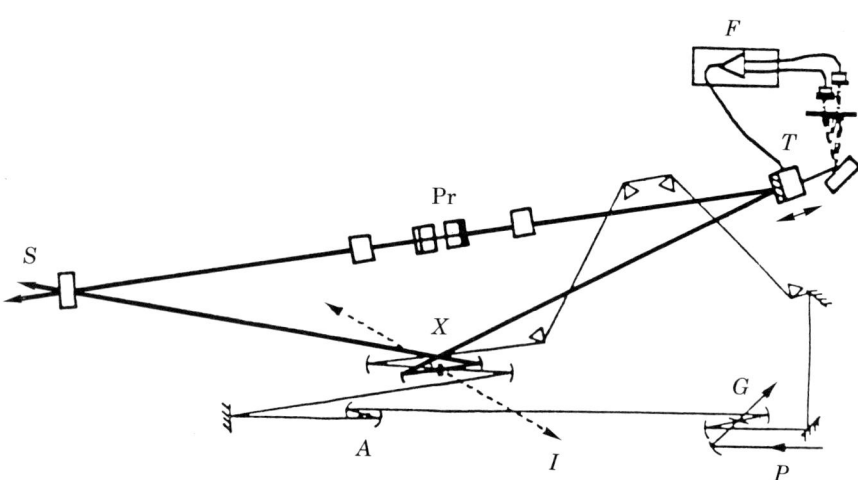

Fig. 3.5. – Layout for the dye-laser-pumped femtosecond OPO. The optics and beam line for the signal (S) are shown in bold. The dotted lines are idler beam (I) which exit the crystal as shown. One flat mirror is mounted on a piezoelectric transducer (T) for length adjustment. This is driven by a feedback loop (F) to maintain proper synchronism. Dispersion compensation is provided by four flint glass prisms (Pr) in the vertical plane. The oscillator is pumped by the intracavity beam (fine line) of a CPM dye laser. P, G and A are the Ar$^+$ laser pump, the gain jet and the absorber jet, respectively.

in excess of 10^{10} W/cm^2 are possible at an intracavity focus of such a dye laser.

Placing the nonlinear crystal inside the mode-locked femtosecond pump laser cavity, the pump laser cavity and that of the OPO become closely coupled together, so that adjustment of one cavity often affects the alignment of the other, which makes the intracavity-pumped OPO exceedingly difficult to operate. The configuration of the first operational intracavity-pumped dispersion-compensated KTP fs OPO is shown in fig. 3.5. A ring cavity rather than a conventional linear cavity is used to ensure that the crystal is always pumped each time the signal pulse passes through it when unidirectional pump pulses are used. The KTP crystal is 1.4 mm long. The pump is a standard Rh6G-DODCI colliding-pulse mode-locked (CPM) laser with the following modifications: the output coupler is replaced by a high reflector, a third intracavity focus is formed by two $r = 20$ cm high reflectors, and the dispersion compensation quartz prism pairs are separated by an additional 14 cm to compensate for the group velocity dispersion in 1.4 mm of KTP at 620 nm ($d^2n/d\lambda^2 = = 0.723$ μm^{-2}).

The crystal is hydrothermally grown KTP cut for type-II (o → e + o) phase matching near normal incidence with a single layer of MgF$_2$ antireflection coating on each surface. Hydrothermally grown KTP crystals appear to have a higher optical-damage threshold than the flux (phosphate or tungstate)-grown

types of KTP. Because of better detector sensitivity, the preference is to resonate the signal wave, which is an extraordinary wave, in the wavelength
range below 1 μm. With an extraordinary signal wave and an ordinary pump
wave in a type-II interaction, normal incidence with antireflection on the intracavity nonlinear crystal to minimize the losses for both waves must be used instead of cutting the crystal at Brewster angle for either the pump or the signal
wave, for the intracavity loss for the orthogonally polarized component would
be too high.

The phase-matched interaction is slightly noncollinear with an internal
2.8° angle between the pump and signal wave vectors. At this angle, the corresponding Poynting vectors are collinear. It was an experimentally established
fact that, when curved focusing mirrors are used as is shown in fig. 3.5, the
OPO optimizes itself by choosing the interaction to be in such way that the
Poynting vectors of the pump and signal waves are in the same direction, thus
ensuring maximum overlap of these two beams. This is the reason for the slight
noncollinearity in the pump and signal wave vectors.

This empirically observed fact [30] has important ramifications for the design of practical singly resonant OPO's quite generally. Normally, to minimize
walk-off in anisotropic crystals, one resonates the parametrically generated
wave of the same polarization as the pump in a collinear type-II phase-matching
geometry. This results in collinear Poynting vectors as well as the *k*-vectors of
the pump and resonated waves. The above observation shows that an analogous
reduction in walk-off can be achieved by requiring collinearity of the Poynting
vectors only. This means that it is possible to resonate the wave of polarization
orthogonal to that of the pump in a noncollinear phase-matching geometry (the
internal noncollinear angle equal to the walk-off angle). Since this conclusion is
equally applicable to both type-I and type-II configurations, it greatly increases the range of choices available for singly resonant OPO design operating in
any time domain.

For the intracavity-pumped OPO, the initial alignment of the device is exceedingly difficult due to the total number of intracavity optical elements and
the fact that the pump laser cavity and the OPO cavity are directly coupled. The
key to the alignment procedure is to make sure that all the elements are first
positioned and oriented to very close tolerances, so that it will be possible to de-
tect spontaneous parametric emission and optimize its amplification to achieve
oscillation. With all the difficulties, the intracavity-pumped KTP OPO was successfully operated as the first broadly tunable femtosecond. Tuning was demonstrated from approximately 700 nm to 3.5 μm [26, 30] (fig. 3.6).

A major breakthrough came with the development of the c.w. mode-locked
titanium-doped sapphire laser. The average power level of the mode-locked
Ti:sapphire laser could be in the 2 to 3 W range with pulse widths less
than 100 fs at 10^8 Hz rate. The average pump power required to achieve

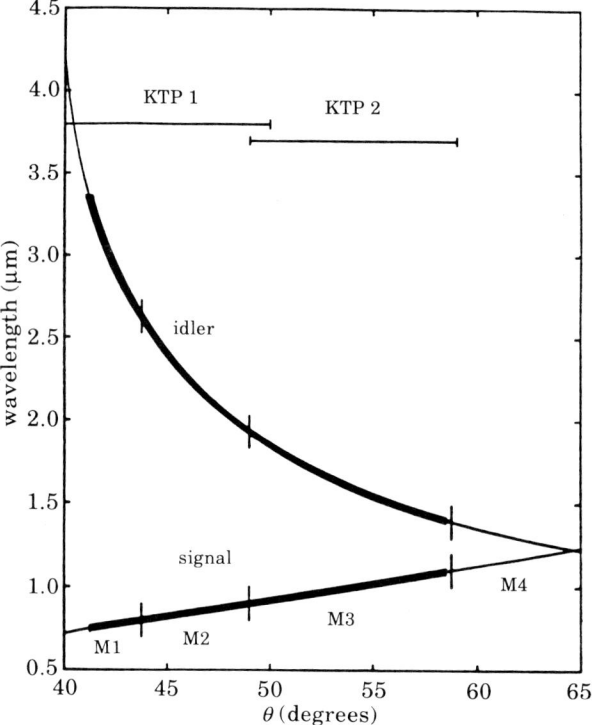

Fig. 3.6. – Tuning curves of Rh6G dye-laser-pumped femtosecond KTP OPO. To cover the entire tuning range, two KTP crystals (#1 and #2) and four sets of miorrors (M1-M4) are needed. Thick solid portions of the tuning curves correspond to demonstrated range. The remaining gap near the degeneracy point at 65° is due to the lack of suitable optics at the time the experiment was done (ref. [30]) and does not correspond to any inherent limitations in the device. 615 nm pump, 1.0° noncollinear.

oscillation in an externally pumped angle-tuned KTP fs OPO is on the order of 200 mW, which is easily exceeded by the TI:sapphire fs laser.

Operation of Ti:sapphire laser-pumped fs KTP OPO's was first reported in 1992. This was quickly followed by fs OPO's using lithium triborate (LBO, LiB_3O_5) [31] and potassium titanyl arsenate (KTA, $KTiOPO_4$) [29].

Figure 3.7 shows the schematic of the first Ti:sapphire laser-pumped KTP OPO with intracavity dispersion-compensation prisms as reported in ref. [25]. These prisms are important. Without them (ref. [32]), the pulse width will be wide and there will be substantial chirping on the wings out to hundreds of femtosecond even using extracavity pulse compression and dispersion compensation to reduce the chirp, even though the width of the coherent peak part of the pulse may be as short as 62 fs. With the intracavity prisms, ideal chirp-free pulses down to 57 fs have been obtained (fig. 3.8).

The efficiency and output power of Ti:sapphire laser-pumped fs OPO are

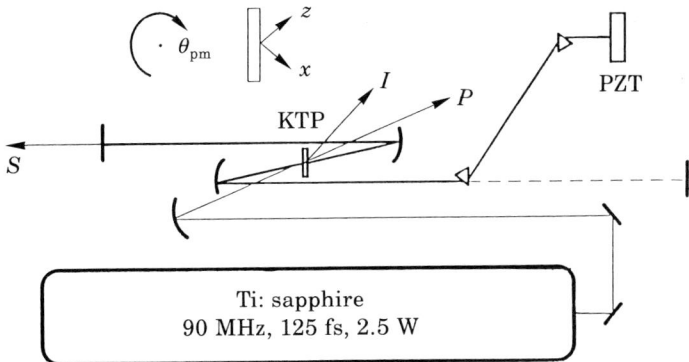

Fig. 3.7. – Ti:sapphire laser-pumped KTP femtosecond OPO with intracavity dispersion compensation prisms. P, S, I = pump, signal and idler beams. θ_{pm} = phase-matching angle. PZT = PZT scanned mirror.

surprisingly high. Pump depletion over 50% due to parametric conversion and total average signal power nearly 700 mW have been observed. An additional remarkable feature of the OPO is that precisely synchronized pulse trains at seven different wavelengths are present simultaneously, which include the signal, the idler, the pump, the phase-unmatched second harmonic of the signal and idler, and the phase-unmatched sum of the pump and the signal, and the pump and the idler. The phase-unmatched second harmonic can be very intense. In fact, almost 100 mW was generated when the total signal power generated was approximately 450 mW. The simultaneous availability of all these

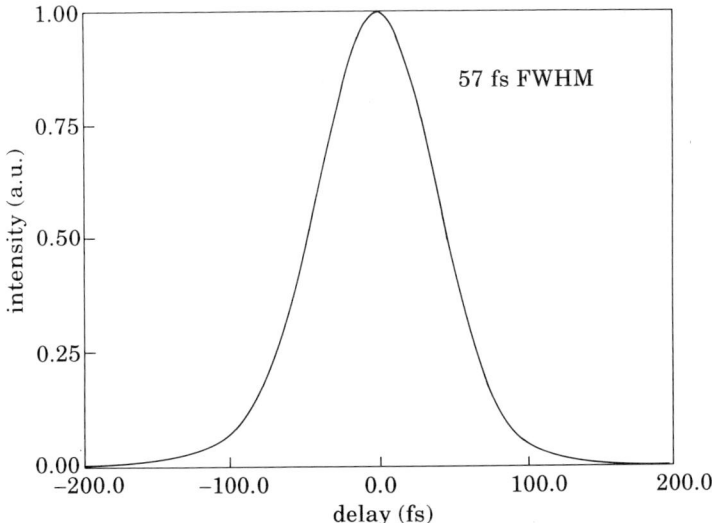

Fig. 3.8. – Intensity autocorrelation trace for a 57 fs KTP OPO pulse at 1295 nm.

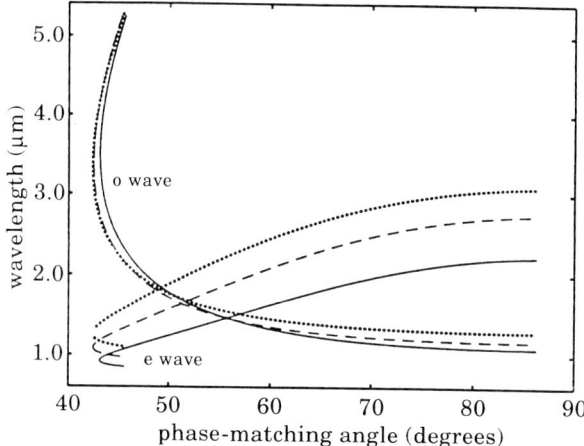

Fig. 3.9. – Tuning curves of Ti:sapphire laser-pumped KTA femtosecond OPO; pump wavelength: ····· 910 nm, --- 820 nm, —— 730 nm.

wavelengths makes it a particularly useful femtosecond source for doing pump-and-probe experiments in studies of ultrafast processes.

KTA is a relatively new crystal[12]. It is an arsenate isomorph of KTP. Its properties are similar to those of KTP but without the orthophosphate overtone absorption at 3.5 μm, as such its infrared absorption edge is red-shifted by about a micrometer and the tuning range of the KTA OPO extends to the 3 to 5 μm region where there are many interesting molecular transitions. The Ti:sapphire laser-pumped KTA fs OPO is, therefore, of particular interest. The tuning curves of the KTA OPO are shown in fig. 3.9. An interesting feature of the KTA tuning curves is that, for some angles, there are two pairs of phase-matched signal and idler wavelengths instead of one each as in the usual case.

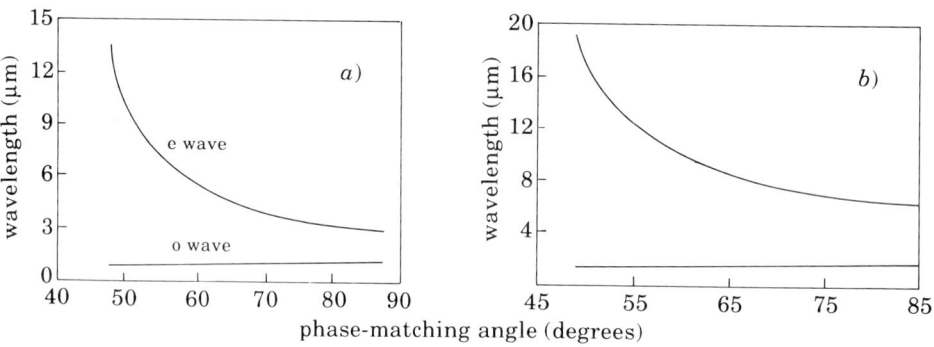

Fig. 3.10. – a) Tuning range of AgGaS$_2$ OPO pumped at 800 nm with type-II interaction (e → e + o). b) Tuning range of AgGaSe$_2$ OPO pumped at 1.3 μm with type-I interaction (e → o + o).

To go further into the infrared, crystals such as AgGaS$_2$ or AgGaSe$_2$ can be pumped at suitable wavelengths for parametric interaction out to 12 µm or 20 µm, respectively. Possible tuning curves are shown in fig. 3.10. The pump requirements in wavelength and in power can be met easily with Ti:sapphire laser-pumped OPO's. For femtosecond generation in the infrared, it is also possible to use difference-frequency generation in these crystals using the OPO as the source.

For extension of the output of the Ti:sapphire laser itself[27] and Ti:sapphire laser-pumped KTP OPO[27] to the visible and to the ultraviolet, a variety of harmonic or sum-frequency generation processes can be used. One can use intracavity or external-cavity doubling or summing of the output of OPO, or the doubled or tripled output of a laser as the pump source for the OPO. Some of these have already been demonstrated with impressive results. Particularly successful examples are the intracavity BBO doubled Rh6G dye laser to 315 nm [26], the Ti:sapphire laser to the blue[27], and the KTP OPO to cover most of the visible-to-near-i.r. range[28].

The key issue in frequency-doubling of the the femtosecond sources is the need to avoid broadening of the femtosecond pulses at the fundamental wavelength and the second-harmonic wavelength. This requires that the doubling crystal must be very thin. For very thin crystals, the single-pass conversion efficiency is invariably very low even with very tight focusing, which means that intracavity doubling is often necessary. Because of its UV transparency and phase-matching properties, the crystal of choice for this purpose is β-barium borate (BBO).

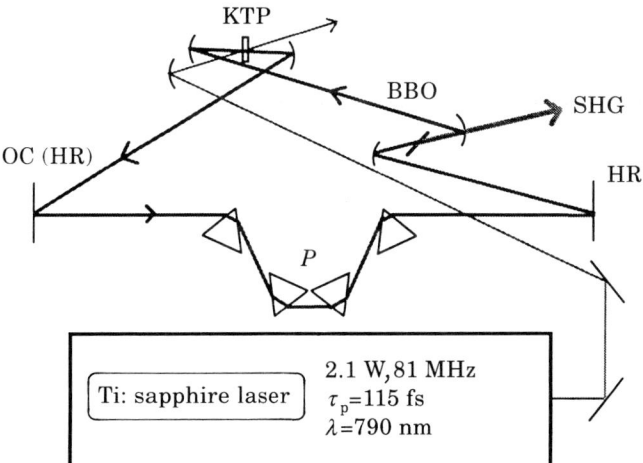

Fig. 3.11. – Intracavity-doubled KTP OPO setup. SHG = doubled second-harmonic output. OC = output coupler. HR = high reflecting. 1.5 mm KTP crystal; cut for type-II phase-matching in the (x, z)-plane. 47 µm thick BBO crystal; cut for Brewster-angle phase matching of SHG at $\theta_{pm} = 27°$. 60° SF-14 prisms spaced 20 cm tip to tip.

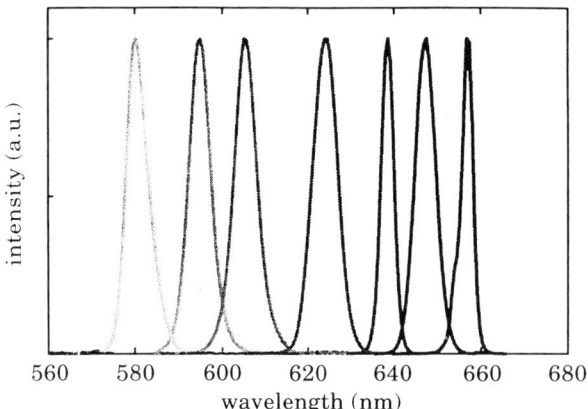

Fig. 3.12. – Spectra of intracavity-doubled KTP OPO. Demonstrated tuning range of the intracavity frequency-doubled OPO from 580 to 657 nm. Tuning is accomplished by rotating phase-matching angle of KTP, and optimizing output coupler and cavity length. Total power generated in second harmonic varies between ~ 80 and 240 mW over range shown.

   In the experiment on intracavity doubled KTP fs OPO reported in ref. [28], an intracavity BBO crystal with a thickness of only 47 μm was used. A schematic of such an arrangement is shown in fig. 3.11. An additional focus to accommodate the BBO crystal is introduced into the OPO ring cavity. A ring cavity rather than a linear cavity is used in this case, so that the OPO would oscillate in one direction and a single unidirectional second-harmonic beam is emitted. Due to the BBO crystal's large second-harmonic generation (SHG) phase-matching bandwidth around the zero group velocity mismatch point at 1.47 μm, tuning the frequency-doubled OPO output in the range of ~ 1.1 to ~ 1.6 μm (SHG ~ 550 nm to ~ 800 nm) requires no adjustment of the BBO phase-matching angle. Continuous tuning of the SHG from 550 to 660 nm limited only by optics available at the time of the experiment has been demonstrated (fig. 3.12). Potential tuning range is from ~ 500 nm to near 1 μm where the KTP fs OPO kicks in. The second-harmonic pulse train exhibits excellent stability, and as demonstrated by the real-time interferometric autocorrelation the pulses are chirp-free. Regardless of the transverse-mode structure of the Ti:sapphire pump laser, one can achieve an exceptionally clean TEM_{00} mode for the OPO which is imparted to the intracavity frequency-doubled beam.

   In conclusion, by using different crystals, pump sources and frequency conversion techniques, OPO's will ultimately extend the range of broadly tunable sources of coherent radiation, including high-repetition-rate broadly tunable femtosecond pulses, even further into the i.r. and the UV. Such sources will have a great impact on spectroscopic and other applications.

APPENDIX

The results in this lecture are given in the CGS Gaussian system as were in many of the pioneering papers on nonlinear optics. The MKS system tends, however, to be more commonly used in the more recent literature. In addition, different conventions and definitions of the nonlinear optical coefficients are used in the literature by different authors. These different choices often lead to confusion when comparing quantitative results. In this appendix, we give a few key results to facilitate comparison of the results using different conventions and units.

First, in the MKS system, the displacement vector D is related to the E-field and the induced polarization P in the medium as follows:

$$(A.1) \qquad D = \varepsilon_0 E + P = \varepsilon_0 E + P^{(1)} + P^{(2)} + \dots .$$

The corresponding wave equation is

$$(A.2) \qquad \frac{\partial^2}{\partial z^2} E_i(z, t) - \frac{1}{c^2} \frac{\partial^2}{\partial t^2} E_i(z, t) = \frac{1}{c^2 \varepsilon_0} \frac{\partial^2}{\partial t^2} P_i(z, t).$$

For the second-order polarization and the corresponding Kleinman d-coefficients, two definitions are in use. A more popular definition in the current literature is the following:

$$(A.3) \qquad P^{(2)} = \varepsilon_0 d_2 EE .$$

In an earlier widely used textbook, YARIV[33] defined his d-coefficient as follows:

$$(A.4) \qquad P^{(2)} = d_2^{(\text{Yariv})} EE .$$

The numerical values of $d_2^{(\text{Yariv})}$ in this reference (e.g., table 16.2) are given in $(1/9) \cdot 10^{-22}$ MKS units. The numerical value of ε_0 in the MKS system is $10^7 \cdot (1/4\pi c^2)$ in MKS units. Thus, for example, a tabulated value of $d_2^{(\text{Yariv})} = 0.5 \cdot (1/9) \cdot 10^{-22}$ MKS units in ref.[33] converts into numerical value of $d_2 = 0.628$ pm/V in MKS units.

In the CGS Gaussian system, the displacement vector D is related to the E-field and the induced polarization P in the medium as follows:

$$(A.5) \qquad D = \varepsilon_0 E + 4\pi P = \varepsilon_0 E + 4\pi P^{(1)} + 4\pi P^{(2)} + \dots .$$

The corresponding wave equation is

$$(A.6) \qquad \frac{\partial^2}{\partial z^2} E_i(z, t) - \frac{1}{c^2} \frac{\partial^2}{\partial t^2} E_i(z, t) = \frac{4\pi}{c^2} \frac{\partial^2}{\partial t^2} P_i(z, t).$$

The conventional definition of d_2 is as follows:

$$(A.7) \qquad P^{(2)} = d_2 EE .$$

The numerical value of d_2 in CGS Gaussian units is, therefore, equal to $3 \cdot 10^4 / 4\pi$ times the numerical value of d_2 in MKS units. Thus, continuing with the numerical example given in the above paragraph, $d_2 = 0.628$ pm/V is equal to $1.5 \cdot 10^{-9}$ cm/stat–volt or $1.5 \cdot 10^{-9}$ e.s.u.

As a final check, the expression eq. (3.13) for the parametric gain in the CGS Gaussian system becomes, in the MKS system,

$$(A.8) \qquad g = \frac{d_{\text{eff}} \, |E_{\text{p}}| \, \sqrt{k_2 k_3}}{2 n_2 n_3} \, .$$

The E-field refers to that inside the medium.

* * *

This work was supported by Joint Services Electronics Program and the National Science Foundation.

REFERENCES

[1] N. KROLL: *Phys. Rev.*, **127**, 1207 (1962).

[2] J. A. GIORDMAINE and R. C. MILLER: *Phys. Rev. Lett.*, **14**, 973 (1965).

[3] See, for example, C. L. TANG: *Spontaneous and stimulated parametric processes*, in *Treatise in Quantum Electronics*, Vol. 1, *Nonlinear Optics*, Parts A and B, edited by H. RABIN and C. L. TANG (Academic Press, New York, N.Y., 1975), p. 419.

[4] Q. HUANG and J. JIANG: *Acta Phys. Sin.*, **30**, 559 (1981); C. CHEN, B. WU, A. JIANG and G. YOU: *Sci. Sin. B*, **28**, 235 (1985), and references therein.

[5] C. CHEN, Y. WU, A. JIANG, B. WU, G. YOU, R. LI and S. LIN: *J. Opt. Soc. Am. B*, **6**, 616 (1989); S. LIN, Z. SUN, B. WU and C. CHEN: *J. Appl. Phys.*, **67**, 634 (1990); S. ZHAO, C. HUANG and H. ZHANG: *J. Cryst. Growth*, **99**, 805 (1990).

[6] F. C. ZUMSTEG, J. D. BIERLEIN and T. E. GIER: *J. Appl. Phys.*, **47**, 4980 (1976); H. VANHERZEELE: *Appl. Opt.*, **29**, 2246 (1990).

[7] *Nonlinear Optical Properties of Organic and Polymeric Materials*, edited by D. WILLIAMS (American Chemical Society, Washington, D.C., 1983).

[8] *Nonlinear Optical Properties of Organic Molecules*, edited by D. CHEMLA and J. ZYSS (Academic Press, New York, N.Y., 1987).

[9] D. EIMERL, S. VELSKO, L. DAVIS, F. WANG, G. LOIACONO and G. KENNEDY: *J. Quantum Electron.*, **25**, 179 (1989).

[10] S. K. KURTZ, J. JERPHAGNON and M. M. CHOY: in *Landolt-Boerstein Numerical Data and Functional Relationships in Science and Technology*, New Series, edited by K. H. WELLWEGE, Group III, Vol. 11 (Springer-Verlag, Berlin, 1979), p. 167.

[11] I. BIAGGIO, P. KERKOC, L. S. WU, P. GUNTHER and B. ZYSSET: *J. Opt. Soc. Am. B*, **9**, 507 (1992); B. ZYSSET, I. BIAGGIO and P. GUNTHER: *J. Opt. Soc. Am. B*, **9**, 380 (1992).

[12] L. K. CHENG, J. D. BIERLEIN and A. A. BALLMAN: *J. Cryst. Growth*, **110**, 697 (1991).

[13] L. K. CHENG, E. M. McCARRON III, J. CALABRESE and J. D. BIERLEIN: *J. Cryst. Growth* (to be published).

[14] J. A. ARMSTRONG, N. BLOEMBERGEN, J. DUCUING and P. S. PERSHAN: *Phys. Rev.*, **127**, 1918 (1962).

[15] T. G. GIALLORENZI and C. L. TANG: *Phys. Rev.*, **166**, 225 (1968).

[16] D. A. KLEINMAN: *Phys. Rev.*, **174**, 1027 (1968).

[17] P. P. BEY and C. L. TANG: *IEEE J. Quantum Electron*, QE-8, 361 (1972).

[18] P. F. BORDUI, G. D. CALVERT and R. J. BLACHMAN: *J. Cryst. Growth*, **129**, 371 (1993).

[19] L. K. CHENG, W. R. BOSENBERG and C. L. TANG: *Appl. Phys. Lett.*, **53**, 175 (1988); W. R. ROSENBERG: Ph. D. thesis, Cornell University (1990).

[20] Y. X. FAN, R. C. ECKARDT, R. L. BYER, C. CHEN and A. JIANG: *CLEO'86*, post deadline paper ThT4; Y. X. FAN, R. C. ECKARDT, R. L. BYER, J. NOLTING and R. WALLENSTEIN: *Appl. Phys. Lett.*, **53**, 2014 (1988).

[21] W. R. BOSENBERG, L. K. CHENG and C. L. TANG: *Appl. Phys. Lett.*, **54**, 13 (1989).

[22] W. R. BOSENBERG and C. L. TANG: *Appl. Phys. Lett.*, **56**, 1819 (1990).

[23] D. C. EDELSTEIN, E. S. WACHMAN and C. L. TANG: *Appl. Phys. Lett.*, **54**, 1728 (1989).

[24] See, for example, D. E. SPENCE, P. N. KEAN and W. SIBBETT: *Opt. Lett.*, **16**, 42 (1991).

[25] W. S. PELOUCH, P. E. POWERS and C. L. TANG: *Opt. Lett.*, **17**, 1070 (1992).

[26] D. C. EDELSTEIN, E. S. WACHMAN and C. L. TANG: *Appl. Phys. Lett.*, **52**, 2211 (1988).

[27] R. J. ELLINGSON and C. L. TANG: *Opt. Lett.*, **17**, 343 (1992).

[28] R. J. ELLINGSON and C. L. TANG: *Opt. Lett.*, **18**, 438 (1993).

[29] P. E. POWERS and C. L. TANG: paper CThK2, *CLEO*, Baltimore, Md. (May, 1993); *Opt. Lett.*, **18**, 1171 (1993).

[30] E. S. WACHMAN, D. C. EDELSTEIN and C. L. TANG: *Opt. Lett.*, **15**, 136 (1990); *J. Appl. Phys.*, **70**, 1893 (1991).

[31] J. D. KAFKA, M. L. WATTS and J. W. PERTERSE: post deadline paper, *CLEO*, Baltimore, Md. (May, 1993).

[32] Q. FU, G. MAK and H. VAN DRIEL: *Opt. Lett.*, **17**, 1006 (1992).

[33] A. YARIV: *Quantum Electronics* (J. Wiley, New York, N.Y., 1975), p. 410.

The Growth and Perfection of Crystals of Nonlinear Optical Materials.

P. J. HALFPENNY, J. N. SHERWOOD and G. S. SIMPSON

Department of Pure and Applied Chemistry, University of Strathclyde
295 Cathedral Street, Glasgow G1 1XL, U.K.

1. – Introduction.

The rapid development of optical communications systems has led to a demand for nonlinear optical materials of high performance for use as components in optical devices. The search for new materials has identified novel organic and inorganic systems of considerable potential and high performance[1]. To use these effectively requires that they are prepared in the form of single crystals of high structural and optical quality. The present lecture addresses the problems involved in the crystal growth of these materials and the factors which influence the quality of the final product. It is proposed to concentrate particularly on the growth of organic materials which by their nature present most problems in growth. The methods to be described and the techniques which are used for seed development are readily applicable to the growth of most inorganic nonlinear optical materials either directly or by a translation in temperature. For some inorganic materials the need to grow them at very high temperatures requires the more specialized technique of high-temperature top-seeded solution growth. Growth by this technique and the problems involved will be briefly addressed.

2. – Organic crystals.

2`1. *Purification and analysis*. – Impurities in organic materials are of considerable importance, not only because of their influence on the physical and chemical properties of the resulting crystals, but also since they can play a dominant role in controlling crystal growth behaviour. In the latter case impurities can modify crystal morphology[2] and growth rates as well as altering the stability of growth through constitutional supercooling.

The methods used to achieve ultra-purity are as many and as varied as organic materials themselves. The main sources of problems are:

a) the presence of close homologues, derivatives and optical isomers;

b) the lack of stability in both the melt phase and the solid state at elevated temperatures.

The purification schemes used in this laboratory are based on the use of a combination of some or all of the following standard techniques:

a) recrystallization,

b) continuous-column chromatography,

c) sublimation,

d) melt phase zone refining.

An important aid to purification is the identification of probable impurities, a knowledge of which can often be gained from the synthetic route used to achieve the material being purified.

Recrystallization can be applied to most soluble materials, the only real problem being the choice of a suitable solvent phase from which the recrystallization takes place. The choice of solvent is made such that at room temperature the material has little or no solubility, but at or near the solvent boiling point the material is readily soluble. The recrystallization should be carried out two or three times to ensure optimum purification. The solvent for such recrystallization should be distilled to prevent contamination.

The second technique, continuous-column chromatography [3], scales up the commonly practised small-scale chromatography technique. Relatively large quantities, (10–100) g, of material are passed through a stationary phase which facilitates purification by the differences in adsorption of the impurities and the material on the stationary phase. The choice of stationary phase and solvent for elution can be assessed using small-scale test columns. The stationary phases used are silica gel (mesh size 60–100) and alumina. These can be used separately or occasionally can be combined to achieve optimum separation of impurities. Distilled solvent is again used to eliminate any contamination. The apparatus used for such purification is shown schematically in fig. 1. The starting material is placed in the flask fitted with a glass frit at the top of the column. The distilled solvent is placed in the flask at the bottom of the column and heated until boiling. The solvent vapour travels via the system of side arms and condenses and drips onto the material in the glass frit. The solution then passes through the frit and onto the stationary phase. The rate of condensation of vapour can be matched to the rate of solution flow through the PTFE tap at the base of the column so that the system remains balanced. Eventually all the material in the top

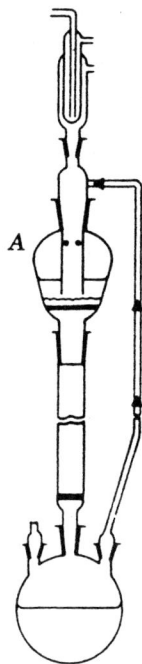

Fig. 1. – Schematic diagram of continuous-column chromatography apparatus.

flask is passed down through the stationary phase where the impurities are «trapped» usually at the top of the column. Appropriate choice of a stationary phase/solvent system can lead to very effective purification.

The two techniques described above both use the presence of a solvent phase to achieve the desired purification. The presence of such solvent can lead to problems in the later stages of purification or crystal growth. The removal of solvents is often most effectively achieved by the use of vacuum sublimation. This process is additionally a highly efficient purification technique, removing both highly volatile and nonvolatile impurities from the system. The technique, of course, depends on the material having a reasonable vapour pressure without which the separation is both time consuming and inefficient. The apparatus used to achieve sublimation is shown in fig. 2. The solvent and volatile impurities can be trapped out and the nonvolatile impurities remain in the first cham-

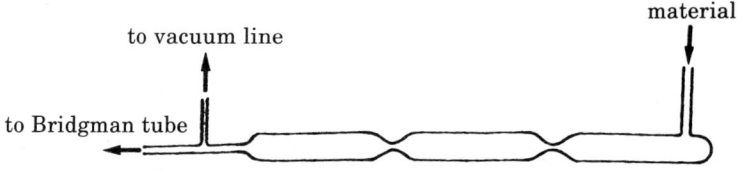

Fig. 2. – Apparatus for purification by vacuum sublimation.

ber of the tube. This process can be repeated again to achieve better purification.

The final technique of zone refining [4-6] allows the most direct and efficient means of achieving ultra-pure material. The procedure is based on the principles of fractional crystallization which utilizes the difference in impurity concentration in the liquid and solid phases. This difference is represented by the distribution coefficient k_0 which is given by

$$k_0 = C_s / C_l ,$$

where C_s is the concentration of impurities in the solid phase and C_l the concentration of impurities in the liquid phase. The rate at which purification be achieved depends on the extent to which k_0 differs from unity. If $k_0 = 1$, no purification can be obtained by zone refining.

The practical aspects of zone refining are based on the automation of the repetitive passage of a zone (or zones) of molten material from one end of a solid charge to the other. The type of zone refiner used in this laboratory is a twelve-zone, multipass system operated by a cam which moves the molten zone from one heater zone to the next until it reaches the bottom of the tube. The charge material is sealed in a thick-walled glass tube which is evacuated to prevent oxidation and decomposition reactions from occurring during zone refining. The resulting distribution of impurities after zone refining depends on the distribution coefficients k_0 of the impurities. Those with $k_0 < 1$ concentrate at the bottom of the ingot of material, while those with $k_0 > 1$ concentrate at the top end of the ingot. Since the exact nature of the remaining impurities and their distribution coefficients may be unknown, it is customary to assume that either type may be present and to use only the centre section of ingot. Zone refining cannot be used for materials which are unstable in the melt or in the solid state near the melting temperature.

The purity achieved by such processes as recrystallization, column chromatography, sublimation and zone refining should, of course, be monitored by a suitably sensitive analytical technique to assess the progress towards the goal of ultra-pure material. The most generally useful technique used to monitor improvements in purity is gas liquid chromatography (glc). This standard analytical technique is capable of achieving sensitivities of 1 to 100 p.p.m. (0.1 p.p.m. in the very best cases) impurity content in a 0.1 mg sample. The main problems encountered with this technique are masking of the impurity peaks by the large main peak and material instabilities under the conditions experienced during injection. The first problem can be overcome by appropriate adjustment of the following parameters:

$a)$ oven temperature,

$b)$ gas flow rates,

c) injection block temperature,

d) method of injection,

e) stationary phase,

f) type of detection system.

The use of capillary columns can also alleviate this problem but with a reduction in sensitivity. The second problem, if it occurs, usually precludes the use of glc as a means of analysis. The exact conditions and nature of the column stationary phase have to be optimized for each individual material. Alternative but less generally useful techniques are absorption and emission spectroscopy and differential thermal analysis.

It is informative to highlight some actual purification routes used in this laboratory to achieve the necessary purity for crystal growth. The two examples given illustrate all the techniques described above.

2˙1.1. 3-nitroaniline (mNA). This material was one of the first organic nonlinear optical materials to be studied [7, 8]. The purification of this material illustrates the benefit of a knowledge of the synthetic route. The method of synthesis [9] is to selectively reduce under very mild conditions (using sodium sulphide, for example) 1,3-dinitrobenzene (DNB) to give mNA. The possible impurities are unreacted DNB and 1,3-diaminobenzene (mDAB). The first stage in the purification is the recrystallization of mNA from boiling water. This choice of solvent is ideal since only at the boiling point does mNA have an appreciable solubility in water. Another advantage of the system is the complete lack of solubility of mDNB in aqueous solution. In fact the presence of mDNB (melting point 90 °C) can be detected as a brown liquid in the flask. The recrystallized material was then sublimed to remove any residual water. Analysis using glc showed that the purity of mNA had been raised from an initial purity of (95–98)% to 99.1%. The final stage of the purification relies on the excellent melt stability which allows zone refining of mNA. The mNA was subjected to 50 zone passes in the zone refiner. The zone-refined ingot has a bright yellow, crystalline appearance with visible segregation of impurities to the lower end. The central portion was analysed using glc and found to have a purity of 99.92%. This material has been used to grow large ((2–3) cm diameter) crystal boules and large-area single crystalline thin-film samples for optical studies.

2˙1.2. 2-methyl-4-nitroaniline (MNA). MNA has been extensively studied because of its extremely high linear electro-optic coefficient [10]. The initial purity ((94–96)%) of the starting material was obtained using glc analysis. This also showed that there were three principal impurities present. Each impurity was not identified, but after a series of purification

TABLE I. – *The analysis of purity of* MNA *using* glc.

Purification history	MNA purity (%)
crude	94.02
column chromatography/sublimation	99.75
column chromatography/sublimation/zone refining	99.35
column chromatography/sublimation/zone refining/ /column chromatography/sublimation	99.96

steps two of these could be removed completely leaving a final purity of 99.96%. The five-stage route to this final purity is as follows.

The first stage involved column chromatography of the MNA on column made up of silica gel with alumina placed on the top. The eluting solvent was distilled dichloromethane. The alumina band trapped a dark impurity band at the top of the column, whereas with silica gel alone this impurity band was found to partially spread down the column. The material collected in the flask at the column base was vacuum sublimed to remove solvent and was analysed. Two of the impurities had been removed, but no removal of the third impurity was achieved. The melt stability of this purified material was good enough to allow the use of zone refining. The material in the ingot was subjected to 35 passes on the zone refiner. A fluorescent purple impurity was observed along the length of the tube, while a dark impurity band was segregated to the lower end of the ingot. The analysis of the central portion of the ingot showed that, although the two impurities removed by column chromatography/sublimation had been partially regenerated, the removal of the third impurity had been achieved. The final steps in the process were to repeat the first two purification steps on the central portion of the zone-refined material leading to the final purity of 99.96%. Table I shows the progress towards the final purity after each step.

This material has been used for both Bridgman and vapour growth studies.

The other organic nonlinear materials studied in this laboratory have also been purified using a combination of the techniques described. The purities quoted in the two examples can be achieved routinely if careful analysis is carried out.

3. – Crystal growth.

3`1. *General properties.* – There are a number of properties, particularly relevant to crystal growth, which are common to many organic materials. Firstly, the intermolecular forces are comparatively weak, being predominantly Van der Waals or permanent dipole-dipole interactions. This typically results in low melting points and relatively high vapour pressures. Also, mechanical properties are, in general, rather poor with most organic solids being relatively soft. Thermal instability is also common with many organic materials undergoing thermal decomposition at or below the melting point. The occurrence of polymorphism is widespread, leading to solid-state phase changes between the melting point and ambient temperature. These factors together with the low thermal conductivities typical of such materials can pose substantial problems in crystal growth, particularly from the melt. In the case of organic nonlinear optical materials, the highly polar nature of the molecules, which gives rise to their highly nonlinear optical response, also complicates growth from solution, since the influences of solvent-solute interactions are greatly enhanced by a large solute ground-state dipole moment.

3'2. *Supersaturation.* – In order to grow crystals of a range of such materials, a broad selection of growth techniques, spanning vapour, melt and solution phase growth, should be available to enable selection of the optimum crystal growth conditions for a given material. In all growth methods the driving force for crystallization is the difference in chemical potential ($\Delta\mu$) of the material in the vapour or liquid phase and that in the solid. For growth from the melt $\Delta\mu$ is proportional to the supercooling

$$\Delta\mu = \Delta H \, \Delta T / T_e \, ,$$

where $\Delta T = T_e - T$, T_e is the equilibrium melting temperature and ΔH is the enthalpy of fusion. For solution or vapour growth, concentration (C) or pressure (P), respectively, are more appropriate. For solution growth

$$\Delta\mu = RT \ln (C/C_e) \sim RT \, \Delta C/C_e \, .$$

The ratio $\Delta C/C_e$ is known as the supersaturation (σ). For vapour growth

$$\Delta\mu = RT \ln (P/P_e) \sim RT \, \Delta P/P_e \, .$$

The ratio $\Delta P/P_e$ is also sometimes known as the supersaturation. Nucleation will not occur until some critical supercooling or supersaturation has been exceeded. The region between equilibrium and the onset of homogeneous nucleation at this critical supersaturation is known as the metastable zone. In the presence of heterogeneous nuclei (*e.g.*, dust, active sites on vessel walls) or of existing crystallites, crystallization may occur at supersaturations less than the critical value, *i.e.* within the metastable zone.

3'3. *Melt growth.* – The most rapid and frequently the simplest approach to the growth of single crystals is through crystallization from the molten state. The onset of nucleation in the melt is in general gradual with wide metastable zone widths, enabling relatively easy control of nucleation and hence single-crystal growth. For this reason a high degree of supercooling can be imposed on the growing crystal, thus yielding large growth rates, typically of the order of several millimetres per hour. Furthermore the absence of a solvent eliminates a major source of impurities. The poor thermal stability, typical of many organic materials, can, however, pose significant problems, often limiting both the purity and structural perfection of the resulting crystals. The high vapour pressures exhibited by many organic solids can also limit the application of melt growth techniques unless a sealed system is employed. In one respect, the growth of organic crystals is considerably simpler than many other classes of materials because of the relatively low melting points of most organics. This allows the use of all glass apparatus, with the resulting ease of observation greatly simplifying monitoring and control of the growth process.

3˙3.1. Bridgman method. Perhaps the simplest of the available melt growth techniques is the Bridgman method in which a melt is directionally solidified by passing it through a temperature gradient, as shown in fig. 3a). The growing crystal is contained in a sealed ampoule, thereby reducing the problems associated with high vapour pressures. Because of the low growth temperatures involved, there is minimal interaction between the melt and its container. This together with the sealing of the ampoule enables the high purity of the starting material to be maintained in the grown crystal. Growth is carried out in a simple two-zone furnace which may be heated either electrically or, as shown in fig. 4, by refluxing solvents. This latter approach, developed at Strathclyde [11], has proved highly successful and versatile for organic melt growth up to temperatures of 300 °C. The refluxing solvents in the outer glass

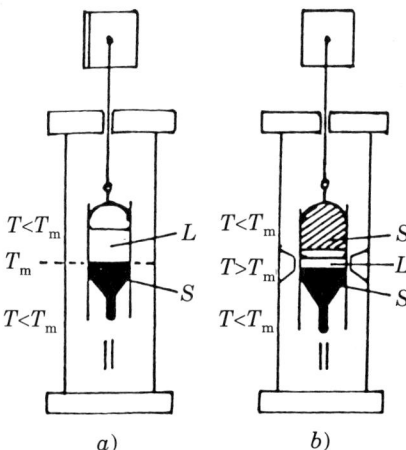

Fig. 3. – Schematic diagram of a) the Bridgman method and b) the modified Bridgman method for partial melting.

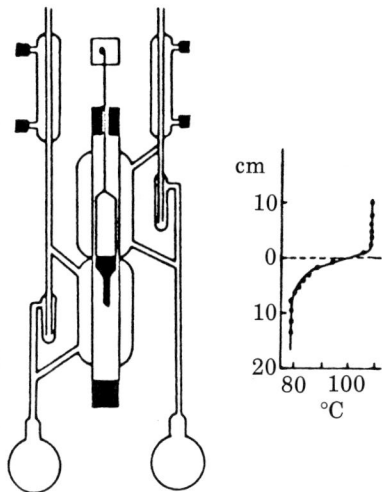

Fig. 4. – Refluxing-solvent heater system for melt growth of organic crystals.

jackets provide very stable and uniform temperature zones without the need for a sophisticated control system. The temperatures of the zones can be varied by appropriate selection from the wide range of organic solvents available. Since the extent of nucleation depends strongly upon the volume of supercooled liquid, the bottom of the growth tube is tapered and ends with a capillary to promote the formation of a single dominant nucleus. The development of single-crystal growth can be further enhanced by the use of kinks in the capillary, double-walled growth tubes or baffles within the tube[12]. Where materials exhibit a very high degree of supercooling in the melt, a small amount of solid can be left unmelted at the bottom of the capillary to provide crude seeding of the melt. In all cases, however, Bridgman growth allows very little control over the orientation of the grown boule.

Figure 5 shows a crystal of 3-nitroaniline prepared by the Bridgman method. The starting material was purified by recrystallization from boiling water, vacuum sublimation and zone refining (80 passes). The crystal was grown under vacuum in a double-walled ampoule with a capillary. A two-zone refluxing-solvent heater system was employed as described above, with butanol (Bpt 116 °C) in the upper zone and heptane (Bpt 96 °C) in the lower.

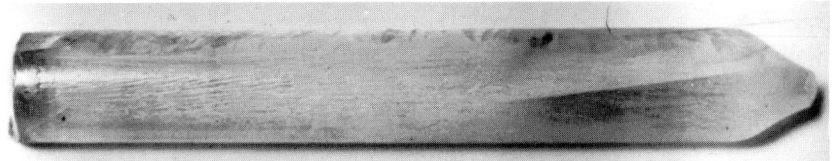

Fig. 5. – Crystal of 3-nitroaniline grown by the Bridgman method.

The ampoule was lowered through the resulting temperature gradient at a rate of 0.45 mmh^{-1}.

While melt growth is most suited to thermally stable materials, a substantial number of organic solids have been grown successfully despite undergoing appreciable decomposition at the melting point. Partial melting of the charge, through the use of a three-zone furnace (fig. 3b)), decreases the length of time for which the material is molten, so greatly reducing thermal decomposition. In a manner akin to zone refining, a molten zone passes along the solid charge as it is traversed through the temperature gradients. With the controlled nucleation provided by the shape and design of the ampoule, single-crystal growth of even quite thermally sensitive materials can be achieved as in the case of 2-(N,N-dimethylamino)-5-nitroacetanilide (DAN)[13]. Of particular importance in such cases is the fact that decomposition products frequently have segregation coefficients substantially less than unity. This influences growth in two contrasting ways. Firstly the decomposition-induced impurities are predominantly rejected into the melt, thus leaving the crystal relatively pure and free from impurity-related defects. As growth proceeds, however, the impurity concentration in the melt rises continuously with a consequent depression of the melting point of the remaining melt. The supercooling of the melt is, therefore, increased. Beyond a critical point this constitutional supercooling leads to instability of the growth front causing poor and often polycrystalline growth. This aspect can limit the size of crystals which can be grown in the case of thermally sensitive materials.

3·3.2. Czochralski method. Although very widely and successfully applied to the growth of inorganic materials, the method has only rarely been used for the growth of organic crystals[14]. The method does, however, offer a number of important advantages over simple Bridgman growth. In Czochralski growth a crucible containing the melt is positioned in a two-zone furnace such that the top of the melt lies just below the level of the melting-point isotherm (fig. 6). A seed crystal is brought into contact with the melt and itself melts up to the level of the melting-point isotherm, contact being maintained by surface tension. The seed is then slowly withdrawn above the melting-point isotherm, resulting in solidification and growth of a crystal boule. Both crystal and crucible can be rotated to provide stirring of the melt and to even out circumferential temperature nonuniformities. Again the low growth temperatures involved enable the use of all glass systems which greatly simplify growth, particularly at the seeding stage. Both electrical heating and the refluxing-solvent system described above have been successfully employed. Figure 7 shows a crystal of benzil grown by the Czochralski method[14]. The relatively high vapour pressure of benzil at the melting point requires the use of the minimum practical temperature gradient between melt and seed to prevent condensation of vapour on the latter and consequent polycrystalline growth. The greater complexity of

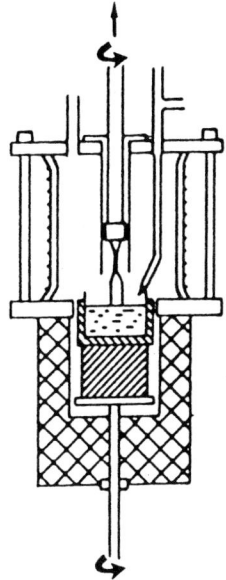

Fig. 6. – Schematic diagram of the Czochralski method.

Fig. 7. – Crystal of benzil grown by the Czochralski method.

Czochralski growth is offset by a number of advantages, particularly the ability to select the orientation of the boule through controlled seeding. Furthermore the lack of mechanical confinement of the growing crystal eliminates the large mechanical stresses which can degrade the structural quality of crystals grown by the Bridgman method. Application of Czochralski growth is, however, limited to materials which are thermally stable in the melt, since partial melting of the charge is not, in general, practical.

3˙3.3. Undercooled melt. Growth of crystals from the undercooled melt employs apparatus essentially identical to that used for low-temperature solution growth discussed below and shown in fig. 8. The growth flask containing the melt is maintained in a thermostat bath. The melt is slightly supercooled ($\Delta T \sim 0.1$ K) and a seed crystal introduced. Provided care is taken to eliminate

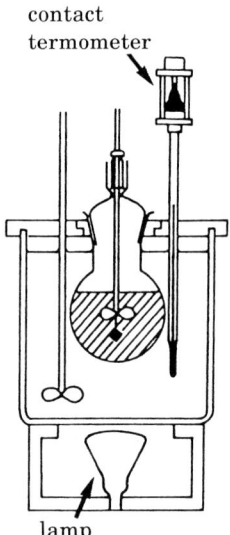

contact
termometer

lamp

Fig. 8. – Solution growth crystallizer.

spurious nucleation, growth then occurs on the seed to yield a facetted crystal. As with the Czochralski method, this approach enables a high degree of control over the seeding process. Moreover, it not only avoids the mechanical stresses associated with confinement of the growing crystal, but also essentially eliminates the thermomechanical stresses which arise from thermal gradients in the case of Czochralski growth.

3`4. *Growth from solution.* – Most organic materials are soluble in a wide range of solvents and can be grown from solution by a variety of methods at temperatures close to ambient. The supersaturation for growth may be achieved in a number of ways, including slow cooling of a saturated solution, evaporation of solvent or by circulating the solution between a saturating tank at temperature T_1 and a growth tank at a lower temperature T_2[15]. Figure 8 shows a system suitable for growth from solution either by slow cooling or solvent evaporation[16]. A flask containing a saturated solution is immersed in a large-capacity water bath which is heated by means of an infrared lamp. Temperature control is achieved by means of an electric-contact thermometer and, under favourable circumstances, control can be maintained to within ±0.005 K over long periods of time. The solution is stirred using a paddle from which a seed crystal is suspended in the case of seeded growth. The contact in the thermometer is driven by a motor to lower the temperature and achieve the desired cooling rate. For growth by solvent evaporation, the temperature is kept constant and a controlled vapour leak included in the system. This may either be open directly to the atmosphere or have a slow gas flow through the flask over

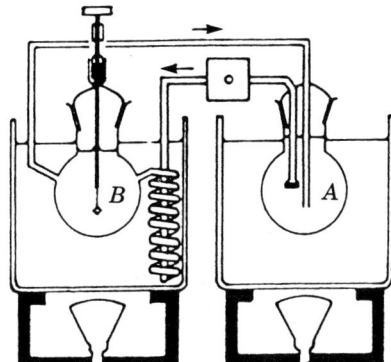

Fig. 9. – Schematic diagram of the temperature difference method.

the solution. In the case of initial trials for the preparation of small seed crystals, this level of sophistication is not employed. Crystals may be obtained by evaporation of solvent from solutions in small, partially covered dishes. The principle of the temperature difference method is shown schematically in fig. 9. The difference in temperature between the two tanks produces the supersaturation for growth. Saturated solution from tank 1 at temperature T_1 is pumped into the growth flask at T_2 ($T_2 < T_1$) at which temperature it is supersaturated. It is then circulated back to tank 1 where it is resaturated. The required supersaturation and growth temperature can be accurately and independently set by selection of temperatures T_1 and T_2. The increasing complexity of the various methods offers different levels of control over the growth conditions and their stability. In the simplest case, growth by solvent evaporation requires only very basic apparatus, but control over supersaturation is extremely limited and sudden fluctuations in growth rate are common. Slow cooling of a saturated solution, on the other hand, provides much greater stability and greatly reduces short-term fluctuations in growth rate. It is, however, difficult to maintain a constant supersaturation throughout a given growth run since the actual supersaturation imposed depends upon the temperature lowering rate and the rate at which the supersaturation is reduced as the growing crystal removes solute from the solution. The supersaturation may, therefore, increase or decrease substantially during growth unless a balance between these two factors is achieved. This possible lack of long-term control over supersaturation can pose significant problems in the crystal growth of some materials. In contrast the temperature difference method provides accurate long-term control over supersaturation during growth and enables growth under constant supersaturation conditions. Being a pumped system, however, it is rather more susceptible to short-term fluctuations in supersaturation and, therefore, growth rate. Figure 10 shows a crystal of the organic secondary explosive cyclotrimethylene

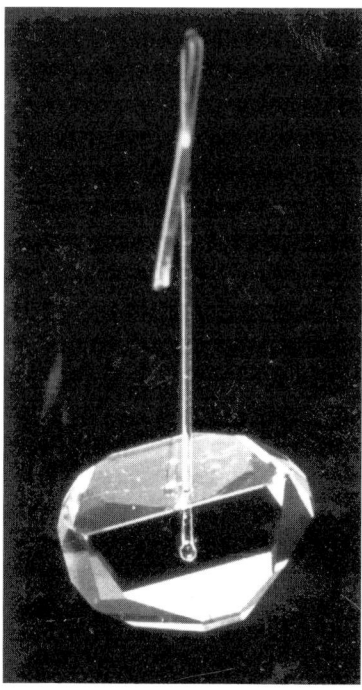

Fig. 10. – Crystal of cyclotrimethylene trinitramine (RDX) grown from acetone solution by temperature lowering.

trinitramine (RDX) grown from acetone solution by temperature lowering [17].

Many of the advantages of growth from solution are obvious from the preceding discussion of melt growth. The low temperatures employed largely eliminate the problem of thermal decomposition. Polymorphic phase transitions can also be avoided in many cases by suitable selection of growth temperature. The lack of confinement of the growing crystal together with the negligible temperature gradients mean that the only major source of mechanical stress is the rod or thread from which the seed crystal is suspended. The principal disadvantage of solution growth is the fact that solvent is inevitably incorporated into the crystal during growth, thereby decreasing the purity and degrading the structural perfection of the grown crystal.

The influences of the solvent are not confined to the effects of incorporation described above. The interactions between solvent and solute molecules in solution and between the solvent and crystal surface play a central role in determining growth mechanisms, growth rates and, therefore, crystal morphology [18]. As a result of the highly polar nature of organic nonlinear optical materials, the effects of these molecular interactions are frequently very pronounced, leading to highly anisotropic growth behaviour. In extreme cases, solvates or complexes may occur with solvent molecules forming part of the crystal lattice. Consequently appropriate selection of solvents is a key factor in the successful growth of such crystals.

The mechanistic influences of the solvent are twofold: firstly by changing the total number of growth sites through modification of the surface energetics and secondly by decreasing the fraction of sites available for growth as a result of solvent adsorption at the surface. The alpha factor, or surface entropy factor, α, is a key factor defining the roughness and, therefore, the growth mechanism of a crystal face [19, 20]:

$$\alpha = \zeta_{hkl} \frac{\Delta H}{RT} ,$$

where

ζ_{hkl} is the anisotropy factor defined as E_{slice}/E_{cr},
E_{slice} is the total energy of the bound within a slice of the face (hkl),
E_{cr} is the total crystallization energy and
ΔH is the enthalpy of the phase change.

For solution growth ΔH is the enthalpy of solution. For values of α less than about 3, the crystal face is rough and molecules can be added to the surface at any point. With increasing values of α, greater than about 3, the growth interface becomes progressively smoother and growth proceeds via a layer mechanism, involving either surface nucleation or dislocation sources. Expressing the enthalpy of solution as

$$\Delta H_s = \Delta H_{fus} + \Delta H_{mix} ,$$

the reason for solvent influences can be seen. For an ideal solution $\Delta H_s = \Delta H_{fus}$, while for solutions exhibiting negative deviations from ideality ΔH_{mix} is negative and so α is decreased. Conversely, positive deviations from ideality result in an increase in α factor. In addition to influencing surface roughness, solvent molecules may also affect growth kinetics through adsorption at and blocking of growth sites. The consequent reduction in the fraction of available growth sites can lead to an observable reduction in growth rate of the face. Since both of these factors depend upon the structural nature of the faces concerned, the solvent-related influences can result in substantial anisotropy of growth from certain solvents and considerable variation in morphology from one solvent to another. It is not surprising, given the highly polar nature of organic nonlinear optical molecules and crystals, that solvent-related effects are particularly pronounced in such systems.

3`4.1. Selection of solvent. A useful first step in the assessment of potential solvents for crystal growth is the examination of UV-visible spectra of a range of solutions. Substantial differences between the spectra can indicate the occurrence of solvate formation in unsuitable solvent systems. In the case of growth from solution by temperature lowering, we have found for a wide range of organic materials, satisfactory growth can be achieved using a simple empirical rule involving the solubility, S, and the gradient dS/dT of the solubility curve[16]. Many materials can be grown successfully at a temperature lowering rate of about 0.01 Kh^{-1} when the ratio $(dS/dT)/S$ lies in the range 0.01 to 0.03. This general rule provides a valuable initial guide to the selection of suitable solvents and approximate growth conditions. Selection of solvents on the basis of the morphology which they yield is performed largely on a trial and error basis with small crystals being grown by solvent evaporation in dishes. For most purposes crystals of an equant, prismatic habit are desirable. It should be noted, however, that, in such simple evaluations, supersaturation is not well controlled, depending both on the solubility of the solute and the volatility of the solvent. Consequently, it is highly probable that crystals from different solvents will have been grown under different supersaturations. Any observed variations may not, therefore, result solely from solvent effects.

Although there are numerous examples of changes in the morphology of crystals grown from different solvents, there have been relatively few quantitative studies of solvent effects on the growth of organic crystals. As far as we are aware, none have been reported to date for organic nonlinear optical materials. The interpretation of such qualitative observations in terms of solvent effects must be made with caution in view of the comments made above con-

TABLE II. – *Changes in morphology with solvent.*

Material	Solvent	Habit
MBA-NP	methanol	prismatic
	toluene	triangular
NMBA	hexane	needles
	ethyl acetate	rhombic plates
DCNP	toluene	needles
	ethyl acetate	rhombic plates
PNP	chloroform	needles
	acetone/water	prismatic
	ethanol/water	prismatic

MBA-NP is $(-)$ 2-N-(α-methylbenzylamino)-5-nitropyridine,
NMBA is 4-nitro-4'-methylbenzylideneaniline,
DCNP is 3-(1,1-dicyanoethenyl)-1-phenyl-4,5-dihydro-1H-pyrazole,
PNP is 2-(N-prolinol)-5-nitropyridine.

cerning supersaturation. Table II shows some examples of variations in morphology of several organic nonlinear optical materials grown from different solvents.

TANAKA and MATSUOKA[21] have suggested for 3-chloronitrobenzene (mCNB) and 2-methyl-4-nitroaniline (MNA) that solvent/solute combinations which have similar dipole moments and which exhibit near-ideal behaviour can be expected to yield crystals of a less anisotropic habit. In general, however, the morphological influences of solvents will depend upon the detailed chemistry of the solution and the specific molecular interactions between the solvent and the crystal surface. Thus consideration of dipole moments alone may prove insufficient.

As the final form of a crystal, a needlelike habit is undesirable. In several cases, however, such a crystal habit provides the starting point for the growth of large crystals of more equant habit which cannot be prepared directly. The case of 4-nitro-4'-methylbenzylideneaniline (NMBA) illustrates this well[22]. Crystals of NMBA can be grown from toluene or ethyl acetate solutions by solvent evaporation to yield crystals of a prismatic habit. Further growth by temperature lowering using such crystals as seeds results in rapid growth of the (010), (102) and (100) faces with near-zero growth of {111} and {110} facets. Consequently this yields octahedral crystals bounded by the latter two forms, with largest dimension only slightly larger than the original seed crystal. Little further growth occurs after this point. Larger dimensions can be achieved through the use of needlelike seeds. Growth of NMBA from n-hexane solution yields crystals with their needle axis along [100]. Growth by temperature lowering from toluene or ethyl acetate using such crystals as seeds again results in the development of {111} and {110} faces to yield rhombic plates, limited in size to the maximum dimension of the seed. In this case, however, the needles can be grown to many centimetres in length, thus enabling the preparation of large crystals

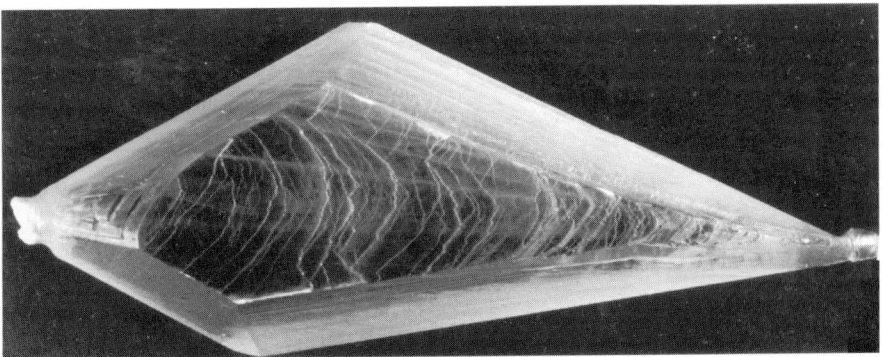

Fig. 11. – Crystal of 4-nitro-4'-methylbenzylideneaniline (NMBA) grown from ethyl acetate.

of NMBA (fig. 11). The crystal thickness can be increased by using seeds
cut from the previously grown crystals.

3'4.2. Supersaturation effects—MBA-NP. Crystals of $(-)2$-N-$(\alpha$-
methylbenzylamino)-5-nitropyridine (MBA-NP) prepared by spontaneous nucleation from saturated methanol solutions are tabletlike in form, bounded principally by $\{001\}$, $\{100\}$ and $\{011\}$ faces as shown in fig. 12a). The growth sectors corresponding to the (012) and (01$\bar{2}$) faces are highly defective producing a
distinctive semi-opaque triangle within the crystal. The apex of this triangle

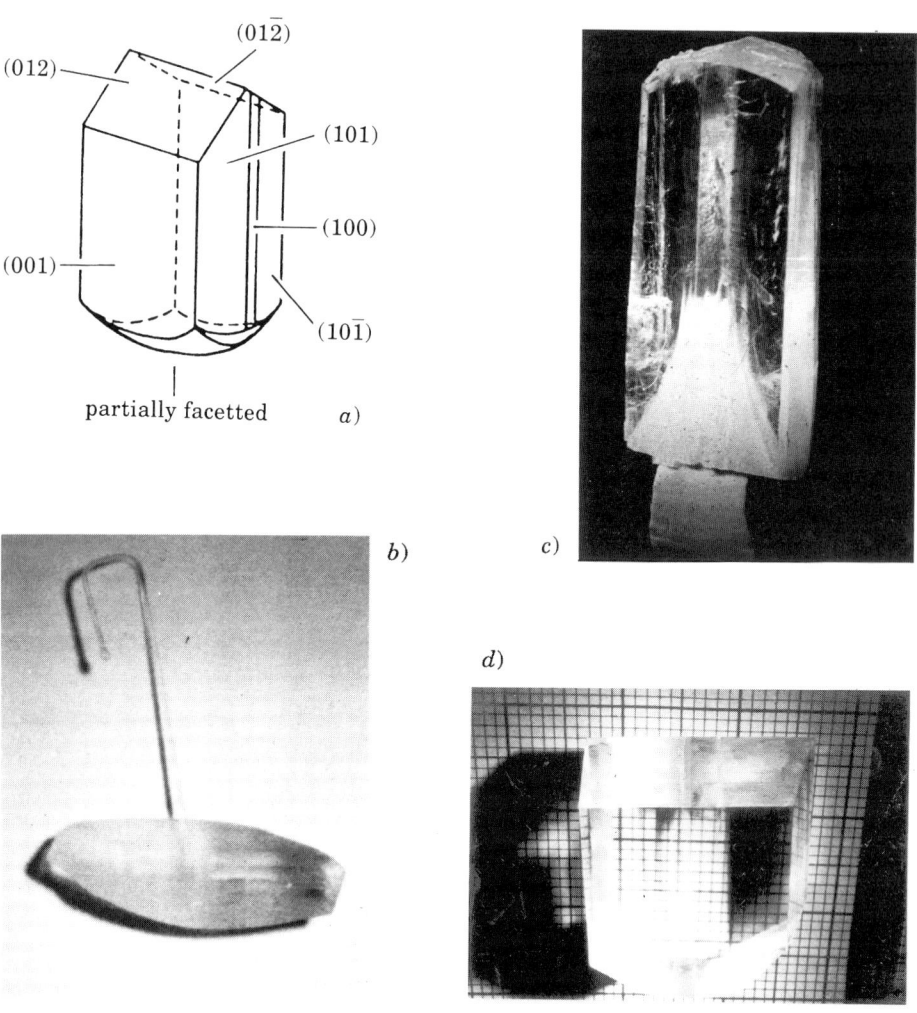

Fig. 12. – a) Crystal morphology of $(-)2$-N-$(\alpha$-methylbenzylamino)-5-nitropyridine
(MBA-NP), b) lozenge-shaped crystal of MBA-NP grown under low supersaturation,
c) crystal of MBA-NP with inclusions grown under high supersaturation, d) large high-
quality MBA-NP crystal.

defines the nucleation point N of the crystal which lies very close to the $[0\bar{1}0]$ end of the crystal. It is apparent, therefore, that little or no growth has occurred in the $[0\bar{1}0]$ direction. Further development of such crystals in methanol solutions at low supersaturations yields growth in the $\langle 001 \rangle$ and $\langle 100 \rangle$ directions together with the appearance of additional facets. No detectable growth occurs, however, in either of the $\langle 010 \rangle$ directions. As growth proceeds at low supersaturation some of the faces become rough and irregular, the crystal assuming a lozenge shape (fig. 12b)). Once the crystal becomes completely bound by rough faces, no significant further growth takes place. At substantially higher supersaturations rapid, but imperfect, growth occurs in the $[010]$ direction (fig. 12c)), but no detectable development in the $[0\bar{1}0]$ direction is observed. The $[0\bar{1}0]$ end of the crystal remains rough and irregular, but the fast growing (012) and $(01\bar{2})$ facets become reasonably planar. Despite the poor crystal quality along $\langle 010 \rangle$, the lateral sectors grow well. At still higher supersaturations nucleation occurs on the static $[0\bar{1}0]$ end of the crystal to yield 180° twins.

This supersaturation dependence of growth in the $\langle 010 \rangle$ direction at one extreme prevents the growth of large crystals of MBA-NP, but at the other holds the key to its successful growth. Growth from solution is generally carried out at relatively low supersaturations with moderately slow growth rates, since such conditions ususally favour higher structural perfection of the resulting crystals. These conditions are clearly inappropriate for the growth of large crystals from small seeds in the case of MBA-NP. The solution to this problem lies in the preparation of seed crystals of suitable length in the $\langle 010 \rangle$ direction which can subsequently be grown under low supersaturation to yield large, high-quality crystals. Small seed crystals are developed in methanol solution at high supersaturation yielding the rapid, imperfect growth along $[010]$ described above. The lateral sectors, however, grow well to produce narrow regions of material of high quality. From the outermost parts of these regions can be cut seeds whose length along $\langle 010 \rangle$ is many times greater than the initial seed crystal. These larger seeds can be developed and cut again to produce successively larger seeds of high quality. Once the desired length has been achieved, crystals are grown from the elongated seeds under low supersaturation. Although no further growth occurs along $\langle 010 \rangle$, slow growth in the lateral directions yields large crystals (greater than $(7 \times 4 \times 3)$ cm^3) of good quality as shown in fig. 12d) [23].

3'5. *Vapour growth.* – The relatively high vapour pressures of many organic solids offer the possibility of growth from the vapour at low temperatures which can eliminate the problems associated with thermal decomposition. Additionally, this flexibility in the choice of growth temperature enables the preparation of crystals of polymorphic materials by growth below the temperatures of polymorphic phase transitions. Growth under vacuum is also inherently much cleaner than either melt or solution growth and can yield crystals of the very highest purity.

The major factor responsible for the very limited application of vapour growth techniques is undoubtedly the difficulty in controlling nucleation from the vapour phase. Unlike melt and solution growth, the onset of nucleation in vapour growth occurs very rapidly above the critical supersaturation and metastable-zone widths are generally rather narrow. This difference in behaviour results primarily from the strong dependence of nucleation rate upon diffusion, which in turn varies substantially with the viscosity of the medium. To avoid extensive nucleation and consequent polycrystallinity, growth must in general be carried out at quite low supersaturations. Low growth rates are, therefore, typical of vapour growth methods. A wide variety of methods have been devised for vapour growth of organic crystals. All achieve the supersaturation for growth by means of a temperature differential between the source material and the growing crystal. The major differences in the techniques arise from the alternative means employed to control nucleation.

3˙5.1. Continuous pulling[24]. In many respects this technique bears a close resemblance to Bridgman growth from the melt. The source material, contained in a sealed growth tube evacuated to about 10^{-5} torr, is placed in a three-zone furnace as shown in fig. 13. The source end of the tube lies within temperature zone A, while the tip of the tube is initially positioned within zone B at a higher temperature. The tube is then translated through the furnace with the temperature of the tip decreasing as it passes through the temperature gradient BC. When the temperature of the tip drops below that of the source, material will sublime along the tube. At some lower temperature still, nucleation will take place at the tip and growth will occur. Several factors are important in controlling nucleation in this method.

1) An undercritical supersaturation is employed to limit nucleation to a few active sites on the walls of the tube. In this context the pre-treatment of the tube is of the utmost importance. Care must be taken to minimize the number of

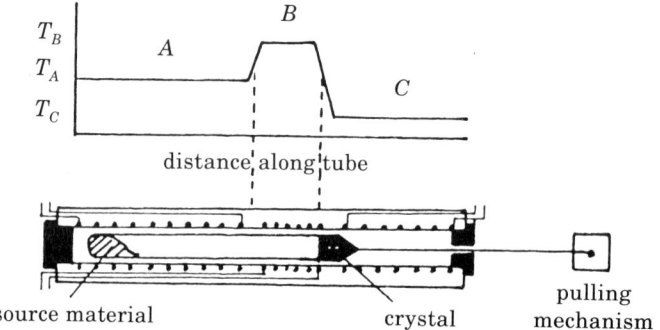

Fig. 13. – The continuous-pulling vapour growth technique.

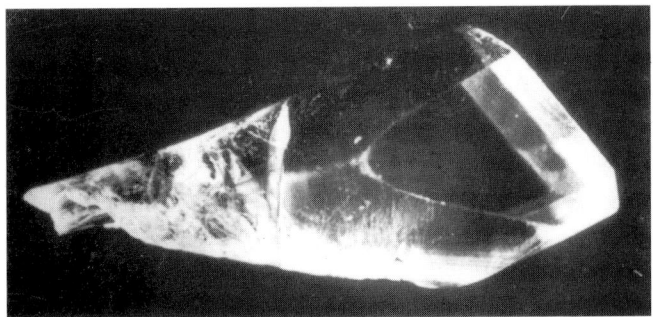

Fig. 14. – Crystal of anthracene grown from the vapour phase.

active sites on the walls. Careful cleaning and flame polishing can be effective in reducing nucleation at surface imperfections on the glass. Additionally, traces of source material, often introduced during filling, must be excluded from the growth end of the tube. Imposing the maximum achievable reverse temperature gradient immediately prior to growth (*i.e.* source colder than the growth end) can achieve this, although for thermally sensitive materials this may prove difficult.

2) As in the case of Bridgman growth, the use of a tapered growth tube and a capillary restricts nucleation and promotes the formation of one crystal from the tip of the tube.

3) To achieve practical growth rates and yet prevent spurious nucleation on the walls of the growth tube, a balance must be maintained between the crystal growth rate and the tube translation rate such that the growth front remains at a temperature near the lower end of the metastable zone.

Figure 14 shows a crystal of anthracene grown from the vapour phase using this method ($T_A = 383$ K, $T_C = 293$ K) at a growth rate of 0.001 mmh^{-1}.

3˙5.2. Sloan method. Substantially greater control over nucleation can be achieved using a method developed by SLOAN[25], based on the early work of HONIGMAN[26] and HONIGMAN and HEYER[27]. The apparatus, shown in fig. 15, consists of a well-insulated heater block into which is placed a sealed glass growth ampoule. Through the side of the heater block, and insulated from it, there is a sliding brass or copper rod which has a small-diameter sliding rod within. This assembly forms a heat leak, the temperature of which is controlled independently of the main block. Opposite the heat leak is an observation window.

During growth the ampoule is heated to a suitable temperature, T_B, to achieve an adequate vapour pressure over the solid contained within. The inner rod of the heat leak, at a temperature T_L (where $T_B - T_L \geqslant 2$ K), is moved into contact with the wall of the ampoule. This results in multiple nucleation of small

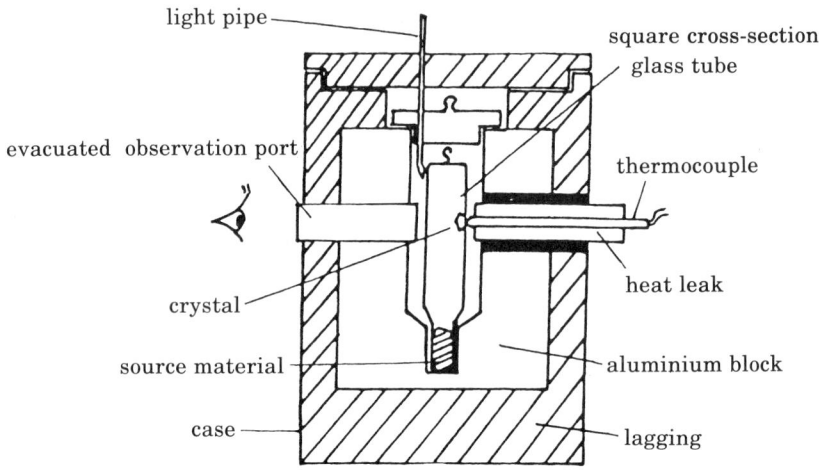

Fig. 15. – The Sloan vapour growth method.

crystallites on the wall of the ampoule. The temperature of the heat leak is then increased to about $T_B + 1$ K resulting in partial evaporation of the crystallites. These are inevitably of different sizes and, by virtue of the Gibbs-Thompson relation, those of smaller dimensions have higher vapour pressure. They, therefore, evaporate more rapidly. This oscillation of the heat leak temperature is repeated causing alternate growth and evaporation until only one crystal nucleus remains. The temperature T_L is then set to a value providing a supersaturation high enough for adequate growth rates but low enough to prevent further nucleation. As growth proceeds, the outer section of the heat leak is advanced to allow growth of a larger crystal. The use of square-section glass tubing for the ampoule ensures good thermal contact between the heat leak and the ampoule over a large area. The ultimate size of the crystal which can be grown depends upon the temperature distribution around the heat leak.

The plate sublimation method, developed by KARL [28], although similar in operation to the Sloan technique, employs a large-diameter cylindrical glass ampoule with large metal heater blocks above and below. Nucleation and growth supersaturations are controlled by adjustment of the temperatures of the two blocks in a manner similar to that described above. The material sublimes from the lower to the upper planar face of the ampoule to yield crystals up to ~ 1 cm^3 in size. This arrangement is particularly suited to the growth of mixed crystals and the introduction of dopants [29].

4. – Polymorphism.

The problems associated with polymorphism, which is so common amongst organic materials, are twofold. Firstly the existence of a polymorphic phase transformation between the growth temperature and ambient temperature frequently results in a loss of crystallinity. Secondly, even if the grown crystal remains single, different polymorphs frequently display substantial differences in chemical or physical properties. Consequently, the stable polymorphic form at room temperature may not exhibit the desired properties. This can be a particular problem in the case of materials for second-order nonlinear optical effects, for which a noncentrosymmetric structure is an essential pre-requisite.

Both of these aspects of polymorphism are illustrated by the behaviour of the material 1-(4-nitrophenyl)-2-(4-methoxyphenyl)-1-cyanoethene (CMONS). CMONS is known to exist in four polymorphic forms[30]. All four can be obtained by growth at different temperatures from the vapour phase. At temperatures above $\sim 160\ ^\circ$C a deep orange-coloured modification, designated (O_1), is obtained in the form of needles. Below 160 $^\circ$C two other forms are obtained; firstly a yellow form (Y_1) and another orange form (O_2) though the latter is lighter in colour than O_1 and occurs as rectangular platelets. Finally at growth temperatures below $\sim 140\ ^\circ$C a second yellow form (Y_2) is obtained. The high-temperature orange form (O_1) is found to transform readily to the yellow form (Y_1) either spontaneously or as a result of an applied stress, as in cleaving, for example. This transformation is accompanied, in most instances, by the loss of single crystallinity. An assessment of the frequency-doubling properties indicates that two of the four forms, O_1 and Y_1, exhibit significant SHG activity, 8 and 10 times MNA, respectively. This observation must, however, be interpreted with care because of the ready transformation of the O_1 to Y_1 form. A small amount of the latter in a crystal of the unstable (O_1) form could give a misleading indication of the SHG activity of the O_1 form. It is unclear at present whether there are two active forms (O_1 and Y_1) or just one (Y_1). Growth of CMONS from dioxane solution by solvent evaporation at $\sim 25\ ^\circ$C yields prismatic yellow crystals which show no SHG activity and probably correspond to the Y_2 form obtained by vapour growth below 140 $^\circ$C. Using growth from the melt by the Bridgman method, large (30 mm $\times$ 8 mm diameter) deep orange CMONS crystal boules can be obtained. In all cases the crystals change to yellow over a period of several days with a consequent loss of crystallinity. The yellow form is active (Y_1).

The behaviour of 3-acetamido-4-pyrrolidinonitrobenzene (PAN)[31] illustrates another important aspect of polymorphism, namely solvent-mediated transformations[18,32]. PAN exists in two polymorphic forms, one active and the other inactive. Rapid recrystallization under high supersaturation from ethanol/water solutions yields needlelike crystals of the active form, while slower growth from the same solvent system results in platelike crystals of the inactive form. Additionally, when the needlelike crystals are maintained in contact with the solution from which they were recrystallized, platelike crystals form and grow at the expense of the needles. It is apparent that the platelike, inactive form is the more stable. The fact that, for two polymorphic phases I and II, the more stable form (II) has the lowest chemical potential (μ) implies that it will always be the less soluble.

I.e. since for the solid phases

$$\mu_{II} < \mu_{I} .$$

For each polymorph in contact with its saturated solution, the chemical potential of the solid phase equals that of the material in solution. Thus

$$\mu_{II} = \mu^0 + RT \ln a_{II}$$

and

$$\mu_{I} = \mu^0 + RT \ln a_{I} .$$

Therefore,

$$a_{II} < a_{I} ,$$

where a_{I}, a_{II} are the activities of the two polymorphic forms.

Below the polymorphic-phase-transition temperature, crystallization occurs through nucleation and growth of the high-temperature form, since this is the more soluble phase and, therefore, experiences a higher supersaturation for a given temperature decrease. Because of the resulting differences in nucleation and growth rates, the high-temperature form will dominate if crystallization is carried out rapidly. Formation of the more stable phase, which is limited by the kinetics of either nucleation or growth, occurs upon slow growth or subsequent aging of the unstable form in its saturated solution.

5. – Inorganic crystals.

5˙1. *Purification.* – Inorganic materials present less problems in this respect than organic materials. Source materials or their components of good and well-defined quality are readily available from general supply houses. For the purposes of top-seeded solution growth (TSSG), which inevitably involves the mixing of the components to form the flux from which the growth occurs, can be used directly without further purification. The crystals which can be obtained are of good quality and show little evidence of impurity incorporation.

5˙2. *Top-seeded solution growth.*

5˙2.1. Fluxes. A number of different flux systems have been used to grow KTP. We have used the classic K_6 flux which is formed by using potassium carbonate, potassium dihydrogen phosphate and titanium dioxide according to the reaction shown below:

$$K_2CO_3 + KH_2PO_4 + TiO_2 \rightarrow KTiOPO_4 + K_6P_4O_{13} + H_2O + CO_2 .$$

This has the advantage that no ions foreign to KTP are present in the flux. The water and carbon dioxide are removed from the charge in the crucible prior to insertion in the furnace. This is done to prevent foaming of the flux which causes overspill and consequent damage to the furnace liner from the corrosive flux. The quantities used at the saturation temperature of approximately 900 °C give 0.5 g KTP/g K_6.

Other fluxes which have been used are molybdate[33] and tungstate[34] fluxes. Tungstate fluxes were used due to the less viscous nature of the flux, but the crystals produced were found to contain optical inhomogenities caused by the incorporation of tungsten[35].

5˙2.2. Equipment. For the most part TSSG is carried out at temperatures approaching 1000 °C. The core of the growth system is, therefore, a vertical three-zone, high-temperature furnace of standard construction. The furnace is lined with silica and alumina cladding and the temperature of each of the three zones is controlled by platinum/rhodium thermocouples controlled by Eurotherm controllers. The sensing heads are placed adjacent to the centres of each of the zones. The temperature of each zone can be adjusted to give an axial temperature stable to better than 1 °C and no radial gradient.

5˙2.3. Modes of operation. Such furnaces operate in two modes, depending on the size and type of samples required. The first growth method uses

TABLE III. – *A comparison of the growth conditions used during spontaneous and seeded growth runs.*

	Spontaneous	Seeded
charge mass (g)	500	500
soak time	3 h at 950 °C	12 h at 1000 °C
saturation temperature	904 °C	904 °C
crucible rotation cycle	0 to 30 r.p.m. in 25 s	0 to 40 r.p.m. in 25 s
	5 s hold at 0 and 30 r.p.m.	5 s hold at 0 and 40 r.p.m.
seed rod rotation rate	—	0 to 90 r.p.m. in 25 s
cooling rate	1.13 °C/h (904 to 800 °C) —————————— 60 °C/h (800 to 25 °C)	hold at saturation temperature for 10 h 0.6 °C/h to 800 °C
		hold for 12 h and raise crystal from flux 20 °C/h ((800–500) °C) then 60 °C/h ((500–25) °C)
crystal size and quality	∼ (3 × 2 × 2) mm³	(7 × 15 × 9) mm³
	good	good

spontaneous nucleation to produce small crystals which can be used as seeds in the second type of growth run.

Table III shows the different conditions used during both types of operation and notes the results obtained.

The seed crystals were mounted on the end of a platinum-tipped alumina rod using platinum wire and inserted into the crucible containing the flux. The growth on the seed is initiated by the supersaturation generated by the cooling of the flux.

6. – Structural characterization.

6˙1. *Defects in organic crystals.* – In common with all crystalline materials, no matter how carefully they are prepared, organic crystals inevitably contain imperfections. These defects may be chemical in nature, as in the case of impurity atoms or precipitates, or may involve a purely structural departure from the perfect crystal. The structural defects which can occur in crystalline solids may be divided into four categories:

1) point defects such as vacant lattice sites, interstitials and impurity atoms;

2) line defects such as dislocations;

3) planar defects such as stacking faults, grain boundaries, twins and growth sector boundaries;

4) volume defects such as impurity inclusions.

The importance of crystal imperfections should not be underestimated. They exert substantial influences over virtually every property of crystalline materials. For example, mechanical properties depend upon the motion and multiplication of dislocations through glide or climb. Point defects such as vacancies and interstitials are central to the mechanisms of impurity and self-diffusion in solids. Impurities and structural defects can control electronic and optical behaviour by generation of charge carriers, by acting as carrier traps or by modifying the perfect crystal band structure. Dislocations having a screw component influence crystal growth by providing a continuous source of surface steps, thereby eliminating the need for two-dimensional nucleation. Chemical as well as physical properties can also be affected, since many crystal defects exhibit enhanced chemical reactivity over that of the surrounding crystal and so act as preferential sites for chemical reactions. These few examples not only illustrate the breadth of the influence of defects upon solid-state properties, but also confirm the need for a detailed understanding of the nature, properties and origins

of crystal imperfections. Of the various defect types, point defects and dislocations are by far the most extensively studied in organic materials and most of
the following discussion will be confined to these.

6'1.1. Point defects. Being equilibrium defects, intrinsic point defects
such as vacancies and interstitials are especially important since they and many
of their effects cannot be eliminated. Much of the information on intrinsic point
defects in organic crystals [36] is derived from X-ray expansivity, self-diffusion, excess specific heat and positron annihilation studies. In particular the
difference between X-ray and bulk expansivities reveals the dominant point defects to be vacancies. Furthermore, comparison of calculated point defect formation and migration energies with experimental self-diffusion coefficients indicates that diffusion mechanisms in organic crystals are in general vacancy related. One would intuitively expect vacancies to dominate the point defect
structure of many organic crystals, where the incorporation of a large, irregu-
larly shaped molecule in interstitial sites is likely to be energetically un-
favourable due to steric interactions.

The incorporation and behaviour of extrinsic point defects (*i.e.* impurities)
in organic crystals are of particular scientific interest and technological importance. They provide, in addition to molecular-engineering methods, a further
means by which optical, electrical, mechanical and many other properties may
be modified and tailored. A major difficulty with intentional doping is the generally rather limited degree of impurity incorporation in organic crystals. For
appreciable dopant acceptance there must be a close geometrical fit between the
dopant molecule and the unrelaxed vacancy produced by removal of a host
molecule from the lattice [37]. The extent to which doping can be achieved depends, therefore, not only on the dopant size in relation to the host molecule,
but also on molecular shape and the nature of intermolecular bonding. There are
many similarities between the behaviour of impurities in organic and inorganic
materials, including the modification of mechanical properties through lattice
hardening by impurities and the generation of charge carriers by impurity ionization. Significant differences, however, also exist. Low charge carrier concentrations and mobilities are a direct consequence of the wide band gaps and narrow energy bands typical of most organic materials. Under these circumstances
the influence of both impurities and intrinsic point defects upon electronic properties, particularly acting as carrier traps, is very pronounced [38]. In the case
of highly conducting charge transfer salts such as TTF-TCNQ energy bands
exist as a consequence of overlap of orbitals on neighbouring molecules. Impurities within the lattice can have a pronounced influence on the conductivity of
these materials, the extent of the effects depending upon geometrical factors
which disrupt the regular molecular packing and affect the degree of orbital
overlap of the impurity molecule.

6'1.2. Dislocations. Because of the low configurational entropy associated with a line of displaced atoms, dislocations, unlike point defects, are not equilibrium defects. An important consequence of this is that it is in principle possible to prepare crystals free from all dislocations. A situation which can never be achieved for point defects. Dislocations may be conveniently grouped into two broad categories: growth-induced dislocations and mechanical dislocations. The former occur during growth as a consequence of disruptions to the regular arrangement of molecules, for example as the crystal grows over a solvent inclusion. They are, in general, straight and lie along directions close to the growth normal of the growth sector in which they occur. These directions, which are rarely low-indexed, correspond to those of minimum elastic line energy of the dislocations [39]. They depend upon the Burgers vector of the dislocation, the growth sector in which it lies and upon the elastic constants of the material. Mechanical dislocations, in contrast, are introduced as a result of mechanical stress either during or subsequent to growth. The factors governing the character of mechanical dislocations and the active slip systems in organic crystals are rather more complex than in the case of metals and simple ionic compounds. The mechanical properties of organic crystals are determined by three principal factors: a) the nature of intermolecular bonding, b) the molecular packing and crystal symmetry and c) the size and shape of the molecules. In many organic crystals the cohesive forces are predominantly weak Van der Waals forces. In such cases the materials are relatively soft. Their crystal structures are determined by the molecular packing which occupies the minimum volume. The mechanical properties are then defined by this packing and by steric interactions between molecules [40]. In contrast, the presence of hydrogen bonding or other directional bonding will result in considerable anisotropy of mechanical properties which may dominate geometrical and steric effects. In the simplest case the slip system of a given crystal is defined by two geometrical rules. Firstly, the slip plane expected to dominate is that with the highest atomic packing density and, therefore, the largest interplanar spacing. Secondly, since the energy of a dislocation depends upon the square of the Burgers vector length, the shortest lattice translations are the most favourable slip directions. These simple rules are not always applicable, however, to organic compounds particularly when they show pronounced anisotropy of bonding as discussed above or have complex molecular shapes. The shortest lattice translation may not be an observed slip direction because of intermolecular contacts which occur when that translation is performed.

6'2. *Characterization methods.* – There is in general a very wide range of techniques for the assessment of perfection and examination of defects in crystalline materials. Certain properties of organic materials, however, restrict the applicability of some of these, notably the electron beam techniques such as transmission electron microscopy (TEM). High vapour pressures and susceptibility to electron beam damage prevent the examination of many organic materials using these techniques. X-ray diffraction methods do not suffer from such drawbacks and, together with defect etching techniques, provide a powerful array of tools with which the defect properties of organic crystals can be examined.

6.2.1. Defect etching. A crystallographic defect and its immediate environment differ from the surrounding crystal not only as a result of lattice strain but also because of accumulation of impurities at the defect. Both of these factors can result in substantially altered chemical reactivity in the vicinity of defects. Upon treatment with a suitable reagent or solvent, pits or other etch features are formed on the surface of a crystal at the points where defects intersect that surface. This delineation of crystal defects by virtue of their altered reactivity is known as defect etching [41]. Etching provides a simple, rapid means of examining defects and their behaviour, yielding information about defect density and distribution. In the case of dislocations it allows determination of slip planes, dislocation line directions and velocities. It is, however, a surface technique and care must be exercised in the interpretation of etching studies, since defect densities and properties in the surface and bulk regions of crystals frequently show marked differences. Since etching is most commonly used to delineate dislocations, we will confine our discussion to this aspect. For any defect etchant, it is essential to assess the extent to which a correspondence exists between etch features and dislocations. In general, it should be noted that not all etch pits are necessarily formed at emergent dislocations and that not all dislocations give rise to etch features. A true dislocation etchant, however, should satisfy the following requirements:

1) Since cleavage of a crystal will intersect dislocation lines, the pattern of etch pits on conjugate cleaved surfaces should be related by mirror symmetry.

2) Continued etching of a surface should in general result in no net change in the number of pits other than where pits are associated with, for example, dislocation loops. In such cases the pits should appear or disappear in pairs.

3) Mechanically induced dislocations lie along characteristic slip planes which are typically low-indexed planes. This should be reflected in crystallographic alignments of the associated etch pits.

Ideally a defect etchant should be «calibrated» by demonstrating a correspondence between etch features and defect images obtained by either TEM or X-ray topography. These diffraction techniques enable detailed characterization of the defects by analysis of their diffraction contrast. The specific application of dislocation etching methods to organic crystals has been discussed in ref. [42].

Figure 16 shows etch pits associated with mechanically induced dislocations around a Knoop microhardness indentation, revealed by etching in dilute sulphuric acid. Note the alignments of pits along the interactions of the dislocation slip planes with the surface. These observations confirm the correspondence of etch pits with dislocations in this system [43]. Figure 17 shows an etched surface of pentaerythritol tetranitrate (PETN) [44] revealing the location of emergent dislocations by the presence of pits. Three different etch pit shapes are visible in this micrograph. These variations in shape arise as a consequence of the different character of the dislocations responsible for their formation. As discussed above, the occurrence of etch pits is an illustration of the enhanced chemical reactivity of dislocations, while in this

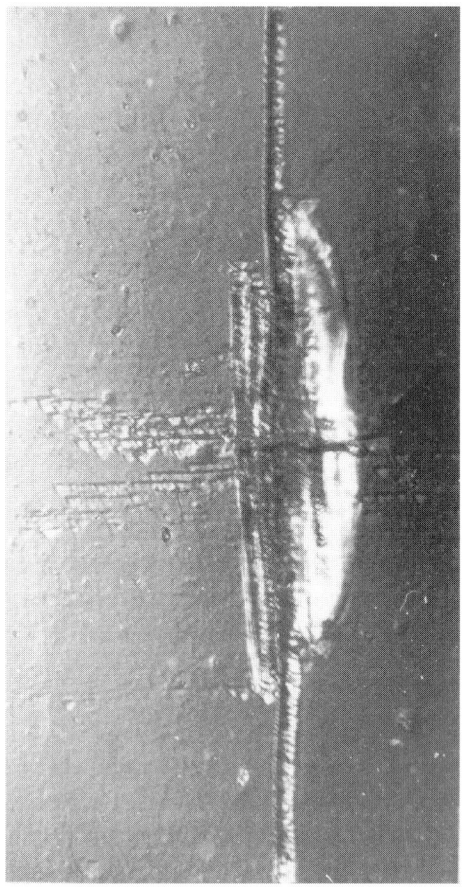

Fig. 16. – Etched microhardness indentation on the (001) cleavage face of MBA-NP.

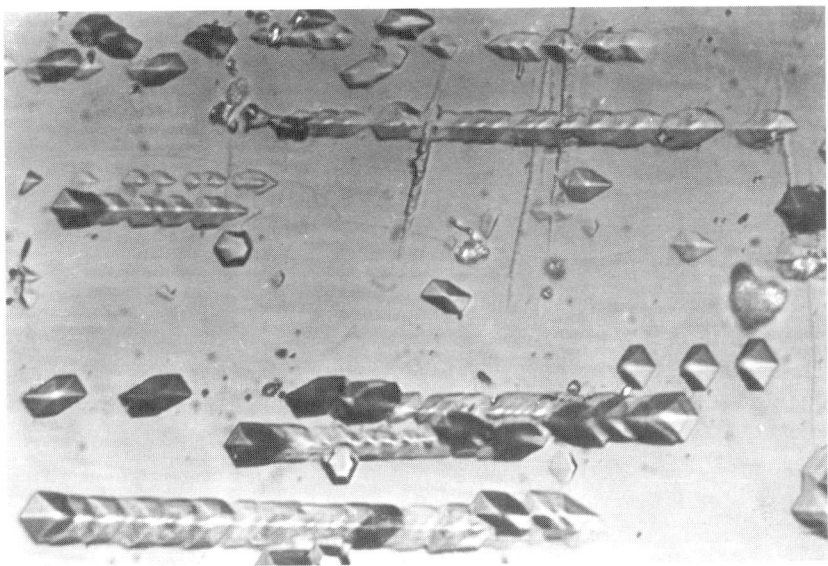

Fig. 17. – Etched (110) surface of pentaerythritol tetranitrate (PETN).

case the different pit shapes on the same crystal face demonstrate the
dependence of this behaviour upon dislocation character.

The dislocation etching technique is rapid, requires only very basic equipment and yields much valuable information, particularly about mechanical
properties. It is, however, destructive and not well suited to the examination of
bulk defect configurations and properties.

6˙2.2. X-ray diffraction methods. X-ray diffraction, on the other hand,
provides a powerful, nondestructive means of assessing structural perfection
and of studying the defect properties of organic crystals. These techniques comprise two main categories:

1) The Laue method and related film techniques, together with goniomet-
ric methods which provide an overall indication of structural quality and
uniformity.

2) A range of defect imaging techniques under the general heading of
X-ray diffraction topography [45].

a) *The Laue method.* Although most widely employed in the determination of crystal orientation, transmission and back reflection Laue pho-
tographs are also valuable tools for preliminary assessments of structural
perfection. Lattice strain or bending are manifested in Laue patterns as
radial streaking of the diffraction spots. From the extent of this streaking,
quantitative measurements can be made. The occurrence of multiple patterns

or split spots is indicative of either twinning or polycrystallinity, respectively.

The size of the incident X-ray beam is typically less than 0.5 mm in diameter, so the volume of the sample to which the diffraction patterns relate is very
small. It should be noted, therefore, that an apparently single-crystal Laue pattern is not proof of a single-crystal sample, since the X-ray beam may impinge
on only one crystallite of a polycrystalline sample.

b) *Orientation goniometry.* A more quantitative assessment of overall
crystal quality can be obtained using an X-ray goniometer to determine relative
variations in orientation across a crystal. The essential components of an X-ray
orientation goniometer are shown in fig. 18. Orientation is achieved using a single Bragg reflection from the crystal. A monochromatic collimated beam from
an X-ray tube T hits the crystal surface at point P which coincides with the vertical axis of rotation of the goniometer. A detector (D) is positioned at an angle
2θ to the incident beam, where θ is the Bragg angle of the chosen reflection.
The crystal is rotated about the goniometer axis at P until a diffraction peak is
located by the detector. The angle ω at which the peak occurs, therefore, defines the angular position of the diffracting planes relative to the incident beam.
The sample, mounted on a precision linear bearing, is then translated a short
distance to bring the incident beam to a new position on the crystal surface. The
new angle ω of the diffraction peak is measured and the process repeated across
the crystal. Any changes in this angle ω at different positions are due to
changes in either lattice plane spacing or orientation with the effect of the latter
being dominant. An angular precision of better than 2 arcmin can be achieved
under favourable conditions. Thus even small changes in orientation due to
grain boundaries, twinning, warpage, etc. can be readily detected.

c) *X-ray topography.* Although the previous techniques allow the overall
quality of samples to be assessed, they provide little information as to the nature of the structural imperfections present. X-ray topographic methods, how-

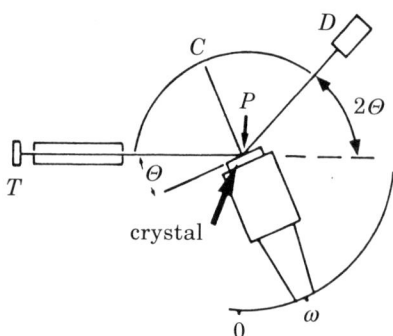

Fig. 18. – Schematic diagram of an X-ray orientation goniometer.

ever, not only enable routine assessment of crystal perfection, but also allow full and unambiguous characterization of defects, including their nature, density and distribution.

X-ray topography in all its various forms has been extensively reviewed in the literature [45-47], together with the mechanisms of defect image formation. All such methods utilize Bragg reflection from a chosen set of atomic planes and rely upon the fact that the resulting diffracted X-ray intensity from a perfect region of the crystal is substantially different from that emanating from the strained and distorted volume around a defect. Recording the diffracted intensity on a photographic film thus yields images of defects within the crystal. A variety of techniques exist to enable imaging of either surface (reflection topography) or bulk (transmission topography) defects and with varying levels of sensitivity to strain and misorientation. In the examples discussed here, two forms of transmission topography have been employed, namely projection and section topography. Projection topographs produce an image of the crystal and its defect content projected through the entire thickness of the sample. A wide parallel beam may be used in the case of a synchrotron source, while for monochromatic laboratory sources the crystal and film are traversed through a narrow (100 to 200 μm) beam. Alternatively, in the case of section topography [48], a very narrow X-ray beam (< 50 μm) can be used to obtain a topograph which is effectively a cross-sectional view through the sample. Two factors limit the sample thickness which can be examined:

   1) loss of information due to overlapping images in high-defect-density samples,

   2) X-ray absorption.

The low absorption coefficients typical of organic solids allow relatively thick samples to be examined ranging from several millimetres for copper K_α radiation to several centimetres for molybdenum K_α. The problems associated with high defect densities, however, limit the application of projection topography to thin crystals ((1–2) mm thick) or slices cut from thicker crystals unless the defect density is very low. Section topograph images, on the other hand, are formed from a very thin volume ((10–50) μm) within the crystal. This, together with the cross-sectional nature of the images, allows topography of thick crystals (up to 2 cm in thickness) having substantially higher defect densities than can be tolerated for projection topography. Thus section topography of whole crystals of organic materials is possible with the maximum thickness limited only by X-ray absorption. Consequently section topography is completely nondestructive. While conventional characteristic X-ray sources are perfectly adequate for topography in many cases, the availability of high-intensity, continuous synchrotron radiation offers a number of advantages. Unlike characteristic radiation topographs, use of the broad synchrotron spectrum allows even quite distorted samples to be imaged in their entirety for both projection and section topography. Furthermore the flexibility of wavelength selection possible with synchrotron radiation allows optimum diffraction conditions to be achieved for section topography of thick crystals [49].

6˙3. Factors influencing crystal perfection. – The final perfection of a crystal depends upon a wide range of parameters related both to the fundamental material and defect properties and to the specific growth conditions employed. The following factors have been identified as being particularly significant:

1) mechanical confinement of the growing crystal,

2) thermomechanical stresses,

3) seeding and refacetting,

4) crystal growth rate and fluctuations in growth conditions,

5) crystal morphology.

As a result of the comparative softness and poor mechanical properties of many organic crystals, factors such as growth temperature, thermal gradients and mechanical confinement of the growing crystal play a dominant role in determining structural quality.

6˙3.1. Mechanical confinement. Confinement of the growing crystal was discussed earlier when considering the relative merits of the various melt growth techniques. Predictably the deleterious effects of confining the crystal are most pronounced in the case of softer materials and at higher temperatures. Figure 19 shows a synchrotron topograph of a crystal of n-eicosane grown by the Brigdman method [50]. The image of this very soft material is highly distorted and reveals no detail, indicating a high level of lattice strain. In contrast the topograph shown in fig. 20 demonstrates the much higher perfection which can be achieved by Bridgman growth of m-nitroaniline—a considerably harder

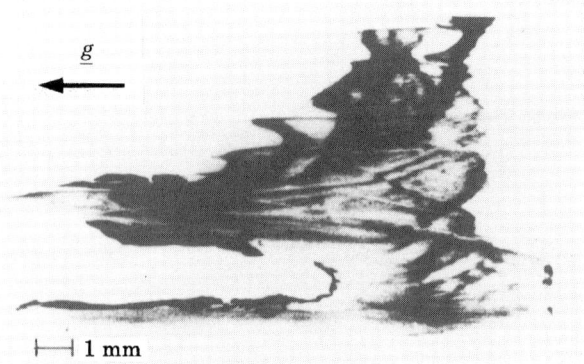

Fig. 19. – White-radiation synchrotron topograph of a highly strained crystal of n-eicosane grown by the Bridgman method.

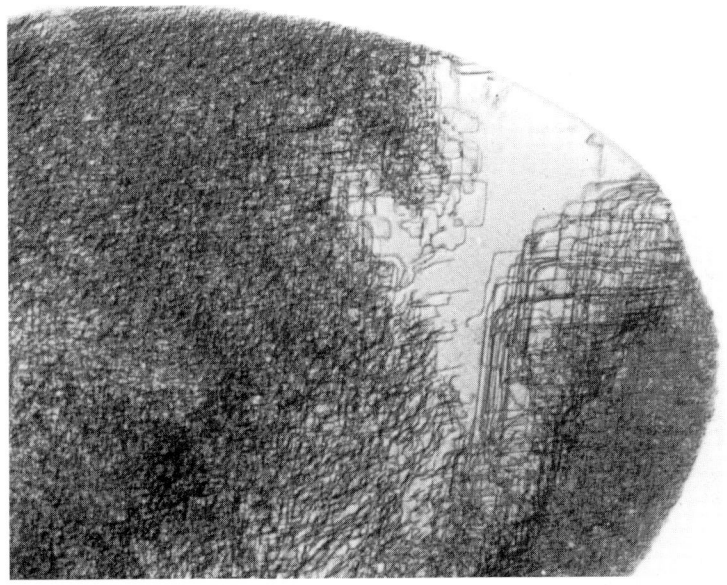

Fig. 20. – X-ray topograph of a cleaved slice from a Bridgman boule of mNA.

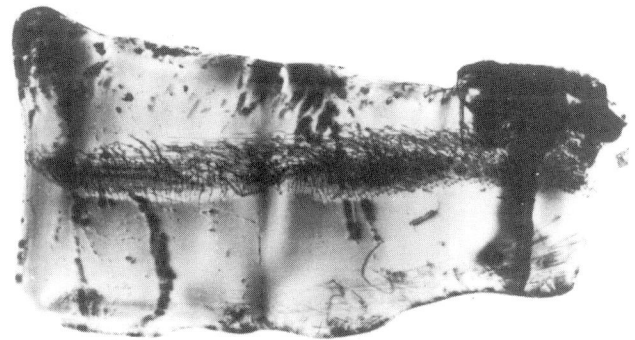

Fig. 21. – X-ray topograph of a longitudinal slice from a Czochralski-grown crystal of benzil.

material[51]. The substantial improvement obtainable when the growing crystal is unconfined is illustrated in fig. 21. This topograph shows a slice cut from a crystal of benzil grown by the Czochralski technique[14]. The limited defect generation arises principally from the seed-crystal interface and is confined to the central region of the boule.

6˙3.2. Thermomechanical stresses. Figure 22 shows a topograph of another slice cleaved from a different Bridgman-grown boule of mNA. In this case, however, the temperature gradient used during growth was substantially reduced, thereby decreasing the thermomechanical stresses imposed upon the

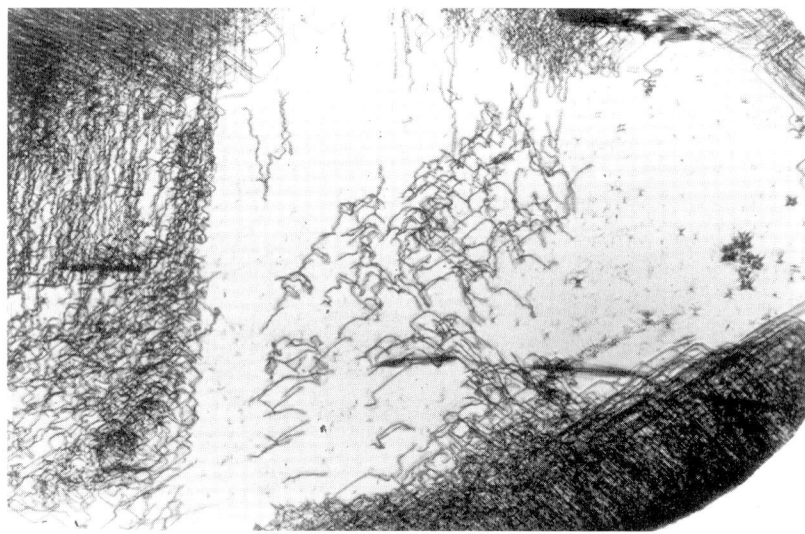

Fig. 22. – X-ray topograph of a slice of mNA cleaved from a boule grown under low-thermal-gradient conditions.

growing crystal. The resulting reduction in dislocation density is obvious with large volumes of the crystal essentially dislocation free. Finally, the elimination of thermomechanically induced stresses by growth from the undercooled melt enables crystals of relatively hard materials to be grown with very low defect densities and yields substantial improvements in the quality of softer crystals. Figure 23 shows a dislocation-free slice from a benzophenone crystal grown

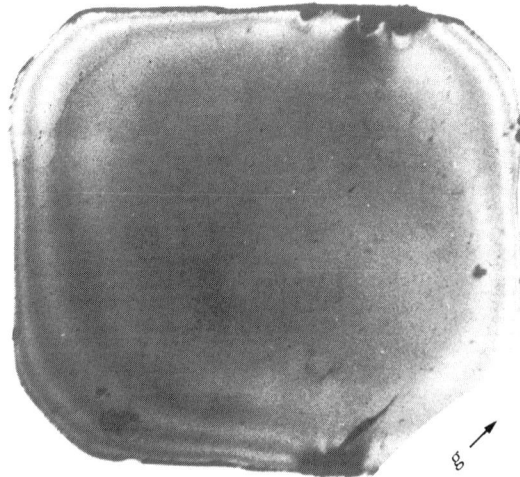

Fig. 23. – An X-ray topograph of a dislocation-free (110) slice from a crystal of benzophenone grown from the undercooled melt.

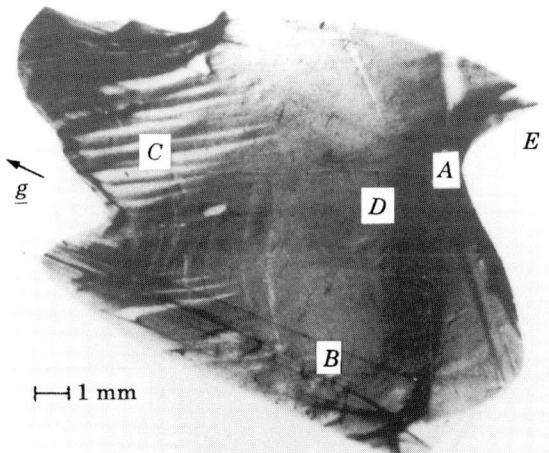

Fig. 24. – A synchrotron white-radiation topograph of a crystal of n-eicosane grown from the undercooled melt.

from the undercooled melt [52], while fig. 24 reveals the considerable improvement in the quality of the much softer n-eicosane grown by this method [52].

Growth from solution offers the greatest potential for preparation of very-high-quality crystals of many materials. Growth temperatures are close to ambient. Thermal gradients and, therefore, thermomechanical stresses are negligible. Furthermore crystals can be grown in an unconstrained manner, virtually eliminating mechanical stresses during growth. The principal cause of imperfections in solution growth are the development of dislocations on refacetting of the seed/crystal interface and the inclusion of solvent with its consequent generation of growth defects. Three factors are particularly important in the introduction of solvent inclusions: a) seeding and refacetting, b) fluctuations in the crystal growth conditions and c) growth instability. In each case, as the crystal grows over a volume of included solvent, lattice closure mistakes can occur, leading to the formation of growth dislocations, twins or, in extreme cases, polycrystalline growth.

6˙3.3. Seeding and refacetting. For solution growth, as well as most other crystal growth techniques, the most critical stages defining the crystal perfection are the seeding and initial growth on the seed leading to refacetting. The nature and quality of the seed itself are also important. It should be a single, untwinned crystal free from mechanical damage. The dislocation density should be low, though this is of secondary importance, since, in the absence of mechanically induced defects, it is generally found that most of the dislocations present in a given crystal are nucleated at the seed-crystal interface during refacetting. The quality of this interface depends upon the nature of the seed

Fig. 25. – X-ray section topograph of a crystal of MBA-NP grown from solution.

surface and upon the growth rate (defined by the supersaturation) at which refacetting occurs. Figure 25 shows a section topograph through the centre of a crystal of MBA-NP grown from solution[49]. The seed, visible in the centre of the image, was initially bounded by (001) and larger (100) surfaces formed by perfect and imperfect cleavage, respectively. The irregularity caused by the poor cleavage on (100) is clearly visible in the topograph. During initial growth this irregular surface leads to trapping of solvent, particularly under conditions of rapid growth. As the crystal grows over these inclusions to form planar facets, lattice closure mistakes occur, causing the nucleation of dislocations. The extensive solvent inclusions and resulting high dislocation density are clearly visible. Dislocations produced in this manner are referred to as growth dislocations as distinct from mechanically induced dislocations.

6˙3.4. Fluctuations in growth conditions. Fluctuations in growth conditions, particularly those leading to dissolution followed by rapid growth, generally result in a degradation of perfection. Dissolution typically leads to rounding of the edges of the crystal which upon rapid regrowth can result in the inclusion of solvent. This is an identical situation to the initial refacetting of the seed crystal. Once again the major consequence of solvent inclusions is the nucleation of growth dislocations. Less severe fluctuations in growth rate can give rise to differing levels of incorporation of individual solvent or other impurity molecules. Variations in the degree of impurity incorporation can lead to corresponding changes in lattice parameter. These growth striations are revealed in X-ray topographs as bands of varying contrast parallel to the growth facets of the crystal as shown in fig. 26 for a crystal of pentaerythritol tetranitrate (PETN)[51].

The pronounced improvement in crystal quality with increased control of the growth conditions is well illustrated by the series of topographs of PETN shown in fig. 27[44,51]. Under conditions of spontaneous nucleation and growth by rapid, uncontrolled solvent evaporation a highly strained and defec-

Fig. 26. – Monochromatic X-ray topograph illustrating growth striations in a crystal of PETN.

tive crystal (*a*)) is obtained. The defect density in this case is so high that individual imperfections cannot be resolved in the topograph. Slower growth by solvent evaporation at constant temperature yields a considerable improvement in quality (*b*)). The dislocation density in this case is low enough for individual defects to be detected and quite large volumes of the sample have very low dislocation densities. The fluctuations in growth rate resulting from the poor control achievable using the solvent evaporation method are revealed by the presence of strongly contrasted growth striations. A further reduction in strain and dislocation density, together with the elimination of detectable growth striations, is achieved through slower, well-controlled growth by temperature lowering (*c*)). Much of the sample is essentially free from dislocations and solvent inclusions. There is also a significant reduction in the level of strain associated with the growth sector boundaries.

The same can be seen in the KTP crystal slice depicted in fig. 28. The very few dislocations D in this section spring from the inclusion I which has probably resulted from a growth fluctuation and from impurity inclusions in the growth sector boundary B_1. Otherwise, the section is remarkably free from dislocations for a material prepared under such adverse conditions.

6˙4. *Crystal morphology and growth sector boundaries.* – Any crystal which exhibits facetted growth is divided into separate volumes known as growth sectors, each of which is associated with one of the crystal faces. The boundaries between these volumes, known as growth sector boundaries, chart the position

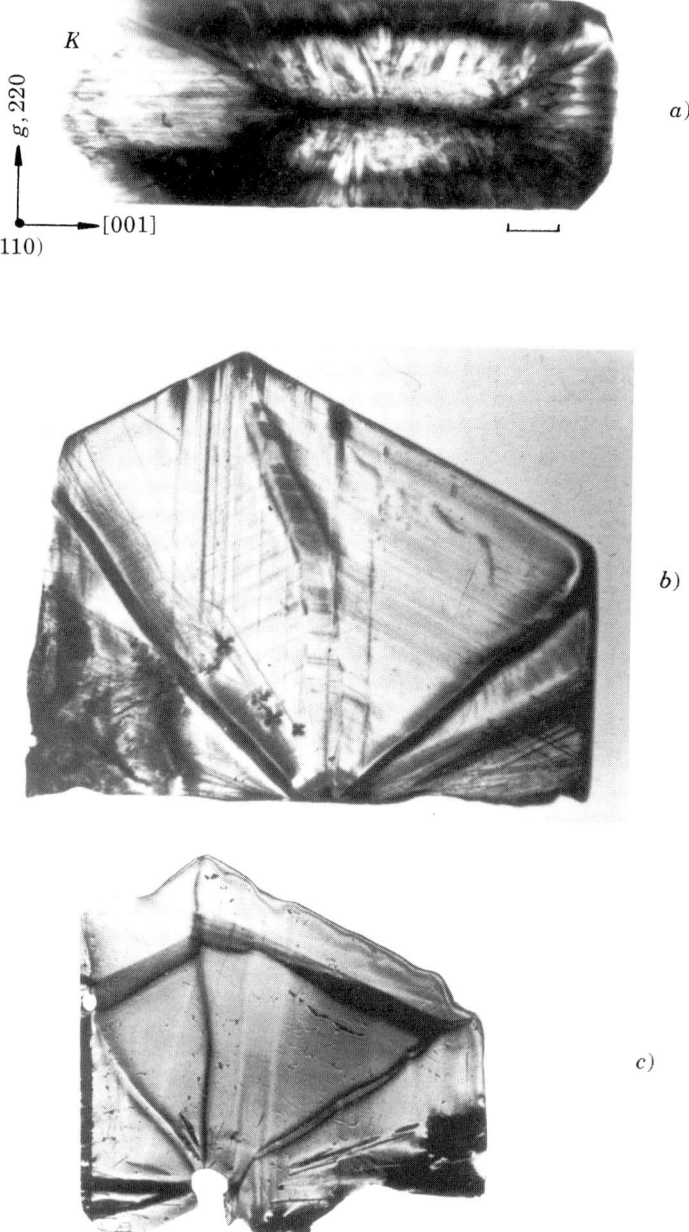

Fig. 27. – Monochromatic X-ray topographs illustrating the increase in perfection result-ing from improved control of growth conditions in the case of growth of PETN from sol-ution: a) uncontrolled nucleation and growth by solvent evaporation, b) slower growth at constant temperature using solvent evaporation and c) careful seeded growth under opti-mum conditions using temperature lowering.

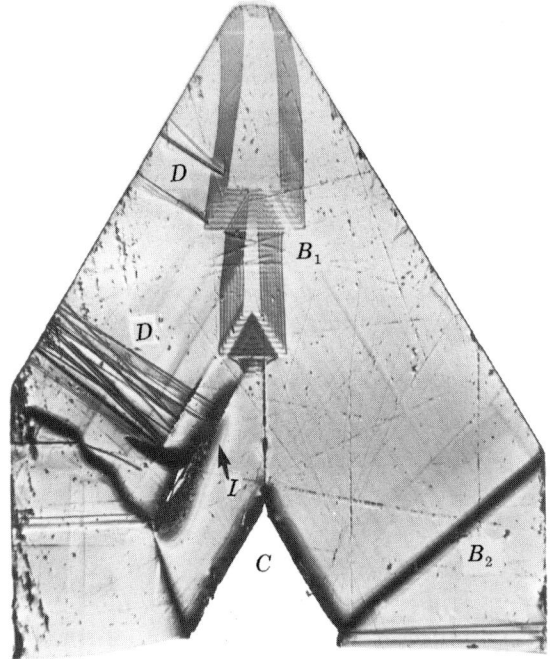

Fig. 28. – An X-ray topograph of a (001) slice of TSSG KTP taken using a 040 reflection at 0.5 Å.

of the crystal edges throughout growth. Two typical growth sector boundaries, B_1 and B_2, can be seen in fig. 28. Growth sector boundaries may comprise discontinuities in the lattice, or may be strained and can often act as preferential sites for impurity incorporation. In addition the lattice may be misoriented either side of such a boundary. Those in the topograph of KTP shown in fig. 28 show well the different characteristics which reflect the degree of strain across the boundary. B_1 exhibits fringe contrast indicating a lower level of strain compared with that of B_2 which shows a much stronger contrast for a high strain variation misorientation. As noted above, the dislocation D probably initiates at impurity clusters trapped in the boundary B_1. These defects are unavoidable in any facetted crystal. There are indications that growth sector boundaries exert significant influences over crystal properties, for example acting as preferential sites for thermal decomposition [53] and affecting the pyroelectric properties of triglycine sulphate (TGS) [54]. It has, however, been demonstrated that slower

Fig. 29. – X-ray section topograph of a higly defective crystal of urea grown from methanol.

growth can reduce the level of strain associated with growth sector boundaries in some instances [53]. Furthermore, since the distribution of growth sector boundaries depends directly upon the external morphology of the crystals, modifications of the latter by solvent [18], impurity [2] or supersaturation effects can yield a more favourable arrangement of growth sector boundaries to enable cutting of samples free from such defects. A similar situation exists with growth dislocations. Since, in facetted crystals, these generally propagate outwards from the centre of the crystal in directions close to the normals to the crystal faces, the final dislocation distribution will exhibit a marked dependence upon crystal morphology.

6˙5. *Growth instability.* – There are a number of solution growth systems which, like growth from the vapour, require rather precise control of supersaturation to achieve high-quality single-crystal growth. One such example is urea which shows considerable instability in the growth along the needle direction of the crystals. If too high a supersaturation is imposed, growth in the needle direction becomes very rapid with extensive inclusion of solvent. This unstable growth is particularly common during refacetting of the seed crystal. Figure 29 shows a section topograph recorded longitudinally through the centre of a urea crystal grown from methanol which illustrates the situation [49]. It is only during the latter stages of growth where the supersaturation is lower that a coherent growth front forms and stable, good-quality growth is established. A similar situation, although less sensitive to supersaturation changes, occurs in the case of MBA-NP as discussed previously.

7. – Conclusion.

The major techniques for the purification, crystal growth and structural characterization of organic crystals have been described. These have been illustrated with examples from a range of organic materials, including a number of technologically important nonlinear optical crystals. Crystals of many of the materials have been grown to large sizes and with exceptionally high structural quality despite numerous difficulties associated with the specific chemical and physical properties of organic materials. In favourable cases dislocation-free samples have been prepared.

The importance of this advanced crystal growth capability is considerable, not only in the context of fundamental research into the properties of organic crystals and their defects, but also for the exploitation of technologically important organic materials such as those for nonlinear optics.

* * *

The work described here represents the accumulated experience of a number of years' work in this area. The work has had the financial support of a number of bodies. Of those we wish particularly to acknowledge the support of the Science and Engineering Research Council and Imperial Chemical Industries.

REFERENCES

[1] D. S. CHEMLA and J. ZYSS: *Non-linear Optical Properties of Organic Molecules and Crystals* (Academic Press, New York, N.Y., 1987).

[2] H. E. BUCKLEY: *Crystal Growth* (Wiley, New York, N.Y., 1951).

[3] R. C. SANGSTER and J. W. IRVINE: *J. Chem. Phys.*, **54**, 670 (1954).

[4] W. G. PFANN: *Zone Melting*, 2nd edition (Wiley, New York, N.Y., 1966).

[5] J. S. SHAH: in *Crystal Growth*, 2nd edition, edited by B. PAMPLIN (Pergamon Press, Oxford, 1980), Chapt. 8, p. 301.

[6] E. F. G. HERINGTON: *Zone Melting of Organic Materials* (Blackwell, Oxford, 1963).

[7] A. CARENCO, J. JERPHAGNON and A. PERIGAUD: *J. Chem. Phys.*, **66**, 3806 (1977).

[8] P. D. SOUTHGATE and D. S. HALL: *Appl. Phys. Lett.*, **18**, 456 (1971).

[9] A. STREITWIESER and C. H. HEATHCOCK: *Introduction to Organic Chemistry* (McMillan, New York, N.Y., 1976), p. 962.

[10] B. F. LEVINE, C. G. BETHEA, C. D. THURMOND, N. T. LYNCH and J. L. BERNSTEIN: *J. Appl. Phys.*, **50**, 2523 (1979).

[11] B. J. MCARDLE, J. N. SHERWOOD and A. C. DAMASK: *J. Cryst. Growth*, **22**, 193 (1974).

[12] J. N. SHERWOOD and S. J. THOMSON: *J. Sci. Instrum.*, **37**, 42 (1960).

[13] J. N. SHERWOOD: in *Organic Materials for Non-linear Optics*, edited by D. BLOOR and R. A. HANN (Royal Society of Chemistry, London, 1988), p. 71.

[14] J. C. BLEAY, R. M. HOOPER, R. S. NARANG and J. N. SHERWOOD: *J. Cryst. Growth*, **43**, 589 (1978).

[15] C. FORNO: *J. Cryst. Growth*, **21**, 64 (1974).

[16] R. M. HOOPER, R. S. NARANG, B. J. MCARDLE and J. N. SHERWOOD: in *Crystal Growth*, 2nd edition, edited by B. PAMPLIN (Pergamon Press, London, 1980), p. 395.

[17] P. J. HALFPENNY, K. J. ROBERTS and J. N. SHERWOOD: *J. Cryst. Growth*, **69**, 73 (1985).

[18] R. J. DAVEY: in *Current Topics in Materials Science*, Vol. 8, edited by E. KALDIS (North-Holland, Amsterdam, 1982), Chapt. 6.

[19] K. A. JACKSON: in *Liquid Metals and Solidification* (American Society of Metals, 1958), p. 174.

[20] P. BENNEMA and G. H. GILMER: in *Crystal Growth: an Introduction*, edited by P. HARTMAN (North-Holland, Amsterdam, 1973), p. 274.

[21] Y. TANAKA and M. MATSUOKA: *J. Cryst. Growth*, **99**, 1130 (1990).

[22] E. E. A. SHEPHERD, J. N. SHERWOOD, G. S. SIMPSON and C. S. YOON: *J. Cryst. Growth*, **113**, 360 (1991).

[23] R. T. BAILEY, F. R. CRUICKSHANK, S. M. G. GUTHRIE, B. J. MCARDLE, H. MORRISON, D. PUGH, E. E. A. SHEPHERD, J. N. SHERWOOD, C. S. YOON, R. KASHYAP, B. K. NAYAR and K. E. WHITE: *Opt. Commun.*, **65**, 229 (1988).

[24] W. W. PIPER and S. J. POLICH: *J. Appl. Cryst.*, **32**, 1278 (1961).

[25] G. J. SLOAN: *Mol. Cryst.*, **2**, 323 (1967).

[26] B. HONIGMAN: *Z. Elektrochem.*, **58**, 322 (1954).

[27] B. HONIGMAN and H. HEYER: *Z. Kristallogr.*, **106**, 199 (1955).

[28] N. KARL: in *Crystals, Growth Properties and Applications*, Vol. 4, edited by H. C. FREYHARDT (Springer-Verlag, Berlin, 1980), p. 1.

[29] N. KARL: *Mol. Cryst. Liq. Cryst.*, **171**, 157 (1989).

[30] S. GILMOUR, E. E. A. SHEPHERD, G. S. SIMPSON and J. N. SHERWOOD: unpublished work.

[31] S. R. HALL, P. V. KOLINSKI, R. JONES, S. ALLEN, P. GORDON, B. BOTHWELL, D. BLOOR, P. A. NORMAN, M. HURSTHOUSE, A. KARAULOV, M. GOODYEAR and D. BISHOP: *J. Cryst. Growth*, **79**, 745 (1986).

[32] R. J. DAVEY, P. T. CARDEW, D. McEWAN and D. E. SADLER: *J. Cryst. Growth*, **79**, 648 (1986).

[33] L. K. CHENG, J. D. BIERLEIN and A. A. BALLMAN: *J. Cryst. Growth*, **110**, 697 (1991).

[34] A. A. BALLMAN, H. BROWN, D. H. OLSEN and C. E. RICE: *J. Cryst. Growth*, **75**, 390 (1986).

[35] A. YOKOTANI, A. MIYAMOTO, T. SASAKI and S. NAKAI: *J. Cryst. Growth*, **110**, 963 (1991).

[36] J. N. SHERWOOD: in *Point Defects in Solids*, edited by J. H. CRAWFORD jr. and L. SLIFKIN, Vol. 2 (Plenum, New York, N.Y., 1975), p. 441.

[37] A. I. KITIAGORODSKII: *Mixed Crystals* (Springer-Verlag, Berlin, 1984).

[38] N. KARL and K. H. PROBST: *Mol. Cryst. Liq. Cryst.*, **11**, 155 (1970).

[39] H. KLAPPER: in *Characterisation of Crystal Growth Defects by X-ray Methods*, edited by B. K. TANNER and D. K. BOWEN (Plenum, New York, N.Y., 1980), p. 133.

[40] T. WATANABE and K. IZUMI: *J. Cryst. Growth*, **46**, 747 (1979).

[41] S. AMELINCKX: *Solid State Phys.*, **6** (Supplement) (Academic Press, New York, N.Y., 1969).

[42] B. J. McARDLE and J. N. SHERWOOD: in *Advanced Crystal Growth*, edited by P. DRYBURGH, B. COCKAYNE and K. C. BARRACLOUGH (Prentice-Hall, New York, N.Y., 1987), p. 179 and references therein.

[43] M. LEONA, P. J. HALFPENNY and J. N. SHERWOOD: unpublished work.

[44] P. J. HALFPENNY, K. J. ROBERTS and J. N. SHERWOOD: *J. Mater. Sci.*, **19**, 1629 (1984).

[45] A. R. LANG: in *Modern Diffraction and Imaging Techniques in Material Science*, edited by S. AMELINCKX (North-Holland, Amsterdam, 1978), p. 623.

[46] B. K. TANNER: *X-ray Diffraction Topography* (Pergamon, London, 1976).

[47] B. K. TANNER and D. K. BOWEN: *Characterisation of Crystal Growth Defects by X-ray Methods* (Plenum, New York, N.Y., 1980).

[48] A. R. LANG: *Acta Metall.*, **5**, 358 (1957).

[49] P. J. HALFPENNY and J. N. SHERWOOD: *Philos. Mag. Lett.*, **62**, 1 (1990).

[50] K. J. ROBERTS, J. N. SHERWOOD, D. K. BOWEN and S. T. DAVIES: *Mater. Lett.*, **2**, 104 (1983).

[51] P. J. HALFPENNY, J. N. SHERWOOD and G. S. SIMPSON: unpublished work.

[52] J. N. SHERWOOD and C. S. YOON: unpublished work (1991).

[53] P. J. HALFPENNY, K. J. ROBERTS and J. N. SHERWOOD: *J. Cryst. Growth*, **67**, 202 (1984).

[54] P. J. HALFPENNY: Ph. D. Thesis, University of Strathclyde (1982).

Nonlinear Optical Side-chain Polymers for Electro-Optic Switching: Materials, Poling, Test Devices.

G. R. Möhlmann

Akzo Nobel Electronic Products b.v.
P.O. Box 9300, 6800 SB Arnhem, The Netherlands

C. P. J. M. van der Vorst

Akzo Nobel Central Research, Applied Physics Department
P.O. Box 9300, 6800 SB Arnhem, The Netherlands

1. – Introduction.

Generally, there is a worldwide tendency to replace or complement electronic functions, processes and systems by photonic alternatives. Photons are taking over the role of electrons as the active particle or carrier with respect to data transmission and data processing. This is mainly due to limitations currently encountered in purely electronic circuitry and transmission systems. These limitations comprise issues such as transmission or processing bandwidth, speed, power consumption and packing densities. Examples of the introduction of photonics in the previously exclusive world of electronics are, among others, optical communications (optical fibres), optical storage (compact discs) and optical interconnects in and between computers.

The photonics approach and photonic solutions of problems require novel ways of thinking about basic concepts and corresponding architectures. In addition, novel (electro-) optical systems, subsystems, components and finally novel (electro-) optical materials are required. This lecture mainly deals with a novel class of photonic materials in the form of nonlinear optical polymers, also called photonic polymers.

Historically, the electronic industry was used to applying inorganic single crystalline materials for their optical and electro-optical components. However, these materials are difficult to make (grow) and difficult to process if thin films are required. In the 1970's, attention was drawn towards organic materials, first crystals and later on polymers, because these materials showed attractive nonlinear optical properties which are required for applications in the field of photonics.

To fulfil the needs for high-speed and high-capacity information technology, vast research programs at universities, national laboratories and industrial laboratories have been initiated to develop materials, components and systems suitable for these applications. In the case of materials for optical switching and data storage, both organic- and inorganic-material developments are still pursued. Organic materials, however, might provide the systems developer with additional and unique features including, among others, larger degrees of flexibility with respect to architecture, design and processing as compared to inorganic materials[1-6]. Attractively high nonlinear coefficients appear to be achievable, but perhaps of even larger importance is the relatively easy way to process organic polymeric materials into practical devices[7,8].

Contrary to the case of inorganic single crystals, in the case of organic nonlinear materials, the optically nonlinear effects are of a purely electronic-displacement nature (electron polarization) and can be quite fast, owing to the small mass of the electrons that are displaced by external electric or optical fields. In inorganic materials, besides rapid electron displacements, slower (heavy)-ion polarization contributions to the nonlinear effect may occur as well, thus slowing down the overall speed of the practical effect. The purely electronic displacement effect in organics ensures, in combination with a low dielectric constant, the very-high-speed operation features required for future communication systems.

This lecture deals with the development of optically nonlinear (also called electro-optically active, or hyperpolarizable) organic molecules and of the corresponding side-chain polymers incorporating these nonlinear molecules. The specific properties of such materials will be discussed in detail, especially with respect to the processes related to the induction of the required polar order in thin films of such polymers as well as the (undesired) relaxation of this orientation as a function of time. In addition, since the development of photonic polymers is dedicated to applications in nonlinear optical devices, the materials evaluation includes the testing in prototype waveguide devices such as polymeric modulators and switches. The performance of several prototypes of such polymeric passive as well as actively switching or modulating devices will be discussed in this lecture.

2. – Nonlinear optical materials.

2˙1. *Macroscopic optical nonlinearity.* – Nonlinear optical (NLO) materials with a large second-order NLO susceptibility $\chi^{(2)}$ will possibly be used in the ultrafast optical switches that are needed in future telecommunication networks.

The NLO susceptibilities $\chi^{(2)}$, $\chi^{(3)}$, etc. are defined by the general relationship between the components of the induced polarization density P (at angular frequency ω) and those of the electric field(s) E (at angular frequencies $\omega_1, \omega_2, \omega_3, \ldots$) [1-4, 7-13]:

$$(1) \quad P(\omega)_I = \chi^{(1)}(-\omega; \omega_1)_{IJ} E(\omega_1)_J + \chi^{(2)}(-\omega; \omega_1, \omega_2)_{IJK} E(\omega_1)_J E(\omega_2)_K +$$

$$+ \chi^{(3)}(-\omega; \omega_1, \omega_2, \omega_3)_{IJKL} E(\omega_1)_J E(\omega_2)_K E(\omega_3)_L + \ldots .$$

Summation over identical indices is assumed throughout this lecture (Einstein convention). $I, J, K = X, Y$ or Z are in the macroscopic frame of reference. If P was to represent the total polarization density, a zeroth-order term in the field strength should be included in eq. (1), corresponding to the permanent polarization density. For having even-order (like second or zeroth order) terms, the material must be noncentrosymmetric; for odd-order terms no such strict symmetry requirement exists. In general, the series expansion in eq. (1) converges fast, implying that, *if* the material is noncentrosymmetric, then the dominant nonlinear term is the $\chi^{(2)}$ term. The linear susceptibility $\chi^{(1)}$ is related to optical refraction and absorption. The most common effects due to $\chi^{(2)}$ are frequency doubling ($\chi^{(2)}(-2\omega; \omega, \omega)$) and the Pockels effect ($\chi^{(2)}(-\omega; \omega, 0)$) [1, 11]. These are both special cases of three-wave mixing. The $\chi^{(3)}$ term leads to four-wave mixing with, *e.g.*, frequency tripling and the optical Kerr effect as special cases. The two mentioned $\chi^{(2)}$ effects are of technological importance in the near future. Frequency doubling becomes important in the field of optical data storage, where it will allow high-density storage using the frequency-doubled light from cheap laser diodes. The Pockels effect forms the basis for electro-optic switching.

2'2. *The electro-optic effect.* – The electro-optic (EO) effect is defined as the dependence on the field strength of a material's dielectric impermeability ε^{-1} when the material is subjected to an external electric field E. In a series expansion, one can write

$$(2) \quad \delta(\varepsilon^{-1})_{IJ} = r_{IJK} \cdot E_K + s_{IJKL} \cdot E_K E_L + \ldots .$$

The first term in eq. (2) corresponds to the linear EO effect, or Pockels effect, where the change of the impermeability is a linear function of the field strength [13, 14]. The tensor r is defined as the Pockels coefficient. The second term in eq. (2) corresponds to the quadratic EO effect or Kerr effect. The linear EO effect is a direct consequence of a material's second-order NLO properties, whereas the quadratic EO effect is a consequence of the third-order nonlinearity. Attention will be focussed on the noncentrosymmetric materials, where the dominant term in eq. (2) is the linear term.

For small changes of diagonal components of the impermeability tensor, these changes are linearly proportional to changes of the corresponding refractive indices. Choosing the principal axes of the indicatrix, the Pockels effect can be rewritten in terms of the refractive-index changes as[14]

$$(3) \qquad\qquad \delta n_I = -\frac{1}{2}\, n_I^3 \cdot r_{IIK} \cdot E_K\,.$$

The off-diagonal terms of the impermeability tensor correspond to polarization rotation and will not be discussed further in detail. Because r_{IJK} is invariant under permutation of the first two indices, often a condensed notation is used in which r has only two indices. The first index represents the combination IJ and may have the values $1 = XX$, $2 = YY$, $3 = ZZ$, $4 = YZ = ZY$, $5 = ZX = XZ$, $6 = XY = YX$. The second index is K. Instead of labelling the coordinate axes with X, Y and Z, often the notation $1(= X)$, $2(= Y)$ and $3(= Z)$ is used, so $K = 1, \ldots, 3$.

It can be derived that the refractive-index change caused by the Pockels effect is related to the susceptibility $\chi^{(2)}$ by

$$(4) \qquad\qquad \delta n_I = (\chi^{(2)}_{IIK}/n_I) \cdot E_K\,,$$

so that r and $\chi^{(2)}$ are related by the following equation[3, 13]:

$$(5) \qquad\qquad \chi^{(2)}_{IIK} = -\frac{1}{2}\, n_I^4 \cdot r_{IIK}\,.$$

The refractive-index change of a material caused by the Pockels effect can be converted in several ways into a change of the intensity of the light passing through the material. Such a conversion is important for the use of a $\chi^{(2)}$ material in an EO switch, but it also enables measurement of the Pockels coefficient. Different principles or mechanisms can be applied for the conversion of the EO modulation of the refractive index (phase modulation) into intensity modulation, *e.g.* interference between two legs of a Mach-Zehnder interferometer[14, 15], interference between multiple reflected rays[16-18] or use of a polarization analyser in case of relative phase modulation[15, 16]. See also subsect. 4˙5.

3. – Organic nonlinear optical materials.

3˙1. *Advantages of organic materials.* – Organic materials have distinct advantages over inorganic materials for optical-switching applications[1, 2, 9, 19, 20]. Electro-optic coefficients up to $34\,\mathrm{pm/V}$ have been reported in poled polymeric films and devices, which compares favourably to competing technologies such as lithium niobate, with the highest EO coefficient (r_{51}) of $32\,\mathrm{pm/V}$. The values mentioned for polymers cannot be regarded as the upper limit. Enhancement of the EO coefficients can be realized by modifications of the NLO side-groups, increase of the NLO density and/or optimization of the poling process of the polymeric materials.

The electronic nature of the nonlinear effects guarantees a practically unlimited switching speed. As far as integrated EO switches are concerned (in contrast to all-optical switching), the maximum bandwidth is usually dominated by the dielectric constant of the electro-optic material rather than by the intrinsic upper speed limit. At very high speeds, the driving electric field should be phase-matched to the optical wave in the device. The phase velocity difference for the optical and electric fields should be minimized, *i.e.* the dielectric constant of the material and hence the electrical capacity should be as low as possible. In the case of the well-known inorganic crystal $LiNbO_3$, with a permittivity of approximately 30, the maximum length·bandwidth product of an electro-optic switch is 10 GHz · cm [20]. Using polymers, with dielectric coefficients typically in the order of 3 to 3.5, a length · bandwidth product of over 20 GHz · cm has already been demonstrated [21]. The calculated maximum value for polymeric materials is around 120 GHz · cm [20].

Apart from speed considerations, polymeric materials offer more interesting properties. Especially the processing of polymers into integrated optic devices is very attractive, since it requires no high-temperature processing. For this reason, it can be anticipated that polymeric optical devices can be integrated with electronic circuitry. The feasibility of this concept has already been demonstrated by DIEMEER *et al.* [22], who integrated a polymeric waveguide structure with a laser diode on a GaAs substrate.

3˙2. *Molecular optical nonlinearity.* – The macroscopic optical nonlinearities ($\chi^{(2)}$, $\chi^{(3)}$, ...) of solid organic substances are based on the corresponding nonlinearities on a molecular level. The molecular analogues of the macroscopic $\chi^{(2)}$ and $\chi^{(3)}$ are called hyperpolarizability (β) and second hyperpolarizability (γ), respectively. These molecular properties are defined by the relationship between the components of the induced molecular dipole moment p and the components of the *local* electric field(s) E', in a way completely analogous to eq. (1):

$$(6) \quad p(\omega)_i = \alpha(-\omega; \omega_1)_{ij} E'(\omega_1)_j + \beta(-\omega; \omega_1, \omega_2)_{ijk} E'(\omega_1)_j E'(\omega_2)_k +$$

$$+ \gamma(-\omega; \omega_1, \omega_2, \omega_3)_{ijkl} E'(\omega_1)_j E'(\omega_2)_k E'(\omega_3)_l + \dots .$$

Here, α is the linear polarizability. The indices $i, j, k = x, y, z$ are the Cartesian coordinates in the molecular frame, whereas $I, J, K = X, Y, Z$ are in the macroscopic frame. Like in eq. (1), a zeroth-order term in the field strength (the permanent dipole moment μ_0) can be added in eq. (6), if p is to represent the total dipole moment.

Well-known hyperpolarizable molecules such as para-nitro-aniline (pNA) or 4-dimethylamino-4'-nitrostilbene (DANS) are of the donor-acceptor type. They have a strongly polarizable conjugated π-electron system, such as a stilbene group, which is most extended in only one direction; the molecule has a long axis. Consequently, the linear polarizability is anisotropic. Along the long axis, the π-electron system is asymmetrically substituted by an electron donor D at one end and an acceptor A at the other end, thus breaking the centrosymmetry of the unsubstituted π-system. The donor-acceptor nature of the so-called DπA molecules causes both a nonzero permanent dipole moment μ_0 (by definition in the z direction of the molecular frame of reference) and a large hyperpolarizability component (β_{zzz}) along μ_0. The permanent dipole moment points approximately from A to D and is, therefore, approximately along the long axis.

Many AπD groups can be approximated as molecules that are rotationally invariant along their long axis, or quasi-one-dimensional molecules, implying that only the dominant β_{zzz} component is taken into account, whereas the other components are neglected (quasi-one-dimensional or «β_{zzz}-only» approximation).

3˙3. *Relation between* $\chi^{(2)}$ *and* β *in organic materials.* – The macroscopic nonlinear susceptibilities (defined per unit volume) are composed as a sum of all corresponding molecular contributions (also per unit volume), each molecular component being mapped onto the appropriate macroscopic unit vectors. For the second-order terms, the relation can be written as follows:

$$(7) \quad \chi^{(2)}(-\omega;\, \omega_1,\, \omega_2)_{IJK} = Nf(\omega)f(\omega_1)f(\omega_2)\beta(-\omega;\, \omega_1,\, \omega_2)_{ijk} \cdot \langle O_{Ii}O_{Jj}O_{Kk} \rangle .$$

Here, N is the number density of hyperpolarizable molecules. The O's are projections of the molecular axes onto the macroscopic frame and the brackets indicate an averaging over all molecular orientations weighted by the orientational distribution function. The $f(\omega_i)$ terms are local-field correction factors[1,11,13], which relate the local fields E' to the applied external fields E. Depending on the frequency of the fields involved, these are usually approximated by either Lorenz-Lorentz-type corrections (optical frequencies) or Onsager-type corrections (quasi-static fields). In a more careful analysis, local-field correction factors are included inside the averaging brackets[23]. We refer to ref.[24] for an extensive discussion of local-field effects in NLO materials and to the textbooks of BÖTTCHER[25] for a general discussion of the local-field problem.

Equation (7) shows that a nonzero value for $\chi^{(2)}$ in organic materials requires the presence of noncentrosymmetric molecules in a noncentrosymmetrical orientational distribution. The noncentrosymmetry on a molecular level is necessary for β to be nonzero. The noncentrosymmetry of the orientational distribution is necessary to obtain a nonzero value for $\langle O_{Ii}O_{Jj}O_{Kk} \rangle$. This requirement means that the AπD molecules should have the so-called polar order—not to be confused with axial order.

3˙4. *Definition of polar order and axial order.* – Polar order of AπD molecules is the degree of alignment of the AπD's dipole moment—a vector or first-rank tensor—along the direction of another vector. For poled polymers (subsect. 3˙5) this second vector is the poling field. Axial order is the degree of aligment of the AπD molecule's long axis—a second-rank tensor—along another axis, the so-called director $\tilde{n}$. For poled polymers, this axis is parallel to the poling-field direction. In the case of polar order, the orientational distribution function is noncentrosymmetric and the «up-down symmetry is broken», *viz.* more molecules are directed «upward» than «downward». In the case of axial order, the up-down symmetry need not be broken and the orientational distribution function may remain centrosymmetric.

The orientational distribution function $G(\theta)$ is defined in such a way that $G(\theta)\sin\theta\,d\theta = -G(\theta)\,d(\cos\theta)$ represents the number of molecules with a polar angle θ between θ and $\theta + d\theta$. $G(\theta)$ can be developed in a series expansion of Legendre polynomials $P_l(\cos\theta)$ of order l with coefficients $\langle P_l(\cos\theta)\rangle$. Odd-order $\langle P_l\rangle$ represent polar order, even-order $\langle P_l\rangle$ represent axial order. Polar-order parameters that are often used in NLO literature are $\langle\cos^3\theta\rangle$ and $\langle(1/2)\cos\theta\sin^2\theta\rangle$. These order parameters—defined in subsect 4˙3—are linear combinations of the more general polar-order parameters $\langle P_1\rangle$ and $\langle P_3\rangle$, *viz.* $\langle\cos^3\theta\rangle = [3\langle P_1\rangle + 2\langle P_3\rangle]/5$ and $\langle(1/2)\cos\theta\sin^2\theta\rangle = [\langle P_1\rangle - \langle P_3\rangle]/5$. The most important axial-order parameters are $\langle P_2\rangle$ and $\langle P_4\rangle$.

Although the Legendre functions, in particular the ones describing polar order and the ones describing axial order, are independent (orthogonal) functions, this does not mean that polar order and axial order are independent. In practice, polar order is always accompanied by some amount of axial order. The reverse is not true: axial order may exist without any polar order present. Polar- and axial-order parameters often show correlation. If polar order is not present spontaneously, but is induced by an electric field, axial order may enhance the induced polar order. All this can be explained by considering the energy terms leading to axial and/or polar order. More about this subject can be found in sect. 5 and in ref.[26].

3˙5. *Methods for obtaining polar order.* – The presence of hyperpolarizable AπD molecules will only lead to a nonzero value for $\chi^{(2)}$, if the molecules can be polarly aligned (see eq. (7)). In many cases, thermal motion and antiparallel molecular association between permanent molecular dipole moments leads to a centrosymmetric orientational distribution, and thus to $\chi^{(2)} = 0$. A number of methods are available to obtain a noncentrosymmetric orientational distribution of the hyperpolarizable molecules. The main routes are:

1) growth of noncentrosymmetric single crystals[4,10],

2) Langmuir-Blodgett techniques[27,28],

3) electric-field alignment of AπD molecules in polymers[1,2,7,9,11-13,19,22,29-31].

The first route has the advantages of high NLO density (number density N

of nonlinear optical molecules), high stability and high homogeneity (important for frequency doubling). Disadvantages are the low chance for finding a new material that crystallizes in a noncentrosymmetrical crystal structure, the tedious growth of single crystals of sufficient size and the difficulty to process crystals into usable devices.

The second route, although already several decennia old, has regained attention because of its ability to obtain nice polar order in monolayers which can be stacked on top of each other. Progress in this field is hampered by loss of order after stacking many (double) layers and by dust problems, leading to high optical losses in waveguides.

In the third route, the AπD molecules are brought into a polymer, either as a dopant (up to a few tens of percents by weight) or chemically attached. The permanent dipole moments of the donor-acceptor molecules (and, with them, the dominant component of the hyperpolarizability) are oriented by applying a very strong electric field while the polymer is heated up to its glass transition temperature. The field-induced order is subsequently «frozen in» by cooling well below the glass transition temperature, where the polymer is in the glassy state. It has been demonstrated that this last route is very promising for EO switching. Several devices have been realized using poled polymers. Realization and performance will be discussed in sect. 8.

In the following section, NLO side-chain polymers will be discussed in which the AπD groups are attached as pendant groups to a polymer backbone. Also, the poling step will be discussed in more detail.

4. – Nonlinear optical side-chain polymers.

4'1. *General structures.* – NLO side-chain polymers can be represented by a structure consisting of three principal building blocks: a polymer backbone, a pendant NLO side-group and a spacer connecting the previous two components. Figure 1 shows typical examples of NLO side-chain polymers.

The properties of this type of NLO polymers can be altered by changing at least one of the three building blocks. The actual NLO effects of the polymer depend primarily on the NLO side-group, since this limits the maximum obtainable NLO coefficient. However, the stiffness of backbone or spacer group influences the rotational freedom of the NLO group, and consequently has an effect on the maximum polar order after a poling cycle.

The thermal and mechanical properties of the NLO polymer depend on all three components. The backbone is especially important for the glass transition temperature T_g of the polymer. Acrylates exhibit a T_g in the order of 80 °C, but other polymer systems exist which have very high T_g's. A well-known example is polyimide, with T_g's up to 400 °C. Apart from backbone effects, the T_g of a side-chain polymer also depends on the spacer and side-groups. Flexible long

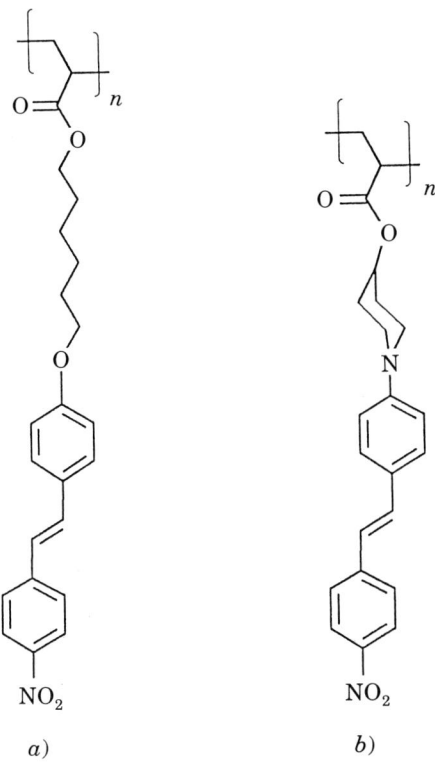

Fig. 1. – Chemical structures of two NLO side-chain polymers: *a*) backbone: acrylate; spacer: C6 aliphatic chain; NLO side-group: 4-methoxy-4′-nitro-stilbene (MONS); *b*) backbone: acrylate; spacer: 6-ring; NLO side-group: 4-amino-4′-nitro-stilbene (DANS).

spacers will in general lead to a lower T_g, and very short or rigid spacers will result in a higher T_g.

The optimization of NLO side-chain polymers for a specific application is a time-consuming process, since not only the thermal and mechanical properties depend on the building blocks, but the optical properties such as refractive index, dispersion and NLO effects depend on the specific combination too. When MONS or DANS type of side-groups are incorporated, these will dominate the optical properties mentioned above. If groups with lower values of the first- and second-order polarizabilities are used, the backbone and the spacer will have an increased influence on the optical properties. For instance, aromatic species will lead to a higher refractive index than aliphatic species in backbone or spacer.

A final aspect of the NLO side-chain polymers is related to their processability. The processability is a key item for the introduction of NLO polymers. Compared to competing technologies in the field of integrated optics, polymer processing is the simplest way to produce devices. In sect. 7 the processing technology will be discussed, but at this point it is important to mention a few characteristics. Polymeric-thin-film deposition requires polymer solutions suited for spin coating or dipping. Moreover, deposited films should withstand other technological procedures required to realize an integrated optic device. This implies chemical resistance to several acids and solvents, no serious outgassing in high-vacuum vessels and adherence to several types of substrates. These properties again are related to the choice of the constituting building blocks of the NLO side-chain polymer and complicate the development of practical classes of these materials.

4'2. *The poling procedure.* – In order to align the permanent molecular dipole moments of the AπD moieties, the polymer has to be exposed to an external electric d.c. field at a temperature at which the dipoles have enough freedom to rotate. Usually this is around the polymer's glass transition temperature T_g. Since at the end of the poling procedure the poling field has to be switched off at room temperature, without any risk of appreciable relaxation, this T_g must be well above ambient temperature. To pole a polymer, it must be brought in the electric field between a pair of electrodes and heated at the same time. Different electrode systems can be used, rigid electrodes, thin metal films, free-standing electrodes or electrodes attached to the sample. Different configurations are possible, *e.g.* coplanar, interdigital or plane parallel. Instead of using metal electrodes, also corona poling is possible. Thin polymer films sandwiched between a pair of plane-parallel metal electrodes on a flat substrate will be considered here (fig. 2). Such polymer samples are very useful for fundamental studies of the poling and relaxation processes. Actual waveguiding

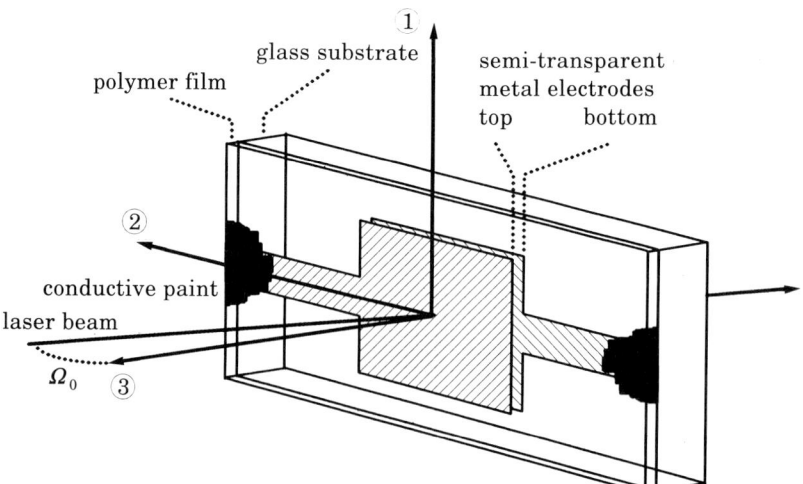

Fig. 2. – The polymer test sample between plane-parallel, semi-transparent metal electrodes. Unit vectors of the coordinate system are shown.

devices only differ from this more simple sample by the use of buffer layers and the patterning of core layers and electrodes. Heating of the samples is easy, *e.g.* using a hot stage or oven.

The poling procedure consists of a temperature profile $T(t)$ (with T the temperature and t the time) and a field profile $E(t)$. For the poling of linear side-chain polymers, these profiles can be kept quite simple. The sample is heated to a maximum poling temperature T_p, which is maintained during poling. At T_p the poling field is switched on either abruptly or (preferably) gradually. Poling is continued during a certain time span t_p, after which the sample is cooled down to room temperature. Then the field is switched off. Practical poling temperatures range from T_g to about $10\,°C$ below T_g. The ultimate value of the Pockels coefficients has been found to hardly differ in this region. The only important temperature dependence is in the time it takes to arrive at this ultimate value. Therefore, t_p has to be varied with T_p. Typical values for t_p in our polymers are from a few minutes at T_g to a few hours at $10\,°C$ below T_g.

4'3. *Relation between $\chi^{(2)}$ and β in poled polymers.* – Figure 2 shows a simple polymer test sample used in our laboratories for EO pre-screening tests in newly developed polymers and for more fundamental poling and relaxation studies in much used polymers. The test sample consists of a polymeric thin film between plane-parallel metal electrodes on a plane substrate (*e.g.*, glass or silicon). The polymeric film is formed by spin coating from a solution, the electrodes are vacuum evaporated. Ignoring side effects due to the finite dimensions of the sample, the poling field is homogeneous and the symmetry of the sample is $C_{\infty v}$. The number of different, nonzero $\chi^{(2)}(\omega; \omega_1, \omega_2)$ components IJK is restricted by symmetry to four. Defining the Z or 3 unit vector of the macroscopic frame of reference along the poling field perpendicular to the film, these components are: ZZZ, $XXZ = YYZ$, $XZX = YZY$ and $ZXX = ZYY$. For the Pockels effect, the first two indices I and J of $\chi^{(2)}(\omega; \omega, 0)$ may be interchanged, so the number of different nonzero components IJK becomes three: ZZZ, $XXZ = YYZ$ and $XZX = YZY = ZXX = ZYY$. If dispersion is also neglected, this number further reduces to two: ZZZ and $XXZ = YYZ = XZX = YZY = ZXX = ZYY$.

Neglecting dispersion and using the quasi-one-dimensional approximation of the AπD molecules (β_{zzz} only), the relationship(s) between macroscopic and molecular second-order NLO properties (eq. (7)) in polymer samples of $C_{\infty v}$ symmetry reduces to the following two relations:

(8)
$$\chi^{(2)}_{ZZZ} = NF\beta_{zzz}\langle\cos^3\theta\rangle,$$

(9)
$$\chi^{(2)}_{XXZ} = \chi^{(2)}_{YYZ} = \chi^{(2)}_{XZX} = \chi^{(2)}_{YZY} = \chi^{(2)}_{ZXX} = \chi^{(2)}_{ZYY} = NF\beta_{zzz}\langle(1/2)\cos\theta\sin^2\theta\rangle.$$

Here, θ is the angle between the permanent dipole moment and the electric poling field. The expressions in brackets describe the degree of polar order obtained during the poling process («polar-order parameters»). All corrections for local-field effects are put in the factor F.

An optimum value for a $\chi^{(2)}$ component requires optimization of N, F, β_{zzz} and the polar-order parameters $\langle \cos^3 \theta \rangle$ or $\langle (1/2) \cos \theta \sin^2 \theta \rangle$. The local-field factors F can hardly be influenced for the AπD materials considered here. Segregation effects often limit the number density N of hyperpolarizable molecules in solid solutions [2,3,9]. Much higher concentrations are feasible by attaching the AπD groups as side-chains to a polymer backbone. The hyperpolarizability can be increased by increasing the length of the π-electron system or by increasing the donating or accepting «strength» of the D and A groups. However, there is a trade-off between hyperpolarizability and optical absorption, since the AπD group usually dominates the optical absorption of such polymer systems [32,33].

Generally, $\langle \cos^3 \theta \rangle$ is larger than $\langle (1/2) \cos \theta \sin^2 \theta \rangle$ (see sect. 5). So, the most efficient EO modulation is obtained using the Pockels coefficient r_{33} (using the same pair of electrodes for poling and for modulation). In this case $\langle \cos^3 \theta \rangle$ must be optimized. Factors that influence polar order, especially $\langle \cos^3 \theta \rangle$, can be found in sect. 5.

4'4. *The Pockels effect in poled polymers.* – Consider the sample between plane-parallel electrodes shown in fig. 2. If the same pair of electrodes that was used for poling (by definition in Z or 3 direction) is also used for electro-optic modulation, the modulating field is in the 3 direction. Applying a voltage V over the electrodes at a separation d, the strength of the modulating field is $E_3 = V/d$. The variations δn of the refractive index (eq. (3)) can be written as

(10) $$\delta n_e = - n_e^3 r_{33} E_3 / 2 \,,$$

(11) $$\delta n_o = - n_o^3 r_{13} E_3 / 2 \,.$$

Because of the symmetry of the sample, the indicatrix (refractive-index ellipsoid) is uniaxial with the unique (here slow) axis n_e (extra-ordinary index) in the poling-field direction and fast axes n_o (ordinary index) in the plane of the film. In a poled sample, the birefringence $\Delta n = n_e - n_o$ is nonzero because of the induced order. Since n_e and n_o are modulated to a different extent, also the birefringence is modulated in a modulating field (relative phase modulation). The relationship between the $\chi^{(2)}$ components and the Pockels coefficients, as given by eq. (5), can be rewritten as

(12) $$\chi^{(2)}_{ZZZ} = - n_e^4 r_{33} / 2 \,,$$

(13) $$\chi^{(2)}_{XXZ} = - n_o^4 r_{13} / 2 \,.$$

4'5. *Measurement of the Pockels effect in poled-polymer test samples.* – In the sample geometry as depicted in fig. 2, the Pockels coefficients can be measured in transmission, in which case the electrodes have to be semi-transparent. With gold, silver or aluminium electrodes of a few tens of nanometres thickness, a good compromise can be obtained between electrical conductivity (square resistance of a few Ω) and optical transmission ((10–40)%). Waveguiding experiments are also possible, but then optical buffer layers have to be used between NLO core layer and metal electrodes (see sect. 7 and 8). This section is restricted to transmission experiments, of which a number of types can be used: crossed-polarizer measurements [15,16], angle-tuned Fabry-Perot measurements [16,17] and Mach-Zehnder interferometric measurements [14,15]. The first two measurements can be performed at the same sample spot, using a common experimental set-up [16] shown in fig. 3.

Because of the high reflectivity of the metal electrodes, the sample forms a Fabry-Perot interferometer or etalon. The tuned Fabry-Perot type(s) of measurement is (are) based on interference of multiple reflected rays, converting EO modulation of the phase difference between different rays into intensity modulation. In the angle-tuned Fabry-Perot measurement [16,17], the angle of incidence is fine-tuned to give maximum intensity modulation. In the similar wavelength-tuned Fabry-Perot method [18], maximum modulatiom is obtained by tuning the wavelength of the incident light. With the laser beam polarized in the plane of the film, we thus observe modulation of n_0 via the Pockels coefficient r_{13} (eq. (11)).

In the crossed-polarizer method, modulation of the birefringence is first converted into a modulation of the polarization state of the transmitted light. This is only possible if the incident, linearly polarized light has components both par-

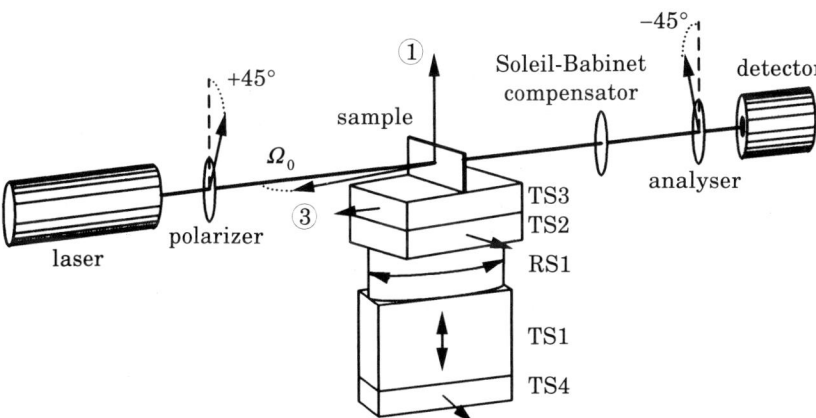

Fig. 3. – Crossed-polarizer set-up with the polymer test sample between plane-parallel electrodes. Modulation of the birefringence is converted into intensity modulation of the laser beam. Translation stages TS1-TS4 and rotation stage RS1 are for sample adjustment. By rotating the polarizer to $0°$, and removing the analyser and compensator, the sample can be used as a tunable Fabry-Perot etalon [16].

allel and perpendicular to the 3 direction. Therefore, the sample is rotated so that the angle of incidence is not zero. Also, the initial polarization direction is not taken parallel to the rotation axis of the sample. The angle of incidence is fine-tuned so that the transmission is at a maximum and interferometric intensity modulation is suppressed[15, 16]. Next, the modulation of the polarization state is converted into intensity modulation by a polarization analyser. The measured electro-optic effect is enhanced, because the light has traversed the polymer film several (f) times. The enhancement factor f can be measured[15, 16].

In the Mach-Zehnder interferometric experiment, the sample is placed in one arm of a Mach-Zehnder interferometer. With the laser beam polarized in the plane of the film, the electro-optic modulation of n_o through r_{13} is converted into intensity modulation by interference between the beams that have passed through the two arms of the interferometer. Since the sample by itself is also an interferometer, any Fabry-Perot interferometric intensity modulation must be suppressed. This is accomplished by fine-tuning the angle of incidence and/or the chosen sample spot, making use of a small variation of the layer thickness over the sample. The correction factor f has also to be taken into account[15].

In all three types of measurement, the electro-optic intensity modulation is very small, owing to the short optical-path length, which is determined by the layer thickness (a few micrometres). Therefore, phase-sensitive detection is necessary, using a.c. modulation. Especially the first two methods are powerful measuring techniques for studying poled polymers. Not only can both methods be applied to measure the Pockels coefficients in already poled polymers, *e.g.* as a function of time (relaxation studies), but measurements can also be performed during the poling process. This is called d.c.-induced Pockels-effect measurements[8, 16], in analogy to d.c.-induced second-harmonic generation (DCSHG or EFISH) measurements. To this end, the sample is placed in a metal block the temperature of which can be controlled. The block is mounted in the sample holder on the optical bench. Two voltages are applied to the electrodes: 1) a d.c. or very-low-frequency voltage for aligning the dipoles and inducing the electronic $\chi^{(2)}$, 2) a weaker a.c. voltage for measurement of the Pockels effect due to this d.c.-induced $\chi^{(2)}$. The entire process is a combined rotational electronic $\chi^{(3)}$ process like DCSHG. The d.c.-induced Pockels-effect technique is very useful to study the process of electric-field-induced alignment, *e.g.* as a function of time and temperature.

5. – Molecular statistical theories for poling.

5`1. *Introduction*. – The fraction of one-dimensional AπD molecules with a polar angle between θ and $\theta + \mathrm{d}\theta$ is given by $G(\theta)\sin\theta\,\mathrm{d}\theta = -G(\theta)\,\mathrm{d}[\cos\theta]$, in which $G(\theta)$ is the orientational distribution function. As already mentioned in subsect. 3˙4, this function can be developed in Legendre polynomials $P_l(x)$ of order l in $x = \cos\theta$, $l = 0, 1, 2, \ldots$. The coefficient, or moment, of $P_l(x)$ in the expansion is its orientational average, $\langle P_l \rangle$[13]. Axial-order parameters are the even moments, polar-order parameters are the odd moments. The polar-order parameters, that are most important for poled polymers (see eqs. (8) and (9)), are linear combinations of the more general polar-order parameters $\langle P_1 \rangle$ and $\langle P_3 \rangle$ (subsect. 3˙4).

For the calculation of order parameters, a number of simple theoretical models have been described in the literature on electric-field alignment:

the isotropic model[2,9],

the Ising model[2,9],

the model of Singer, Kuzyk and Sohn (SKS model or linear model)[13],

the self-consistent Maier-Saupe model as extended by VAN DER VORST and PICKEN (MSVP model)[26,34,35].

These models do not describe the time dependence of the poling process; they are not dynamical models. Only the ultimate values of the order parameters, that are approached after a very long poling cycle, can be calculated. These are steady-state values corresponding to the thermodynamically stable state in the presence of the poling field. Since entropy terms are omitted, the models are not real thermodynamical models, but molecular-statistical models. The four models use the mean-field approximation, *i.e.* an effective single-particle potential $U(\theta)$ is used to describe the interaction of a molecule with its environment. Two-particle interactions like dipole-dipole interactions are not taken into account explicitly. In all models it is assumed implicitly that the dipoles are perfectly rotatable at the poling temperature T_p (around T_g).

In the models mentioned above, average values of any function $A(\theta)$, either corresponding to polar order or axial order, are calculated using the orientational distribution function $G(\theta)$ as a weight factor:

$$(14) \qquad \langle A \rangle = \int_{-1}^{1} d[\cos\theta] A(\theta) G(\theta).$$

Using the mean-field approximation, the orientational distribution function is given by

$$(15) \qquad G(\theta) = Z^{-1} \cdot \exp[-U(\theta)/k_B T_p].$$

Here, k_B is Boltzmann's constant, T_p is the absolute poling temperature, $U(\theta)$ is the (mean-field) potential in the presence of an electric field E_p and Z is the partition function:

$$(16) \qquad Z = \int_{-1}^{1} d[\cos\theta] \exp[-U(\theta)/k_B T_p].$$

The main difference between the four models is in the choice of the expression for the single-particle energy $U(\theta)$. In a generalized notation, $U(\theta)$ can be written as a sum of terms, each with a different dependence on the poling field[34,35]:

$$(17) \qquad U(\theta) = U_0(\theta) + U_1(\theta) + U_2(\theta).$$

The terms $U_i(\theta)$ ($i = 1, \ldots, 3$) in eq. (17) are proportional to the i-th power of the local-field strength E_p. Expressions for the energy terms, used in the different models can be found in table I.

The U_1 term corresponds to the energy of the permanent dipole moment μ_0 in the electric field E_p. This term is taken into account in all four models and nearly identical expressions are used for $U_1(\theta)$ (the only differences arising from differences in the treatment of the local-field problem). U_1 is a linear function of the field strength. Since the angular dependence is described by $\cos\theta$, $U_1(\theta)$ is minimal for $\theta = 0$, maximal for $\theta = \pi$. This means that U_1 by itself induces polar order. However, also axial order is induced by U_1 alone. In the isotropic model, intended to describe poling in an initially isotropic medium ($\langle P_2 \rangle = 0$ at $E_p = 0$), the only energy term considered is U_1 (table I). Using the isotropic model, $\langle P_2 \rangle$ can be calculated exactly[35]:

$$(18a) \qquad \langle P_2 \rangle_{\text{isotropic}} = 1 - 3(\text{ctg}\,a)/a + 3/a^2,$$

$$(18b) \qquad \langle P_2 \rangle_{\text{isotropic}} \simeq a^2/15.$$

The variable $a \equiv \mu_0 E_p/k_B T_p$ is the reduced dipole energy. Equation (18b)

TABLE I. – *Energy expressions in the four molecular statistical models mentioned in the text.*

Energy term	Isotropic model[2,9]	Ising model[2,9]	SKS model[13]	MSVP model[34,35]
$U_0(\theta)$	0	0 (for $\theta = 0, \pi$)	(no analytic expression)	$-\varepsilon\langle P_2 \rangle P_2(\cos\theta)$
		∞ (for $\theta \neq 0, \pi$)	(no analytic expression)	$-\varepsilon\langle P_2 \rangle P_2(\cos\theta)$
$U_1(\theta)$	$-\mu_0 E_p \cos\theta$	$-\mu_0 E_p \cos\theta$	$-\mu_0 E_p \cos\theta$	$-\mu_0 E_p \cos\theta$
$U_2(\theta)$	0	0	0	$-\Delta\alpha E_p^2 P_2(\cos\theta)/3$

is the first (quadratic) term in a Taylor series expansion of the exact solution in eq. (18a).

The U_2 term corresponds to the energy of the linearly induced dipole moment in the poling field and is a quadratic function of E_p. Although normally this term can be neglected at moderate fields, its contribution can be significant for (local) poling-field strengths above 10^8 V/m. The U_2 term is taken into account only in the MSVP model, which is intended for the strong-field regime [35]. Because the AπD molecules have a strongly anisotropic polarizability, U_2 is angle dependent. This angle dependence takes the simple shape as given in table I, if the one-dimensional approximation is used for the AπD molecules [34, 35]. See table I: $\Delta\alpha$ is the anisotropy of the molecular polarizability and $P_2(x) = (1/2)(3x^2 - 1)$ is the second-order Legendre polynomial of $x = \cos\theta$. Assuming $\Delta\alpha$ positive, $U_2(\theta)$ is minimal for $\theta = 0$ and π and maximal for $\theta = \pi/2$. Therefore, this term by itself tends to give rise to axial order, not polar order.

U_0 is an effective mean-field potential which takes the mutual interactions between strongly anisotropic, rodlike molecules into account. It describes the tendency of such molecules to align their long axes mutually parallel. If this tendency is large enough to overcome thermal motion, a liquid crystalline phase may result with spontaneous (*viz.* at $E_p = 0$) axial order. The expression in the MSVP model for U_0 was used originally by MAYER and SAUPE [36-38] in their theory for the axial order in the nematic phase and for the nematic-isotropic phase transition. The parameter ε determines the absolute strength of the potential. The θ dependence is approximated by (again) the second-order Legendre polynomial of $\cos\theta$ (table I). The tendency for the particle to align also depends on the axial order of the environment. This is described by the factor $\langle P_2 \rangle$. For an isotropic material (at zero field strength) $\langle P_2 \rangle = 0$. This means that the potential U_0 is zero and thus no axial order is induced by the U_0 term. For perfect molecular alignment $\langle P_2 \rangle = 1$. The theory of Maier and Saupe predicts that the temperature at which the nematic-isotropic phase transition takes place, T_c («clearing» temperature in the absence of a field), is related to the parameter ε through $k_B T_c \simeq 0.22\,\varepsilon$ [36-38]. VAN DER VORST and PICKEN have combined the Maier-Saupe expression for U_0 (without explicit field dependence) with the two field-dependent terms U_1 and U_2. If the tendency towards cooperative axial order is not large enough to overcome thermal motion without an electric field, it can be «helped» by the field (through U_1 and U_2) and a field-induced isotropic-nematic phase transition is predicted by the MSVP model [26, 34, 35].

The isotropic model is intended for isotropic systems with no axial order in the absence of a poling field. Even though axial order can be induced by a field in this model (eq. (18a), (18b)), any change on the cooperative effect is ignored: $U_0 \equiv 0$ (table I). The Ising model, on the other hand, is intended for ideal liquid crystalline poling media with perfect axial order ($\langle P_2 \rangle = 1$) with or without a field. This means that the molecules are oriented either «up» ($\theta = 0$) or «down» ($\theta = \pi$). Such an ideal situation can be described formally without using any U_0 term, simply by allowing only these two orientations in the calculations [2, 9]. The molecules can also be forced in either one of the two orientations by means of the U_0 term as given in table I [34].

In the SKS model [13], an analytical expression for $U_0(\theta)$ is not given (table I). Instead, in the calculation of order parameters $\langle A(\theta) \rangle$ (for $A(\theta)$ equal to $\cos^3 \theta$ or $(1/2) \cos \theta \sin^2 \theta$), the product $A(\theta) \exp[-U_0(\theta)/kT]$ is developed in terms of the $\langle P_2 \rangle$ and $\langle P_4 \rangle$ order parameters of a liquid crystalline host. For the SKS model these are input parameters which have to be determined experimentally.

A second important difference between the four models is wether or not exact solutions are derived for the order parameters, either analytically or numerically. In the isotropic, Ising and SKS models, approximate analytical solutions are derived, which are linear functions of the poling-field strength [2, 9, 13]:

$$(19) \qquad \langle \cos^3 \theta \rangle_{\text{Ising}} \simeq a \,,$$

$$(20) \qquad \langle \cos^3 \theta \rangle_{\text{isotropic}} \simeq a/5 \,,$$

$$(21) \qquad \langle \cos^3 \theta \rangle_{\text{SKS}} \simeq a(1/5 + 4\langle P_2 \rangle/7 + 8\langle P_4 \rangle/35) \,.$$

In eq. (21), $\langle P_2 \rangle$ and $\langle P_4 \rangle$ are the axial-order parameters of a liquid crystalline host for the AπD molecules. Their numerical values are input parameters that have to be determined experimentally.

The linearization is a consequence of the truncation of Taylor series expansions (of intermediate results in the derivations) after the linear term. In the SKS model, linearization is necessary to obtain analytical expressions for the polar-order parameters. In the isotropic and Ising models, such linearization is not strictly necessary, since exact analytical solutions can be found as well [34, 35]:

$$(22) \qquad \langle \cos^3 \theta \rangle_{\text{Ising}} = \frac{e^a - e^{-a}}{e^a + e^{-a}} = \text{tgh}\, a \,,$$

$$(23) \quad \langle \cos^3 \theta \rangle_{\text{isotropic}} = \frac{e^a(a^3 - 3a^2 + 6a - 6) + e^{-a}(a^3 + 3a^2 + 6a + 6)}{(e^a - e^{-a})a^3} \,.$$

Of course, linear solutions only apply for moderate poling fields, where, for instance, saturation effects do not occur.

By using a quasi-one-dimensional description of the acceptor-donor molecules, the MSVP model is capable of going beyond the (usual) linear approximation, which implies that it is suitable for the strong-field regime. The model predicts deviations from linear behaviour in both directions (super- and sublinear). In systems that are isotropic without a field, $\langle \cos^3 \theta \rangle_{\text{MSVP}}$ may show a linear regime, followed by regimes with, respectively, a superproportional dependence on the field strength (enhancement by field-induced axial order) and a subproportional dependence (saturation)[35]. Other situations are also possible[26]. The model does not provide analytical expressions for the order parameters. Instead, first a self-consistent value of $\langle P_2 \rangle$, satisfying eqs. (14)-(17) for $A = P_2$ and using the energy expressions of table I, must be calculated by numerical methods. After substituting the obtained $\langle P_2 \rangle$ in the expression for $U_0(\theta)$, other averages $\langle A \rangle$ can be calculated, such as the polar-order parameter $\langle \cos^3 \theta \rangle$. The entire calculational procedure must be performed for each specific (redundant) set of 5 input parameters (μ_0, $\Delta\alpha$, ε, T_p and E_p), which set can also be reduced to a nonredundant set of 3 parameters[34,35]. The MSVP model is intended for the strong-field regime, which differs from the moderate-field regime in a number of respects:

1) truncation after the linear term is no longer allowed,

2) the energy of the induced dipole moment becomes significant,

3) field-induced axial order must be incorporated, which enhances field-induced polar order and *vice versa* (next subsection). This is especially necessary in the vicinity of a field-induced phase transition.

For a more detailed discussion of the above-mentioned models as well as other models, the reader is referred to [26].

5˙2. *Influence of axial order on polar order.* – It has been recognized that the polar order of AπD molecules obtained from alignment by an electric field can be enhanced by axial order[2,9]. Since substantial axial order can exist spontaneously, *viz.* without a field, in liquid crystals ($\langle P_2 \rangle = 0.4$–$0.9$), it has been considered advantageous to use liquid crystals as a poling medium[2,9,13,39-41]. Liquid crystals consist of strongly anisotropic, often rodlike molecules, which tend to align their long axes mutually parallel. If the anisotropic interactions between the molecules are strong enough to overcome thermal motion, a liquid crystalline phase (nematic phase) results with cooperative axial order. Above a certain temperature T_c, thermal motion becomes too strong, and a transition takes place to an isotropic phase with no axial order (without a field). Since the AπD molecules are also strongly anisotropic, rodlike molecules, they are similar to molecules forming a liquid crystalline phase. A first step to utilize the spontaneous axial order in liquid crystals for the enhancement of field-induced polar order of AπD molecules is to dope the AπD molecules in an existing liquid crystal. To freeze in field-induced order, this is preferably a liquid crystalline *polymer* with $T_g < T_c$, *e.g.* a liquid crystalline side-chain polymer with the mesogenic groups (= groups tending to form a liquid crystalline phase) attached as side-chains to a polymeric backbone. As the next step, side-chain *copolymers*[39] can be made with two types of side-

chains: side-chains containing AπD groups and side-chains containing the mesogenic groups. In both cases of «mixed-rod systems», it is hoped that the spontaneous axial order of the liquid-crystal molecules is transferred to the AπD
molecules, by molecular alignment. In the ideal case, the hyperpolarizable and
mesogenic properties are united in one single type of side-group («single-rod
system»). This is indeed possible, since many AπD molecules also possess mesogenic properties [40, 41].

The enhancement of polar order by axial order is clearly suggested by the
fact that the value of the polar-order parameter $\langle \cos^3 \theta \rangle$, calculated in the Ising
model (with perfect axial order), is about five times as high as the correspond-
ing value in the isotropic model (with no axial order at $E_p = 0$): compare eqs.
(19) and (20). This enhancement of $\langle \cos^3 \theta \rangle$ can be understood qualitatively by
comparing the Ising and the isotropic model. In the Ising model, two states are
allowed $(\theta = 0, \pi)$, each with an extreme value of $\cos^3 \theta$, but of opposite sign
$(+1$ and -1, respectively). At zero field strength, both states have equal en-
ergy and are equally populated, so that $\langle \cos^3 \theta \rangle_{E=0} = 0$. When an electric field is
present, the energy levels are separated so that the lower state $(\theta = 0)$ is more
densely populated than the higher one. This leads to a nonzero value for $\langle \cos^3 \theta \rangle$.
In the isotropic model, many states of intermediate energy have to be taken
into account. For every state (θ, ϕ) there is a state $(\pi - \theta, \phi)$ with opposite
$\cos^3 \theta$. In a field, these intermediate states with $|\cos^3 \theta| < 1$ are not so far apart
in energy, and their difference in population is smaller than in the Ising model.
This, combined with the smaller value of $|\cos^3 \theta|$ for these intermediate states,
leads to a smaller value for $\langle \cos^3 \theta \rangle$ in the isotropic model. (See also [26].)

The enhancement of polar order by axial order only concerns the polar-order
parameter $\langle \cos^3 \theta \rangle$. The second polar-order parameter $\langle (1/2) \cos \theta \sin^2 \theta \rangle$ apparently is negatively influenced by perfect axial order as the (linearized) solution
in the isotropic model is $a/15$, whereas it is zero in the Ising model [2, 9].
The other two models, the SKS and the MSVP model, are both able to de-
scribe intermediate ranges of axial order $(0 \leqslant \langle P_2 \rangle \leqslant 1)$ and thus link the
isotropic to the Ising model. Both the SKS and the MSVP model predict that
axial order enhances $\langle \cos^3 \theta \rangle$, whereas only substantial axial order diminishes
$\langle (1/2) \cos \theta \sin^2 \theta \rangle$. See also below.

The SKS model takes only the spontaneous axial order of a liquid crystalline
host into account, which is transferred to AπD dopant molecules (mixed-rod
system). The order is supposed to be field-independent, as field-induced axial
order is not considered. Together with the linearization mentioned in the preceding subsection, this limits the use of the SKS model to the region of intermediate field strengths.

The MSVP model is intended to describe the electric-field poling of mesogenic, hyperpolarizable AπD groups in concentrated single-rod systems. It considers all types of axial order, *viz.* spontaneous liquid crystalline axial order,
field-induced liquid crystalline axial order and field-induced nonliquid crys-

talline axial order. All this axial order is beneficial for the polar-order parameter $\langle \cos^3 \theta \rangle$.

Some calculational results of the MSVP model have been reported in ref. [35] for a specific choice of some of the input parameters: $\mu_0 = 7\,\mathrm{D}$, $\Delta\alpha = 47\,\text{Å}^3$, $T_p = T_g = 380$ K, $T_c = 0.22\,\varepsilon/k_B = 340$ K and 420 K, $E_p = (0\text{--}1000)\,\mathrm{V}/\mu\mathrm{m}$. Most polymers have a dielectric strength of a few hundred volt per micrometre (externally applied field strength). Using an Onsager local-field correction factor (approximately 2.5) to convert external fields into local fields, the maximum value for the local-poling-field strength, $E_p = 1000\,\mathrm{V}/\mu\mathrm{m}$, seems an appropriate practical limit. The values of dipole moment and polarizability anisotropy could represent the DANS molecule. The relative importance of the U_2 energy term with respect to the U_1 term for inducing axial order (approximately quadratically in the field strength) can be estimated by the ratio $\Delta\alpha k_B T_p/\mu_0^2$. This is the ratio of the linear term in U_2 and the quadratic term in U_1 in a Taylor series expansion of the distribution function. For the present numerical example, it amounts to about 5%. It is assumed that poling is performed at the polymer's glass transition temperature, $viz.$ $T_p = T_g$. To show the influence of (initial) liquid crystallinity on poling, two values of the (zero-field) clearing temperature are considered: one value of T_c that is higher than T_g—in which case the hypothetic sample is liquid crystalline before poling—and one value of T_c lower than T_g—the sample is isotropic before poling. During the cooling period after poling (field still on), the liquid crystalline order of the sample with $T_c > T_g$ is frozen in. The sample with $T_c < T_g$ remains nonliquid crystalline on cooling be-

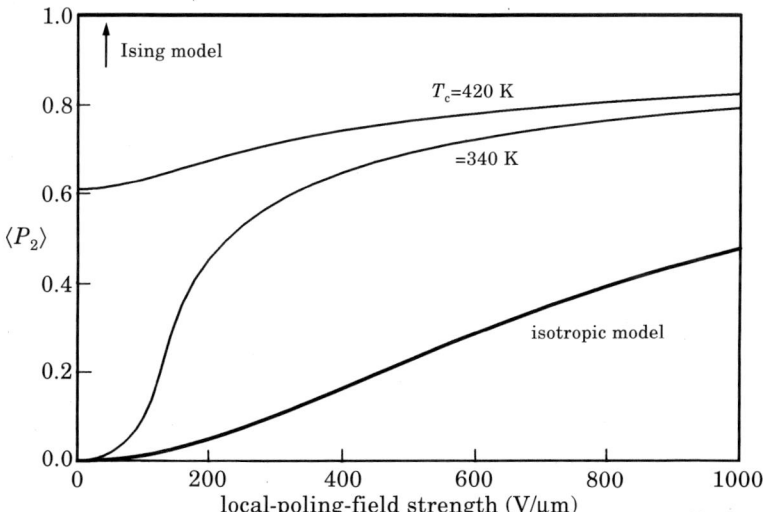

Fig. 4. – The axial-order parameter $\langle P_2 \rangle$ using the MSVP model, as a function of the local poling field for $\mu_0 = 7\,\mathrm{D}$, $\Delta\alpha = 47\,\text{Å}^3$, $T_g = 380$ K and T_c as indicated in the figure. The exact solution for $\langle P_2 \rangle$ in the isotropic model is also shown.

low T_c, as molecular rearrangements are not possible below T_g. Results of calculations with the MSVP model using the above-mentioned set of parameters are presented in fig. 4-7. Exact results from the isotropic and the Ising model are also included. The value of the axial-order parameter $\langle P_2 \rangle$, calculated self-

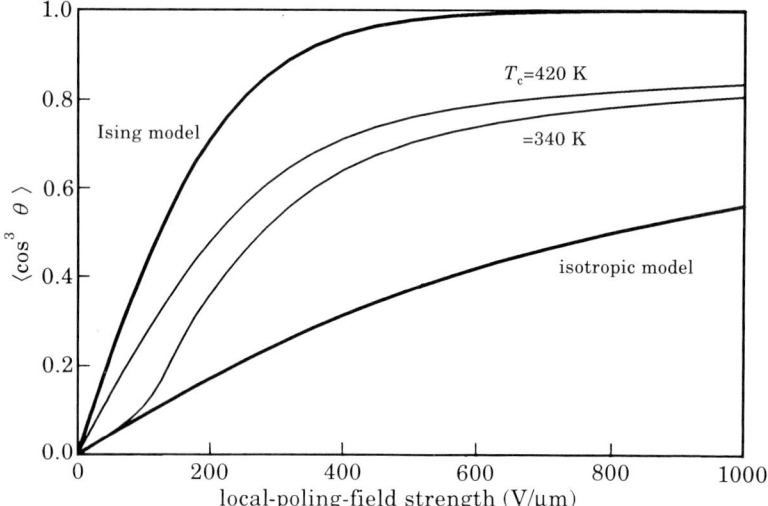

Fig. 5. – The polar-order parameter $\langle \cos^3 \theta \rangle$ as a function of the electric field E using the same values of μ_0, $\Delta \alpha$, T_g and T_c as for fig. 4. The thick solid lines are from the exact solutions of the Ising and isotropic models as indicated. The thin solid lines are from the MSVP model.

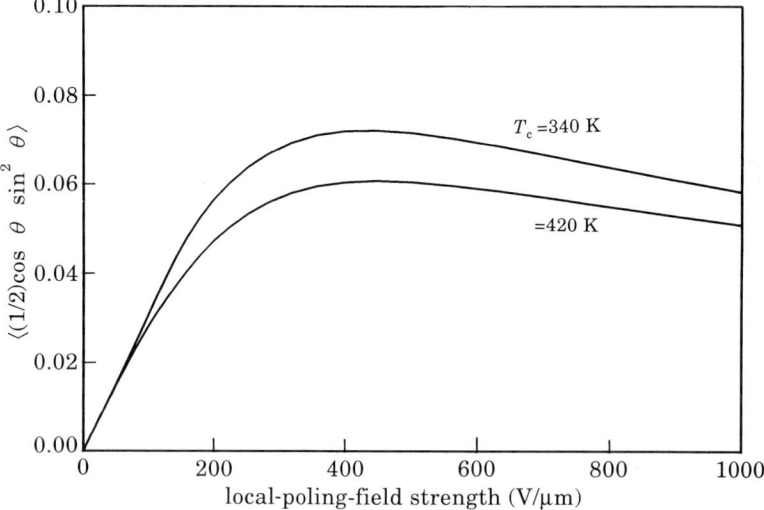

Fig. 6. – The polar-order parameter $\langle (1/2) \cos \theta \sin^2 \theta \rangle$, using the MSVP model, as a function of the local poling field E, for the same values of μ_0, $\Delta \alpha$, T_g and T_c as in fig. 4 and 5.

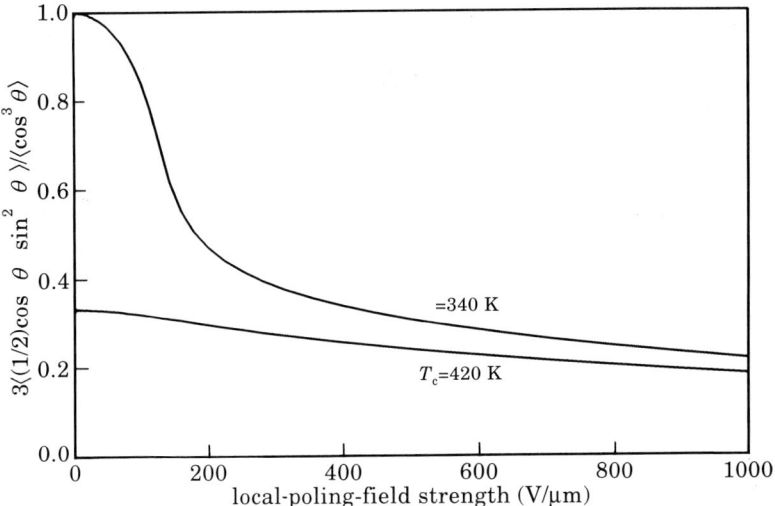

Fig. 7. – The ratio of polar-order parameters $3/v = 3\langle(1/2)\cos\theta\sin^2\theta\rangle/\langle\cos^3\theta\rangle$ $(v \equiv \langle\cos^3\theta\rangle/\langle(1/2)\cos\theta\sin^2\theta\rangle)$ as a function of the field strength, as calculated from the MSVP model for the same parameter setting as in fig. 4-6.

consistently, is shown in fig. 4; the polar-order parameter $\langle\cos^3\theta\rangle$ is given in fig. 5; $\langle(1/2)\cos\theta\sin^2\theta\rangle$ in fig. 6; and the ratio $3\langle(1/2)\cos\theta\sin^2\theta\rangle/\langle\cos^3\theta\rangle$ is shown in fig. 7.

In the low-field regime, *viz.* $E_p < 75$ V/μm for the present parameter settings, there is a large difference in axial order between the initially isotropic-model case and the initially liquid-crystalline-model case. Axial order is still insignificant in the isotropic case—although it grows quadratically with the poling-field strength—but the (spontaneous) axial order in the liquid crystalline case is already large (fig. 4). As a consequence, the polar-order parameter $\langle\cos^3\theta\rangle$ of the liquid-crystalline-model case is enhanced (fig. 5) with respect to the solution of the isotropic-model case—which behaves fully according to the isotropic-poling model. Both model cases show a linear field dependence, but with an enhanced slope for the liquid crystalline case. The enhancement is smaller than by a factor of 5 (Ising limit), since the liquid crystalline axial order is not perfect.

For field strengths beyond 75 V/μm, field-induced axial order starts to become appreciable in the initially isotropic case $(T_c < T_g)$ (fig. 4). This induced axial order grows very steeply with increasing field strength—up to approximately 140 V/μm. At this field strength, a field-induced transition to a liquid crystalline phase is predicted. From there on, axial order grows less steeply. In the initially liquid crystalline case $(T_c > T_g)$, the axial order increases as well, but to a much smaller extent, due to the absence of a phase transition and more pronounced saturation. As a result, the axial-order parameters of the initially isotropic and the initially liquid-crystalline-model cases approach each other with increasing poling-field strength (fig. 4). This has important consequences for the polar-order parameter $\langle \cos^3 \theta \rangle$ (fig. 5). The field-induced axial order of the initial isotropic-model case enhances $\langle \cos^3 \theta \rangle$ and causes a higher slope. With growing field-induced axial order, this slope becomes increasingly higher: $\langle \cos^3 \theta \rangle$ shows a superproportional dependence on E and approaches the solution of the initial liquid-crystalline-model case. After the field-induced phase transition, there is only an insignificant difference between the initial liquid crystalline and the initial isotropic-model cases. The MSVP model, therefore, predicts that the influence of initial liquid crystallinity on $\langle \cos^3 \theta \rangle$ vanishes at very high poling fields.

Whereas axial order, either initially present or induced by the field, clearly has an enhancing effect on $\langle \cos^3 \theta \rangle$, the relation between axial order and $\langle (1/2) \cos \theta \sin^2 \theta \rangle$ is less obvious. For field strengths smaller than $75 \, \text{V}/\mu\text{m}$, where no significant axial order is induced yet, $\langle (1/2) \cos \theta \sin^2 \theta \rangle$ is still linear in E_p and the lines of the initially isotropic and the initially liquid-crystalline-model cases coincide (fig. 6). Apparently, spontaneous liquid crystalline order has no influence on this second polar-order parameter. Above $75 \, \mu\text{m}$, where axial order is induced by the field (most prominently in the initially isotropic case) and where also saturation effects start (most pronounced in the initially liquid crystalline case), the parameter $\langle (1/2) \cos \theta \sin^2 \theta \rangle$ levels off, most prominently in the initially liquid crystalline case. Although induced axial order and $\langle (1/2) \cos \theta \sin^2 \theta \rangle$ are negatively correlated, saturation plays a bigger role in determining $\langle (1/2) \cos \theta \sin^2 \theta \rangle$. For increasing E_p, this polar-order parameter goes through a maximum and finally, for E_p approaching infinity, $\langle (1/2) \cos \theta \sin^2 \theta \rangle$ goes to zero (Ising limit).

From the behaviour of $\langle \cos^3 \theta \rangle$ (fig. 5) and $\langle (1/2) \cos \theta \sin^2 \theta \rangle$ (fig. 6), the behaviour of the ratio $3 \langle (1/2) \cos \theta \sin^2 \theta \rangle / \langle \cos^3 \theta \rangle$ is evident and presented in fig. 7. For intermediate field strengths, the value of this ratio is around 1 for the initially isotropic sample and much smaller for the initially liquid crystalline sample (though not 0 as in the Ising case). When appreciable axial order is induced, the model solutions of the initially isotropic and of the initially liquid crystalline cases approach each other, and both go slowly to zero for infinite E_p (Ising limit).

All molecular statistical models considered in subsect. 5'1 neglect dipole-dipole interaction. In a qualitative analysis, VAN DER VORST and PICKEN [26] have argued that antiparallel dipole-dipole correlation (which lowers $\langle \cos^3 \theta \rangle$) is more prominent in axially ordered systems than in isotropic systems. This introduces a new mechanism through which axial order has an adverse instead of an enhancing effect on $\langle \cos^3 \theta \rangle$! Dipole-dipole may become important in highly concentrated, liquid crystalline «single-rod» systems such as side-chain homopolymers with (hyper)polarizable and mesogenic AπD side-chains. The new mechanism may well explain the observation in a MONS homopolyester that the field-induced EO coefficient was apparently smaller in the liquid crystalline phase than in the isotropic phase [26,42].

5'3. *Comparison between experiment and theory.* – A comparison between absolute values of experimental and theoretical values of Pockels coefficients is complicated by many factors [8]. Theoretical values can be obtained from eqs. (8), (9) and the specific theoretical poling model by inserting the necessary input parameters. These input parameters, however, may not be precisely known. Since they are the result of other experiments, they contain experimental errors. The theoretical calculation of Pockels coefficients is complicated furthermore by the local-field problem. Four local-field factors, which only can be approximated, enter the calculations, three from the relation between macroscopic and molecular NLO properties (eq. (7)) and one for the poling field used in the poling models (subsect 5'1). Usually, Lorentz local-field factors $((n^2 + 2)/3)$ are used at optical frequencies, whereas Onsager local-field factors are more appropriate at lower frequencies. For a general discussion of the local-field problem, the reader is referred to the textbooks of Böttcher [25], for a more specific treatment of the local-field problem in poled polymers to [24]. The measured Pockels coefficients too are hampered by experimental errors, both experimental scatter and systematic errors, *e.g.*, due to piezoelectric contributions to apparent Pockels coefficients. Theoretical and experimental values must be compared under conditions that are «matched» to each other [8]. For instance, possible resonance enhancement of experimental values must be properly taken care of, either by performing the measurements far from resonance, or by applying a correction for resonance enhancement [43,44]. Also, any decay of the EO coefficients during cooling and during or after switching the field off must be prevented, since these effects—though themselves interesting—are not covered by the theoretical models. This is possible by measuring the EO coefficients during poling at T_g, although this may increase a piezoelectric contribution. Part of the «real» discrepancy between experiment and theory can be attributed to sterically hindered rotatability of the side-chains [45-47], but probably also to significant antiparallel dipole-dipole association [26].

A comparison between theory and experiment is easier for relative instead of absolute values of Pockels coefficients. For instance, the value of r_{33} relative to that of r_{13} allows a comparison with predictions from poling models without much complications. The ratio r_{33}/r_{13} is approximately equal to the ratio of the corresponding polar-order parameters ($v \equiv \langle \cos^3 \theta \rangle / \langle (1/2) \cos \theta \sin^2 \theta \rangle$). Another example is the (relative) behaviour of a Pockels coefficient as a function of the poling-field strength or poling temperature.

Of particular importance would be a comparison between experiment and theory for the value of $r_{33} (\langle \cos^3 \theta \rangle)$ measured in a liquid crystalline phase relative to the value measured in the isotropic phase. To suppress temperature effects, the measurements or calculations preferably have to be performed at the same, or nearly the same temperature. Three experimental ways seem possible[26]. One way is to perform the measurements during poling slightly above T_c—in the isotropic phase—and slightly below T_c—in the oriented liquid crystalline phase[42]. Disadvantages are stronger sample degradation and larger piezoelectric contributions[26], because the polymer is usually very soft above T_g ($T_c > T_g$). A second way would be to perform the poling experiments below T_g in the frozen-in isotropic and the frozen-in single-domain liquid crystalline phase. The above-mentioned problems would be avoided this way. However, it remains to be investigated whether a suitable poling temperature below T_g can be found where poling in the frozen-in isotropic phase is already possible, whereas formation of an ordered liquid crystalline phase is still too slow. This second route might also be used to investigate the influence of liquid crystalline axial order on relaxation at a temperature below T_g. A third and slightly different route might be to perform the measurements in the single-domain liquid crystalline phase and in the polydomain liquid crystalline phase, since the latter is expected to behave macroscopically as an isotropic medium in some respects—not as far as the influence of dipole-dipole interaction on the static values of the EO coefficients is considered, nor considering the kinetics of poling and relaxation. Here, it must be prevented that the polydomain sample becomes single domain, while the AπD moieties become oriented by the electric field. This third route has been applied succesfully by STUPP and co-workers[48,49] in a guest-host system using a strong magnetic field for domain prealignment and a weak electric field for dopant alignment. The electric field had to be weak to prevent electrical-domain alignment in the polydomain sample.

Hereafter, some experimental results in different polymer systems will be compared with theory. First, dilute systems with a low AπD density and negligible pair interaction will be treated. Concentrated systems with AπD pair interaction follow. The concentrated systems are subdivided in isotropic («almost» liquid crystalline) and liquid crystalline systems.

5˙3.1. Dilute systems. Liquid crystalline systems in the literature with a low hyperpolarizability density N are either guest-host systems with a liquid crystalline polymer host or liquid crystalline copolymers. Both categories are examples of mixed-rod systems. In the guest-host systems, the hyperpolarizable molecules are doped into the liquid crystalline host, which is either a side-chain polymer[2,9] or a main-chain polymer[48,49]. In the liquid crystalline side-chain copolymers[39,50], separate hyperpolarizable and mesogenic groups are attached as side-chains to the backbone.

The first mention in the literature of enhanced poling by liquid crystallinity was by MEREDITH, VAN DUSEN and WILLIAMS[2,9]. Their suggestion of enhanced poling, based on a comparison between the isotropic and Ising models, was accompanied by experimental results, which seemed to support the theoretical predictions. A liquid crystalline polymer was doped with 2 wt./wt.% DANS and poled with a field strength of $1.4\,\mathrm{V}/\mu\mathrm{m}$. $\chi^{(2)}_{333}$ was determined from SHG experiments and compared with a theoretical prediction using the Ising model. The result, $\chi^{(2)}_{\mathrm{exp}} \approx 0.5\chi^{(2)}_{\mathrm{Ising}} = 2.5\chi^{(2)}_{\mathrm{isotropic}}$, claimed to be consistent with an axial-order parameter determined from dichroism measurements ($\langle P_2(\cos\theta)\rangle \approx$ ≈ 0.3 or $\langle\cos^2\theta\rangle \approx 0.5$), was regarded as experimental proof for the enhancement[2,9]. A direct comparison between experimental and theoretically calculated NLO coefficients is, however, very dangerous to draw conclusions from, as has been remarked at the beginning of this subsection. This first experimental indication for enhanced poling should, therefore, be treated with caution.

STUPP and co-workers[48,49] studied dopes of the well-known chromophore Disperse Red 1 (4-[ethyl(2-hydroxyethyl)amino]-4′-nitroaxobenzene) in a solidified liquid crystalline main-chain polymer. The host contained three different structural units, all aperiodically arranged and connected by ester bonds. The semi-flexible chains formed a nematic phase in the temperature range between the melting point of 185 °C and 220 °C. The nematic phase may extend below 185 °C. Above 220 °C, a nematic-isotropic biphasic fluid was observed. Solvation of the dye was possible for concentrations up to 15 wt.%. The samples were poled by an electric field below the melting point of the host using moderate field strengths which were too weak to align the domains and form a single-domain phase. Single-domain formation was possible before poling by aligning the main chains with a strong magnetic field. The magnetic (pre-)orientation only introduces macroscopic axial order, not polar order. The second-order NLO properties induced by the electric field were studied in magnetically aligned (single-domain) and nonaligned (polydomain) samples using SHG measurements. In the samples loaded with 5 wt.% dye, it was found that magnetic alignment increased the SHG intensity by a factor of 6–9. Taking into account optical scattering effects, this increase could be ascribed to an enhancement of the field-induced polar order in the monodomain nematic phase by a factor of about 3 with respect to the polydomain phase. In addition, the authors found that poling proceeded more rapidly in the macroscopically aligned samples than in the nonaligned samples; thermal relaxation of polar order, on the other hand, was slower in the magnetically ordered system. The found enhancement of field-induced polar order is qualitatively in agreement with the poling models, considering that the multidomain phase can be treated as isotropic in a molecular-statistical-model calculation taking into account only the energy of the permanent dipoles in the electric field (isotropic model). Considering also dipole-dipole interaction, or considering the kinetics of poling and relaxation, the polydomain nematic phase is expected to show more resemblance to the monodomain nematic phase than to a real isotropic phase. It is, therefore, difficult to draw conclusions from the above-mentioned observations with regard to the influence of liquid crystallinity on the kinetics of poling and relaxation.

5'3.2. Concentrated isotropic systems. The mesogenic properties of the AπD rods will only give rise to a liquid crystalline phase in the single-rod systems, if their concentration is high enough, *viz.* if they are not diluted too much by the polymer. To achieve high concentrations without recrystallization problems, the AπD rods have to be attached chemically, *e.g.*, as side-chains to a polymer backbone. Experimentally, the important single-rod systems are mainly side-chain homopolymers and also some copolymers, some of which indeed show a liquid crystalline phase [40-42]. Here, the *isotropic* side-chain homo- and copolymers will be reviewed, which do not show spontaneously a liquid crystalline phase, but which do show behaviour related to strong field-induced axial order.

A field-induced phase transition from an isotropic to a liquid crystalline phase—as predicted by our model—has been reported by GONIN *et al.* [51, 52] in an isotropic copolyether and also in an isotropic homopolymethacrylate, both containing OCB (oxycyanobiphenyl) in the side-chain (next two paragraphs).

A number of copolyethers were prepared by chemical modification of atactic polyepichlorohydrin with the sodium salt of 4-4'-hydroxycyanobiphenyl [51, 52]. The CH_2Cl side-chains were partly substituted by the classical mesogenic group OCB, which is also slightly hyperpolarizable. The OCB group is attached to the backbone via a —CH_2— spacer group. Although these polymers are copolymers and not homopolymers, they can be classified as single-rod systems, since the OCB moiety is the only rodlike group having hyperpolarizable or mesogenic properties. Substitution percentages larger than approximately 65% resulted in the formation of a liquid crystalline (nematic) phase between T_g and $T_c > T_g$. For instance, 80% substitution gave a polymer «A» with a nematic phase between a T_g of 79.5 °C and a T_c of 146.5 °C. A second copolyester «B» with a substitution percentage of 55% had a T_g of 56 °C and was isotropic without a field [51, 52]. In addition to the OCB grafted copolyethers, also commercially available OCB side-chain homopolymers were studied with OCB connected via a —$(CH_2)_3$— spacer to the backbone (Merck). Polymer «C» was a polyacrylate with a smectic S_A phase between a T_g of 51.5 °C and a T_c of 97 °C. The analogous polymethacrylate («D») had a T_g of 97 °C and was isotropic at rest. Here, only results in the isotropic polymers B and D are reviewed. Results in the liquid crystalline polymers A and C are mentioned below in the subsection on concentrated liquid crystalline systems.

The location of the *virtual* zero-field clearing temperature T_c of polymer B was estimated in the region (40–56) °C by extrapolation of the measured T_c values of A-B mixtures, using the mutual-miscibility method [53]. Samples were poled with corona poling, which permits extremely high poling-field strengths approaching the intrinsic dielectric strength—usually a few hundred $V/\mu m$ in polymers. The value of the axial-order parameter $\langle P_2 \rangle$ in the poled polymer was determined from dichroitic absorption measurements and was found to be large, about 0.45. Also, the value of the ratio of second-harmonic co-efficients (about equal to the ratio of the Pockels coefficients) was found to be much larger than the value of 3 in the isotropic model: $d_{33}/d_{13} = 11$. These results strongly suggest a transition to the nematic phase during poling. It was concluded that the enhancement of the NLO performance by the field-induced liquid crystalline character is by a factor (slightly) larger than 3 compared to predictions by the isotropic model. Similar results were also obtained in the isotropic side-chain homopolymethacrylate *D*.

A superproportional behaviour of the EO effect as a function of poling field strength—theoretically predicted in isotropic concentrated polymers—has been observed in several occasions.

A first indication of superproportional behaviour can be seen in the poling results of Hoechst Celanese polymers as shown in the plenary papers by STA-MATOFF [54] and HAAS *et al.*[55]. Figure 1 in the paper by STA-MATOFF [54] shows the Pockels coefficient as a function of the poling voltage. After a linear region (isotropic), one can observe a superproportional region (phase transition) which is followed again by a second linear region with a steeper slope (liquid crystalline). Figure 4 in the paper by HAAS *et al.*[55] shows similar, though less detailed, poling results in a polymer HCC-1232. The authors do not give any experimental details and do not comment on the shown behaviour.

Superproportional behaviour has also been observed by our group in a DANS side-chain polymer with $T_g = 140\,°C$ [45, 46]. Test samples consisting of a spuncoat single layer of the polymer in between semi-transparent, plane-paral-lel metal electrodes were studied during poling at $130\,°C$ with the d.c.-induced Pockels-effect technique, where a superposition of a d.c. poling voltage and an a.c. modulation voltage is applied to the sample electrodes. The EO effect was measured between crossed polarizers in transmission and in reflection. The applicable poling-field strength is limited in the single-layer test samples by dielectric breakdown at weak spots in the sample. Electric discharge at a weak spot is self-healing and only results in elimination of the weak spots (sparkling). In principle, this does not influence the measurements very much, unless the measure spot—where the laser hits the sample—itself is affected. Then, a new measure spot has to be selected, which may lead to a discontinuity (shift) in the field-dependent measurements due to the nonuniformity of the layer thickness. Despite the increasing number of dielectric breakdowns at higher field strengths and occasional repositioning, a superproportional behaviour could be observed for field strengths in excess of $50\,V/\mu m$ (fig. 8)[45, 46].

Very interesting are the recent observations of Meyrueix *et al.* [56] in a

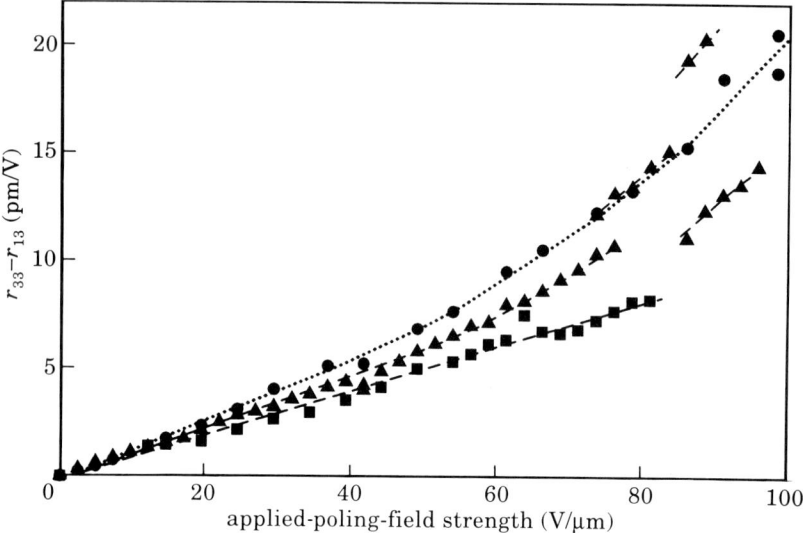

Fig. 8. – Dependence of the EO coefficients on the applied-poling-field strength in a DANS side-chain polymer with $T_g = 140\,°C$. The measurements are between crossed polarizers during poling at $130\,°C$: measurements in trasmission (dashed) and in reflection (dotted). Note the superproportional field dependence, tentatively associated with a field-induced isotropic-to-nematic transition. Discontinuities are due to dielectric breakdown at the measure spot. A new measure spot was selected when the breakdown at the measure spot was self-healing[45].

polyurethane («PU11») with the DANS-like azo chromophore DR17 as active side-group. Not only did these authors find a value for r_{33}/r_{13} larger than 3, which increases with increasing poling-field strength, but also a superproportional dependence of r_{33} and a subproportional dependence for r_{13}, all as predicted by our model.

BUCKLEY[57] reports superproportional behaviour in a PMMA side-chain polymer containing about 75 wt.% DANS as active group.

KELDERMAN[58] mentions a superproportional field dependence during poling at $105\,°C$ of a 75 wt.% film of tetrapropoxytetranitrocalix[4]arene. Disregarding the parabolic (?) curve fitted through the data points in fig. 6.5 of his thesis, one can recognize a linear regime up to 20 V/µm, a superproportional regime between 20 and (30–40) V/µm, levelling off to a second, almost linear regime between (30–40) and 50 V/µm (highest applied field strength).

At field strengths lower than in the superproportional region (of r_{33}), the predictions from our model coincide with those from the isotropic model. Both models then predict a linear field dependence of the polar-order parameters $\langle \cos^3 \theta \rangle$ and $\langle (1/2) \cos \theta \sin^2 \theta \rangle$ and a value of their ratio $\langle \cos^3 \theta \rangle / \langle (1/2) \cos \theta \sin^2 \theta \rangle$ equal to 3. Measurements in several of our polymers indeed showed a linear field dependence of the Pockels coefficients at moderate field strength and a ratio r_{33}/r_{13} near the value 3 [15, 16, 45, 46]. More recent results of ratio *smaller* than 3 [47] must probably be attributed to piezoelectric contributions to our *apparent* Pockels coefficients. MEYRUEIX *et al.* reported a method for correcting the apparent Pockels coefficients for the piezoelectric effect [56].

5˙3.3. Concentrated liquid crystalline systems. Liquid crystalline side-chain homopolymers with the mesogenic and hyperpolarizable properties united in a single type of rodlike side-group (single-rod system) were expected to provide the largest Pockels coefficients as they combine a high number density of hyperpolarizable groups with the liquid crystallinity that would enhance polar order. Several groups have started synthesis along this route. Liquid crystalline side-chain homopolymers with hyperpolarizable side-chains have indeed been obtained in this way [40-42, 51, 52].

The nematic OCB grafted side-chain copolyether A and the smectic OCB homopolyacrylate C of Gonin *et al.* [51, 52] have already been mentioned above. Only minor experimental results were reported on polymer C. The ratio d_{33}/d_{13} in polymer A was determined to be 15 ± 3 from SHG measurements. For $\langle P_2 \rangle$, a value was obtained of 0.55 ± 0.02 from dichroism measurements. These experimental results were compared with the results of the theoretical models mentioned in subsect 5˙1. Reasonable agreement was obtained using either the SKS or the MSVP model. The ratio d_{33}/d_{13} significantly larger than 3 indirectly confirms that the polar-order parameter $\langle \cos^3 \theta \rangle$ can be enhanced by liquid crystallinity, whereas $\langle (1/2) \cos \theta \sin^2 \theta \rangle$ is reduced.

Direct experimental evidence on enhanced poling in concentrated liquid crystals is scarce—if not nonexistent—in the literature, which is rather strange considering that several groups ventured to study the liquid crystalline option of poled polymers already in the second half of the eighties. Of course, the enhancement of poling in a field-induced liquid crystalline phase and also the superproportional behaviour accompanying the field-induced transition, both discussed above and in subsect. 5˙2, can be regarded as positive evidence for enhanced poling in liquid crystals.

Results in our own laboratories in spontaneous liquid crystalline side-chain homopolymers suggested quite the contrary. Instead of an enhancing effect on poling from liquid crystallinity, we found indications of an adverse effect in two liquid crystalline side-chain homopolymers [42]. An adverse effect cannot, however, be concluded without ambiguity, because of certain experimental problems [26].

From the above, it can be inferred that the theoretical models neglecting dipole-dipole interaction apply for dilute systems and even for concentrated isotropic systems—at least in their qualitative predictions. For concentrated liquid crystalline systems, more research is needed. Looking more quantitatively, there are indications that an antiparallel dipole-dipole association does exist in the concentrated isotropic systems. For instance, experimental values of Pockels coefficients are often appreciably smaller than theoretically predicted values using eqs. (8), (9), (12), (13) and the poling models for calculating the order parameters. VAN DER VORST and HUYTS [15] found a discrepancy by a factor of 5 between experiment and theory (isotropic model) in a MONS side-chain polyacrylate poled in the linear region. In a moderately poled DANS side-chain polymer, a discrepancy by a factor 3 was found [16]. If resonance enhancement of the experimental Pockels coefficients is (better) taken into account, this discrepancy may become even larger. Experimentally, there are indications for sterical hindrance during poling, which may account in part for the discrepancy between experiment and theory [45, 46]. It is, however, unlikely that sterical hindrance alone is responsible for such a large difference. Some degree of antiparallel dipole-dipole association might well explain the remainder, perhaps the largest part, of the discrepancy between experiment and theory.

6. – Dynamics of poling and relaxation.

6˙1. *The «fast» process*. – From initial poling and relaxation studies at Akzo in DANS and MONS polymers, it was found that the linear EO effect consists of (at least) two different contributions with a completely different time and temperature behaviour. These have been termed by us as the «fast» contribution and the «slow» contribution [8].

The «fast» contribution is quasi-instantaneous on the time scale of our experiments using phase-sensitive detection (typical modulation frequency 275 Hz, typical time constant of lock-in amplifier 0.3 s). When the d.c. poling field is switched on/off, the «fast» contribution develops/relaxes far within one second, whereas the measurement of an EO coefficient takes about (1–15) min, depending on the method and the accuracy. Measurements at Lockheed [59] have revealed that the «fast» process develops even within 10 ms. We found that both speed and magnitude (at constant poling-field strength) are hardly temperature dependent far below T_g, the first observation implying that the activation energy of the «fast» process is very low. The magnitude itself is approximately a linear function of the poling-field strength. After the «fast» decrease—when the poling field has been switched off at the end of poling—the Pockels effect can be reinstalled to its original value by switching the poling field on again—with the same polarity. However, if the polarity is reversed, the Pockels effect decreases for a second time by approximately the same amount. In unpoled samples, a Pockels effect can be induced at ambient temperature by applying a d.c. poling field. Upon switching this field off, the Pockels effect disappears again completely. Applying equal poling-field strengths, the induced effects seem to be slightly larger in unpoled samples than in poled samples. All these related observations are quasi-instantaneous at ambient temperature and can also be observed at elevated temperatures—then superimposed on slower poling and relaxation phenomena («slow» process).

We attribute the above-mentioned quasi-instantaneous effect(s) to orientational polarization of AπD moieties that only exists in the presence of the poling field and that cannot be «frozen in». In an oversimplified picture of a hyperpolarizable AπD molecule in its polymer environment, the AπD group might be represented by a *rod rattling in a cage*. Now, two types of nonfreezable orientational polarization can be envisioned: 1) polarization within the cage itself (free volume) without touching the walls of the cage, 2) additional rotation where the AπD rod is pressed into the soft cage wall through the torque exerted by the poling field. The second movement has energetic consequences in the form of elastic deformation energy of the cage wall, the first one has not (or a to lesser extent). Both types of orientational polarization can be expected to be very fast. The speed of the free-volume contribution 1) will be very insensitive to temperature, but the speed of the elastic part 2) should show a temperature dependence similar to that of the elastic modulus. Considering the magnitude of the polarization, one should certainly expect a temperature dependence of the elastic part 2). The reason why no significant dependences upon temperature could be detected far below T_g might be due to the small sizes of the effects and the rather large experimental errors. With regard to the relative size of contributions 1) and 2) to the fast process, we believe that contribution 2) is the dominant one. Three arguments in favour of this interpretation are given below.

The field strength dependences of both contributions will be very different. For low field strengths, polarization will be mainly in the free volume 1), characterized by a linear field dependence with a large slope determined by the balance between the driving ($\mu_0 E_p$) and counteracting forces ($k_B T_p$). With increasing field strength, contribution 1) becomes frustrated when the rod touches the wall. Then, the elastic contribution 2) takes over and a second linear region is expected with a smaller slope because both thermal motion ($k_B T_p$) and elastic forces now counteract the rotation. Experimentally, we found a linear dependence on the field strength in the whole investigated range from 0.2 to 26 V/μm (DANS polymer, $T_g = 140$ °C). Instead of a kink between two linear regions, we found an apparent small and positive offset, which—in our opinion—corresponds to free-volume polarization already saturated at the lowest field strength of 0.2 V/μm. The linear increase then corresponds to elastic deformation 2).

A second argument in favour of a dominant elastic contribution has been formulated by VAN DER VORST *et al.* [47], who compared the poling and relaxation behaviour of three similar side-chain polymers differing mainly in the stiffness of the connection between side-chain and main chain. If the instantaneous decay was mainly relaxation within the free volume, it would reflect the size of the free volume—which is expected to correlate positively with the speed of the (slow) relaxation process. Experimentally, the opposite was found. A third and similar argument could be deduced from annealing experiments [45, 46]. Annealing a polymer sample after poling with the poling field still on was found to affect the initial part of the slow relaxation process, but its effect on the size of the fast (quasi-instantaneous) decay was negligible. If this decay was mainly relaxation within the free volume, the fast decay should have become smaller after annealing.

The slightly larger effect of switching the poling field on or off in an unpoled sample (see above) can be explained as a statistical effect. If all AπD molecules rotate towards the field over a certain angle, the effect on $\langle \cos^3 \theta \rangle$ will be largest in an unpoled sample, since most molecules are then at right angles with respect to E_p where the slope $d[\cos^3 \theta]/d\theta$ is largest.

If the interpretation of the quasi-instantaneous effects ($\ll 1$ s) in terms of «rattling rods» is correct, one would expect to see an orientational Kerr effect in an electric field. Indeed, this effect has been observed by us. Applying a strong a.c. voltage over the electrodes, an intensity modulation at the double frequency could be detected. The measurement of an orientational Kerr effect is expected to be an interesting technique for gaining information on the orientational freedom of the rattling rods as a function of frequency and temperature.

6`2. *The «slow» process.* – The slow response during poling corresponds to molecular rearrangements within the AπD rod's deformed environment, which relax the associated elastic deformation energy and also allow further rotation. In other words, it corresponds to a *conversion of elastic into plastic polar order* or to a *rotation of the cages*. Whereas the orientational polarization of the rods within the cages cannot be frozen in, the rotation of the cages themselves, corresponding to plastic deformation of the polymer, can be «frozen in»—but only in a kinetic sense. Because the poled situation is thermodynamically unstable after the poling field has been switched off far below T_g, the oriented dipoles will tend to randomize thermally, thereby using the small but nonnegligible motional freedom within the free volume. This slow relaxation process can be represented as a rotation of the cages back to the nonoriented situation. Time and temperature dependence of the «slow» part of the poling and relaxation processes are studied in our laboratories by measuring the development of the Pockels coefficient $r(t; T)$ in time t at constant temperature T, after the poling field has been switched on (poling) or off (relaxation). Measurements are performed in polymer films between plane-parallel, semi-transparent metal electrodes on glass (or silicon) substrates in a transmission (or reflection) experiment. Both crossed polarizers and Fabry-Perot techniques are used (subsect. 4`5).

Thus far, the results have usually been fitted with a single-exponential expression. For relaxation, this expression is

$$(24) \qquad r(t; T)^{\mathrm{rel}} = r(0; T)^{\mathrm{rel}}_{\mathrm{slow}} \exp[-\alpha(T)^{\mathrm{rel}}_{\mathrm{slow}} t].$$

For poling, the following expression is used:

$$(25) \qquad r(t; T)^{\mathrm{pol}} = r(\infty; T)^{\mathrm{pol}}_{\mathrm{slow}} [1 - \exp[-\alpha(T)^{\mathrm{pol}}_{\mathrm{slow}} t]] + r(\infty; T)^{\mathrm{pol}}_{\mathrm{fast}}.$$

The temperature dependences of the frequencies $\alpha(T)^{\mathrm{pol}}_{\mathrm{slow}}$ or $\alpha(T)^{\mathrm{rel}}_{\mathrm{slow}}$, thus obtained from poling or relaxation experiments at different constant temperatures T, were subsequently fitted with an Arrhenius expression:

$$(26) \qquad \alpha(T)_{\mathrm{slow}} = \alpha(\infty)_{\mathrm{slow}} \exp[-A_{\mathrm{slow}}/RT].$$

For a DANS side-chain polymer, measurements were performed at temperatures ranging from 100 °C to 140 °C (T_g). It was found that the Arrhenius plot of the poling process and that of the relaxation process overlap [16]. The activation energy for both was about 300 kJ/mole («slow» contribution). At each temperature, the poling and relaxation frequencies do not differ significantly; $\alpha(T)^{\mathrm{pol}}_{\mathrm{slow}}$ and $\alpha(T)^{\mathrm{rel}}_{\mathrm{slow}}$ were equal within a factor of about 3, which was within the experimental inaccuracy. Measurements below 100 °C were not very practical, because it took too much time to observe any changes. In a sample of the DANS polymer kept at 43 °C, no significant decrease of the EO effect could be detected in a period of 110 days. Information on the stability at lower temperatures can be obtained from relaxation measurements at higher temperature by extrapolation. In this way it can be predicted that a poled sample of the DANS polymer, stored at ambient temperature, will lose 50% of its initial activity in a period of 10^{10} years! Of course, this estimate is very rough, but it means that the lifetime of a device, based on this DANS polymer and kept at ambient temperature, will *not* be limited by relaxation of polar order.

For practical applications of EO devices, however, stability at ambient temperature is not sufficient. EO devices will be mounted together with heat-dissipating electrical equipment and should withstand temperatures of (40–100) °C during a lifetime of several years. Also, the device should survive short excursions to much higher temperatures during assembling. For instance, temperatures may rise to 200 °C or even 300 °C during soldering activities in the vicinity of the device. Requirements that polymer-based EO devices must fulfil are listed in ref. [21]. To meet these requirements, new generations of polymers with high T_g have to be developed.

7. – Polymeric multilayer structures.

7˙1. *Slab waveguides.* – For the testing of the nonlinear optical polymeric materials it is necessary to evaluate their properties under as realistic conditions as possible. Therefore, polymeric waveguide devices have to be made in order to carry out such tests.

In making polymeric (nonlinear) optical test devices, one often starts using a thin-film multilayer structure on a substrate: a so-called «optoboard». The substrate is mostly silicon, but it can consist of other materials such as glass, metals, plastic or III-V semiconductors as well. On top of this substrate, a lower metal electrode (silver, gold or aluminium) is deposited via evaporation or sputtering. Then, a lower polymeric cladding layer with relatively low refractive index is spuncoat onto it, followed by the waveguiding core layer exhibiting a higher refractive index than the cladding layer. Finally, an upper cladding layer and the top metallization (electrode) are successively deposited. A schematic picture of a multilayer thin-film structure which is in fact a slab waveguide is given on the left-hand side of fig. 9. The role of the low-refractive-index cladding layers is to prevent the light travelling in the core layer from «seeing» the metal electrodes, thereby reducing undesired scattering and absorption. The central waveguiding layer (core layer) is often made from NLO polymers.

The metal layers have thicknesses ranging from tens of nanometres to several micrometres, depending on the type of electrical drive system for the device in combination with the type of electrodes (lumped or travelling-wave electrode). The upper metal layers serve three purposes. First, during the electric-field poling to induce the electro-optic effect at a temperature close to T_g of the polymer. Second, to apply the switching voltages to the finished (electro-optic) device. The third application is in between, where the metal layers also serve the purpose of creating the masks used for the realization of the actual waveguide channels via photobleaching as discussed further on. For the first application, the upper metal layer need not be structured. For the other two applications, the upper metal layer is structured, different for each purpose.

Thicknesses of the various polymer layers, deposited by spin coating or dipping, is of the order of micrometres, depending on whether monomode or multimode waveguide devices are desired. Thickness homogeneities of

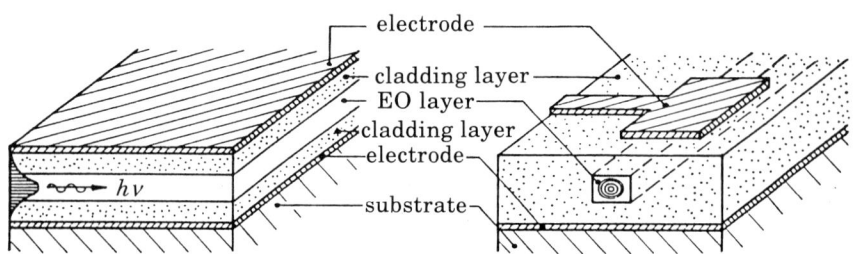

Fig. 9. – Multilayer structures: slab waveguide (left-hand side) and channel waveguide (right-hand side)

better than 0.5% over 3 inch diameter silicon wafer areas have been achieved in spin-coated polymer layers.

The important electric-field poling step is carried out by placing the entire multilayer structure on a hot plate or in an oven heated to temperatures around T_g where the polymer is soft, thus providing sufficient molecular segmental mobility for reorientation (poling) of the permanent electric dipoles of the optically nonlinear side-groups by the strong electric field. Field strengths in excess of $200\,V/\mu m$ have been applied to the electrodes during the poling step[60]. Optically nonlinear slab waveguides (optoboards) can thus be made; such optoboards are the starting structures for further processing into devices.

7'2. *Channel waveguides.* – Starting with the optoboard and after having carried out the electric-field poling step (poled optoboard), channel waveguides have to be realized in the central (core) layer via the UV bleaching technique and applying the appropriate mask exhibiting the desired waveguide layout[61]. As a consequence of the bleaching process, the refractive index of the selectively (via the mask) irradiated areas decreases, thus creating laterally confined optical pathways (see also on the right in fig. 9). The type of mask used during bleaching determines the optical pathway of the channel waveguide and therewith the type of device.

The applied bleaching time combined with the irradiation power of the applied (mercury) lamp determines the magnitude of the locally induced refractive-index change (decrease) in the central layer. The combination of refractive-index contrast between core and cladding as well as that of the bleached and unbleached areas in the core together with the waveguide dimensions determine whether the waveguides are monomode or multimode. By selecting the appropriate refractive-index difference combination of bleached core, unbleached core and (unbleached) cladding, the desired modal properties of the waveguide can be realized.

Following the electric-field-assisted poling step and the channel waveguide definition by the UV bleaching technique in the multilayer structures, the realization and processing of the electrode structure, to drive the device, must be carried out. The processing is done by photolithographic techniques and wet chemical or dry (reactive ion) etching steps.

7'3. *Passive and active waveguide features.* – The quality and performance of the polymeric channel waveguides made as described above strongly depend on the quality of the mask used for bleaching and on the accuracy of the various processing steps. To test the functions and performance of the produced waveguiding structures, light must be injected into the channel waveguides. Polishing of the end-faces of the multilayer structure—or cleaving, if a crystalline (*e.g.*, silicon) substrate is used—provides flat enough end-faces to permit endfire coupling or fibre butt coupling for coupling the light into the waveguide.

End-fire coupling of laser light from a fibre via a lens into the channel waveguides is the easier method, since there is no physical contact required between the light source (fibre) and the device chip. In the future, when pigtailed and packaged devices are required, permanent fibre to chip butt coupling or other hybrid coupling solutions need to be developed.

In fig. 10 is shown how the quality of a 1×2 or Y-branch splitter has been improved over time by improving mask quality and corresponding processing steps. The shown 3D plots have been obtained by imaging with an infrared camera the exit face of the polymeric multilayer structure comprising the 1×2 channel waveguide pattern. After digitization and computer processing of the obtained infrared camera signals, the pictures can be generated and show actually measured intensity distributions of the light at the end-face of the channels. In the left-hand side in fig. 10, a lot of undesired stray light propagating in the core layer between the two output channels can be observed. The right-hand side of fig. 10 shows a more recent result obtained after additional optimization of mask making and optoboard processing and here almost no undesired stray light or interference features can be observed. So, really high-quality waveguides with respect to light confinement can be made and optical propagation losses of 0.5 dB/cm at 1330 nm wavelength have been measured in the case of polymeric passive waveguides in multilayer structures made via the photobleaching method.

It must be realized that by electric-field poling the material in the waveguide has become birefringent. The poling-induced polar order of the hyperpolarizable AπD moieties causes second-order nonlinearity on a macroscopic scale, whereas the induced axial order of the anisotropically polarizable AπD groups results in the creation of an anisotropically polarizable (birefringent) bulk material. Generally, the refractive-index component coinciding with the direction

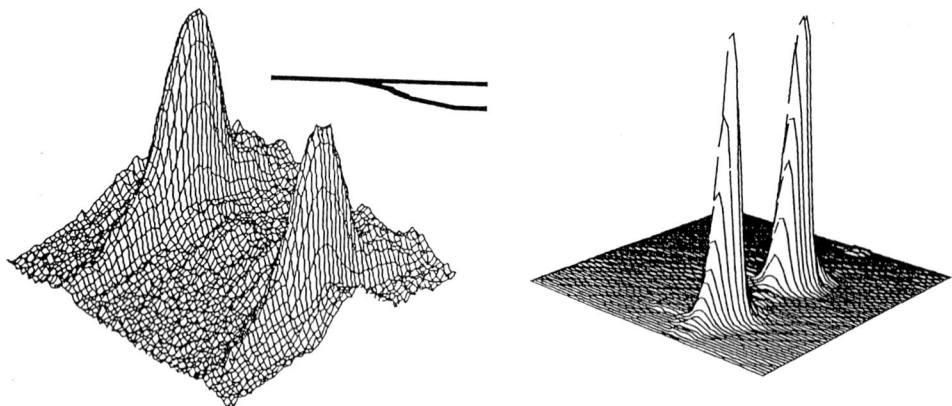

Fig. 10. – Two stages showing improved channel waveguide quality of passive polymeric optical 1×2 splitter waveguides.

of the electric poling field, which is usually applied perpendicular to the plane of the multilayer film structure, has the higher value. Consequently, the refractive-index component in the plane of the film (and perpendicular to the poling field) has the lower value.

Therefore, light propagating in a waveguide and having its electric-field vector perpendicular to the plane of the film (transverse magnetic or TM) experiences the higher refractive index. Components of the optical electric field in the plane of the film (transverse electric or TE) experience the lower refractive index. Light injected in the waveguide under an angle not equal to 0° and 90° experiences polarization rotation.

In the case of active waveguides, besides the optical coupling of the waveguide with the outside world (fibre, etc.), electrical coupling is required in order to get the electrical driving signals to the electrodes to operate and test the devices. In the case of lumped electrodes, simple electrical connections between the powering wires and electrodes are sufficient. For very-high-speed devices, travelling-wave electrode systems, combined with appropriately terminated power lines, are required.

8. – Polymeric modulators and switching devices.

8`1. *Introduction*. – Modulation and switching in waveguides can be brought about in several ways by applying passive (nonelectro-optical) as well as active (electro-optic) devices. In passive waveguide structures such as directional mode couplers, the switching depends on, among others, the wavelength or frequency of the light used. By varying the wavelength of the propagating light, the coupling efficiency can be changed. Switching functions can be obtained in such passive systems by varying the properties of the propagating light, although the waveguiding structure itself is passive. Another way to vary the coupling efficiency is by changing the temperature of the waveguide thereby changing the refractive indices and the coupling characteristics. Other examples of such thermo-optic switching effects are given further on in this lecture.

Opto-optic switching effects, in which the switching is caused by optically induced refractive-index variations, is not discussed in this lecture but will be dealt with in those lectures of this volume that are related to third-order NLO effects.

Electro-optic switching phenomena in polymers have been recently demonstrated and show promising and attractive features for application in electro-optic devices. Several examples will be discussed in the subsections hereafter.

8`2. *Polymeric integrated phase modulators*. – A basic polymeric electro-optic device structure is the phase modulator, or an array of such phase modula-

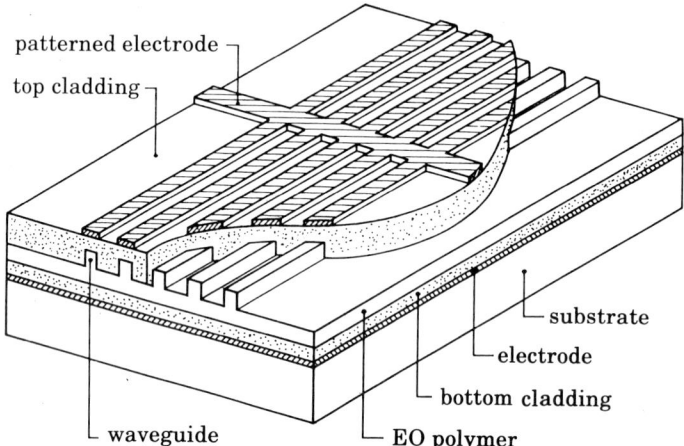

Fig. 11. – Structure of polymeric integrated phase modulator array.

tors. It comprises one or more straight-channel waveguides realized in a poled polymeric multilayer structure. A schematic representation of an array of polymeric integrated phase modulators is given in fig. 11.

On a silicon substrate, a lower gold electrode has been deposited via evaporation or sputtering, followed by a 2.4 μm thick spin-coated passive polymer cladding of relatively low refractive index[19]. The successively deposited central (core) layer, comprising a DANS-containing side-chain polymer with a T_g of about 140 °C, was 1.4 μm thick and has a higher refractive index than the cladding. This was followed by the upper (low index) polymeric cladding which is again 2.4 μm thick. With this procedure, the central layer has initially everywhere the highest refractive index of the three layers, thus forming at first a slab waveguide.

Following the deposition of the top metallization, the entire sandwich was brought to 130 °C and was poled with an electric-field strength of 100 V/μm for about 10 min. After cooling down to room temperature, the electric field was removed from the sample and the induced electro-optic effect appeared to be frozen in and permanent.

The channel waveguides were then realized via UV bleaching. With this bleaching process, the refractive index of the polymer in the central waveguide layer is locally reduced according to an optical-mask pattern, thus forming the desired waveguide structure. The optical mask for bleaching was made using the already present top metallization in combination with a mother mask and photolithographic steps in photoresist, followed by the necessary etching steps. In the case of the phase modulator array, the obtained waveguide pattern is an array of linear monomode channel waveguides.

For longer bleaching times, the refractive-index contrast will further increase; this may lead to the undesired formation of multimode channel waveguides and deterioration of the performance of the phase modulators. This can be understood as follows. In a multimode channel waveguide, each mode has its own propagation characteristics and is influenced differently (different phase shift) by an applied external electric field for modulation, giving rise to a less «clean» phase-shifted wave front as compared to monomode systems.

The device waveguide dimensions were: length 25 mm, width 6 μm and a bleached waveguide ridge 0.5 μm high (only partial bleaching through the central-layer thickness); the refractive-index difference between bleached and unbleached material at 1.32 μm wavelength was equal to 0.03.

Finally, after realizing the channel waveguides, the waveguide mask is removed and the mask for the the top electrode structure is applied on a newly metallized top layer, similarly as was described for making the waveguide pattern, thus realizing the top electrode structure. One now has achieved a processed optoboard. After cutting and polishing of the end-faces a so-called «optochip» results.

Characterization of the performance of phase modulators can be carried out by placing the obtained optochip in one arm of an external bulk Mach-Zehnder interferometer or by placing it between crossed polarizers; both methods lead in fact to intensity modulation. For the current modulator, the Mach-Zehnder approach for evaluation was taken and laser light of 1.32 μm wavelength has been end-fired via a lens into the channels under investigation. In one of the earlier experiments, in the case of TM radiation, the V_π obtained in the device was 9.7 V; this corresponds to a r_{33} of 8 pm/V. For the TE mode in the same device, a r_{13} value of 2.3 pm/V has been measured. The r_{33}/r_{13} ratio of about 3 agrees with theory. The frequency response of this phase modulator has been measured in the range (0–5) MHz (instrumentation limited) and was found to be flat.

Phase modulators with improved performance have been made using basically the same optically nonlinear polymers and processing techniques but with improved poling and handling procedures [19]. By applying a stronger electric poling field (156 V/μm) in 2.6 μm thick cladding layers and a 2.4 μm thick core, combined with optimized combinations of poling times and poling temperatures, a r_{33} value of 28 pm/V at 1.34 μm wavelength in the monomode channels of the phase modulator array was obtained. The measured frequency response (instrumentation limited) was flat in the range 20 Hz-65 MHz. Further improvements could be obtained by again applying DANS polymers with a T_g equal to 140 °C combined with more accurate processing and induced r_{33} values of 34 pm/V were shown [60]. In this case, the multilayer was poled for 10 min at 130 °C at the higher electric-poling-field strength of 170 V/μm. The active part of the device was 38.6 mm long. The measured d.c. V_π was 1.75 V at 1.3 μm; for higher driving frequencies, the V_π increased to 3.4 V. The rise in V_π as a function of increasing modulation frequency may be due to conductive impurities such as inorganic ions in the polymer. The exact reason is being investigated.

Besides the modulation voltages, the optical losses in poled polymer waveguides are important. At a wavelength of 1.32 μm, the measured waveguide propagation loss was less than 1 dB/cm. This value has to be improved to preferably below 0.3 dB/cm, which is believed to be feasable in the future.

8'3. *Polymeric integrated Mach-Zehnder interferometers.* – An intensity modulator can be made using an integrated Mach-Zehnder interferometer structure (fig. 12). By applying virtually the same techniques and procedures as in the case of the phase modulators, (arrays of) Mach-Zehnder intensity modulators have been produced. Also here, large improvements in performance have been achieved owing to improved design, mother mask and processing conditions.

Mach-Zehnder structures have been obtained by poling for 10 min at 135 °C with a field strength of 205 V/μm using a polymer with $T_g = 140$ °C. The V_π was equal to 8 V for 14 mm long electrodes; the observed extinction ratio was in excess of 20 dB at 1.3 μm wavelength; the corresponding calculated r_{33} was equal to 18.5 pm/V.

Processing improvements yielded still better devices. A recently induced d.c. r_{33} value achieved in a polymeric integrated Mach-Zehnder device was equal to 32 pm/V and corresponds to $V_\pi = 4.4$ V [60]. For higher modulation frequencies (100 Hz), the V_π value, however, increased to 9 V; the reason for this nonflatness of the frequency response is not clear at this moment but may be due to electrical imperfections in the core and/or buffer layers.

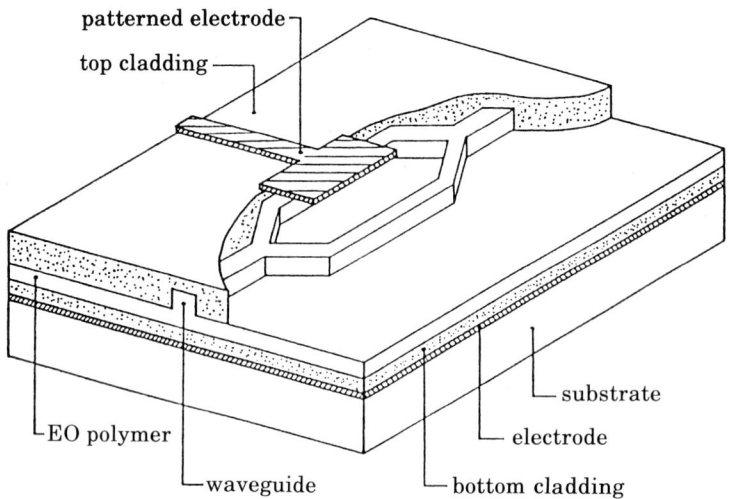

Fig. 12. – Structure of polymeric integrated Mach-Zehnder interferometer.

High-speed Mach-Zehnder interferometers have recently been reported[62] with modulation frequencies up to 20 GHz, applying Akzo's NLO polymers. This latter device has been poled with an electric-field strength of 70 V/μm for 1 min at 120 °C. The active length of the device and of the corresponding electrode was equal to 17 mm. The electrodes were so-called travelling-wave electrodes with matched impedances permitting high-speed drive signals. The measured V_π was 9 V only and intensity modulation ratios of about 10 dB were observed. This result shows the attractive features of polymeric integrated electro-optic devices. It has been theoretically shown[62] that polymeric electro-optic devices can probably be made for modulation frequencies of about 150 GHz for 1 cm long devices before walk-off between the electrical and optical signals occurs.

8˙4. *Polymeric integrated directional mode couplers*. – Another type of device is the integrated polymeric 2×2 space, or directional, switch in the form of a directional mode coupler[60]. The same techniques and procedures as mentioned in the cases of phase modulators and Mach-Zehnder modulators were applied to produce the mode couplers. These mode couplers appear to be rather difficult to produce because of the dimensional features of waveguide and electrode structures. The channel waveguides, each a few micrometres wide and high, are parallel to each other over a distance of centimetres while separated by a few micrometres only.

The devices thus made showed good switching from the cross to the bar state with a modulation depth of $- 17$ dB by applying a switching potential difference of 7.5 V over the 14 mm long active electrodes. By adding a bias voltage, the working point could be varied over the modulation curve; the switching potential difference was not affected in this way.

8˙5. *Polymeric integrated Mach-Zehnder switch*. – The 1×2 Mach-Zehnder switch[63] comprises: a Y-branched optical power splitter, first followed by two parallel arms with at least in one arm an electro-optic phase shifter, then followed by a symmetric 3 dB directional coupler.

In case of a perfect 3 dB coupler, for an induced phase difference of $\pi/2$, all optical power is propagated through one of the output arms; for a phase shift of $-\pi/2$, the light is shifted entirely into the other output arm. The voltage needed to switch the light completely from one arm to the other arm at the wavelength of 1.3 μm was 10 V for TM radiation leading to r_{33} equal to 16 pm/V. Recently, improved 1×2 switches have been made by optimizing design and processing steps[64]. Now, both arms of the phase shift section are equipped with electrodes. By proper electrical biasing of the voltages on the electrodes, the optical intensities in the two output arms can be set to any ratio. The device can be biased before all the light is initially in one of the output arms; after applying $V_\pi = 8.5$ V to the electrode, the light is switched to the other channel with an extinction ratio of better than 20 dB.

8˙6. *Thermo-optic digital switch.* – Polarization/wavelength-insensitive polymeric switches comprising asymmetric Y-junctions have been realized [63]. The switching properties are based on refractive-index modulations causing variations in the mode evolution and mode propagation in such asymmetric Y-junctions [65]. This type of switch is often called «digital optical switch» owing to its digital transfer function with respect to the applied switching signal. Switches made in GaAs and based on this principle of operation have recently been demonstrated [66].

In the present polymeric-device case, the required asymmetry has been induced by exploiting the large refractive-index variation of the polymer upon heating; the refractive index decreases with increasing temperature. The variation of the refractive index in amorphous polymers is of the order of $10^{-4}/°C$ and is generally isotropic and thus polarization independent for light propagating through the structure.

The branch angle of the waveguide at the Y-junction was 1.5°. One of the output branches was covered by a gold stripe heater; the gold heater was separated from the DANS polymer by a polymeric cladding preventing absorption or scattering of the propagating light by the electrode. The temperature of the polymer under the heater stripe in the arm of the Y-junction could be easily raised, by passing a current through the stripe, to about 140 °C which is close to the T_g of the polymer. The corresponding decrease of the refractive index was in excess of 10^{-2}, thereby exceeding the UV bleaching induced (effective) index difference of the ridge waveguide. In this latter state, the waveguide no longer exists. The operation of the switch is, therefore, based on a transition from a mode evolution regime (small temperature increase) to a cut-off regime (large temperature increase). In both regimes, the switch exhibits polarization and wavelength insensitivity, and digital behaviour. In principle, the large thermally induced refractive-index changes would permit wide branch angles of the junction section, leading to shorter devices, without excessive cross-talk.

In the case of unpolarized 1300 nm wavelength light from a butt-coupled optical fibre and without any electrical activation (heating), equal power output in both branches was obtained. If a current was passed through the heater stripe covering one arm, the optical power through the heated arm decreased, while that in the other (unheated) arm increased. For increasing current in the heated arm (further lowering the refractive index), all optical power could be directed into the other (unheated) arm. This state remained unchanged on increasing the current further. However, ultimately, the power in both arms starts to decrease owing to the spreading of the heat zone causing both arms to enter the cut-off mode regime. The switching time was measured to be of the order of milliseconds. For driving electrical powers of several tens of milliwatt, the extinc-

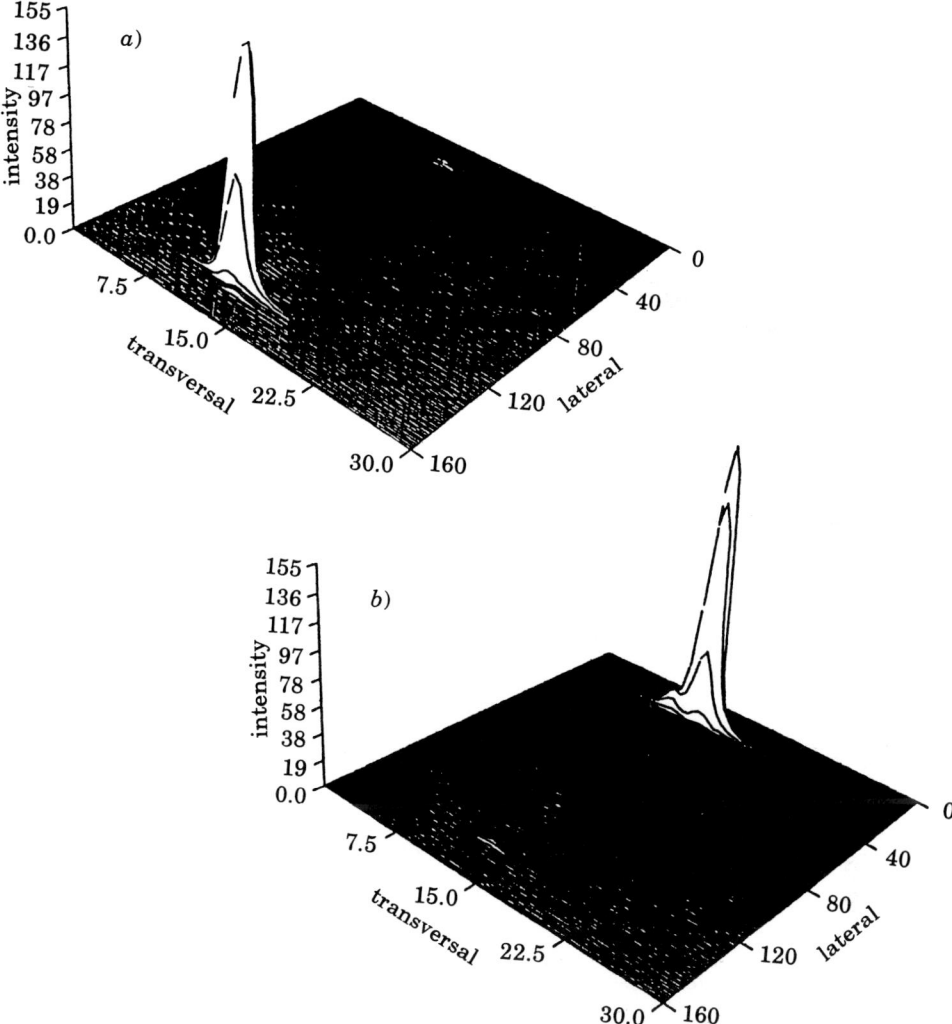

Fig. 13. – Thermo-optic switching in a 1×2 Y-branch splitter structure.

tion ratio was in excess of 20 dB. A 3D representation of the intensity distribution as measured with the i.r. camera is given in fig. 13.

Recently, a thermo-optic switch with optimized thermal design including two inputs and two active outputs has been developed; the polarization and wavelength insensitivity will be further investigated. This type of device is attractive in those places in the network where slower switches are permitted.

Experiments have been performed to set the bias of an integrated polymeric Mach-Zehnder interferometer by means of the thermo-optic effect applied to one of the two arms, whereas the other arm was driven electro-optically. The advantage of this way of operation is the drift-free character of the thermo-optic effect as compared to the sometimes observed effect of d.c.-induced bias drift. A Mach-Zehnder interferometer has been made with a procedure similar to that described earlier in the present lecture. Each arm is equipped with an electrode, one for the electro-optic switching voltage and the other to carry the current for heating the underlying polymer. The required electrical power to switch thermo-optically π radians was equal to 0.5 mW. Simultaneously, the device could be modulated with an a.c. voltage on the other arm; the electro-optic V_π was 20 V. The reason why the Mach-Zehnder structure requires only 0.5 mW for thermo-optic switching as compared to the tens of millivolt needed in the case of the thermo-optic 1×2 splitter is due to the fact that interferometric processes are more sensitive to refractive-index changes than mode evolution processes.

9. – Conclusions.

The results thus far obtained combined with the progress achieved in the field of polymeric (electro-optic, thermo-optic) devices have been very promising in the past few years. The feasibility of these materials for applications in optical information technology has been demonstrated. However, critical questions still need to be answered. The thorough understanding of the thermal relaxation of the electric-field-induced polar order is essential in order to develop new classes of polymeric materials with high-temperature stability. Although results on more stable NLO polymers have been reported[67-69], much work need to be conducted to solve the problems related to high-temperature operation.

In addition, questions related to long-term electrical, optical and environmental stability of EO devices are also of extreme importance in relation to the potential application of these devices. To answer these questions, extensive testing of polymeric EO components in test systems comparable with practical applications is required.

* * *

The authors wish to thank Dr. C. T. J. WREESMANN, Dr. U. E. WIERSUM, Ing. B. H. M. HAMS, E. W. P. ERDHUISEN, T. BOONSTRA and D. VAN OLDEN for the organic syntheses; Prof. L. W. JENNESKENS, Dr. J. W. HOFSTRAAT, Ing. J. P. B. VAN DEURSEN and J. W. VAN VELDEN for the polymer characterization; Miss C. J. M. VAN WEERDENBURG, Miss M. VAN RHEEDE and Ing. R. VAN GASSEL for the preparation and EO characterization of the EO test samples; Dr. Ir. H. M. M. KLEIN KOERKAMP, Drs. J. L. P. HEIDEMAN, Ing. J. W. MERTENS, Dr. Ir. W. H. G. HORSTHUIS and Ing. R. RAMSAMOEDJ for the preparation and testing of the EO devices; and Dr. S. J. PICKEN for the self-consistent field calculations.

REFERENCES

[1] D. J. WILLIAMS: *Angew. Chem. Int. Ed. Engl.*, **23**, 690 (1984).

[2] G. R. MEREDITH, J. G. VAN DUSEN and D. J. WILLIAMS: *Macromolecules*, **15**, 1385 (1982).

[3] A. F. GARITO, K. D. SINGER and C. C. TENG: in *Nonlinear Optical Properties of Organic and Polymeric Materials, ACS Symposium Series*, Vol. **233**, edited by D. J. WILLIAMS (American Chemical Society, Washington, D.C., 1983), Chapt. 1, p. 1.

[4] J. ZYSS: *J. Non-Cryst. Sol.*, **47**, 211 (1982).

[5] S. MIYATA, Editor: *Nonlinear Optics: Fundamentals, Materials and Devices, Proceedings of the Fifth Toyota Conference on Nonlinear Optical Materials, Aichi-ken, Japan, 6-9 October 1991* (North-Holland, Amsterdam, 1992).

[6] G. ZERBI, Editor: *Organic Materials for Photonics: Science and Technology, European Materials Society Monographs*, Vol. **6** (North-Holland, Amsterdam, 1993).

[7] G. R. MÖHLMANN and C. P. J. M. VAN DER VORST: in *Side Chain Liquid Crystal Polymers*, edited by C. B. MCARDLE (Blackie, Glasgow, 1988), Chapt. 12, p. 330.

[8] C. P. J. M. VAN DER VORST, W. H. G. HORSTHUIS and G. R. MÖHLMANN: in *Polymers for Lightwave and Integrated Optics: Technology and Applications*, edited by L. A. HORNAK (Marcel Dekker Inc., New York, N.Y., 1992), Chapt. 14, p. 365.

[9] G. R. MEREDITH, J. G. VAN DUSEN and D. J. WILLIAMS: in *Nonlinear Optical Properties of Organic and Polymeric Materials, ACS Symposium Series*, Vol. **233**, edited by D. J. WILLIAMS (American Chemical Society, Washington, D.C., 1983), Chapt. 5, p. 109.

[10] J. ZYSS and G. TSOUCARIS: *Mol. Cryst. Liq. Cryst.*, **137**, 303 (1986).

[11] J. D. LE GRANGE, M. G. KUZYK and K. D. SINGER: *Mol. Cryst. Liq. Cryst.*, **150b**, 567 (1987).

[12] K. D. SINGER, J. E. SOHN and S. J. LALAMA: *Appl. Phys. Lett.*, **49**, 248 (1986).

[13] K. D. SINGER, M. G. KUZYK and J. E. SOHN: *J. Opt. Soc. Am. B*, **4**, 968 (1987).

[14] M. SIGELLE and R. HIERLE: *J. Appl. Phys.*, **52**, 4199 (1981).

[15] C. P. J. M. VAN DER VORST and R. A. HUYTS: *SPIE Proc.*, **1126**, 6 (1989).

[16] C. P. J. M. VAN DER VORST and C. J. M. VAN WEERDENBURG: *SPIE Proc.*, **1337**, 246 (1990).

[17] C. A. ELDERING, A. KNOESEN and S. T. KOWEL: *SPIE Proc.*, **1337**, 348 (1990).

[18] T. KOBAYASHI, H. UCHIKI and K. MINOSHIMA: *SPIE Proc.*, **971**, 59 (1988).

[19] G. R. MÖHLMANN, W. H. G. HORSTHUIS, C. P. J. M. VAN DER VORST, A. MCDONACH, J. M. COPELAND, C. DUCHET, P. FABRE, M. B. J. DIEMEER, E. S. TROMMEL, F. M. M. SUYTEN, P. VAN DAELE, E. VAN TOMME and R. BAETS: *SPIE Proc.*, **1147**, 245 (1989).

[20] R. S. LYTEL: *SPIE Proc.*, **1216**, 30 (1990).

[21] G. F. LIPSCOMB, R. S. LYTEL, A. J. THICKNOR, T. E. VAN ECK, S. L. KWIATKOWSKI and D. G. GIRTON: *SPIE Proc.*, **1337**, 23 (1990).

[22] M. B. J. DIEMEER, F. M. M. SUYTEN, E. S. TROMMEL, G. R. MÖHLMANN, W. H. G. HORSTHUIS, C. P. J. M. VAN DER VORST, A. MCDONACH, J. M. COPELAND, C. DUCHET, P. FABRE, S. SAMSO, E. VAN TOMME, P. VAN DAELE and R. BAETS: in *Proceedings of the 15th European Conference on Optical Communications, ECOC'89*, p. 425.

[23] A. F. GARITO, K. Y. WONG, Y. M. CAI, H. T. MAN and O. ZAMANI-KHAMIRI: *SPIE Proc.*, **682**, 2 (1986).

[24] G. R. MEREDITH: *SPIE Proc.*, **824**, 126 (1987).

[25] C. F. J. BÖTTCHER: *Theory of Electric Polarization*, volumes **1** and **2** (Elsevier, New York, N.Y., 1973).

[26] C. P. J. M. VAN DER VORST and S. J. PICKEN: in *Polymers as Electro-Optical and Photo-Optical Active Media*, edited by V. P. SHIBAEV (Springer-Verlag, Heidelberg, in print 1994).

[27] D. B. NEAL, M. C. PETTY, G. G. ROBERTS, M. M. AHMAD, W. J. FEAST, I. R. GIRLING, N. A. CADE, P. V. KOLINSKY and I. R. PETERSON: *Electron. Lett.*, **22**, 460 (1986).

[28] J. P. CRESSWELL, J. TSIBOUKLIS, M. C. PETTY, W. J. FEAST, N. CARR, M. GOODWIN and Y. M. LVOV: *SPIE Proc.*, **1337**, 358 (1990).

[29] M. A. HUBBARD, N. MINAMI, C. YE, T. J. MARKS, J. YANG and G. K. WONG: *SPIE Proc.*, **971**, 136 (1988).

[30] S. ESSELIN, P. LE BARNY, P. ROBIN, D. BROUSSOUX, J. C. DUBOIS, J. RAFFY and J. P. POCHOLLE: *SPIE Proc.*, **971**, 120 (1988).

[31] M. EICH, A. SEN, H. LOOSER, D. Y. YOON, G. C. BJORKLUND, R. TWIEG and J. D. SWALEN: *SPIE Proc.*, **971**, 128 (1988).

[32] L. T. CHENG, W. TAM, G. R. MEREDITH, G. L. J. A. RIKKEN and E. W. MEIJER: *SPIE Proc.*, **1147**, 61 (1989).

[33] L. T. CHENG, W. TAM, A. FEIRING and G. L. J. A. RIKKEN: *SPIE Proc.*, **1337**, 203 (1990).

[34] C. P. J. M. VAN DER VORST and S. J. PICKEN: *SPIE Proc.*, **866**, 99 (1987).

[35] C. P. J. M. VAN DER VORST and S. J. PICKEN: *J. Opt. Soc. Am. B*, **7**, 320 (1990).

[36] W. MAIER and A. SAUPE: *Z. Naturforsch. A*, **13**, 564 (1958).

[37] W. MAIER and A. SAUPE: *Z. Naturforsch. A*, **14**, 882 (1959).

[38] W. MAIER and A. SAUPE: *Z. Naturforsch. A*, **15**, 287 (1960).

[39] P. LE BARNY, G. RAVAUX, J. C. DUBOIS, J. P. PARNEIX, R. NJEUMO, C. LEGRAND and A. M. LEVELUT: *SPIE Proc.*, **682**, 56 (1986).

[40] A. C. GRIFFIN, A. M. BHATTI and R. S. L. HUNG: *SPIE Proc.*, **682**, 65 (1986).

[41] J. B. STAMATOFF, A. BUCKLEY, G. CALUNDANN, E. W. CHOE, R. DEMARTINO, G. KHANARIAN, T. LESLIE, G. NELSON, D. STUETZ, C. C. TENG and H. N. YOON: *SPIE Proc.*, **682**, 85 (1986).

[42] C. P. J. M. VAN DER VORST and S.J. PICKEN: *SPIE Proc.*, **2025**, 243 (1993).

[43] D. M. BISHOP: *Phys. Rev. Lett. B*, **61**, 322 (1988).

[44] C. R. MOYLAN, S. A. SWANSON, R. J. TWIEG, C. A. WALSH, J. I. THACKARA, R. D. MILLE and V. Y. LEE: in *Organic Thin Films for Photonic Applications Topical Meeting, Toronto, October 5-7, 1993*, Optical Society of America, *Technical Digest Series*, Vol. **17** (1993), p. 77.

[45] C. P. J. M. VAN DER VORST and M. VAN RHEEDE: *SPIE Proc.*, **1775**, 186 (1992).

[46] C. P. J. M. VAN DER VORST, M. VAN RHEEDE and C. J. M. VAN WEERDENBURG: in *Proceedings of the 25th Europhysics Conference on Macromolecular Physics, St. Petersburg, July 1992, Int. J. Polym. Mater.*, **22**, 113 (1993).

[47] C. P. J. M. VAN DER VORST, M. VAN RHEEDE and B. H. M. HAMS: *SPIE Proc.*, **2025**, 137 (1993).

[48] S. I. STUPP, H. C. LIN and D. R. WAKE: *Chem. Mater.*, **4**, 947 (1992).

[49] S. I. STUPP: in *Proceedings of the 25th Europhysics Conference on Macromolecular Physics, St. Petersburg, July 1992, Int. J. Polym. Mater.*, **22**, L19 (1993).

[50] G. BERCOVIC, V. KRONGAUZ and S. YITZCHAIK: *SPIE Proc.*, **1442**, 44 (1990).

[51] D. GONIN, C. NOËL, A. LE BORGNE, G. GADRET and F. KAJZAR: *Makromol. Chem. Rapid Commun.*, **13**, 537 (1992).

[52] D. GONIN, G. GADRET, C. NOËL and F. KAJZAR: *SPIE Proc.*, **2025**, 129 (1993).

[53] A. SAUPE: in *Liquid Crystals and Plastic Crystals*, Vol. **1**, edited by G. GRAY and P. A. WINSOR (Horwood, Chichester, 1974), Chapt. 2.

[54] J. B. STAMATOFF: in *International Photonics Research Topical Meeting, Hilton Head, S. Carolina, March 1990, Technical Digest Series*, **5**, paper MA2.

[55] D. HAAS, C. C. TENG, H. YOON, H. T. MAN and K. CHIANG: in *International Photonics Research Topical Meeting, Hilton Head, S. Carolina, March 1990, Technical Digest Series*, **5**, paper MF2.

[56] R. MEYRUEIX, G. TAPOLSKY, M. DICKENS and J. P. LECOMPTE: *SPIE Proc.*, **2025**, 117 (1993).

[57] A. BUCKLEY: *Adv. Mater.*, **4**, 153 (1992).

[58] E. KELDERMAN: *Molecules for second order non-linear optics*, thesis, Twente University, Enschede (1993).

[59] J. VALLEY: private communication. See also J. F. VALLEY and J. W. WU: *SPIE Proc.*, **1337**, 226 (1990).

[60] G. R. MÖHLMANN, W. H. G. HORSTHUIS, A. McDONACH, M. J. COPELAND, C. DUCHET, P. FABRE, M. B. J. DIEMEER, E. S. TROMMEL, F. M. M. SUYTEN, E. VAN TOMME, P. BAQUERO and P. VAN DAELE: *SPIE Proc.*, **1337**, 215 (1990).

[61] M. B. J. DIEMEER, F. M. M. SUYTEN, E. S. TROMMEL, A. McDONACH, J. M. COPELAND, L. W. JENNESKENS and W. H. G. HORSTHUIS: *Electron. Lett.*, **26**, 379 (1990).

[62] D. G. GIRTON, S. L. KWIATKOWSKI, G. F. LIPSCOMB and R. LYTEL: *Appl. Phys. Lett.*, **58**, 1730 (1991).

[63] G. R. MÖHLMANN, W. H. G. HORSTHUIS, J. W. MERTENS, M. B. J. DIEMEER, F. M. M. SUYTEN, B. HENDRIKSEN, C. DUCHET, P. FABRE, C. BROT, J. M. COPELAND, J. R. MELLOR, E. VAN TOMME, P. VAN DAELE and R. BATES: *SPIE Proc.*, **1560**, 426 (1991).

[64] H. M. M. KLEIN KOERKAMP, J. L. P. HEIDEMAN and W. H. G. HORSTHUIS: in *Proceedings of the European Conference on Integrated Optics (Post Deadline Paper), Neuchatel, Switzerland April 18-22, 1993*.

[65] Y. SILBERBERG, P. PERLMUTTER and J. E. BARAN: *Appl. Phys. Lett.*, **51**, 1230 (1987).

[66] H. YANAGAWA, K. UEKI and Y. KAMAKATA: *IEEE J. Lightwave Technol.*, **8**, 1192 (1990).

[67] B. RECK, M. EICH, D. JUNGBAUER, R. J. TWIEG, C. G. WILLSON, D. Y. YOON and G. C. BJORKLUND: *SPIE Proc.*, **1147**, 74 (1989).

[68] M. EICH, B. RECK, D. Y. YOON, C. G. WILLSON and G. C. BJORKLUND: *J. Appl. Phys.*, **66**, 3241 (1989).

[69] W. H. G. HORSTHUIS, P. M. VAN DER HORST and G. R. MÖHLMANN: in *Proceedings of the First Topical Meeting on Integrated Photonics Research, Hilton Head, Technical Digest Series OSA*, **5**, 23 (1990).

Molecular Design of Nonlinear Optical Molecules and Polymers by Lambda (Λ)-Type Conformation.

S. Miyata, T. Watanabe and H. Singh Nalwa (*)

Department of Material Systems Engineering, Faculty of Technology
Tokyo University of Agriculture and Technology - Koganei, Tokyo 184, Japan

1. – Introduction.

Organic materials with unusual second-order nonlinear optical properties are attracting a great deal of attention because of their potential applications in photonic devices. For the appearance of second-order nonlinear optical effects, it is necessary that organic molecules must lack a centre of symmetry. Because of this prerequisite, most of the organic materials fail to exhibit second-harmonic generation since they crystallize in a centrosymmetric space group. Therefore, an important point in designing new organic materials for second-harmonic generation (SHG) is to introduce the noncentrosymmetry in crystal structures. One way is to incorporate centrosymmetric materials into organic-polymer backbones which generate second-order nonlinear optical effects after electrical poling is applied to them. Organic polymers possessing NLO chromophores are an important class of materials due to their large nonlinear optical (NLO) coefficients, fast response time, intrinsic tailorability and processability for integrated optics [1-4]. The NLO chromophores can be used either in guest-host systems [5-7] or by covalently attaching them to polymer backbones. The NLO chromophores may be chemically imparted into the polymer backbone either as side-chains [8-12] or making them a segment of the polymer main chain [13-15]. This chemical strategy, not only allows a higher density of NLO chromophores, but also provides materials with good processability. By appropriate chemical modifications, second-order nonlinear optical properties, mechanical and thermal properties of NLO-dye grafted polymers can be tailored.

(*) Hitachi Research Laboratory, Hitachi Ltd., 7-1-1 Ohmika-cho, Hitachi City, Ibaraki 319-12, Japan.

The main-chain NLO polymers are attracting much interest because of their better temporal stability and a variety of head-to-tail main-chain NLO polymers have been reported by several groups [16-19]. Unfortunately, the poling of the head-to-tail NLO polymer is not so easy because a complete reorientation of the polymer chain is difficult to achieve. We have focused our research activities on organic molecules with Λ-type charge transfer conformation because they easily form a noncentrosymmetric crystal structure. These Λ-shaped molecules can be also incorporated into the polymer backbone forming a new class of main-chain NLO polymers. Interestingly, the off-diagonal components of the hyperpolarizability (β) of Λ-shaped molecules are large and this increases the effective phase-matched harmonic generation. Like other organic NLO single crystals, the crystal growth of Λ-shaped molecules with large size and good optical quality is not so easy. Therefore, Λ-shaped molecules designed and synthesized by us were incorporated into the polymer main chain to study their second-order NLO properties. This lecture describes the progress in novel NLO molecules and polymers developed using a Λ-shaped charge transfer conformation.

2. – Strategies for noncentrosymmetric crystal structures for second-harmonic generation.

As pointed out earlier, a prerequisite for the appearance of second-harmonic generation is the absence of a centre of symmetry in organic crystals. However, the majority of organic materials ($\sim 70\%$) crystallize in centrosymmetric space groups, as a consequence they do not show SHG optical properties. In particular, most achiral organic molecules possessing large ground-state dipole moments acquire a centrosymmetric crystal structure that restricts their use for second-order nonlinear optics. Different chemical and physical strategies have been implemented to introduce noncentrosymmetric crystal structures capable of displaying SHG activity through intermolecular forces. The various strategies that have been applied for forming an acentric crystal structure are as follows:

1) chirality [20],

2) hydrogen bonding [21],

3) steric hindrance [22],

4) guest-host systems [23, 24],

5) electrical poling [25],

6) reduced dipole-dipole interaction [26].

The chemical and physical approaches listed above have been found useful for ensuring a dipolar alignment favourable for SHG activity. It is necessary to consider other possibilities of introducing a noncentrosymmetric structure in crystals while designing new materials for second-harmonic generation and Pockel's electro-optical effect. Electrical poling has been often used for preparing noncentrosymmetric structures either from guest-host systems or NLO chromophore grafted polymers. In the following section, we discuss a new molecular-engineering approach for introducing noncentrosymmetry in organic molecules using Λ-type conformation. These Λ-shaped organic molecules and their based polymers constitute a new class of organic materials usable for second-harmonic generation and electro-optical studies.

3. – Introducing noncentrosymmetry by lambda (Λ)-type conformation: theoretical and experimental evidences.

Many approaches have been suggested for generating a noncentrosymmetric crystal structure either by modifying the chemical structures or by physical means. It is difficult to predict the crystal packing merely by applying these approaches and, as a result, very often organic crystals are obtained which show no second-harmonic generation. The important point is achieving the desired optimal molecular orientation in a crystal for efficient phase-matched SHG. ZYSS and CHEMLA[27] developed a relationship between the microscopic optical nonlinearity β of the molecule and the macroscopic optical nonlinearity $\chi^{(2)}$ of a molecular crystal and indicated the optimization of the molecular orientation for bulk phase matching through the oriented-gas model. In this model, the molecular orientation is defined by an angle α between the molecular charge transfer axis and the crystal principal dielectric axis. For one-dimensional charge transfer molecules, the optimal angle for crystal phase-matched optical nonlinearity must be 54.74°. In this case, about 38% of the β_{xxx} can be utilized for the bulk phase matching. The oriented-gas model suggests that molecular structures should be designed taking into account the desired noncentrosymmetric arrangement. Therefore, it is of significant importance to apply new molecular-design strategies to introduce a noncentrosymmetric crystal structure. The molecular design of Λ-shaped charge transfer molecules for SHG discussed here is an entirely new approach that can be applied to both organic molecules and polymers. To substantiate this new molecular-design concept, we have performed theoretical calculations on model compounds as well we have practically synthesized these Λ-shaped molecules in our laboratory.

MOPAC AM1 calculations were performed on a variety of methanediamine derivatives which attain a Λ-type conformation. The first hyperpolarizability (β) of Λ-shaped molecules is listed in table I. Figure 1 shows an optimized conformation of N, N'-bis(p-nitrophenyl)-methanediamine (p-NMDA) which acquires an Λ-shape with permanent dipole moment μ parallel to the molecular y-axis[28]. Interestingly, our results demonstrate that the Λ-shaped molecules

TABLE I. – *Dipole moment, first hyperpolarizability at frequency $\omega = 0$ and lambda angle between two aromatic rings in Λ-shaped molecules.*

Molecule	Lambda angle (degrees)	Dipole moment (Debye)	β_{xxx} (10^{-30} e.s.u.)	β_{xyy} (10^{-30} e.s.u.)
p-NMDA	120	5.4	1.7	12.2
o-NMDA	125	3.0	0.211	2.7
MNPMDA	110	0.6	2.1	2.7
MCPMDA	125	1.7	2.1	10.2
ECPMDA	120	1.5	1.1	10.2

form a noncentrosymmetric crystal structure by staking over and over along one direction. The angle between two charge transfer axes of each of the p-nitroaniline parts of p-NMDA is about $120°$, close to an optimal orientation of two p-nitroaniline molecules for crystal phase-matched optical nonlinearity. For example, the best condition is $2\alpha = 109.48°$. Of considerable interest are the different β components because here the predominant β component of p-NMDA is β_{xyy} resulting from a two-dimensional charge transfer interaction. The β tensors

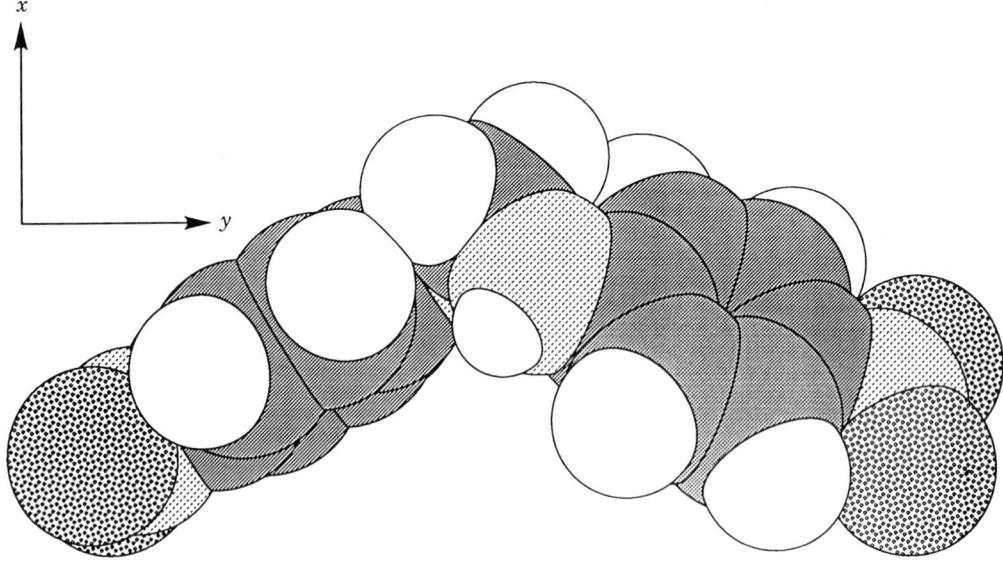

Fig. 1. – Optimized conformation of N, N'-bis(p-nitrophenyl)-methanediamine (p-NMDA) that acquires a Λ-shape with permanent dipole moment μ parallel to the molecular x-axis.

for Λ-shaped molecules were calculated as a function of the dihedral angle θ by the finite-field method. For p-NMDA, the β_{xyy} is $12.2 \cdot 10^{-30}$ e.s.u., which is about 12 times larger than that of the diagonal β_{xxx} component as estimated from the finite-field method. On the other hand, p-nitroaniline is a typical one-dimensional charge transfer molecule possessing β_{xxx} as the largest hyperpolarizability component. Therefore, these Λ-shaped two-dimensional charge transfer molecules are characteristically different from those of one-dimensional charge transfer systems. The p-NMDA and other Λ-shaped molecules were chemically synthesized and they all crystallize in noncentrosymmetric space groups.

4. – Lambda (Λ)-shaped NLO molecules and polymers.

4`1. Λ-shaped molecules. – A variety of methanediamine derivatives attaining Λ-shaped conformations by bonding together two aniline derivatives via a methylene ($-CH_2-$) bridge were synthesized. Table II lists the melting points, cut-off wavelengths and powder SHG of aniline derivatives and methanediamine derivatives derived from them. The eight analogues selected in this study were p-nitroaniline, 2-methyl-4-nitroaniline, N-methyl-4-nitroaniline, 2-methoxy-4-nitroaniline, 2-amino-4-nitrotoluene, 4-amino benzoic acid methyl ester, 4-amino benzoic acid ethyl ester and o-nitroaniline. The Λ-shaped molecules were prepared by the reacting the corresponding aniline with formaldehyde in methanol at room temperature. The powder SHG efficiency of the Λ-shaped molecules was determined at 1.064 µm from a Q-switched pulse of a Nd:YAG laser and compared with the reference urea. Out of 8 aniline derivatives, only 2 were SHG active, whereas almost all Λ-shaped molecules prepared from these aniline derivatives exhibit SHG activity, indicating that they attain noncentrosymmetric crystal structures. Such a high probability (87.5%) of forming noncentrosymmetric crystal structures is the result of the new molecular-design approach to Λ-shaped conformation. To the best of our knowledge, there is no chemical strategy which can offer such a high percentage of noncentrosymmetric molecules. Therefore, our approach to Λ-type conformation is unique and very useful in designing novel SHG active molecules. The powder SHG efficiency of p-NMDA was found to be 80 times larger than that of the urea standard. The cut-off wavelengths of Λ-shaped molecules having an electron acceptor nitro group are within (450–500) nm, which is relatively shorter than in aniline derivatives. The Λ-shaped molecules having ester group show shorter cut-off wavelengths around 330 nm, indicating that these materials are attractive for frequency doubling in the blue region. These Λ-shaped molecules also have higher thermal stability than aniline derivatives because their melting point is as high as 200 °C.

TABLE II. - SHG powder efficiencies of aniline and methanediamine derivatives. T_m melting point, λ measured in 10^{-5} mol/l EtOH solution.

No.	Aniline derivative	T_m (°C)	Cut-off λ (nm)	SHG (for urea)	Methanediamine derivative	T_m (°C)	Cut-off λ (nm)	SHG (for urea)
1	O_2N-⬡-NH_2	147	470	0	O_2N-⬡-$(NH$-$CH_2)_2$	240	460	80
2	O_2N-⬡-NH_2 (CH_3)	131	480	60–150	O_2N-⬡-$(NH$-$CH_2)_2$ (CH_3)	260	450	6.3
3	O_2N-⬡-NH_2-CH_3	106	490	0	O_2N-⬡-$(N$-$CH_2)_2$ (CH_3)	242	470	6
4	O_2N-⬡-NH_2 (OCH_3)	140	490	0	O_2N-⬡-$(NH$-$CH_2)_2$ (OCH_3)	254	480	2.5
5	(NH_2) O_2N-⬡-CH_3	118	510	0	(O_2N) ⬡-$(NH$-$CH_2)_2$ (CH_3)	241	480	3
6	CH_3O-$C(=O)$-⬡-NH_2	114	330	0	$(CH_3O$-$C(=O)$-⬡-$NH)_2$-CH_2	215	330	$\left\{\begin{array}{l}30\\0\end{array}\right.$
7	C_2H_5O-$C(=O)$-⬡-NH_2	90	330	8	$(C_2H_5O$-$C(=O)$-⬡-$NH)_2$-CH_2	195	330	25
8	⬡-NH_2 (NO_2)	71	500	0	⬡-$(NH$-$CH_2)_2$ (NO_2)	195	490	0

Scheme 1.

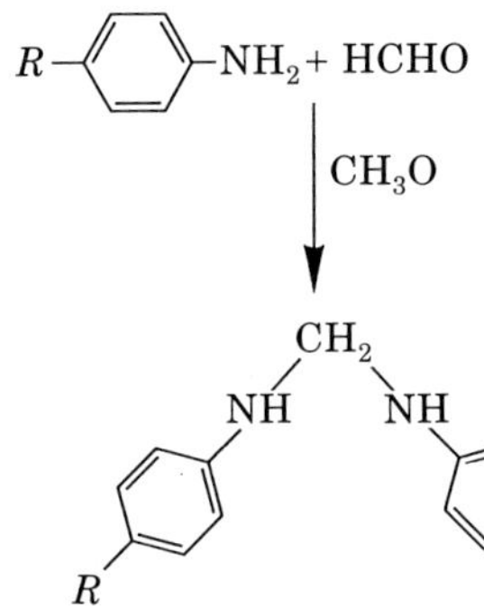

R=NO$_2$, CO–OCH$_3$, CO–OC$_2$H$_5$

TABLE III. – *Crystal structure data on Λ-shaped molecules.*

Crystal data	*p*-NMDA	*o*-NMDA	MNPMDA	MCPMDA	ECPMDA
crystal solvent	NMF	acetone	acetone	acetone	acetone
crystal size (mm^3)	$0.1 \times 0.04 \times 0.7$	$0.1 \times 0.01 \times 2.0$	$0.2 \times 0.1 \times 0.2$	$0.2 \times 0.4 \times 0.8$	$0.2 \times 0.2 \times 0.8$
crystal class	trigonal	monoclinic	monoclinic	monoclinic	tetragonal
space group	*P*3	*Cc*	*C*2	*Cc*	I4$_1$*cd*
unit-cell parameter	$a = 18.791$ Å	$a = 23.237$ Å	$a = 16.795$ Å	$a = 16.275$ Å	$a = 20.472$ Å
	$b = 18.788$ Å	$b = 4.111$ Å	$b = 5.233$ Å	$b = 7.779$ Å	$b = 20.475$ Å
	$c = 4.147$ Å	$c = 14.635$ Å	$c = 9.802$ Å	$c = 14.029$ Å	$c = 8.488$ Å
	$\beta = 120°$	$\beta = 111.4°$	$\beta = 120.6°$	$\beta = 119.5°$	
molecules/cell	$Z = 3$	$Z = 4$	$Z = 2$	$Z = 4$	$Z = 8$
heat of formation (kcal)	75.1	84.9	56.7	− 79.5	− 108.5
density (g/cm^3)	1.36	1.47	1.42	1.35	1.28
R (%)	5.8	6.0	3.9	5.1	5.8
refractive indices			$n_y = 1.630$ (ω)		$n_y = 1.761$ (ω)
			$n_y = 1.687$ (2ω)		$n_y = 1.791$ (2ω)
NLO coefficient (pm/V)			$d_{yyy} = 9.3$		$d_{zzz} = 4.7$
			$d_{yzz} = 13.1$		$d_{zxx} = 20.6$

Single crystals of Λ-shaped molecules were grown from solution at room temperature. Single crystals of p-NMDA were grown from acetone, methylethylketone, methanol, ethanol, tetrahydrofuran (THF), N,N-dimethylacetamide (DMA), N,N-dimethylformamide (DMF), N-methylformamide (NMF), dimethylsulfoxide (DMSO) and hexamethylphosphorictriamide (HMPA). Yellow single crystals with dimensions $(0.1 \times 0.04 \times 0.7)$ mm^3 were grown from NMF for X-ray analysis. The p-NMDA crystallizes in the very rare $P3$ space group. Single crystals of other Λ-shaped molecules were obtained from acetone. The crystal size and crystal data of Λ-shaped molecules are listed in table III. Figure 2 shows the packing pattern of a MNPMDA crystal and the Λ-shaped molecules stack along one direction parallel to the B-axis. Hydrogen bonds formed between amino groups and nitro groups on neighbouring 2_1 screw-related molecules lead to the formation of a noncentrosymmetric structure. Therefore, the interactions due to hydrogen bonds are responsible for introducing a noncentrosymmetric crystal structure in MNPMDA. MCPMDA exhibits polymorphism and the SHG active form belongs to the Cc space group. Figure 3 shows the crystal structure of ECPMDA and here again the Λ-shaped molecules are stacked in one direction every 90° by the 4_1 screw symmetry. The packing mode tends to be dominated by a preferential Λ-shaped stacking though there is no strong hydrogen bonding. The angle between two aniline parts of ECPMDA is about 125° close to the optimum orientation for bulk phase matching. The largest β_{xyy} component effectively contributes to the phase-

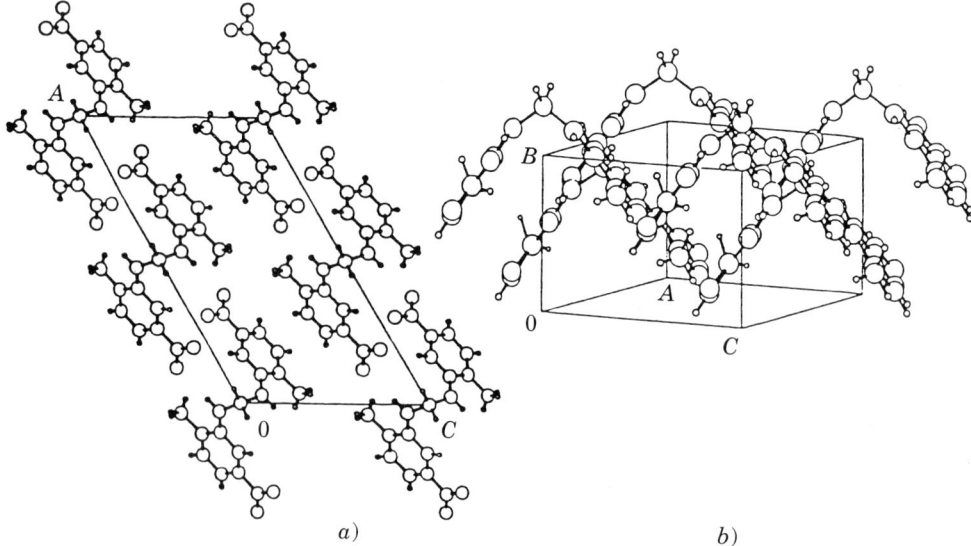

$a)$ $b)$

Fig. 2. – Crystal structure of MNPMDA $a)$ as viewed along the B-axis, $b)$ unit-cell packing pattern.

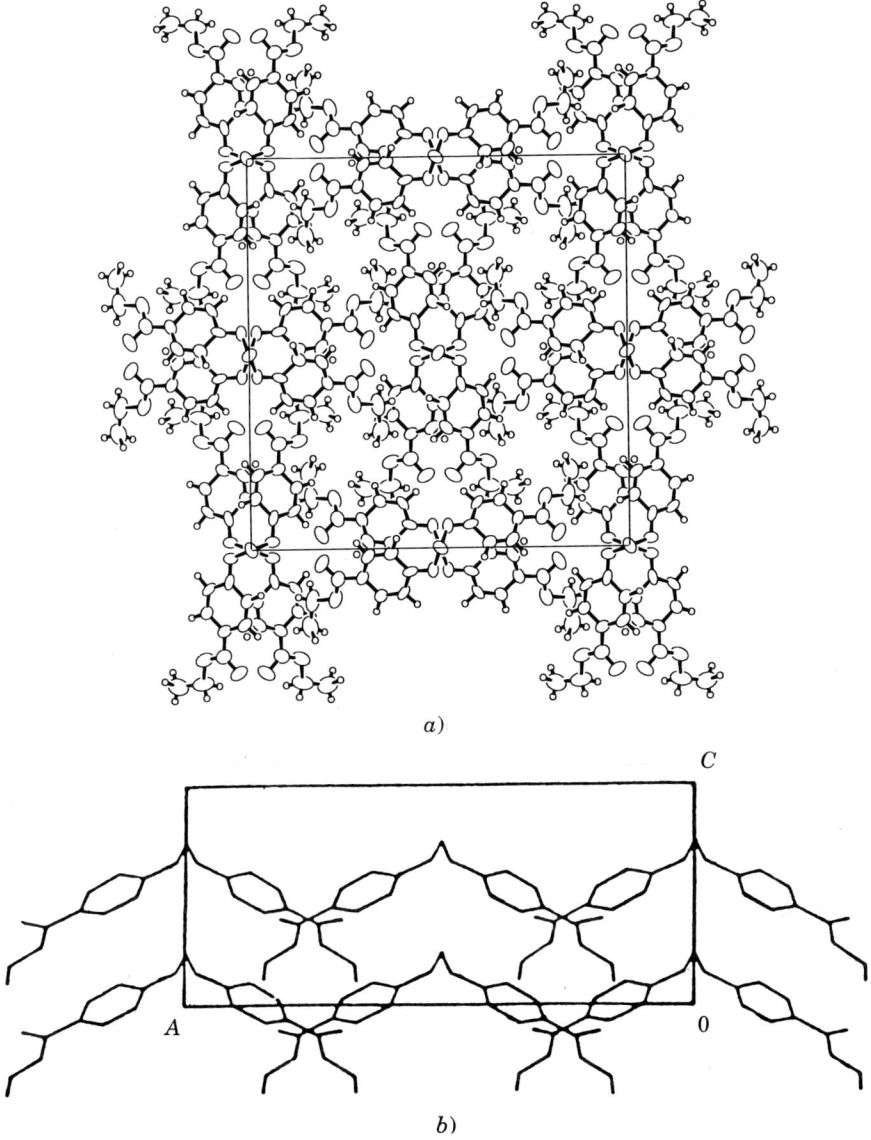

Fig. 3. – Crystal structure of ECPMDA *a*) as viewed along the *C*-axis, *b*) along the *B*-axis.

matchable d_{31} coefficient. The d_{31} coefficient of ECPMDA determined from the oriented-gas model was 20.4 pm/V [28]. ECPMDA shows the possibility of noncritical phase matching by tuning the crystal temperature.

Using the Λ-type conformation initially developed by Miyata's research team, new Λ-shaped nonlinear optical chromophores have been developed by

other researchers. ERMER *et al.* [29] prepared a series of photobleachable
Λ-shaped chromophores based on the (dicyanovinyl)pyran acceptor group.
The   {2,6-bis[3-(9-ethyl)carbazolyl)ethenyl]-4*H*-pyran-4-ylidene)propanedinitrile}
DADC chromophore shown below has $\mu\beta$ 1.5 times that of Disperse Red 1
(DR1) and shows an absorption maximum at 459 nm. The DADC is stable in air
and nitrogen up to 340 °C as shown by thermogravimetric analysis. A Mach-
Zehnder modulator was fabricated with a 15% concentration of DADC dissolved
in a polyimide matrix.

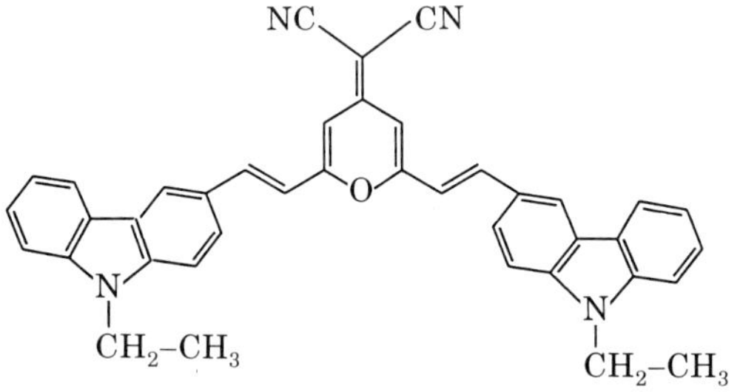

4˙2. *Polymers having* Λ-*shaped* NLO *chromophores as side-chains.* – The
reaction of Λ-shaped molecules with 4,4′-diphenyl methanediisocyanate in
dimethylformamide yields a novel series of polyureas having Λ-shaped chromophores partly in the main chain and mainly as the side-chains. Scheme 2
shows the synthetic route of polyureas having Λ-shaped molecules. Table IV
lists the thermal, optical and NLO coefficients of polyureas derived from Λ-
shaped molecules [30]. The optical transparency and NLO coefficients of these

TABLE IV. - *Chemical structures, cut-off wavelength, initial decomposition temperature*
(IDT) and d_{33} coefficients of aromatic polyureas having Λ-*shaped molecules as side*
chains.

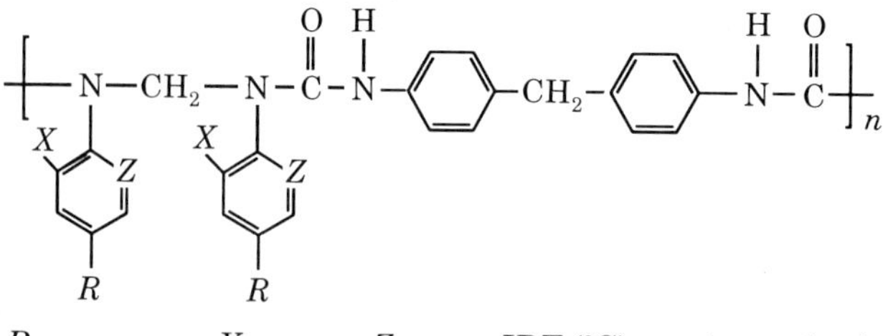

Polyurea	R	X	Z	IDT (°C)	$\lambda_{cut-off}$ (nm)	d_{33} (pm/V)
PURE-Λ1	NO$_2$	H	CH	300	450	10.5
PURE-Λ2	COOCH$_3$	H	CH	220	340	7.5
PURE-Λ3	NO$_2$	CH$_3$	CH	245	450	11.5
PURE-Λ4	NO$_2$	H	N	285	460	7.1

polyureas depend upon the chemical structure. For example, the Λ-shaped molecules with electron acceptor nitro group show large second-order optical nonlinearity and longer cut-off wavelengths. On the other hand, Λ-shaped molecules with electron acceptor ester group have shorter cut-off wavelengths as well as lower NLO coefficients. Furthermore, the Λ-shaped molecules with a pyridine ring have lower NLO coefficient but longer cut-off wavelengths, indicating that the benzene ring is more advantageous for large optical nonlinearity. These polyureas can be easily processed into thin films and show good thermal stability up to 300 °C.

Scheme 2.

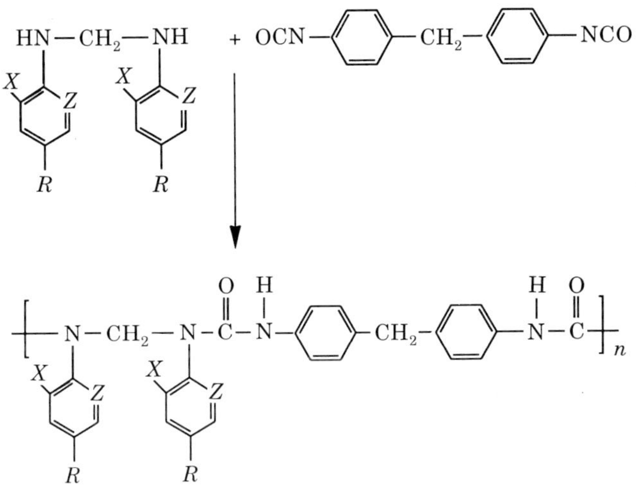

HN—CH₂—NH + OCN—⟨benzene⟩—CH₂—⟨benzene⟩—NCO

$$\left[\!-N-CH_2-N-\overset{O}{\underset{}{C}}-N-\text{⟨benzene⟩}-CH_2-\text{⟨benzene⟩}-N-C\!-\right]_n$$

4'3. *Polymers having* Λ-*shaped* NLO *chromophores in the main chain.* – Polyarylamines having Λ-shaped molecules in their main chain were synthesized by nucleophilic displacement polymerization by reacting bis(4-fluoro-3-nitrophenyl) sulfone (BFNPS) with aliphatic and aromatic diamines in aprotical solvents [31]. The schematic route is shown in scheme 3, here X is a hydrogen or an alkyl group, and R is a flexible spacer group. Polyarylamines are yellow powders and can be easily dissolved in aprotic solvents such as DMSO, DMF and NMP. Polymer films for refractive indices and SHG Maker fringe measurements were prepared by spin coating on glass slides from a NMP solution. These main-chain polymers were poled by the corona poling method. In the experiment, (5-8) kV positive electric voltage was set to the tungsten needle which was kept about 1 cm above the grounded aluminium electrode and the corona current was fixed at 2 µA. The poling temperature was kept near the glass transition temperature (T_g) of the polymers. The poling voltage, temperature and time were varied to have the best poling conditions. The temporal stability of poling-induced

SHG at elevated temperatures was studied by the *in situ* SHG measurements.

Scheme 3.

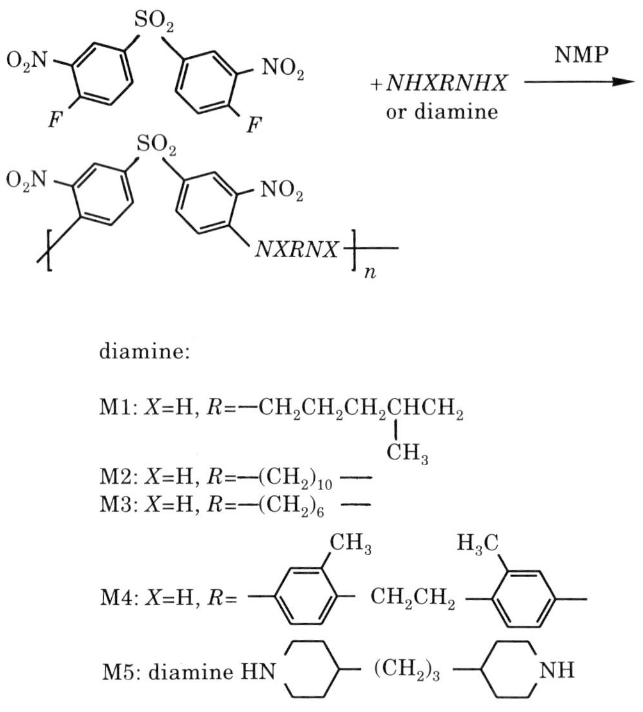

diamine:

M1: X=H, R=—CH$_2$CH$_2$CH$_2$CHCH$_2$
                              |
                              CH$_3$

M2: X=H, R=—(CH$_2$)$_{10}$ —

M3: X=H, R=—(CH$_2$)$_6$ —

M4: X=H, R= (with CH$_3$ and H$_3$C substituents) CH$_2$CH$_2$

M5: diamine HN — (CH$_2$)$_3$ — NH

The absorption spectra of polyarylamine films spin coated on indium-tin oxide (ITO) glass were measured with a UV-visible spectrophotometer at a wavelength of (300–800) nm. After poling, a decrease in absorbance (hypochromic shift) was observed in all polymers, though the absorption spectrum before and after poling was similar, indicating that no chemical changes occurred during poling. The optical absorption characteristics are related to the strength of donor groups and π-conjugation length. In M1 and M2 polymers, the donor is a N-substituted amino group and their cut-off wavelengths are about 500 nm, whereas, in the case of the M3 and M5 polymers, the donor group is a bis-N-substituted amino group and their cut-off wavelengths are around 510 nm. The longest cut-off wavelength is 540 nm for the M4 polymer since an aromatic diamine was used as the spacer unit that forms a longer π-conjugated system over other polyarylamines.

The glass transition temperature of these polyarylamines varies depending on the spacer groups. In the case of longer spacer units, the T_g is lowered since the polymers with long flexible spacer groups are easy to rotate. DTA and TGA

TABLE V. – *Glass transition temperature* (T_g), *poling temperature, refractive indices and NLO coefficients of polyarylamines having* Λ-*shaped molecules in the main chains.*

Polymer	$\lambda_{\text{cut-off}}$ (nm)	T_g (°C)	Poling temperature (°C)	Refractive index (1064 nm)	Refractive index (532 nm)	NLO coefficients	
						d_{33} (pm/V)	d_{31} (pm/V)
M1	500	130	100	1.6483(TM) 1.6474(TE)	1.7485(TM) 1.7427(TE)	12.0	4.2
M2	510	78	70	1.6385(TM) 1.6381(TE)	1.7103(TM) 1.7073(TE)	7.6	2.5
M3	500	105	130	1.6340(TM) 1.6386(TE)	1.7038(TM) 1.7027(TE)	17.6	5.4
M4	540	192	170	1.6541(TM) 1.6654(TE)	1.7059(TM) 1.7289(TE)	9.2	3.0
M5	510	148	135	1.6164(TM) 1.6268(TE)	1.6868(TM) 1.6966(TE)	8.0	2.4

studies showed no significant difference in the decomposition temperature recorded in air and nitrogen atmospheres. Before decomposition no melting point was observed in these polymers, which indicates that these polymers could be amorphous. The amorphous state of these polymers was further supported by the wide-angle X-ray diffraction. The refractive indices of these polymers measured by the *m*-line method in a slab waveguide configuration [32] are listed in table V. The refractive indices of these polyarylamines depend on spacer group lengths, chromophore density and charge transfer absorption wavelengths. For example, the M2 polymer has 10 methylene units, the refractive index of n_{TM} at 1.064 μm is 1.6385 after poling. On the other hand, the M1 polymer has 5 methylene units that lead to a larger density of chromophores and yield higher refractive indices compared with the M2 polymer. The M4 polymer, which has a spacer group with two methylenes and two substituted benzene rings, shows the highest refractive index and the longest cut-off wavelength among these polymers. A very interesting behaviour is the appearance of the positive birefringence before poling, which is governed by both the chromophore density and the steric constraints. For example, the M2 polymer has a long and flexible spacer group $(CH_2)_{10}$ and the birefringence $(n_{\text{TE}} - n_{\text{TM}})$ is 0.0036 at 1.064 μm before poling. Under identical conditions, the birefringence of M4 was found to be 0.0217 because this polymer has a more rigid spacer group. When M1, M2 and M3 were corona poled at a temperature below than their T_g's, large order parameters are introduced by poling and the refractive index $n_\perp$ becomes larger than $n_\parallel$. When the spacer group is more rigid such as in the M4 and M5 polymers, the refractive index $n_\parallel$ is still larger than $n_\perp$ even after poling. The large positive birefringence may be associated with the special molecular and morphological structures of these polymers. Such a kind of birefringence has also been observed in the polyimide films [33]. In polymers with flexible spacer groups, corona poling induces a better alignment and the chromophores get oriented in the direction perpendicular to the film, as a result the refractive index $n_\perp$ becomes larger than $n_\parallel$. Contrary to this, polymers with very rigid space groups have very large steric constraints, the poling-induced orientation is not large enough and the chromophores remain preferentially oriented along the film plane. This study on imparting Λ-shaped charge transfer chromophores in the polymer backbone, not only opens an avenue to new second-order NLO polymers, but also indicates the new possibility of phase matching in polymer films by using positive birefringence.

Figure 4 shows the dependence of SH intensity as a function of the incidence angles in the range of $-70°$ to $70°$ for P and S polarizations of the fundamental beam (1.064 μm) for the M4 polymer. Table V summarizes the calculated nonlinear optical coefficients of polyarylamines. The difference in d_{33} values may be attributed to the degree of molecular alignment induced by poling. All polymers have a common electron acceptor nitro group and the main difference is in their spacer units and donor groups. In polyarylamines prepared from aliphatic diamines with BFNPs such as the M1, M2 and M3 polymers, the spacer units are long flexible methylene chains which can be easily poled by corona discharge, therefore the poling-induced order parameters of these polyarylamines

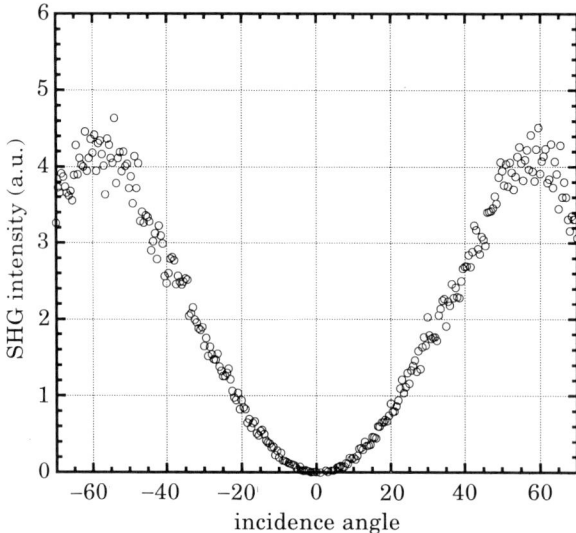

Fig. 4. – Dependence of SH intensity on the incidence angles in the range of $\pm 70°$ for P and S polarizations of the fundamental beam (1.064 μm), respectively, for the M4 polymer.

may be larger than those of the M4 and M5 polymers. Furthermore, the M3 polymer has a stronger donor group than the M1 and M2 polymers. Another factor that could affect the magnitude of NLO coefficients is the NLO chromophore density. For example, M2 has the lowest chromophore density and also shows smaller NLO coefficients than the M1 and M3 polymers. M5 also has a stronger donor group like M3 polymers, but its piperidyl group is rigid, therefore the extent of alignment induced by poling is smaller than in other polyarylamines prepared from aliphatic diamines. The M4 polymer has two methylene units and two substituted benzene rings, therefore it has a relatively longer π-conjugation length. This gives rise to large NLO coefficients in the M4 polymer. M3 has the largest d_{33} value because its poling temperature was higher than its T_g, whereas other polymers were poled below their T_g's.

As seen in the UV-visible spectra data, the absorption at second-harmonic wavelength (2ω) of these polymers is negligible, therefore the NLO activity is not enhanced by a resonance effect. However, the role of resonance enhancement of NLO activity in the M4 polymer cannot be ruled out since it has a considerable red-shifted cut-off wavelength compared with other polyarylamines. The NLO data on these polyarylamines indicate that the main-chain polymer having Λ-shaped chromophores can be successfully aligned by corona poling. To investigate the NLO stability, the poled M4 polymer film, which has stable SHG at room temperature, was heated to 100 °C. Its SHG intensity was monitored continuously at a fixed angle ($\Phi = 60°$). Figure 5 shows the NLO stability of M4 polymers. During the first 50 min of the experiment, the SHG intensity remained approximately 60% of the original value, indicating that even at high temperature the temporal stability of poling-induced SHG of a main-chain polymer with Λ-shaped chromophores is better than that of some of the NLO dye

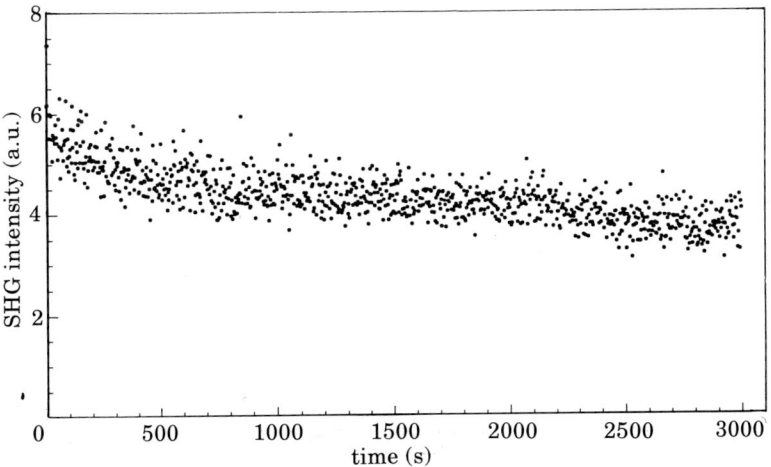

Fig. 5. – Temporal stability of SHG for the M4 polymer at 100 °C.

grafted polymers. This novel series of second-order NLO main-chain polymers having Λ-shaped chromophores shows large birefringence, indicating a preferential orientation in the polymer systems. The d_{33} coefficients of these polymers were found to be in the range (7–18) pm/V and they show good poling-induced temporal stability at elevated temperatures. The SHG properties of Λ-shaped molecules were also investigated in a polymer host system [34]. Both poling and physical aging processes proved to be advantageous for improving the temporal stability of SHG activity. We have demonstrated that a variety of novel nonlinear optical molecules and polymers useful for photonic devices can be synthesized using Λ-shaped chromophores.

REFERENCES

[1] D. S. CHEMLA and J. ZYSS, Editors: *Nonlinear Optical Properties of Organic Molecules and Crystals* (Academic Press, New York, N.Y., 1987).

[2] S. MIYATA, Editor: *Nonlinear Optics, Fundamentals, Materials and Devices* (North Holland, Amsterdam, 1992).

[3] H. S. NALWA, T. WATANABE and S. MIYATA: in *Progress in Photochemistry and Photophysics*, edited by J. F. RABEK (CRC Press, Boca Raton, Fla., 1992), Vol. 5, Chapt. 4, p. 103.

[4] S. R. MARDER, J. E. SOHN and G. D. STUCKY: *Materials for Nonlinear Optics*, ACS Symposium Series, Vol. 455 (American Chemical Society, Washington, D.C., 1991).

[5] G. R. MEREDITH, J. G. VAN DUSEN and D. J. WILLIAMS: *Macromolecules*, 15, 1385 (1982).

[6] T. WATANABE and S. MIYATA: *Proc. SPIE*, 1147, 101 (1989).

[7] K. D. SINGER, J. E. SOHN and S. J. LALAMA: *Appl. Phys. Lett.*, 49, 248 (1988).

[8] K. D. SINGER, M. G. KUZYK, W. R. HOLLAND, J. E. SOHN, S. J. LALAMA, R. B. COMIZZOLLI, H. E. KATZ and M. L. SCHILLING: *Appl. Phys. Lett.*, 53, 1800 (1988).

[9] D. H. CHOI, H. M. KIM, W. M. K. P. WIJEKOON and P. N. PRASAD: *Chem. Mater.*, 4, 1253 (1992).

[10] A. HAYASHI, Y. GOTO, M. NAKAYAMA, K. KALUZYNSKI, H. SATO, T. WATANABE and S. MIYATA: *Chem. Mater.*, 3, 6 (1991).

[11] D. R. ROBELLO: *J. Polym. Sci., Polym. Chem. Ed.*, 28, 1 (1991).

[12] M. A. MORTAZAVI, A. KNOESEN, S. T. KOWEL, R. A. HENRY, J. M. HOOVER and G. A. LINDSAY: *Appl. Phys. B*, 53, 287 (1991).

[13] D. JUNGBAUER, I. TERAOKA, D. Y. YOON, B. RECK, J. D. SWALEN, R. J. TWIEG and C. G. WILLSON: *J. Appl. Phys.*, 69, 8011 (1991).

[14] I. TERAOKA, D. JUNGBAUER, B. RECK, D. Y. YOON, R. J. TWIEG and C. G. WILLSON: *J. Appl. Phys.*, 69, 2568 (1991).

[15] J. T. LIN, M. A. HUBBARD, T. J. MARKS, W. P. LIN and G. K. WONG: *Chem. Mater.*, 4, 1148 (1992).

[16] G. D. GREEN, H. K. HALL, J. E. MULVANEY, J. NOONAN and D. J. WILLIAMS: *Macromolecules*, 20, 716 (1987).

[17] G. A. LINDSAY, J. D. STENGER-SMITH, R. A. HENRY, J. M. HOOVER and R. F. KUBIN: *SPIE-Int. Soc. Opt. Eng. Proc.*, 1497, 418 (1991).

[18] C. S. WILLAND and D. J. WILLIAMS: *Ber. Bunsenges. Phys. Chem.*, **91**, 1304 (1987).

[19] G. D. GREEN, H. K. HALL, J. E. MULVANEY, J. NOONAN and D. J. WILLIAMS: *Macromolecules*, **20**, 716 (1987).

[20] J. L. OUDAR and R. HIERLE: *J. Appl. Phys.*, **48**, 2699 (1977).

[21] J. ZYSS and G. BERTHIER: *J. Chem. Phys.*, **77**, 3635 (1982).

[22] B. F. LEVINE, C. G. BETHEA, C. D. THURMOND, R. T. LYNCH and J. L. BERNSTEIN: *J. Appl. Phys.*, **50**, 2523 (1979).

[23] S. TOMARU, S. ZEMBUTSU, M. KAWACHI and M. KOBAYASHI: *J. Inclusion Phenom.*, **2**, 885 (1984).

[24] T. MIYAZAKI, T. WATANABE and S. MIYATA: *Jpn. J. Appl. Phys.*, **27**, L1724 (1988).

[25] G. R. MEREDITH, J. G. VAN DUSEN and D. J. WILLIAMS: *Macromolecules*, **15**, 1385 (1982).

[26] J. ZYSS, D. S. CHEMLA and J. F. NICOUD: *J. Chem. Phys.*, **74**, 4800 (1981).

[27] J. ZYSS and J. L. OUDAR: *Phys. Rev. A*, **26**, 2016 (1982).

[28] Y. YAMAMOTO, S. KATOGI, T. WATANABE, H. SATO and S. MIYATA: *Appl. Phys. Lett.*, **60**, 935 (1992).

[29] S. ERMER, D. S. LEUNG, S. M. LOVEJOY, J. F. VALLEY and M. STILLER: *ACS/OSA Topical Meeting, October 5-7, Toronto, Canada, 1993, Technical Digest Series*, Vol. **17**, p. 50.

[30] H. S. NALWA, T. WATANABE, A. KAKUTA, A. MUKOH and S. MIYATA: *ACS/OSA Topical Meeting, October 5-7, Toronto, Canada, 1993, Technical Digest Series*, Vol. **17**, p. 85.

[31] T. WATANABE, X. T. TAO, D. C. ZHOU, S. SHIMODA, H. USUI, H. SATO, S. MIYATA, C. CLAUDE and Y. OKAMOTO: *SPIE. Proc.*, **2025**, 429 (1993).

[32] P. K. TIEN, R. ULRICH and R. J. MARTIN: *Appl. Phys. Lett.*, **14**, 291 (1969).

[33] S. HERMINGHAUS, D. BOESE, D. Y. YOON and B. A. SMITH: *Appl. Phys. Lett.*, **59**, 1043 (1991).

[34] S. C. LEE, A. KIDOGUCHI, T. WATANABE, H. YAMAMOTO, T. HOSOMI and S. MIYATA: *Polym. J.*, **23**, 1209 (1991).

Photorefractive Effects and Materials.

C. Medrano and P. Günter

Institute of Quantum Electronics, Swiss Federal Institute of Technology
ETH-Hönggerberg - CH-8093 Zürich, Switzerland

Introduction.

The first observation of the photorefractive effect was reported by Ashkin *et al.* [1] in 1966 while investigating electro-optic and nonlinear properties of $LiNbO_3$. It was first considered as a laser damage effect, undesiderable in nonlinear optical and electro-optical applications. Later it was found that the crystals could recover their original state by briefly heating them above 200 °C. Nowadays this effect is widely used in dynamic holography by writing a phase hologram by interference of two or more waves and inducing space-charge-induced refractive-index changes.

This process in contrast to the one employed in conventional holography allows real-time read-out, which in addition can be used for beam amplification, phase conjugation and optical real-time processing, for example, in optical novelty filters, Fourier filtering, associative memories, optical logic operations and optical phase conjugation.

Since the discovery of the photorefractive effect, a great deal of effort has been oriented to investigate the origin of this effect as well as the microscopic processes involved. These studies have also led to the discovery of additional new effects in electro-optic materials, namely the bulk photovoltaic effect [2] and excited-state polarization.

The fundamental theory published by Kukhtarev [3] describes the nonlinear interaction between the incident light and the nonlinear material response to illumination. The beams will change the refractive index in the electro-optic crystal. The resulting space charge field (E_{SC}) influences the charge migration and the change in refractive indices, which gives rise to interactions between recording beams and these refractive-index variations. Up to now, no analytical solutions of the band transport equations and of the coupled wave equations have been reported, however several limiting cases have been calculated and verified experimentally.

So far the list of materials where the photorefractive effect has been observed includes $LiNbO_3$, $LiTaO_3$, $BaTiO_3$, $KNbO_3$, $K(Ta_xNb)O_3$, $Ba_2NaNb_5O_{15}$, $Ba_{1-x}Sr_xNb_2O_6$, $Bi_4Ti_3O_{12}$, $Bi_{12}(Si, Ge)O_{20}$, KH_2PO_4, CdS, $RbZnBr_4$, $Pb_5Ge_3O_{11}$ among electro-optic materials; $(Pb, La)(Zr, Ti)O_3$ among ceramics; GaAs, InP, CdTe among semiconductors and very recently COANP:TCNQ among organic materials.

Most of the experimental work on the fundamentals and applications of the photorefractive effect has been realized in oxygen octahedra ferroelectrics. It has been also observed that the presence of Fe impurities in these materials is strongly connected with the enhancement of the photorefractive effect. Experimental results can be very well explained with the band transport model introduced first by KUKHTAREV and co-workers. This model is in wide use in the literature nowadays and with small changes it can give answers to new experimental results. However, the more complex problems are described by the Kukhtarev equations, the more complicated is the derivation of analytical solutions.

The highest photosensitivity of oxide and sillenite materials extends from the visible up to the near-infrared spectral range. These materials present the strongest electro-optic effects, which lead to the largest photorefractive-index changes, but for maximum photosensitivity it is necesary to maximize the ratio of the electro-optic coefficients to the dielectric constant and to have a photocarrier drift or diffusion length comparable to the fringe spacing of the phase hologram.

The first observation of the photorefractive effect in doped GaAs at 1.06 μm has been reported in 1984 [4, 5], and so far several other semiconductors with interesting photorefractive properties have been investigated. Some of the important features are: semiconductors can be used in the near infrared, therefore they are important for optical data processing and phase conjugation at diode and Nd:YAG laser wavelengths. They present large carrier mobilities, which leads to an enhancement in photoconductivity and speed. They are widely available in good quality and large size due to well-established techniques to prepare these materials. In addition the basic physical properties are well known in the literature. These facts introduce semiconductor materials as promising ones in the field of photorefractive applications.

We will review the main features which characterize the photorefractive effect in oxygen compounds such as oxygen octahedra ferroelectrics and sillenites, and the models which have been developed in order to explain the different experimental results and the basic mechanism of the photorefractive effects. A description of wave interaction including two-wave mixing, four-wave mixing and self-pumped phase conjugation is also presented.

1. – Photorefractive effects in electro-optic crystals.

1`1. *Generation and transport of charge carriers.* – For a large photorefractive efficiency, suitable donors or traps and efficient charge migration are necessary[6]. In undoped crystals the traps are provided by small traces of impurities and, in the majority of the oxygen octahedra ferroelectrics with a pronounced photorefractive effect, Fe impurities are the most important donor and trapping centres. Iron acts as a donor acceptor trap via intervalence exchanges such as $Fe^{2+} \leftrightarrow Fe^{3+}$. After iron doping and reduction treatment a characteristic band at about 2.55 eV is observed in the absorption spectrum. This band gives rise to photoconductivity upon optical excitation. This band has been assigned to intervalence charge transfer which can be also considered as photoionization of the Fe^{2+} ion; the final state is a Fe^{3+} ion and a mobile electron in the conduction band. This phenomenon occurs when an electron from a $d\varepsilon$ orbital of a Fe^{2+} ion is transferred to the niobium orbital. Upon oxidizing the crystals, the intervalence transfer band decreases and the crystal becomes transparent at $\lambda = 490$ nm when all the impurities are trivalent. The susceptibility to light-induced index changes is directly correlated with the strength of the intervalence band and thus with the Fe^{2+} concentration. Trivalent ions are important since they act as electron acceptors and conduction electrons can be trapped by Fe^{3+} ions, which in this form convert to Fe^{2+} ions. The photorefractive effect in Fe-doped oxygen octahedra ferroelectrics is based on the $Fe^{2+} \leftrightarrow Fe^{3+}$ interconversion, and the space charge set up by the action of light is attributed to a charge redistribution between divalent and trivalent impurities. However, in recent years it has been shown that the simple model of $Fe^{2+} \leftrightarrow Fe^{3+}$ interconversion is unable to explain all phenomena connected with the photorefractive effect and that also other defects such as oxygen vacancies, self-trapped electrons, shallow traps and colour centres are important[7-14]. Moreover other multivalent transition metal ions have also been used as activators, but Fe ions are the most efficient centres in most of these materials[15-17].

Periodic gratinglike excitation is particularly well suited to the experimental observation of the photorefractive effect and also for its mathematical description. The spatial modulation of the light intensity gives rise to a corresponding modulation of the electron and ionized donor centres. Initially the negative and positive charges of the electrons and ionized donors compensate, so that there is no net space charge. However, the electrons move by diffusion under the action of an external field or due to the photovoltaic effect. The electrons are subsequently trapped by empty donor centres and, due to the movement of electrons, there will be a spatial difference in the excitation rate of the ionized electric charge. In this form an electric field builds up which modulates the refractive index via the electro-optic effect.

The space charge fields responsible for hologram formation by the photorefractive effect can be derived from the photocurrent data by taking into account

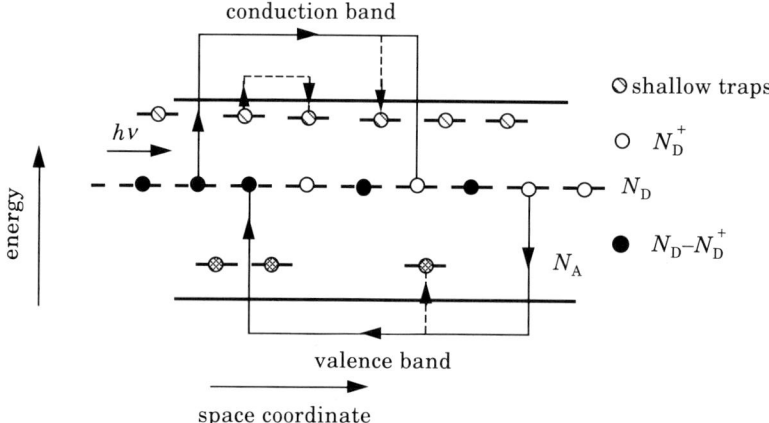

Fig. 1. – Simultaneous electron and hole transport with one set of recombination centres and shallow traps. Electrons are photoionized from filled donors and retrapped from ionized traps. Trapping and recombination in shallow traps is also indicated.

the spatial modulation of photocurrents by the nonuniform illumination. Figure 1 illustrates the situation for the excitation and recombination of electrons in KNbO$_3$ assuming a single donor centre with concentration N_D, a trap centre with concentration N_A and electrons with concentration n, being excited from the donors to the conduction band. An additional level corresponding to shallow traps is also indicated (N_{ST}) with a corresponding energy of 0.3 eV [14].

1'2. *Band transport equations: the Kukhtarev model.* – The movement of the photoexcited free carriers can be affected by three different mechanisms: diffusion, drift with an applied electric field and the photovoltaic effect. To describe how light creates a space charge field E_{SC}, the band transport equations formulated in the Kukhtarev model are the most appropriate. In the charge transport model, KUKHTAREV assumes one species of donor (with density N_D, filled or empty) from which photoelectrons (or holes) are ejected into the conduction (or valence) band. In the following discussion we will not consider the additional shallow-trap level. The donors are partially emptied by compensating acceptors with density N_A. The light distribution leads to a charge distribution among ionized donors with density N_D^+ by photoexcitation, migration of electrons in the conduction band and subsequent retrapping at ionized donor sites. The moving charges constitute an electric-current density j. If the illumination is spatially modulated, this current $j(z)$ is oscillating in space and contains contributions from diffusion, drift and photovoltaic effect. The mobile electrons, with density n, charge e and mobility μ, move under the action of a diffusion field E_D, a drift field E_A if there is an applied electric field, and the photovoltaic effect E_{PV}. In the most general case the transport equation, therefore, contains diffusion, drift and photovoltaic contributions, and the free-carrier current density j is given by

$$j = j_D + j_A + j_{PV} \, . \tag{1}$$

The diffusion current j_D ($j_D = eD\nabla n$, where D is the diffusion constant) results from a concentration gradient of the mobile charges. In the steady state it is compensated by the diffusion field E_D. The drift current j_A has two contributions: movement of carriers through the photorefractive crystal under the action of a constant externally applied electric field E_A, drift of these charge carriers due to the growing space charge field E_{SC}. The bulk photovoltaic current j_{PV} results from anisotropic photoexcitation of charge carriers in piezoelectric crystals. Due to the three mechanisms, diffusion, drift and photovoltaic effect, charges are separated in the photorefractive crystal and form the space charge field E_{SC}, which modulates the refractive index via the linear electro-optic effect. In the steady state E_{SC} compensates all spatially oscillating currents. Therefore, it is periodic itself and creates a refractive-index grating leading to light diffraction.

The total concentration of charge carriers is given by $n(z, t) = n_d + n_1(z, t)$, where n_d is the thermally excited free-carrier concentration in the dark, and $n_1(z, t)$ the excess free-carrier concentration due to illumination given by $n_1(z, t) = \phi \partial n_1 / \partial t = \phi \alpha I(z, t) / h\nu$, where α is the absorption constant, ϕ the quantum efficiency for exciting a charge carrier, $h\nu$ the photon energy and $I(z, t)$ the light intensity distribution. The photogenerated carrier density n_1 and the photoinduced space charge field are nonlinearly coupled and can affect each other specially in the steady-state stage of hologram formation. Depending on the material, different terms will dominate. The photovoltaic term will dominate in materials like $LiNbO_3$ or $LiTaO_3$, whereas the photoconductivity term will dominate in $Bi_{12}SiO_{20}$, $Bi_{12}GeO_{20}$, $K(NbTa)O_3$ and reduced $KNbO_3$. The diffusion term depends on the mobility and is only important for intensity patterns with high spatial frequencies. The effect of each of this terms was considered for $KNbO_3$ in [18].

If we assume that only electrons are photoexcited, then the band transport model can be described by the following equations.

1) The continuity equation for the immobile ionized donors:

$$\frac{\partial N_D^+}{\partial t} = (N_D - N_D^+)(sI + \beta_0) - \gamma_R N_D^+ n \, . \tag{2}$$

The temporal evolution of N_D^+ has been set equal to the ionization rate, or free-electron generation rate $(N_D - N_D^+)(sI + \beta_0)$, minus the ionization rates $\gamma_R N_D^+ n$, where s is the photoexcitation constant, and thus sI is the probability rate of photoexcitation, β_0 is the probability rate of thermal excitation, and γ_R is the recombination constant.

2) The continuity equation for the mobile conduction band electrons:

(3)
$$\frac{\partial n}{\partial t} = \frac{\partial N_D^+}{\partial t} + \nabla \cdot \left(D \nabla n + \mu n \left(E_{SC} + \frac{V}{L} \right) - \frac{\beta}{eI} \right),$$

where n is the free-electron number density, D is the diffusion coefficient ($\mu k_B T$), V is the applied voltage across the crystal length L, and β is the photovoltaic tensor component.

3) Poisson's equation for the space charge field E_{SC}:

(4)
$$\nabla \cdot (\varepsilon_0 \varepsilon_{eff} E_{SC}) = e(N_D^+ - N_A - n),$$

where ε_0 is the vacuum dielectric constant, ε_{eff} is the effective partially clamped dielectric constant. The term N_A accounts for the effect of immobile compensative acceptors which are supposed to be completely ionized (negatively charged), and which do not take part in the photoexcitation process. Accordingly, their presence guarantees that a large number of donors can be vacant even in the dark.

These three band transport equations describe the charge transport in a photorefractive material, and are known as Kukhtarev's equations. They must be supplemented by the appropriate initial and boundary conditions, and will describe the material response to inhomogeneous illumination. The space charge fields responsible for hologram formation by the photorefractive effect can be obtained from photocurrent data by taking into account the spatial modulation of photocurrents by the nonuniform illumination. Let us consider a material with a single donor (with concentration N_D) and trap centre (with concentration N_A), the total concentration being $N = N_D + N_A = $ const. The electrons (with concentration n) are being excited into the conduction band. Photoexcited holes can also contribute to the photoconductivity, as has been demonstrated in different electro-optic crystals [19-22], and, in this special case, the transport equation (1) must be modified to include contributions from photoexcited free holes.

The rate per unit volume for electron excitation into the conduction band is $(sI(z, t) + \beta_0) N_D$, and, for trapping, $\gamma_R n N_A$. Ionized donors act as acceptors, and other compensative acceptor levels are assumed not to be involved in the phototransition. The total change in n resulting from photoionization of the donors, charge transport and trapping is determined by the continuity equation

(5)
$$\frac{dn(z, t)}{dt} = \frac{dN_A(z, t)}{dt} - \frac{1}{e} \frac{\partial j(z, t)}{\partial z},$$

where

(6) $\dfrac{dN_A(z, t)}{dt} = [sI(z, t) + \beta_0][N(z, t) - N_A(z, t)] - \gamma_R n(z, t) N_A(z, t).$

We consider the case in which only a small fraction of the donors will be ionized ($N_A \ll N$). Assuming a linear recombination of photocarriers, we can introduce a free-carrier lifetime independent of n given by $\tau = 1/\gamma_R N_A$, and $\beta_0 N = n_d/\tau$. In this case we will have, in thermal equilibrium, a thermal-excitation rate equal to the recombination rate, and so the time evolution of N_A becomes

(7) $$N_A(z, t) = g(z) - \frac{n(z, t) - n_d}{\tau},$$

where $g(z)$ is the generation rate, and is proportional to the light intensity.

1·3. *Photoinduced space charge fields.* – We now discuss the mechanism of photorefractive hologram recording by considering the formation of an elementary holographic grating. As before the photorefractive grating wave vector is directed along the z-axis and the x-axis is perpendicular to the crystal surface. The interference pattern of the hologram can be described by

(8) $$I(x, z) = I_0 \exp[-\alpha x/\cos \theta](1 + m \cos Kz),$$

where α is the light absorption constant, $K = 2\pi/\Lambda$ is the spatial frequency, in which $\Lambda = \lambda/2 \sin \theta$ is the fringe spacing. The angle between the two writing beams (outside the crystal) is 2θ, and λ is the light wavelength. $I_0 = I_{+1} + I_{-1}$ is the sum of the incident intensities of the two interfering beams with modulation index given by

(9) $$m = 2 \frac{(I_{+1}I_{-1})^{1/2}}{I_{+1} + I_{-1}} \cos(2\theta p).$$

For polarization directions within the plane of incidence, $p = 1$, and for light polarized perpendicularly to the plane of incidence, $p = 0$. Assuming that the hologram is so thin that beam coupling effects can be neglected, the attenuation factor αx in eq. (8) becomes negligible since $\alpha x \ll 1$. The generation of photoelectrons can be assumed to be uniform over the crystal length, and the generation rate can be written as

(10) $$g(z) = g_0(1 + m \cos Kz),$$

where $g_0 = \phi \alpha I_0/h\nu$.

We assume that all the processes, drift, diffusion and photovoltaic effect, are allowed to contribute to the photocurrents, and also that the diffusion length $L_D = (D\tau)^{1/2}$, where D is the diffusion coefficient, the drift length $L_E = \mu\tau E$, where μ is the mobility, and the photovoltaic drift length L_{ph} can be arbitrary, and in particular comparable to or larger than the fringe spacing of the grating.

With the results obtained, one will be able to predict and describe the photorefractive characteristics of electro-optic crystals, understand the primary differences between short- and long-transport-length recording, and show that long
migration lengths can produce a phase shift of the refractive-index pattern with
respect to the light interference pattern thus causing transfer of optical energy
between the recording beams[23]. Such beam coupling effects are very important in dynamic holography, image amplification and self-pumped phase
conjugation.

We will analyse in the following the constant-applied-voltage configuration.
The short-circuit situation is a special case of this configuration, when the applied voltage is zero; whereas the open-circuit case, that is with zero current, is
in practice difficult to achieve. The accumulation rate of the space charge density ρ_{SC} at any point and time is given by the one-dimensional continuity
equation

(11)
$$\frac{\partial \rho_{SC}(z, t)}{\partial t} = \frac{\partial j(z, t)}{\partial z} .$$

Combined with Poisson's equation, this gives

(12)
$$\frac{\partial E_{SC}}{\partial z} = \frac{\rho_{SC}(z, t)}{\varepsilon \varepsilon_0} = -\frac{1}{\varepsilon \varepsilon_0} \int_0^t \frac{\partial j(z, t')}{\partial z} \, dt' + G(t),$$

where ε is the static dielectric constant of the material and $G(t)$ is determined
from the boundary conditions. A constant applied voltage requires

(13)
$$-\int_0^L E_T(z, t) \, dz = V,$$

where $E_T(z, t) = E_{SC}(z, t) - V/L$ is the total electrostatic field. This means
that

(14)
$$\int_0^L E_{SC}(z, t) \, dz = 0 .$$

If the electron transport length is very small $n_1(z) = \tau g(z)$, the carrier concentration is simply proportional to the generation rate. When we analyse the
expression for the total change of n (eq. (6)) and the expression for the temporal
behaviour of N_A, we find that, if $n_1(z) \approx \tau g(z)$, then $\partial j/\partial z \approx 0$. This means that
there is a very slow spatial change in the charge distribution. In the present
analysis, the continuity equation is solved directly for $n(z, t)$ rather than assuming $n_1(z) = \tau g(z)$. Since the time scale of the change in $n_1(z, t)$ is very slow

compared to τ, then with $\mathrm{d}n_\mathrm{d}/\mathrm{d}t = 0$

$$\frac{\partial n_1(z, t)}{\partial t} \ll g(z) - \frac{n_1(z, t)}{\tau}$$

and

(15) $$-\frac{1}{e} \frac{\partial j}{\partial z} \approx g(z) - \frac{n_1(z, t)}{\tau} .$$

Then the difference between g and n/τ represents the rate of change of the space charge and of the grating formation[24]. At saturation, when the change in the space charge field stops ($\partial j/\partial z = 0$), $n_1 = \tau g$.

The coupled system of differential equations (2), (6), (13) subject to the condition expressed in eq. (15) cannot be solved analytically in general. Nevertheless, analytical solutions can be obtained for $n(z, t)$ and $E_{\mathrm{SC}}(z, t)$ with the initial condition $E_{\mathrm{SC}}(z, 0) = 0$ in the initial short-time approximation and in the steady-state saturation time limit[23].

1`4. *Steady-state approximation.* – Let us assume the number density of the free charge carriers to be at any time much smaller than the trap density $n \ll N_\mathrm{A}$ and than the density of unionized donor centres $n \ll N_\mathrm{D} - N_\mathrm{A}$. In addition let us assume a linear generation and recombination rate so that $N_\mathrm{D}^i \approx N_\mathrm{A}$. Then in the steady state, eq. (3) reduces to

(16) $$n = n_0[1 + M \cos(K_g y)],$$

where K_g is assumed to point into the y-direction and n_0 is the mean charge carrier density

(17) $$n_0 = g_0 \tau_\mathrm{R},$$

where the generation rate g_0 is given by

(18) $$g_0 = (N_\mathrm{D} - N_\mathrm{A})(sI_0 + \beta),$$

and the free-carrier lifetime

(19) $$\tau_\mathrm{R} = (N_\mathrm{A} \gamma_\mathrm{R})^{-1};$$

M is the reduced fringe contrast given by

(20) $$M = \frac{m}{1 + \beta/sI_0},$$

which is reduced due to the presence of free carriers even in the dark. In the steady state and neglecting the photovoltaic effect, eq. (3) reduces to

(21) $$\frac{\partial j}{\partial y} = \frac{\partial}{\partial y}\left(D \frac{\partial n}{\partial y} - \frac{e}{q} \mu n E\right) = 0 .$$

The integration of this equation together with eq. (17) finally gives

$$(22) \qquad E = \frac{j}{e\mu n_0} \frac{1}{1 + M\cos(\boldsymbol{K}_g y)} - \frac{q}{e} \frac{DK_g}{\mu} \frac{M\sin\boldsymbol{K}_g y}{1 + M\cos(\boldsymbol{K}_g y)} \, ,$$

where j is the free-carrier current density.

For a boundary condition of constant applied voltage V over the crystal length L we must have

$$(23) \qquad E_0 = \frac{V}{L} = \frac{1}{L} \int_0^L E \, dy \, .$$

Integrating eq. (22) over a large number of grating periods, we are able to find an expression for the current density j:

$$(24) \qquad j = \sqrt{1 - M^2}\, e\mu n_0 \, .$$

The conductivity for sinusoidal illumination is reduced by a factor $\sqrt{1 - M^2}$ compared to the conductivity for uniform illumination at the same average intensity. Finally (22) can be rewritten in the form

$$(25) \qquad E = E_0 \frac{\sqrt{1 - M^2}}{1 + M\cos(k_g y)} - E_d \frac{M\sin(k_g y)}{1 + M\cos(K_g y)} \, ;$$

E_d is the diffusion field given by

$$(26) \qquad E_d = \frac{q}{e} \frac{DK_g}{\mu} = \frac{q}{e} \frac{k_B T}{e} K_g \, ,$$

where the last equation is obtained by the Einstein relation, k_B is the Boltzmann constant and T the temperature. The final equation is an expression for the photoinduced space charge field:

$$(27) \qquad E_{SC} = E - E_0 = E_0 \left(\frac{\sqrt{1 - M^2}}{1 + M\cos(K_g y)} - 1 \right) - E_d \frac{M\sin(K_g y)}{1 + M\cos(K_g y)} \, .$$

Due to the presence of dark conductivity even for a modulation ratio $m = 1$ the reduced fringe contrast M is always smaller than unity, therefore the space charge field E_{SC} physically can never reach infinity.

For later investigations of first-order Bragg diffraction by such photoinduced phase gratings it is useful to expand the space charge field given in the above expression (25) as a Fourier series:

$$(28) \qquad E_{SC} = 2\sqrt{E_0^2 + E_d^2} \sum_{j=1}^{\infty} \left[\sqrt{\frac{1}{M^2}} - 1 - \frac{1}{M} \right]^j \cos(jK_g y - \phi)$$

with

$$\text{(29)} \qquad \operatorname{tg} \phi = \frac{E_{\mathrm{d}}}{E_0} \, .$$

Although a general analytic solution of the system of coupled differential equations (2)-(4) is not possible for sinusoidal illumination, a solution can be found in a linear approximation. For small modulation ratios M and for dominant photoconductivity we can obtain the following expression for the space charge field[25]:

$$\text{(30)} \qquad E_{\mathrm{SC}}^0 = - M \frac{E_0 E_q^2}{(E_q + E_{\mathrm{d}})^2 + E_0^2} \cos (\boldsymbol{K}_g \cdot \boldsymbol{r}) -$$

$$- M \frac{E_{\mathrm{d}}^2 E_q + E_q^2 E_{\mathrm{d}} + E_q E_0^2}{(E_q + E_{\mathrm{d}})^2 + E_0^2} \sin (\boldsymbol{K}_g \cdot \boldsymbol{r}),$$

where E_q is given by

$$\text{(31)} \qquad E_q = \frac{q}{\varepsilon \varepsilon_0} \frac{1}{K} \frac{N_{\mathrm{A}}}{N_{\mathrm{D}}} (N_{\mathrm{D}} - N_{\mathrm{A}}).$$

For typical photorefractive parameters, fig. 2 illustrates the dependence of the photoinduced space charge field on an externally applied electric field E_0, according to eq. (30).

1˙5. *Transient gratings and time-resolved experiments.* – Several optical data processing operations have been performed in photorefractive crystals. Since the refractive-index grating written in them can be phase shifted relative to the optical-intensity grating leading to an intensity redistribution between writing beams, it is possible to use photorefractive materials as optical beam and image amplifiers (see subsect. 1˙6). Most of the optical writing, reading and

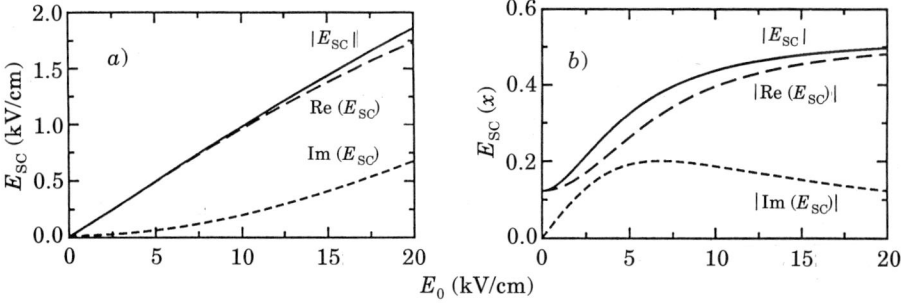

Fig. 2. – Dependence of the photoinduced space charge field on an externally applied electric field E_0 for typical photorefractive parameters[19]: *a)* $\Lambda = 10 \, \mu\mathrm{m}$, $N_{\mathrm{A}} = 10^{16} \, \mathrm{cm}^{-3}$, $m = 0.1$; *b)* $\Lambda = 1 \, \mu\mathrm{m}$, $N_{\mathrm{A}} = 10^{16} \, \mathrm{cm}^{-3}$, $m = 0.1$.

erasing of refractive-index gratings in photorefractive crystals have employed c.w. lasers with writing times between milliseconds and seconds. These times are slower than the ones required for many applications. This has led to an investigation of writing and erasing gratings with pulsed lasers of higher intensity showing nanosecond response [26-28]. Response times in the nanosecond range were already observed in several photorefractive materials like $LiNbO_3$ [26, 29, 30], $BaTiO_3$ [28, 31, 32], $KNbO_3$ [33] and $Bi_{12}SiO_{20}$ [27, 34, 35]; however, for most of the materials the high-speed performance has not been yet optimized.

Investigation of grating dark decays in photorefractive $Bi_{12}SiO_{20}$ crystals [36] showed a dependence of the dark storage time on the light intensity which is used to write the grating. In this paper it was suggested that nonexponential dark decays of photorefractive gratings and their dependence on the light intensity, which is used to record the grating, can be explained by using an extension of the one-charge-carrier Kukhtarev model which considers an additional shallow-trap level. This shallow trap was identified as a shallow hole trap.

In a recent paper [14], transient photorefractive gratings in reduced $KNbO_3$

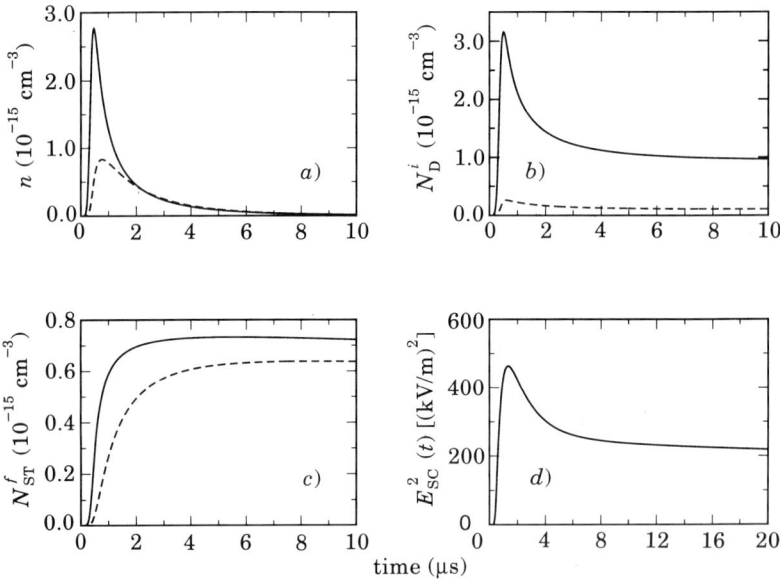

Fig. 3. – a) Time dependence of the number of electrons in the conduction band, b) time dependence of the ionized-donor density, c) time dependence of density of filled shallow traps in the crystal along c, d) time dependence of the square of the first harmonic of the space charge field as calculated from the model. Curves are shown at $z = 0$ and $z = \Lambda/2$ for the following material parameters: donor density $N_D = 4 \cdot 10^{15}$ cm^{-3}, density of shallow traps $N_{ST} = 8.1 \cdot 10^{14}$ cm^{-3}, recombination rates for donors γ and shallow traps γ_{ST}, $\gamma = 5 \cdot 10^{-10}$ cm^{3}s^{-1}, $\gamma_{ST} = 9 \cdot 10^{-10}$ cm^{3}s^{-1}, $\mu = 0.008$ cm^{2}V^{-1}s^{-1}.

were investigated in a time regime from 1 μs to 10 ms for various writing intensities and grating wave vectors. The observed grating build-up time using a pulsed laser was found to be faster than 1 μs, while during the grating decay two different time regimes could be distinguished with typical time constants of a few microseconds and a few milliseconds. The interpretation of these experimental results was done using the Kukhtarev model with the addition of a shallow-trap level near the conduction band. This additional level modifies the basic Kukhtarev equations, and the obtained agreement between theory and experiment is excellent. Figures 3a), b) and c) show the time dependence of the number of electrons in the conduction band, the ionized-donor density and the density of filled shallow traps in the crystal along c, for the following material parameters: donor density $N_D = 4 \cdot 10^{15}$ cm^{-3}, density of shallow traps $N_{ST} = 8.1 \cdot 10^{14}$ cm^{-3}, recombination rates for donors γ and shallow traps γ_{ST}, $\gamma = 5 \cdot 10^{-10}$ cm^3 s^{-1}, $\gamma_{ST} = 9 \cdot 10^{-10}$ cm^3 s^{-1}, $\mu = 0.008$ cm^2 V^{-1} s^{-1}. Figure 3d) shows the time dependence of the square of the first harmonic of the space charge field as calculated from the model for the same above-mentioned material parameters.

1`6. Wave interaction.

1`6.1. Two-wave mixing. In order to describe the nonlinear mixing of light beams in a photorefractive material, it is necesary to consider two parts: the action of the light on the material described by the Kukhtarev equations, and a second part related to the response of the material to the action of light; this is described by the Maxwell's equations of the electrodynamic theory. Let us consider the mixing of two optical waves in a photorefractive crystal in the case in which waves are nearly copropagating or counterpropagating. A typical two-wave mixing configuration is shown in fig. 4.

In this situation the waves will vary along their directions of propagation, so that $E(\mathbf{x}) = E_j(z)$, leading to the exact solution of the coupled-wave equations

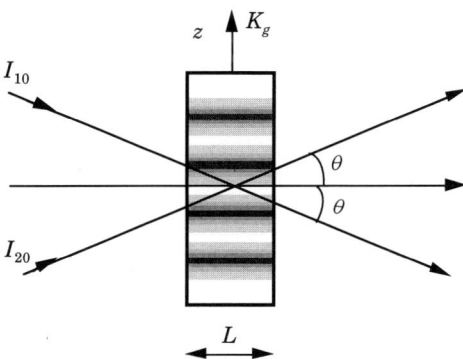

Fig. 4. – Configuration for two-wave mixing experiments.

for the beam amplitudes and phases. For the copropagating waves, the coupled-wave equations are [3, 33, 34]:

$$(32) \quad \frac{dE_1}{dz} = \gamma \frac{E_1 E_2^* E_2}{I_1 + I_2} - \frac{\alpha E_1}{2}, \quad \frac{dE_2}{dz} = \gamma^* \frac{E_2 E_1^* E_1}{I_1 + I_2} - \frac{\alpha E_2}{2},$$

where E_j are the slowly varying amplitudes of the optical electric fields, $I_j = |E_j|^2$ are the respective intensities, α is the intensity absorption per unit length, and γ is the coupling strength per unit legth [35]. For beams with the same frequency and for crystals with very low dark conductivity, γ is independent of the total intensity in the steady state. If we write the field amplitudes as $E_j = \sqrt{I_j} \exp[i\phi_j]$ and collect the real and imaginary parts, eq. (32) can be written as

$$(33) \quad \frac{dI_1}{dz} = 2 \, \mathrm{Re}\{\gamma\} \frac{I_1 I_2}{I_1 + I_2} - \alpha I_1, \quad \frac{dI_2}{dz} = 2 \, \mathrm{Re}\{\gamma\} \frac{I_1 I_2}{I_1 + I_2},$$

$$(34) \quad \frac{d\phi_1}{dz} = \mathrm{Im}\{\gamma\} \frac{I_2}{I_1 + I_2}, \quad \frac{d\phi_2}{dz} = \mathrm{Im}\{\gamma\} \frac{I_1}{I_1 + I_2}.$$

The total intensity $I_0 = I_1 + I_2$ varies with z only by absorption; eq. (33) can be uncoupled and solved. Once the z-dependence of the intensities is known, eq. (34) for the phases can be integrated, and the results are expressed in

$$(35) \quad I_1(z) = I_{10} \exp[-\alpha z] \frac{I_0}{I_{10} + I_{20} \exp[-2 \, \mathrm{Re}\{\gamma\} z]} \equiv G_1(z) I_{10} \exp[-\alpha z],$$

$$(36) \quad I_2(z) = I_{20} \exp[-\alpha z] \frac{I_0}{I_{20} + I_{10} \exp[2 \, \mathrm{Re}\{\gamma\} z]} \equiv G_2(z) I_{20} \exp[-\alpha z],$$

$$(37) \quad \phi_1(z) = \frac{1}{2} \frac{\mathrm{Im}\{\gamma\}}{\mathrm{Re}\{\gamma\}} \ln[G_1(z)],$$

$$(38) \quad \phi_2(z) = \frac{1}{2} \frac{\mathrm{Im}\{\gamma\}}{\mathrm{Re}\{\gamma\}} \ln[G_2(z)],$$

where I_{10} and I_{20} are the incident intensities. If $\mathrm{Re}\{\gamma\} > 0$, beam 1 will be amplified at the expense of beam 2.

Two-wave mixing is a convenient method for measuring the coupling gain $2 \, \mathrm{Re}\{\gamma L\}$ for a particular beam geometry. By measuring the transmitted intensities of beams 1 and 2 with and without coupling, the gain is obtained from

$$(39) \quad 2 \, \mathrm{Re}\{\gamma\} L = \ln\left[\frac{I_{20} I_1(L)}{I_{10} I_2(L)} \right],$$

where L is the interaction length. If I_{10} is made small enough that the depletion of beam 2 is negligible ($I_{10} \ll I_{20} \exp[-2 \, \mathrm{Re}\{\gamma\} L]$), then the ampli-

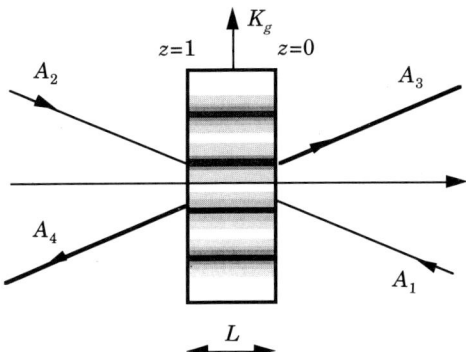

Fig. 5. – Configuration of four-wave mixing experiments. A_1 and A_2 represent the pump beams, A_4 represents the probe and A_3 the phase-conjugate beam.

fication of beam 1 is independent of the intensity of beam 2, and $I_1(L) = I_{10} \cdot$ $\cdot \exp[-\alpha z] \exp[2\,\mathrm{Re}\,\{\gamma\}\,L]$ and (39) becomes $2\,\mathrm{Re}\,\{\gamma\}\,L = \ln[I_1(L)/I_{10}\exp[-\alpha L]]$. In this way the gain can be determined by measuring the intensity of beam 1 after the crystal with and without beam 2.

1‘6.2. **Four-wave mixing.** Complex conjugate wave generation by degenerate four-wave mixing was first reported by KUKHTAREV and ODULOV[36,37] in LiNbO$_3$ and LiTaO$_3$. In these experiments two counterpropagating pump waves with complex amplitudes R_1 and R_2, and a weak signal wave (S_3) interact in the volume of the photorefractive material to produce the fourth wave S_4 complex conjugate to the signal wave. The most simple geometry is the one shown in fig. 5. In this geometry the transmission grating plays the major role, whereas for a crystal orientation with the polar axis parallel to the pump beams the reflection grating would be more important due to the different values of the electro-optic coefficients. In ferroelectric crystals, the four-wave mixing process depends on the direction of the spontaneous polarization. The interference patterns of the beams for two opposite orientations of the sample may either coincide in the volume of the crystal thus increasing the recorded grating, or may cancel each other due to the phase shift equal to π. The phase grating in the materials with diffusion nonlinearity is always $\pi/2$ shifted with respect to the interference pattern, the shift direction depending on polar-axis orientation. On the other hand, the beam diffracted by the phase grating is always $\pi/2$ shifted in phase with respect to the zeroth-order one. This means that the interference pattern of the waves 2 and 4 in transmission geometry is $\pi/2$ shifted with respect to the grating recorded by beams 1 and 3. This last shift does not depend on the c-axis orientation and is summed up or subtracted from the «diffusion» shift.

According to FISHER *et al.* [38], the coupled-wave equations for four-wave mixing are

$$
(40) \qquad \frac{dA_1}{dz} = -\gamma \frac{(A_1 A_4^* + A_2^* A_3) A_4}{I_0} \,,
$$

$$
(41) \qquad \frac{dA_2^*}{dz} = -\gamma \frac{(A_1 A_4^* + A_2^* A_3) A_3^*}{I_0} \,,
$$

$$
(42) \qquad \frac{dA_3}{dz} = +\gamma \frac{(A_1 A_4^* + A_2^* A_3) A_2}{I_0} \,,
$$

$$
(43) \qquad \frac{dA_4^*}{dz} = \gamma \frac{(A_1 A_4^* + A_2^* A_3) A_1^*}{I_0} \,,
$$

where $I_0 = I_1 + I_2 + I_3 + I_4$ and the coupling constant γ is defined by

$$
\gamma = \frac{i\omega n_I \exp[-i\phi_I]}{2c \cos \nu} \,,
$$

ν is the angle of the pump beam with respect to the crystal surface and the complex constant $n_I \exp[i\phi_I]$ characterizes the spatial hologram written by the intensity interference pattern of beams 1 and 4. In these coupled-wave equations (40)-(43) it is assumed that absorption is negligible and that one photorefractive (transmission) grating n_I predominates out of four; this is a good approximation because of the strong dependence of the coupling strength on the orientation of the grating wave vector.

Equations (40)-(43) were first solved in the so-called «undepleted-pump approximation» [38] which assumes that the probe beam 2 is so much weaker than the other beams that the intensities of the pumping beams are unaffected by the interaction. A little later, CRONIN-GOLOMB et al. [39] gave the exact solution to the four-wave mixing problem, by uncoupling equations (40)-(43) noting that the following quantities are conserved:

$$
(44) \qquad d_1 = I_1 + I_4 \,, \qquad d_2 = I_2 + I_3 \,, \qquad c = A_1 A_2 + A_3 A_4 \,;
$$

with these assumptions, the solution becomes

$$
(45) \qquad \frac{A_1(z)}{A_2^*(z)} = -\frac{(\Delta - r) D \exp[-\mu z] - (\Delta + r) D^{-1} \exp[\mu z]}{2c^* [D \exp[-\mu z] - D^{-1} \exp[\mu z]]} \,,
$$

$$
(46) \qquad \frac{A_3(z)}{A_4^*(z)} = -\frac{(\Delta - r) E \exp[-\mu z] - (\Delta + r) E^{-1} \exp[\mu z]}{2c^* [E \exp[-\mu z] - E^{-1} \exp[\mu z]]} \,,
$$

where $\Delta = d_2 - d_1$, $r = (\Delta^2 + 4\,|c|^2)^{1/2}$, $\mu = \gamma r/2I_0$, and D and E are integration constants. Equations (44) to (46) may be solved by applying appropriate boundary conditions for specific situations. For the case of a phase-conjugate mirror externally pumped, the amplitudes of all beams at their respective entrance faces are known, so the starting equations are the values of eqs. (45) and (46) at the boundaries $z = 0$ and $z = 1$.

By solving for D, E and c one can finally obtain the phase-conjugate reflectivity ρ given by the squared modulus of

$$(47) \qquad\qquad \rho = \frac{-2cT}{\Delta T + (\Delta^2 + 4\,|c|^2)^{1/2}} \, ,$$

where $T = \text{tg}\,(\mu l)$. When considering reflectivity as a function of the various input beam intensities, it is convenient to define the probe ratio q:

$$(48) \qquad\qquad q \equiv \frac{I_4(0)}{I_1(0) + I_2(1)} \, ,$$

and the pump ratio R:

$$(49) \qquad\qquad R = \frac{I_2(1)}{I_1(0)} \, .$$

The reflectivity behaviour can be analysed by plotting it as a function of the coupling strength $|\gamma l|$ and for different phase shifts between the grating and the interference fringes. A contour plot of phase-conjugate reflectivity as a function of both pump and probe ratios gives information on the case when the pumping beam 2 is absent, and this has a great importance in the analysis of passive phase-conjugate mirrors[40].

1'6.3. Self-pumped optical phase conjugation. Phase-conjugate mirrors based on degenerate four-wave mixing generally require two additional reference beams, which may complicate the build-up of resonators with phase-conjugate mirrors. It has been shown that there exist mechanisms for self-pumping phase-conjugate mirrors by the self-diffracting part of the wave to be conjugated and by employing four-wave mixing of the incident beam with the self-diffracted and retroreflected self-diffracted beam.

The first observation of a «self-pumped phase conjugator» has been reported by WHITE et al. [41]. It consisted of a photorefractive crystal placed between two mirrors carefully aligned to form a resonator cavity. The input waves produce a pair of self-generated beams which serve as the pumping beams which produce the phase-conjugate signal. The device needs both mirrors to start, but, once the device is working, one of the resonator mirrors can be removed. This second device is known as «passive self-pumped phase-conjugate mirror» [42]. So far, many different configurations of self-pumped phase conjugators relying on four-wave mixing to produce a phase-conjugate wave[41-46] have been reported. The fidelity of phase conjugation and ease of alignment vary considerably from one self-pumped phase conjugator to another. Large optical gains, charge transport processes dominated by diffusion and large electro-optic coefficients are a prerequisite for self-pumped phase conjugation. The quality of the phase-conjugate beam is critically dependent on the alignment, intensities and phase profiles of the pump waves, whereas, in the self-pumped phase conjugators requiring no optical elements, the pump waves are generated by the incident beam itself, making the device performance independent of external mirrors. In this device, an incident laser beam tends to fan out and bends towards the c-axis as it propagates through the crystal. Some of the fanned

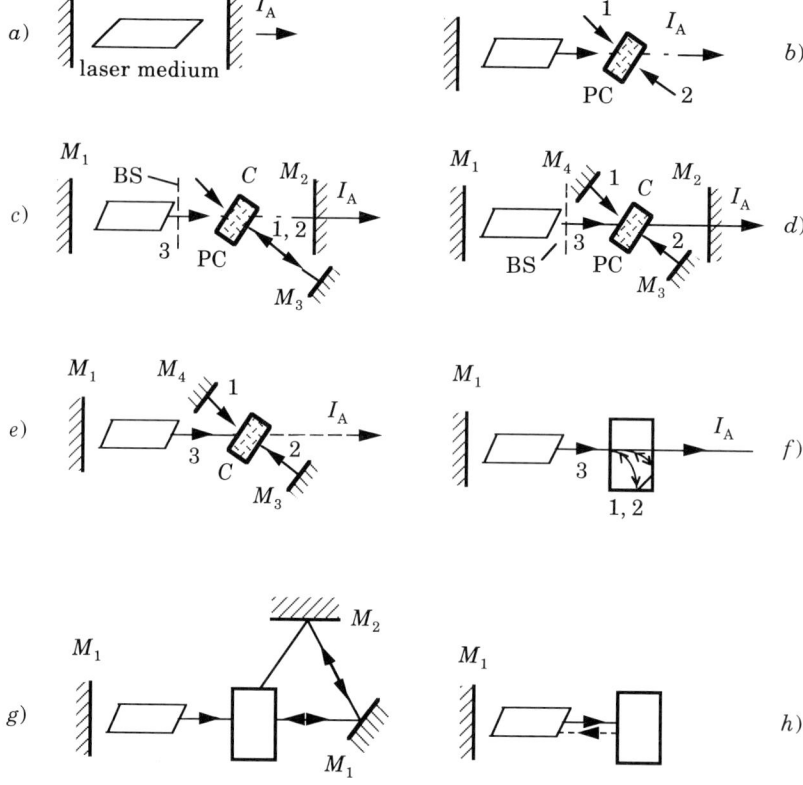

Fig. 6. – Optical resonators: a) conventional two-mirror system, b) system with one phase-conjugate mirror and one conventional one. The phase-conjugate mirror has to be pumped externally by two antiparallel pump beams. c), d) Self-pumped phase-conjugate resonator which is not self-starting. e) Self-pumped phase-conjugate resonator self-starting using two-wave mixing. f) Self-pumped phase-conjugate resonator self-starting using beam bending. g) Ring resonator configuration with input beam redirected back into the crystal. h) Backscattering configuration.

beam of light stays within the crystal and, by total internal reflection from opposing faces and corners of the crystal, forms rings and loops of light within which the four-wave mixing process will self-generate and a phase-conjugate beam will be produced. FEINBERG [43] demonstrated a self-pumped phase conjugator with all the interacting beams contained within a barium titanate crystal. Another phase conjugator without external optical elements was demonstrated by CHANG *et al.* [46]. In this device, phase conjugation is achieved by «stimulated backscattering» in barium titanate due to two-wave mixing between the counterpropagating incident and phase-conjugate waves. Figure 6 shows the different resonator arrangements using photorefractive crystals.

Using a new geometry, a self-pumped phase-conjugate wave was generated by amplified scattered light from the signal wave, propagating along the *c*-axis, and by the wave retroreflected from a coated *c*-face [47]. Very recently [48] a device has been demonstrated where stable phase-conjugate reflectivities as high as 60% have been achieved with back-seeded stimulated photorefractive scattering in a $BaTiO_3$ crystal. In this device the fanned light exiting the crystal strikes a nonspecular, diffusive-type surface, the scattering from which acts as a back-injected seed and lowers the threshold condition for stimulated scattering. This new geometry presents several advantages over previous configurations: *a*) it is more easy to obtain accurate coupling of several beams from independent amplifier legs, since the interaction region in the seeded stimulated photorefractive scattering (SPS) is distributed over a long length rather than over several very localized interaction regions; *b*) since there are no internal loops related to internal corner reflections, higher damage thresholds can be expected; *c*) faster response times can be expected since the seeded SPS geometry involves very-small-period gratings created by counterpropagating beams in the crystal.

REFERENCES

[1] A. ASHKIN, G. D. BOYD, J. M. DZIEDZIC, R. G. SMITH, A. A. BALLMANN and K. NASSAU: *Appl. Phys. Lett.*, **9**, 72 (1966).
[2] A. M. GLASS: *Opt. Eng.*, **17**, 470 (1978).
[3] N. V. KUKHTAREV, V. B. MARKOV, S. G. ODULOV, M. S. SOSKIN and V. L. VINETSKI: *Ferroelectrics*, **22**, 949 (1979); N. V. KUKHTAREV: *Sov. Tech. Phys. Lett.*, **2**, 438 (1976).
[4] M. B. KLEIN: *Opt. Lett.*, **9**, 350 (1984).
[5] A. M. GLASS, A. M. JOHNSON, D. H. OLSON, W. SIMPSON and A. A. BALLMAN: *Appl. Phys. Lett.*, **44**, 948 (1984).
[6] P. GÜNTER: *Phys. Rep.*, **93**, 199 (1982).
[7] K. L. SWEENEY and L. E. HALLIBURTON: *Appl. Phys. Lett.*, **43**, 336 (1983).
[8] J. L. KETCHUM, K. L. SWEENEY, L. E. HALLIBURTON and A. F. ARMINGTON: *Phys. Lett. A*, **94**, 450 (1983).
[9] L. E. HALLIBURTON, K. L. SWEENEY and C. Y. CHEN: *Nucl. Instrum. Methods Phys. Res. B*, **1**, 344 (1984).

262 C. MEDRANO and P. GÜNTER

[10] L. ARIZMENDI, J. M. CABRERA and F. AGULLO LOPEZ: *J. Phys. C*, **17**, 515 (1984).

[11] O. F. SCHIRMER: *J. Appl. Phys.*, **50**, 3404 (1979).

[12] O. F. SCHIRMER and D. VON DER LINDE: *Appl. Phys.*, **33**, 35 (1978).

[13] E. KRÄTZIG: *Ferroelectrics*, **21**, 635 (1978).

[14] I. BIAGGIO, M. ZGONIK and P. GÜNTER: *Opt. Commun.*, **77**, 312 (1990).

[15] K. MEGUMI, H. KOZUKA, M. KOBAYASHI and Y. FURUKATA: *Appl. Phys. Lett.*, **30**, 631 (1977).

[16] R. A. SPRAGUE: *J. Appl. Phys.*, **46**, 1673 (1975).

[17] E. MOYA, L. CONTRERAS and C. ZALDO: *J. Opt. Soc. Am. B*, **5**, 1737 (1988).

[18] P. GÜNTER and F. MICHERON: *Ferroelectrics*, **18**, 27 (1978).

[19] R. ORLOWSKY and E. KRÄTZIG: *Solid State Commun.*, **27**, 1351 (1978).

[20] G. C. VALLEY: *J. Appl. Phys.*, **59**, 3363 (1986).

[21] F. P. STROHKENDL, J. M. C. JONATHAN and R. W. HELLWARTH: *Opt. Lett.*, **11**, 312 (1986).

[22] C. MEDRANO, E. VOIT, P. AMRHEIN and P. GÜNTER: *J. Appl. Phys.*, **64**, 4668 (1988).

[23] D. FLUCK, P. AMRHEIN and P. GÜNTER: *J. Opt. Soc. Am.*, **8**, 2196 (1991).

[24] L. YOUNG, W. K. Y. WONG, M. L. W. THEWALT and W. D. CORNISH: *Appl. Phys. Lett.*, **24**, 264 (1974).

[25] W. KROLIKOWSKI, M. CRONIN-GOLOMB and B. S. CHEN: *Appl. Phys. Lett.*, **57**, 7 (1990).

[26] C. T. CHEN, D. M. KIM and D. VON DER LINDE: *IEEE J. Quantum Electron.*, QE-16, 126 (1980).

[27] L. K. LAM, T. Y. CHANG, J. FEINBERG and R. W. HELLWARTH: *Opt. Lett.*, **6**, 475 (1981).

[28] J. P. HERRMAN, J. P. HERRIAU and J. P. HUIGNARD: *Appl. Opt.*, **20**, 2173 (1981).

[29] C. T. CHEN, D. M. KIM and D. VON DER LINDE: *Appl. Phys. Lett.*, **34**, 321 (1979).

[30] J. K. TYMINSKI and R. C. POWELL: *J. Opt. Soc. Am. B*, **2**, 440 (1985).

[31] P. N. ZANADVOROV, E. L. LEBEDEVA, V. M. MOLDAVSKOYA and E. P. KOKANYAN: *Sov. Phys. Sol. State*, **30**, 1162 (1988).

[32] A. L. SMIRL, G. C. VALLEY, R. A. MULLEN, K. BOHNERT, C. D. MIRL and T. F. BOGESS: *Opt. Lett.*, **12**, 501 (1987).

[33] E. VOIT, M. Z. ZHA, P. AMRHEIN and P. GÜNTER: *Appl. Phys. Lett.*, **51**, 2079 (1987).

[34] G. LE SAUX, G. ROOSEN and A. BRUN: *Opt. Commun.*, **56**, 374 (1986).

[35] G. LE SAUX and A. BRUN: *IEEE J. Quantum Electron.*, QE-23, 1680 (1987); F. P. STROHKENDL: *J. Appl. Phys.*, **65**, 3773 (1989).

[36] N. V. KUKHTAREV and S. G. ODULOV: *JETP Lett.*, **30**, 4 (1979).

[37] N. V. KUKHTAREV and S. G. ODULOV: *Opt. Commun.*, **32**, 183 (1980).

[38] B. FISHER, M. CRONIN-GOLOMB, J. O. WHITE and A. YARIV: *Opt. Lett.*, **6**, 519 (1981).

[39] M. CRONIN-GOLOMB, J. O. WHITE, B. FISCHER and A. YARIV: *Opt. Lett.*, **7**, 313 (1983).

[40] M. CRONIN-GOLOMB, B. FISCHER, J. O. WHITE and A. YARIV: *IEEE J. Quantum Electron.*, QE-20, 12 (1984).

[41] J. O. WHITE, M. CRONIN-GOLOMB, B. FISHER and A. YARIV: *Appl. Phys. Lett.*, **40**, 450 (1982).

[42] M. CRONIN-GOLOMB, B. FISCHER, J. O. WHITE and A. YARIV: *Appl. Phys. Lett.*, **41**, 689 (1982).

[43] J. FEINBERG: *Opt. Lett.*, **7**, 486 (1982).

[44] F. C. JAHODA, P. G. WEBER and J. FEINBERG: *Opt. Lett.*, **9**, 362 (1984).

[45] M. CRONIN-GOLOMB, B. FISCHER, J. O. WHITE and A. YARIV: *Appl. Phys. Lett.*, **42**, 919 (1983).

[46] T. Y. CHANG and R. W. HELLWARTH: *Opt. Lett.*, **10**, 408 (1985).

[47] P. GÜNTER, E. VOIT, M. Z. ZHA and J. ALBERS: *Opt. Commun.*, **55**, 210 (1985).

[48] R. A. MULLEN, D. J. VICKERS and D. M. PEPPER: *Technical Digest: Topical Meeting on Photorefractive Materials, Effects, and Devices II*, January 17-19, 1990, Aussois, France.

MATERIALS FOR CUBIC EFFECTS

Fabrication, Characterization and Below-Band-Gap Optical Nonlinearities of Semiconductor-Doped Glasses.

V. Degiorgio and G. P. Banfi

Dipartimento di Elettronica, Sezione di Fisica Applicata dell'Università
27100 Pavia, Italia

1. – Introduction.

A class of materials potentially interesting for nonlinear optical applications is that of composites which are materials consisting of supermolecules or nanocrystals embedded in a dielectric matrix. We will discuss in this lecture a particular class of composites, semiconductor-doped glasses (SDGs), which are dispersions of II-VI semiconductor nanocrystals in a silicate glass matrix. SDGs containing $CdS_{1-x}Se_x$ crystallites are the basis for commercially available yellow-to-red optical filters presenting a sharp absorption edge at a cut-off wavelength which is adjusted by varying x [1]. The series is extended in the near i.r. with glasses including CdTe crystallites. The discovery that SDGs present significant nonlinear optical properties [2,3], associated with the discussion about the relevance of quantum confinement effects, has stimulated several investigations on their structure. Various nonlinear optical experiments [3] have been performed with both commercial and experimental glasses, the latter glasses tailored to contain very small crystallites.

This lecture contains a short description of the fabrication procedures, a discussion of the techniques used for the determination of the size and the volume fraction of the crystallites (with particular emphasis to the use of small-angle neutron scattering) and an introduction to two-photon absorption experiments probing below-band-gap nonlinearities. The resonant optical nonlinearities are discussed in detail by RICARD in this volume.

Since SDGs present an intrinsic polydispersity in crystallite size, various preparation procedures have been proposed and tested to prepare composites with a narrow distribution of sizes. Some of these techniques, which are for the moment at an early stage of development, are mentioned in the next section.

However, the subsequent treatment concerning structural characterization and optical nonlinearities deals only with SDGs.

2. – Fabrication.

The preparation of semiconductor-doped glasses is as follows[1]: the semiconductor constituents (or their oxides) are added to a melt of alkalisilicate glass at approximately 1300 °C. The melting process is done under reduced conditions in order to avoid oxidation of S, Se and Te. On initial cooling the glass displays only a pale yellow colour indicating a uniform dispersion of the constituents, although it is believed that subcritical nuclei of semiconductor material are already formed at this stage. In a second stage, called the striking process, the glass is annealed by keeping it for several hours at temperatures of (600–700) °C. During this stage nanocrystals can grow on the nuclei by a process of ion diffusion, the average crystal size depending on the type of doping, on the time duration of the treatment and on the furnace temperature.

The base glass composition must satisfy the requirement that submicroscopic small precipitations are formed during the cooling process, and that the precipitated droplets are small, numerous and uniformly distributed all over the glass. This can be controlled by acting on the potassium concentration.

Because of the fact that even small amounts of impurities influence the final colour of the glass, only very pure materials can be used. It is also important to avoid contamination coming from the melting furnace, from the used energy source and from the handling of the materials. The stoichiometry of the nanocrystals may be complicated by the presence of zinc which is usually present in borosilicate glasses. In fact, zinc might substitute cadmium in a small degree[4].

The chemical composition of the coloured glass can differ significantly from the original batch amounts because of the volatility of some components, especially S and Se, at the melting temperature. It should be noted that knowledge of the chemical composition of the coloured glass, obtainable through X-ray fluorescence, wet chemical or microprobe analysis yields the concentration of the semiconductor constituents in the entire glass, and cannot be used to determine the nanocrystal stoichiometry, which can instead be assessed by using X-ray diffraction techniques[4].

A new preparation method which is still in the development stage is that of fabricating semiconductor-doped silica glass thin films by r.f. sputtering[5]. Generally speaking, glass preparation from the gaseous state is superior to the conventional melting and quenching technique as it should allow a better control of the nanocrystal size and concentration, and avoids the reaction between the glass matrix and the semiconductor crystal as a result of the relatively low processing temperature. The glasses used for SDGs contain Na and K which form impurity levels in II-VI semiconductors. The sputtering technique uses pure SiO$_2$ glasses, and gives nanocrystals with few impurity levels. Typically the sputtered films have a thickness of 10 μm, and present a volume fraction of semiconductor crystallites which can be higher than 1%.

Semiconductor crystallites can also be grown in porous materials[6], such as porous glasses or porous polymeric matrices or zeolites: because of the constraints imposed by the porous medium, the size distribution can be narrower than in SDGs. However, the optical quality of the host matrix may represent a problem for optical applications. It is also possible to grow the nanocrystals in a colloidal solution by using inverse micelles or microemulsions. Also in this case the size distribution can be very narrow. Particles prepared in this way can be isolated as a powder, redissolved or chemically attached to a substrate[7].

3. – Structural characterization.

The sizes of crystallites reported in the literature have been obtained with various techniques. The only direct observation technique, which also permits to obtain the size distribution, is transmission electron microscopy (TEM)[4,8]. The recent introduction of filtering techniques to cancel the background signal originating from the glass matrix allows one to obtain rather clear pictures of the crystallite. TEM is, however, a time-consuming and destructive method. A second approach is that of using small-angle scattering of X-rays (SAXS)[9] or of neutrons[10]: this approach, as discussed in detail below, consists in measuring the intensity of scattered radiation as a function of the scattering angle and deriving from such a pattern the size of the scatterer. In some cases, indirect methods, such as Raman scattering[11], were also proposed: here the idea is that Raman scattering can give the frequency of vibrational surface modes of the particles and such a frequency is inversely proportional to the particle size. A comparison among TEM, SAXS and Raman scattering is discussed in a recent publication[12].

In this lecture, we shall discuss in some detail the small-angle scattering techniques. SANS uses typically neutrons from a cold source having kinetic energy in the range (3–30) meV (corresponding to wavelengths in the range (1.5–0.5) nm). SAXS uses X-ray emission lines from metallic targets. The usual line is CuK$_\alpha$ ($\lambda = 0.154$ nm), but in the case of SDGs the absorption of this line is very strong and it can be more convenient to use the MoK$_\alpha$ ($\lambda = 0.07$ nm) line[13]. Both SAXS and SANS are described by the same formalism, the only difference being that the neutron scattering amplitude comes from the interaction of the cold neutron with the nucleus, whereas the X-ray scattering amplitude is related to the interaction between the electromagnetic wave and the electrons. The mismatch in the neutron coherent scattering length between the semiconductor nanocrystals and the glass matrix is large enough to allow a quick and accurate measurement of the scattered intensity as a function of the scattering angle. A feature which emerged from the first SANS experiment with SDGs [10] was that the scattering pattern depends, not only on the crystal size, but also on the size of the depletion region that each crystal creates around itself during growth. When the volume fraction of crystallites is small, one could expect the scattered intensity $I(k)$, observed as a function of the modulus of the scattering vector k, to be proportional to the average form factor of the crystallites, $P(k)$. It was found, however, that this is correct only for values of k somewhat larger than $1/R_0$, where R_0 is the average grain radius. At smaller values of k, $I(k)$ departs considerably from $P(k)$. In fact, the data show that $I(k)$ is close to 0 at very small k and presents a peak at a value $k_p > 0$. The physical origin of such a behaviour lies in the fact that the crystal grows by depleting a surrounding region of size ξ much larger than the crystal radius. The quantity ξ represents the distance over which mass can be transported by diffusion in the annealing time interval t. Since the mass of aggregating material is conserved over a region of size $\geq \xi$, the scattered intensity (which is proportional to the static structure factor) is very low for k smaller than $1/\xi$. By using these considerations, a simple phenomenological model was developed [14, 15], containing as free parameters R_0, ξ and the volume fraction Φ_v of crystallites, which calculates $I(k)$ for a dilute gas of crystallites surrounded by depletion zones. The model can incorporate the size polydispersity of crystallites, and describes very well the experimental data. Note that the peak of $I(k)$ is not observed in the usual small-angle X-ray scattering experiments only because the standard SAXS instruments do not give access to values of k smaller than $0.15\,\mathrm{nm}^{-1}$. However, in the case of glasses with sufficiently small crystallites, a careful analysis of low-k data seems to confirm the presence of the peak [13].

SANS measurements provide also a reliable determination of the volume fraction occupied by the crystallites. This is an important parameter in optical studies, which cannot be derived either from the initial dopant concentration or from a chemical analysis on the fabricated glass. Difficulties in fact do arise from the volatility of S and Se, and also from the fact that not all the dopant atoms end up in the crystallites, but a significant fraction is still dispersed in the matrix in various forms.

Some years ago DURVILLE et al. [16] studied by SANS the nucleation and growth of $MgAl_2O_4$ in a cordierite glass, and subsequently WRIGHT et al. [17] used the same technique for a study of $MgTi_2O_5$ crystallites in a silicate glass. It is interesting to note that also in these cases a peak in $I(k)$ was observed.

3'1. *The small-angle neutron scattering technique.* – The important geometrical quantity in a scattering experiment is the scattering vector, $\boldsymbol{k}$, defined as the difference between the scattered and incident wave vectors, $\boldsymbol{k} = \boldsymbol{k}_d - \boldsymbol{k}_i$. Calling θ the scattering angle, which is the angle formed between $\boldsymbol{k}_d$ and $\boldsymbol{k}_i$, and noting that $k_d \approx k_i$, k is given by the relation $k = 2k_i \sin(\theta/2)$. SANS is used in situations where the relevant structural information occurs on length scales in the range (1–100) nm, much larger than interatomic distances[18]. This means that a detailed description of the structure at the atomic level is not required. A material of given chemical composition is characterized by the scattering length density ρ which represents the scattering amplitude per unit volume. Values of ρ for the relevant semiconductors are given in table II of ref.[19].

It is useful to define, for a material of given composition and structure, the absolute scattered intensity $I(k)$ which represents the fraction of the incident beam scattered, per unit solid angle and per unit length of the sample thickness, at the scattering angle θ. $I(k)$ is measured in sterad^{-1}cm^{-1}. In the simple situation in which monodisperse, spherical and homogeneous particles of radius R_0 and coherent scattering length density ρ_p are embedded in a matrix of coherent scattering length density ρ_0, the absolute scattered intensity can be written as [18]

$$(1) \qquad\qquad I(k) = N\partial_p^2 V_p^2 P(kR_0) S(k),$$

where N is the number of particles per unit volume of the material, $V_p = (4/3)\pi R_0^3$ is the volume of the particle, $\partial_p = \rho_p - \rho_0$ is the mismatch in scattering length density, and $P(kR_0)$ is the particle form factor which depends on k and R_0 only through the product kR_0. $S(k)$ is the structure factor describing the interactions among different particles. Note that, for uncorrelated particles, $S(k)$ becomes identically equal to 1. The measured scattered intensity depends on several experimental quantities. In order to derive the absolute $I(k)$, a calibration procedure is used which essentially consists in a comparison with known scattering standards[18,19].

An adequate physical model for the structure of the doped glass should provide for the conservation of the atomic constituents of the semiconductor crystallites over the distance ξ. This is the distance over which mass can be transported by diffusion in the annealing time interval t. An identical amount of semiconductor must be depleted from the matrix as is found in the nanocrystal. One should, therefore, calculate the form factor of a scattering object made of the crystallite plus the depleted region around the crystallite. This can be modelled with a density profile that is constant and corresponds to a pure semiconductor for $0 \leqslant r \leqslant R_0$, and a depleted zone with its concentration varying linearly from R_0 to the bulk matrix value at ξ, with the constraint that the total amount of material in the core is equal to that missing in the depletion zone. Such a mass conservation condition implies that $I(k)$ is zero for $k \ll 1/\xi$, independently of the mutual arrangement of crystallites. We call $P_m(kR_0, k\xi)$ the modified form factor which is a function of the two variables kR_0 and $k\xi$. Its complete expression can be found in ref.[19]. In the case of SDGs, once the form factor is calculated according to such a model, the choice of $S(k)$ becomes irrelevant, so that one can put $S(k)$ identically equal to 1 in eq. (1) when fitting the experimental data [14].

Taking into account that $\Phi_v = NV_p$, $I(k)$ can be written as

$$(2) \qquad\qquad I(k) = \partial_p^2 \Phi_v V_p P_m(kR_0, k\xi).$$

Note that, for a fixed R_0/ξ ratio, the plot of $I(k)$ vs. k scales with a horizontal scale factor $1/R_0$ and a vertical scale factor ΦV_p. Therefore, the shape of $I(k)$ is determined by R_0 and ξ, whereas its amplitude depends also on Φ_v, the volume fraction occupied by the crystallites.

3'2. *The size and volume fraction of nanocrystals.* – Systematic SANS measurements have been performed on commercial glasses manufactured by Schott Glaswerke (Mainz, Germany). Table I presents the list of the investigated glasses, together with the chemical composition of the nanocrystals and the mismatch in the scattering length densities ∂_p. The number associated with

TABLE I. – *The best-fit values of R_0 and Φ_v for various commercial glasses. N denotes measurements taken at NIST, G at ILL. x is the Se fraction in the $CdS_{1-x}Se_x$ nanocrystal. $|\partial_p|$ is the absolute value of the mismatch in scattering length density between the semiconductor nanocrystal and the glass matrix.*

Glass		x	$|\partial_p| \cdot 10^{-9}$ (cm^{-2})	R_0 (nm)	$\Phi_v \cdot 10^3$
GG 495	N	0.03	19.8	2.1	5.2
OG 530 #1	N	0.13	19.1	3.4	3.6
OG 530 #2	N	0.13	19.1	2.8	4.1
OG 590	N	0.5	16.3	3.0	4.7
OG 610 #1	N	0.58	15.8	3.1	2.9
OG 630	N	0.66	15.2	3.7	3.2
RG 665	N	0.8	14.2	6.5	3.5
RG 695	N	0.92	13.3	4.3	1.6
RG 715	N	1	12.7	5.5	3.2
OG 515	G	0.08	19.4	2.8	4.2
OG 530 #2	G	0.13	19.1	2.7	3.5
OG 570	G	0.37	17.3	3.6	3.5
RG 610 #2	G	0.58	15.8	3.9	4.4
RG 830	G	—	20.5	5.3	1.6
RG 850	G	—	20.5	11.1	1.5

each glass indicates the cut-off wavelength λ_c in nanometres (more precisely, λ_c is the wavelength at which transmission is 50% in a 3 mm thick sample). All the glasses have a nearly similar melt composition, except for the semiconductor anions: the glasses with cut-off wavelength in the visible contain $CdS_{1-x}Se_x$ crystals, the infrared filters contain essentially CdTe.

The composition of a typical doped glass can be found in table I of ref.[19]. Although both boron and cadmium have high absorption cross-sections, their concentrations are sufficiently low that samples of (0.5–2) mm thickness can conveniently be studied with SANS instruments on high-flux reactors.

Most part of the SANS measurements here described were performed at the Cold Neutron Research Facility of the National Institute of Standards and Technology, Gaithersburg, MD, U.S.A.[15,19]. Samples in the form of polished plates of area 2 cm^2 were mounted normal to the incident neutron beam of wavelength $\lambda = 0.55$ nm. The thickness of the sample was 1 mm. Since the experiment was aimed at obtaining the complete shape of $I(k)$, various sample-to-detector distances between 3.5 and 15 m were used. At the shortest distance the measurement time is typically 10 min per sample. The intensity scattered at low k is extremely weak, so that the data taken at the sample-to-detector distance of 15 m required a measurement time of several hours. Data were recorded on a two-dimensional multidetector, and were corrected in the usual manner for background scattering and detector efficiency[18]. The absolute intensity was obtained by calibration with various standards, each appropriate to a different instrument configuration. For each sample, transmission and scattering measurements were made. The typical sample transmission was about 60%.

Some of the SANS measurements were performed at the Institute Laue Langevin, Grenoble, France. The neutron wavelength was set in this case at 1.2 nm, while the detector distance ranged between 1.4 and 2.8 m. At this neutron wavelength, the absorption is stronger than at 0.55 nm[10], and thinner samples (0.5 mm thick) were employed.

We show in fig. 1 and 2 the $I(k)$ obtained at NIST for some commercial glasses. Each plot is obtained by the superposition of data taken at several sample-to-detector distances. Most of the curves of fig. 1 and 2 show a peak at a finite value of k, with $I(k)$ falling to very low values at small k. The observed form of $I(k)$ clearly indicates that semiconductor-doped glasses do not behave as dilute dispersions of homogeneous particles.

In principle, a realistic model should take into account that the crystallites might be nonspherical and show a size distribution. Even if the crystallites show a tendency to grow as hexagonal prisms elongated in the c-direction[8], large deviations from sphericity were not shown by TEM pictures[4], even in high resolution[8], so that we have kept the assumption that the crystallites are spherical. As far as polydispersity is concerned, the TEM observations of Potter and Simmons[20] suggest a size distribution $f(R)$ in agreement with the Lifshitz-Slyozov-Wagner (LSW) theory[21]. We have assumed that the LSW dis-

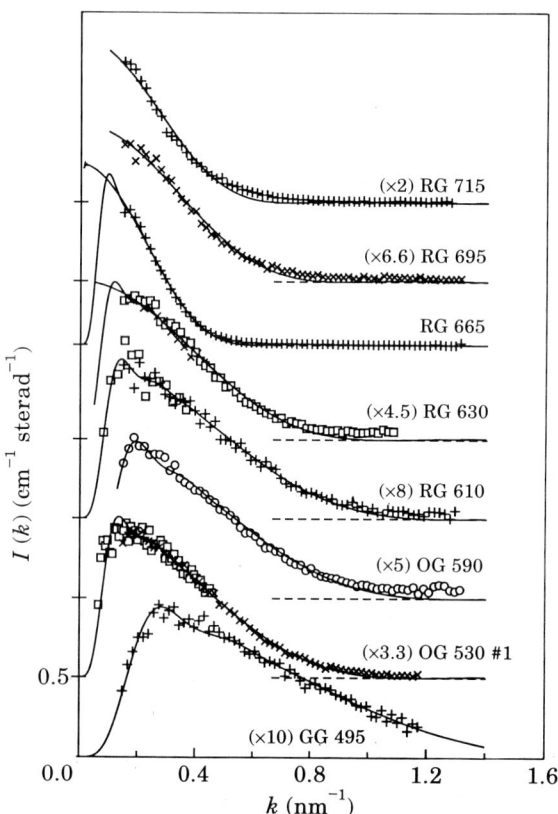

Fig. 1. – The scattered neutron intensity $I(k)$, in absolute units, for various commercial glasses (NIST data). The full lines represent a fit with the model discussed in the text.

tribution applies to all our samples. An indirect, but rather convincing, argument which justifies our assumption will be presented later on in connection with the discussion of the scaling properties of $I(k)$. Since the LSW distribution contains the average radius R_0 as the only free parameter, the insertion of polydispersity within the model does not increase the number of fitting parameters. The shape of $I_{LSW}(k, R_0)$, for an ensemble of particles distributed in size according to the LSW theory around the average radius R_0, is very similar to that of $I(k)$ as calculated for monodisperse particles [15,19]. The comparison between the two expressions suggests the relation

(3) $$I_{LSW}(k, R_0, \xi_0) \approx cI(k, R, \xi),$$

where $R = 1.146 R_0$, $\xi = 1.146 \xi_0$, and $c = 0.94$, if I_{LSW} and I are calculated at the same particle volume fraction. R is larger than R_0 because the scattering technique gives more weight to the largest crystallites. The value of c depends on

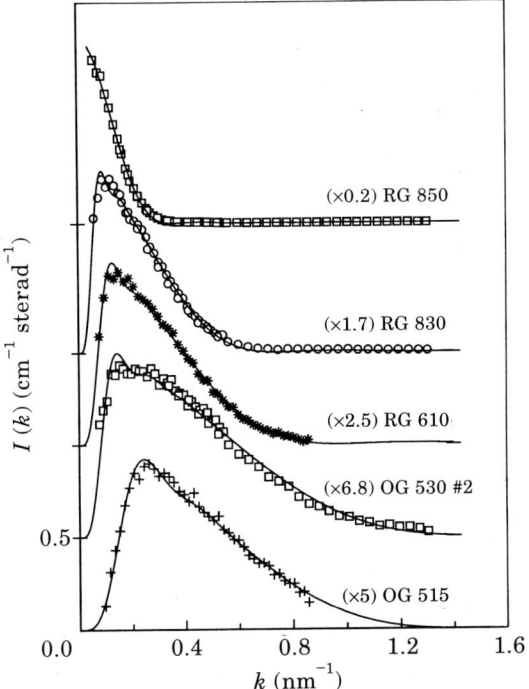

Fig. 2. – The same as fig. 1 for various commercial glasses (ILL data).

the ratio R/R_0 and on the moments M_k of the LSW distribution as follows: $c = (M_6/M_3)(R_0/R)^3$. Using the relations $M_6 = 1.61\,R_0^6$, $M_3 = 1.13\,R_0^3$, we obtain the value of c. The fit performed by using eq. (2) gives a value for the volume fraction which is 6% smaller than the true value.

The considerations developed above indicate the following empirical procedure[19]: by fitting the data with the simple eq. (2) valid for a monodisperse system we obtain R, ξ and a volume fraction Φ'. If we assume a LSW distribution, the average radius R_0, the size of the depleted region ξ_0 and the volume fraction Φ_v can be obtained by scaling the obtained values as $R_0 = 0.873\,R$, $\xi_0 = 0.873\,\xi$, $\Phi_v = 1.06\,\Phi'_v$.

The agreement between the model and the experimental data is remarkable, as shown by the fitting curves in fig. 1 and 2. The best-fit parameters are reported in table I. For some of the glasses, the data extend to sufficiently low k for an unambiguous determination of both R_0 and ξ_0. For other glasses, the data at low k are too poor, or missing. In such a case the fit becomes insensitive to the value of ξ_0, and gives only R_0 and Φ_v. We recall that collecting data at low k requires a much longer measuring time. Samples with small crystallites have also a smaller depletion radius, this is the main reason for which, in spite of similar experimental conditions, it has been possible to obtain ξ_0 values for some commercial glasses and not for others. The determination of R_0 is very accurate, even with a short accumulation time: we estimate that the uncertainty in R_0 is never larger than $\approx 3\%$.

The accuracy in the determination of Φ_v depends, not only on the fitting uncertainty, but also on the accuracy of the absolute calibration and on the precise knowledge of the mismatch δ between crystallites and glass matrix. Taking into account all these contributions, the estimated uncertainty in the evaluation of Φ_v is below 10%. We stress that the obtained values of R_0 and Φ_v are essentially independent of the details of modelling the depletion process because they are mainly determined by the decay of $I(k)$ when k becomes somewhat larger than the peak value k_p.

One of the commercial glasses (OG 530 #2) was tested both at NIST and ILL. The fit parameters are quite similar (4% difference in R_0 and 15% in Φ_v). This is a satisfactory agreement for results taken with two different instruments, each with its own calibration, at two different neutron wavelengths.

Sample-to-sample variation for the same commercial glass can be seen in comparing OG 530 #1 with OG 530 #2, and RG 610 #1 with #2. A slight difference was observed also in their optical absorption edge, which is only guaranteed by the manufacturer to be within 3 nm from the labelling wavelength of the glass. Given allowance for sample-to-sample variation, the radii in table I do agree with literature data for the commercial glasses [22]. A more precise comparison can be made with the results of a recent SAXS experiment [13], where the study of two samples previously investigated with SANS at ILL [10] is presented. As discussed in ref. [13], the agreement in crystallite size is remarkable.

3`3. *The kinetics of crystal growth.* – In this subsection we focus our attention on the kinetics of crystal growth within SDGs. The study [15] was performed by using two series of experimental glasses, both prepared by Schott Glaswerke. Within each series the samples have the same chemical composition but different size of the semiconductor crystallites, because they were annealed for the same time interval but at different temperatures, using a gradient furnace.

The first series of experimental glasses consisted of four samples, called $R1, ..., R4$, which have the same composition of the commercial glass RG 830-850. The samples were annealed for 24 h at the following temperatures T: 1) 923 K, 2) 943 K, 3) 963 K, 4) 983 K. The second series of experimental glasses has the composition of the commercial Schott glass OG 515, containing mixed $CdS_x Se_{1-x}$ crystals with $x \approx 0.1$, and consists of four glasses annealed for 48 h at the following temperatures: 1) 863 K, 2) 883 K, 3) 903 K, 4) 923 K. We call hereon $V1, ..., V4$ the glasses of the second series.

The $I(k)$ data obtained with the R series are reported in fig. 3. The peak

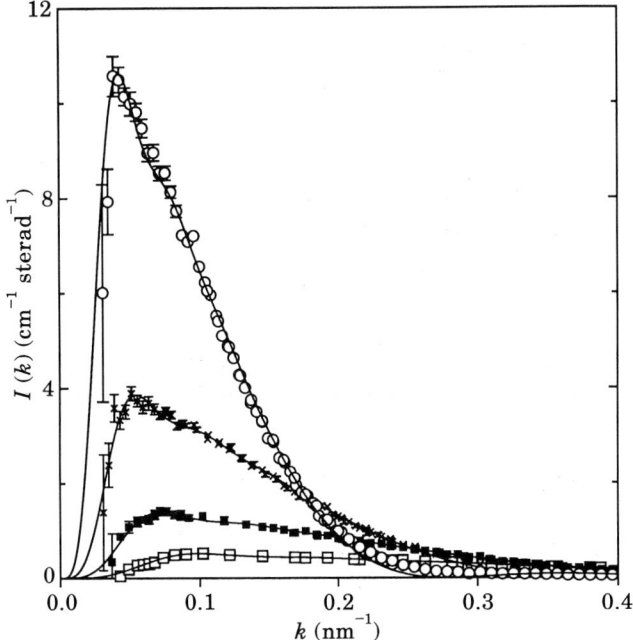

Fig. 3. – The same as fig. 1 for the R series of experimental glasses. The four curves refer to samples annealed at different temperatures: □ $R1$, ■ $R2$, × $R3$, ○ $R4$. The full lines represent a fit with theory.

height increases, and correspondingly the peak position k_p shifts to smaller and smaller values, as the annealing temperature grows. The fit with the model discussed above is very good. The best-fit parameters for both the R and the V series are reported in table II. It should be remarked that the accuracy of ξ_0 becomes very poor for the $V4$ and $R4$ glasses, because for these glasses data at sufficiently low k were not accessible to our apparatus.

It is interesting to check how the position of the optical absorbtion edge of the experimental glasses shifts with the nanocrystal radius. A spectrophotometer transmission measurement, at room temperature, gave for E_c, the photon energy corresponding to λ_c, the values reported in fig. 4. The confinement should increase the energy of the absorption edge of a quantity $\Delta E(R_0)$ which depends on the size of the nanocrystal. In fig. 4 we show an attempt to fit the data by assuming $\Delta E(R_0) \propto 1/R_0^2$, a dependence which should be suitable also for the large radii of our experimental glasses [20, 23]. The agreement is very good for the R series and for the first three glasses of the V series, but a clear deviation appears for the $V4$ sample. A detailed discussion is given in ref. [15] and [19].

Usually the process of phase separation is described as a function of the annealing time interval t. In our case, t is fixed for each family, and the samples

TABLE II. – *Best-fit values of the average nanocrystal radius R_0, of the average radius of the depleted region ξ_0 and of the volume fraction Φ_V. A LSW size distribution has been assumed. The first column lists the mismatch ∂_p in scattering length densities.*

Glass	$|\partial_p| \cdot 10^{-9}$ (cm^{-2})	R_0 (nm)	ξ_0 (nm)	$\Phi_V \cdot 10^3$
V1	19.4	5.2	57	4.0
V2	19.4	7.6	88	3.7
V3	19.4	8.7	104	3.6
V4	19.4	13.5	≈ 140	3.4
R1	20.5	4.8	62	1.8
R2	20.5	6.9	85	1.7
R3	20.5	9.8	113	1.7
R4	20.5	14.0	≈ 150	1.6

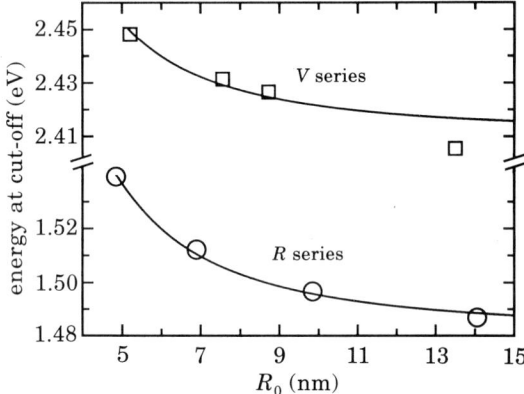

Fig. 4. – Plot of E_c vs. R_0 for the crystallites of the R and V series. $E_c = hc/\lambda_c$, with λ_c the cut-off wavelength. The full curves are a fit assuming the dependence $E_c = a + b/R_0^2$.

are related to different annealing temperatures. Noting that the mass diffusion coefficient D should behave as $D \approx \exp[-A/k_B T]$, where A is an activation energy and k_B is the Boltzmann constant, and that a change of D is equivalent to a change in the time scale of the phase separation process [21], we conclude that time and temperature are connected by the transformation $t \approx \exp[-A/k_B T]$. The transformation is strongly nonlinear, so that small changes of T may correspond to large variations of the corresponding time interval t. We have plotted in fig. 5 the logarithm of R_0 as a function of the reciprocal of T. The fact that the plot is linear confirms the exponential transformation law between annealing time and temperature.

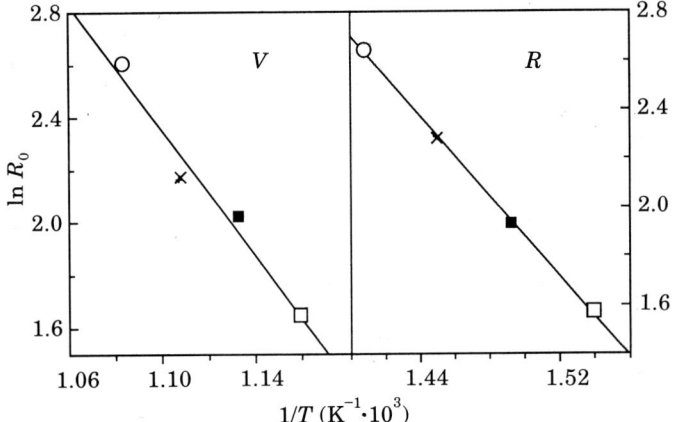

Fig. 5. – The logarithm of the crystallite radius in nanometres as a function of the recipro-cal of the absolute annealing temperature.

The kinetics of nucleation and growth of crystals from a supersaturated solid solution is only partially understood. Recent approaches [24] focus the attention on the structure factor which can be measured by scattering techniques, and in-dicate the existence of a common theoretical basis between spinodal decomposi-tion (SD) and nucleation processes. There is ample experimental evidence of a scaling behaviour of the structure factor in SD [24], but few experiments con-cern nucleation processes [25].

The scaling theory of SD [24] predicts that the scattered intensity $I(k, t)$ is related to the scaled wave vector k/k_p by the equation $I(k, t) = k_p^{-3} F(k/k_p)$,

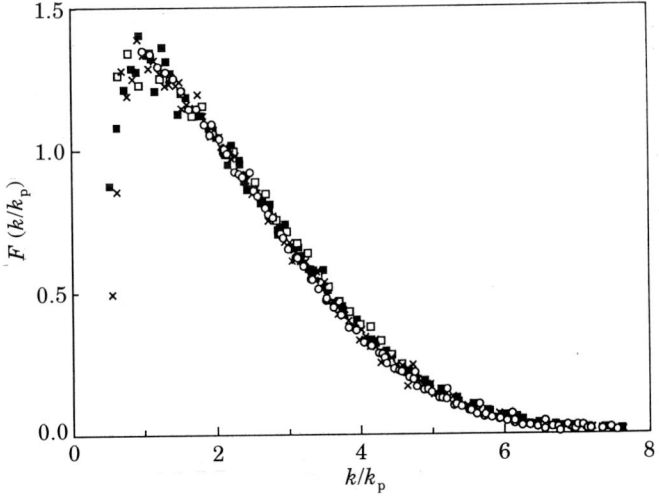

Fig. 6. – Scaling behaviour for glasses of the R series: □ $R1$, ■ $R2$, × $R3$, ○ $R4$.

where $F(k/k_p)$ is a time-independent scaling function. The plots shown in fig. 6 are obtained from those in fig. 3 by multiplying the vertical scale by k_p^3 and dividing the horizontal scale by k_p. We see that the data display scaling behaviour. A similar behaviour is presented by the V series[15].

The fact that scaling applies to our data and that the total amount of material present in the crystallites shows only small changes with temperature (see the values of Φ_v given in table II) indicates that the observed growth is already in the late stage, called coarsening (or ripening), in which, according to the LSW theory, the shape of the size probability distribution is also time-independent, and the only time-dependent parameter is the average crystal size which grows as t^b, with $b = 1/3$. This is in agreement with the electron microscope observations of SDG discussed in ref.[20] Note that, taking $b = 1/3$, the slope of the plots in fig. 5 gives an activation energy $A/k_B = 30\,000$ K.

4. – Two-photon absorption processes.

We discuss in this section very recent measurements of the imaginary part of the third-order optical susceptibility $\chi^{(3)}$ for semiconductor nanocrystals below the band gap[26]. Differently from the resonant case, the optical nonlinearities at light frequencies well below the absorption edge are poorly understood. COTTER et al.[27] have recently presented calculations which predict a strong effect of the crystal size R_0 on the third-order optical susceptibility $\chi^{(3)}$ for semiconductor nanocrystals below the band gap. In particular, the imaginary part of $\chi^{(3)}$, which is directly related to the two-photon absorption (TPA) coefficient β, is predicted to decrease when R_0 decreases. However, the few available experimental results[27-29] would rather indicate the opposite behaviour, with values of β for the crystallites typically five times larger than for the bulk semiconductor. In order to clearly establish the effect of quantum confinement on TPA, it is necessary to obtain accurate optical data over a wide range of nanocrystal sizes and over the whole interval of accessible ratios $y = E_g/\hbar\omega$, where E_g is the energy gap of the SDG and $\hbar\omega$ is the energy of the incident photon. It is also essential to complement the optical measurements with a careful structural characterization of the used SDGs.

Measurements of $\mathrm{Im}\,\chi^{(3)}$ were performed on many of the SDGs listed in table I, covering a range of R_0 between 2 and 11 nm, and a range of y between 1 and 1.9. We find that, in order to extract the imaginary part of $\chi^{(3)}$ from the optical-transmission data in the picosecond time regime, it is necessary to take into account free-carrier absorption (FCA) processes. We have also performed experiments in the femtosecond time domain where the effect of FCA is negligible. The average size and the volume fraction of nanocrystals were measured by small-angle neutron scattering (SANS). The values of $\mathrm{Im}\,\chi^{(3)}$ obtained for the nanocrystals are comparable to those of the bulk semiconductors of similar com-

TABLE III. – TPA *coefficient of the whole semiconductor-doped glass (third column) and* TPA *coefficient of the semiconductor nanocrystals (fourth column).* In the second column, P denotes measurements at 1.06 μm with 30 ps pulses, F denotes measurements at 0.6 μm with 180 fs pulses. When a range of β values is given, the first value is derived by taking $\sigma = 2 \cdot 10^{-18}$ cm^2, and the second by taking $\sigma = 0$.

Sample		$\beta_{\mathrm{SDG}} \cdot 10^3$ (cm/GW)	β_{m} (cm/GW)
OG 570	P	2– 8	1.1– 4.3
OG 590	P	3– 10	1.3– 4.3
RG 610	P	4– 12	2.8– 8.3
RG 630	P	9– 27	6.2– 18
RG 665	P	15– 40	9.5– 25.5
RG 695	P	9– 25	12.5– 34.8
RG 715	P	18 ± 3	12.7 ± 2.1
RG 830	P	17 ± 3	32.5 ± 5.9
RG 850	P	23 ± 2	47 ± 4
GG 495	F	15 ± 3	6.3 ± 1.3
OG 570	F	23 ± 3	12.5 ± 1.6
OG 590	F	20 ± 5	10.1 ± 2.5

position, indicating that quantum confinement effects on TPA, at least in the investigated size range, are very small.

The list of used SDGs, all manufactured by Schott Glaswerke (Mainz, Germany), is given in table III. Nonlinear transmission measurements were performed on all samples by employing 30 ps pulses at 1.06 μm from a Nd-YAG mode-locked laser. The pulse duration was measured by standard correlation techniques. The beam quality factor, obtained by monitoring the transversal intensity profile of the laser beam with a CCD camera, was $M^2 = 1.25$. Typically, the thickness of the samples was (3–5) mm, but for a few SDGs we used also much thicker samples.

The transmission T, measured as the ratio between the transmitted energy and the input energy, is shown in fig. 7 as a function of the peak intensity Φ_0. The decrease of T with Φ_0 is due to TPA processes which become possible when $y \lesssim 2$. We find indeed that the nonlinear absorption of the Nd-YAG laser pulses becomes undetectable for SDGs with $\lambda_c \lesssim 550$ nm. It should be noted that the glass matrix, with its large band gap, plays no role in TPA. In the presence of TPA, the intensity Φ of a pulse propagating along z is given, neglecting diffrac-

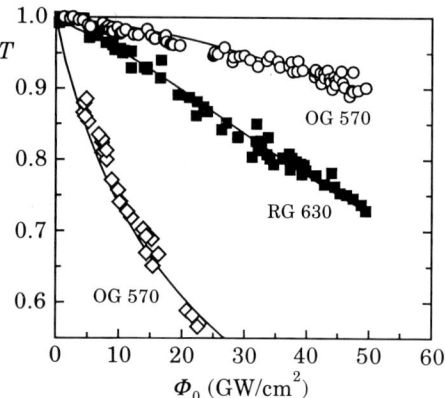

Fig. 7. – Transmission *vs.* peak intensity for some SDGs. The two upper sets of data points, $\circ$ and $\blacksquare$, are taken with 30 ps pulses at 1.06 μm with a sample thickness of 0.8 cm. The lower set of data points, $\diamond$, is taken with 180 fs pulses at 0.6 μm with a sample thickness of 5 cm. Full lines are best-fit curves.

tion, by

$$(4) \qquad \frac{\mathrm{d}\Phi(r, z, t)}{\mathrm{d}z} = -\beta\Phi^2(r, z, t) - \sigma N(r, z, t)\Phi(r, z, t),$$

where r is the radial coordinate, and β is the TPA coefficient which is related to $\mathrm{Im}\,\chi^{(3)}$ by the expression $\beta = \omega(\varepsilon_0 c^2 n^2)^{-1}\,\mathrm{Im}\,\chi^{(3)}$, n being the index of refraction. The second term at left-hand side of eq. (4) accounts for FCA, with N the free-carrier density generated by TPA and σ denoting the related cross-section. Assuming the decay of N to be negligible during the pulse duration (the validity of this assumption is supported by time-resolved degenerate-four-wave-mixing measurements we performed on the samples), N can be calculated through

$$(5) \qquad N(r, z, t) = \int_{-\infty}^{t} \frac{\beta\Phi^2(r, z, t')}{2\hbar\omega}\,\mathrm{d}t'\,.$$

By assigning the spatial intensity profile and the temporal shape of the pulse, the nonlinear dependence of the output energy as a function of the input energy can be calculated by solving numerically eqs. (4) and (5). In principle, the two parameters β and σ could be obtained by a best fit to the data taken with a single sample of thickness L. However, because of the limited range of input energies over which the nonlinear transmission can be investigated and because of the uncertainties associated with the experimental data, it is not possible to reliably extract the two parameters from a single experimental curve. For instance, the data points referring to the OG 570 sample in fig. 7 can be described, with no appreciable difference in the quality of the fit, by adopting any couple of values within the interval $\beta = 8\cdot10^{-3}$ cm/GW, $\sigma = 0$, $\beta = 2\cdot10^{-3}$ cm/GW, $\sigma =$

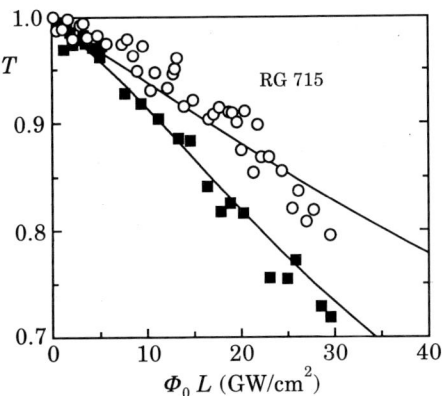

Fig. 8. – Transmission *vs.* $\Phi_0 L$ for two distinct values of L: ■ $L = 0.8$ cm, ○ $L = 5$ cm. Full lines are best-fit curves.

$= 2 \cdot 10^{-18}$ cm^2. It should be noted that, for $\sigma = 0$, the transmission derived from eq. (4) depends on the peak intensity Φ_0 and on L only through the product $\Phi_0 L$. In fig. 8 we show plots of T *vs.* $\Phi_0 L$ for two different sample thicknesses. The two sets of data do not fall on the same curve, which indicates that FCA cannot be neglected. A fit to both sets of data allows one to derive β and σ. Adopting this procedure, we estimated $\sigma \approx 2 \cdot 10^{-18}$ cm^2 for both RG 830 and RG 715. The value is consistent with $\sigma = 1.95 \cdot 10^{-18}$ cm^2 derived for RG 850 in ref. [28].

A simple way to reduce the effect of FCA is that of measuring the nonlinear transmission with ultrashort pulses which can provide a high enough intensity to evidence TPA without exciting too many carriers during the pulse duration. Numerical solutions of eqs. (4) and (5) indicate that, for an extended range of values of L, Φ_0 and σ, the effect of FCA can safely be disregarded with pulses shorter than 200 fs. The ultrashort-pulse measurements were performed at the European Laboratory of Nonlinear Spectroscopy, Florence, Italy. The investigation was limited to GG 495, OG 570 and OG 590, because only the wavelength of 0.6 µm ($\hbar\omega = 2.06$ eV) was available. The 190 fs pulse from the dye-laser amplifier, after passing through a spatial filter, was gently focused to provide ≈ 10 µJ of energy within a smooth Airy disc at the sample. A good accuracy in the measurement of T was achieved by using a differential detector and a reference beam. A cross-check of the absolute calibration (which required measuring energy, beam shape and pulse duration) was given by a separate experiment in which we observed the beam depletion due to second-harmonic generation in a 0.5 mm thick KDP platelet, in type-I phase matching (using KDP as a TPA standard, we assumed d_{36}(KDP) = 0.4 pm/V). Since beam depletion due to second-harmonic generation is formally equivalent to that due to TPA, such a calibration procedure is simple and reliable. We found some evidence of a darkening effect [3] (especially with GG 495): that is, starting with a fresh sample, the linear transmission showed an initial decrease with time before stabilizing after exposure. We relied on pre-darkening to avoid changes of the linear transmission during data collection. A typical transmission curve can be seen in fig. 7.

The values of β_{SDG} derived from the transmission data are reported in table III. For those measurements performed with 30 ps pulses in which it was not possible to derive both β and σ, we fixed σ by considering the two extreme scenarios of $\sigma = 2 \cdot 10^{-18}$ cm^2 and $\sigma = 0$. There is some indication[28] that σ decreases as E_g increases, so that the value $\sigma = 0$ is probably the more appropriate for the largest values of y.

The few published values of β_{SDG} are rather different among each other and are all larger than our values. The origin of the discrepancy with ref.[27] is probably that FCA was not taken into account in the interpretation of transmission data. In the case of ref.[29], the used technique does not allow an easy absolute calibration. The origin of the discrepancy with ref.[28] is unclear.

The relation between the measured $\chi^{(3)}_{SDG}$ of the composite SDG and $\chi^{(3)}_m$ of the nanocrystals is given by[3,30]

(6) $$\chi^{(3)}_{SDG} = \chi^{(3)}_m f^4 \Phi_v \,,$$

where f is the local-field correction factor. Assuming a spherical shape and an isotropic polarizability for the crystallites, f is given by $f = 3n_g^2/(n_m^2 + 2n_g^2)$, where n_g is the index of refraction of the glass matrix and n_m that of the crystallite. At 1.06 µm, f was calculated by taking $n_g = 1.53$, $n_m = 2.84$ (value of bulk

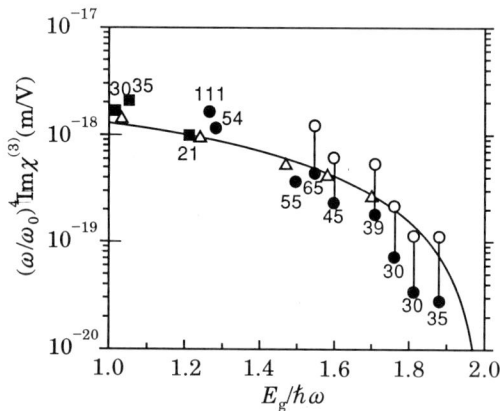

Fig. 9. – Plot of the scaled quantity $(\omega/\omega_0)^4 \, \mathrm{Im}\,\chi^{(3)}$ vs. $E_g/\hbar\omega$. ■ optical measurements at 0.6 µm. Circles refer to optical measurements at 1.06 µm assuming $\sigma = 2 \cdot 10^{-18}$ cm^2 (●), $\sigma = 0$ (○). The numbers associated to the experimental points are the radii of the nanocrystals in ångström. △ Experimental data for bulk semiconductors, taken from ref.[32] and [33], all measured at 1.06 µm, except for the point corresponding to the lowest value of $E_g/\hbar\omega$ which is measured at 0.53 µm. The solid line is the behaviour of $(\omega/\omega_0)^4 \, \mathrm{Im}\,\chi^{(3)}$ predicted for bulk semiconductors.

CdTe) for RG 830 and RG 850, $n_m = 2.33$ (value of bulk CdS) for GG 495, $n_m = 2.54$ (value of bulk CdSe) for RG 715, and then interpolating according to λ_c for the other glasses of the CdSSe series. At 0.6 μm, n_m is about 3% larger and f^4 somewhat lower.

By using eq. (6) one can write $\beta_{SDG} = \beta_m (n_m^2/n_{SDG}^2) f^4 \Phi_v$. This relation allows one to derive β_m from the measured β_{SDG} and Φ_v. The obtained values are given in table III. Taking into account all the uncertainties, the absolute calibration of β_m is estimated to be correct within a factor 2.

The comparison between the TPA coefficient of nanocrystals and that of bulk semiconductors is shown in fig. 9, where the scaled quantity $(\omega/\omega_0)^4 \operatorname{Im} \chi^{(3)}$, with ω_0 the frequency of 1.06 μm radiation, is plotted as a function of $E_g/\hbar\omega$. For the nanocrystals, we set $E_g = hc/\lambda_c$. Such a value of E_g should differ by no more than 20 meV from the value one would derive from the wavelength of the fluorescence peak at room temperature [4], and practically coincides with the energy difference between the higher discrete level of the valence band and the lower one of the conduction band. Due to the blue shift caused by confinement, the nanocrystals and the bulk with the same E_g do not have exactly the same stoichiometry. The distinction is irrelevant for the present purposes, except at the highest $E_g/\hbar\omega$ values. The full curve in fig. 9 gives the bulk values, calculated by using the relation $\operatorname{Im} \chi^{(3)} = \operatorname{const} \cdot (1/\omega E_g^3)(1 - 2\hbar\omega/E_g)^{3/2} (2\hbar\omega/E_g)^5$, which was proposed by VAN STRYLAND and co-workers [31]. The constant in the expression of $\operatorname{Im} \chi^{(3)}$ is fixed by fitting the curve to experimental data for bulk semiconductors [32, 33] which are also reported in fig. 3.

It is found that $\operatorname{Im}\chi_m^{(3)}$ decreases monotonically as y increases. Within experimental errors, TPA in the nanocrystals has the same behaviour as in the bulk semiconductor. Quantum confinement effects on $\operatorname{Im}\chi_m^{(3)}$ must be quite small, if the deviations from the bulk values are not evident even for the small crystallites of the GG 495 glass ($R = 2$ nm). Using the R and y parameters of this sample, the theoretical curves plotted in fig. 1 of ref. [27] predict $\operatorname{Im}\chi_m^{(3)}$ to be at least 4 times smaller than for the bulk.

Our group also performed some preliminary measurements of $|\chi_{SDG}^{(3)}|$ through nearly degenerate three-wave mixing at wavelengths around 1 μm. It was found that the ratio $|\chi_{SDG}^{(3)}|^2/|\chi_{glass}^{(3)}|^2$ is in the range 1-2 for all SDGs. This result indicates a significant contribution of the glass matrix and, making use of eq. (6), appears fairly consistent with the published value of $\chi_b^{(3)}$ of bulk semiconductor.

We can summarize the results of this section by saying that, below the band gap, the third-order susceptibility of the semiconductor nanocrystals with a size ranging from 2 to 11 nm has the same magnitude as for the bulk semiconductor. Such a result calls for a reconsideration of the approximation used in the theoretical treatment of below-band-gap optical nonlinearities in confined systems.

* * *

The experiments performed by our group have received financial support from the Italian Ministry for University and Research (MURST 40% funds) and from Progetto Finalizzato Telecomunicazioni of the Consiglio Nazionale delle Ricerche (Italy).

REFERENCES

[1] K. E. REMITZ, N. NEUROTH and B. SPEIT: *Materials Sci. Eng. B*, **9**, 413 (1991).

[2] R. K. JAIN and R. C. LIND: *J. Opt. Soc. Am.*, **73**, 647 (1983).

[3] C. FLYTZANIS, F. HACHE, M. C. KLEIN, D. RICARD and P. ROUSSIGNOL: *Progr. Opt.*, **29**, 323 (1991).

[4] N. F. BORRELLI, D. W. HALL, H. J. HOLLAND and D. W. SMITH: *J. Appl. Phys.*, **61**, 5399 (1987).

[5] J. YUMOTO, H. SHINOJIMA, N. UESUGI, K. TSUNETOMO, H. NASU and Y. OSAKA: *Appl. Phys. Lett.*, **57**, 2393 (1990).

[6] Y. WANG and W. MAHLER: *Opt. Commun.*, **61**, 233 (1987); Y. WANG and N. HERRON: *J. Phys. Chem.*, **91**, 257 (1987).

[7] A. N. GOLDSTEIN, C. M. ECHER and A. P. ALIVISATOS: *Science*, **256**, 1425 (1992).

[8] M. ALLAIS and M. GANDAIS: *J. Appl. Crystallogr.*, **23**, 418 (1990).

[9] A. I. EKIMOV, A. L. EFROS and A. A. ORUSHCHENKO: *Solid State Commun.*, **56**, 921 (1985).

[10] V. DEGIORGIO, G. P. BANFI, G. C. RIGHINI and A. R. RENNIE: *Appl. Phys. Lett.*, **57**, 2879 (1990).

[11] B. CHAMPAGNON, B. ANDRIANASOLO and F. DUVAL: *J. Chem. Phys.*, **94**, 5237 (1991).

[12] B. CHAMPAGNON, B. ANDRIANASOLO, A. RAMOS, M. ALLIAS, M. GANDAIS and J. P. BENOIT: *J. Appl. Phys.*, **73**, 2775 (1993).

[13] G. FAGHERAZZI, P. RIELLO and G. C. RIGHINI: *J. Non-Crystalline Solids*, **142**, 63 (1992).

[14] A. R. RENNIE, V. DEGIORGIO and G. P. BANFI: *Physica B*, **180&181**, 509 (1992).

[15] G. P. BANFI, V. DEGIORGIO, A. R. RENNIE and J. G. BARKER: *Phys. Rev. Lett.*, **69**, 3401 (1992).

[16] F. DURVILLE, B. CHAMPAGNON, E. DUVAL, G. BOULON, F. GAUME, A. F. WRIGHT and A. N. FITCH: *Phys. Chem. Glasses*, **25**, 126 (1984).

[17] A. F. WRIGHT, A. N. STITCH, J. B. HAYTER and B. E. F. FENDER: *Phys. Chem. Glasses*, **26**, 113 (1985).

[18] See, for instance, the review by S.-H. CHEN: *Annu. Rev. Phys. Chem.*, **37**, 351 (1986).

[19] G. P. BANFI, V. DEGIORGIO and B. SPEIT: *J. Appl. Phys.*, **74**, 6925 (1993).

[20] B. G. POTTER and J. H. SIMMONS: *Phys. Rev. B*, **37**, 10838 (1988).

[21] I. M. LIFSHITZ and V. V. SLYOZOV: *J. Phys. Chem. Solids*, **19**, 35 (1961); C. WAGNER: *Z. Elektrochem.*, **65**, 581 (1961).

[22] P. ROUSSIGNOL, D. RICARD and C. FLYTZANIS: *Appl. Phys. A*, **44**, 285 (1987).

[23] AL. L. EFROS and A. L. EFROS: *Sov. Phys. Semicond.*, **16**, 772 (1982).
[24] See the review papers by K. BINDER and by R. WAGNER and R. KAMPMANN: in *Materials Science and Technology*, Vol. 5, *Phase Transformations in Materials*, edited by P. HAASEN (VCH, Weinheim, 1991), Chapt. 4 and 7.
[25] A. F. CRAIEVICH, J. M. SANCHEZ and C. E. WILLIAMS: *Phys. Rev. B*, **34**, 2762 (1986).
[26] G. P. BANFI, V. DEGIORGIO, M. GHIGLIAZZA, H. M. TAN and S. TOMASELLI: *Phys. Rev. B*, **50**, 5699 (1994). Some preliminary results were presented by G. P. BANFI, M. GHIGLIAZZA, H. M. TAN and S. TOMASELLI: in *Quantum Electronics and Laser Science*, Vol. 13 of the 1992 OSA Technical Digest Series (Optical Society of America, Washington, D.C., 1992), p. 448, paper PTh 48.
[27] D. COTTER, M. G. BURT and R. J. MANNING: *Phys. Rev. Lett.*, **68**, 1200 (1992).
[28] S. M. OAK, K. S. BINDRA, R. CHARI and K. C. RUSTAGI: *J. Opt. Soc. Am. B*, **10**, 613 (1993).
[29] R. TOMMASI, M. LEPORE and I. M. CATALANO: *Solid State Commun.*, **85**, 539 (1993).
[30] S. SCHMITT-RINK, D. A. B. MILLER and D. S. CHEMLA: *Phys. Rev. B*, **35**, 8113 (1987).
[31] M. SHEIK-BAHAE, D. J. HAGAN and E. W. VAN STRYLAND: *Phys. Rev. Lett.*, **65**, 96 (1990).
[32] E. W. VAN STRYLAND, M. A. WOODHALL, H. VANHERZELE and M. J. SOILEAU: *Opt. Lett.*, **10**, 490 (1985).
[33] A. A. SAID, M. SHEIK-BAHAE, D. J. HAGAN, T. H. WEI, J. YOUND and E. W. VAN STRYLAND: *J. Opt. Soc. Am. B*, **9**, 405 (1992).

Optical Nonlinearities of Composite Materials: Metal and Semiconductor Crystallites.

D. RICARD

Laboratoire d'Optique Quantique, Ecole Polytechnique - 91128 Palaiseau, France

1. – Introduction.

This lecture is devoted to the third-order nonlinear optical response and, more specifically, to the optical Kerr effect in which a laser beam at frequency ω_1 modifies the optical properties (embodied in the index of refraction n or the permittivity $\varepsilon = n^2$) at frequency ω_2. As far as this effect is concerned, a given material is characterized (at lowest order) by the third-order susceptibility $\chi^{(3)}(\omega_1, -\omega_1, \omega_2)$. Materials with a large value of $\chi^{(3)}$ are necessary to implement such promising applications as optical bistability, optical phase conjugation, optical switching or, generally speaking, optical data processing. Important special cases correspond to degenerate four-wave mixing when $\omega_1 = \omega_2 = \omega$, a situation also known as laser self-action, and to the static Kerr effect when $\omega_1 = 0$, a situation which may be useful in hybrid devices.

The simplest way of increasing $\chi^{(3)}$ is to get closer to resonance, for example one-photon resonance when ω_1 or ω_2 are close to the frequency of a one-photon allowed transition and the material presented in this lecture pertains to this case: we will be concerned with the resonant Kerr effect and keep only the triply resonant terms. In so doing, we introduce absorption losses and we also modify the population of the levels. It takes some time to these populations to come back to equilibrium. We must keep aware of these two drawbacks: the absorption losses and the finite recovery time.

Another way of increasing the nonlinear response is to tailor the molecules constituting a homogeneous medium so as to increase the nonlinear polarizabilities β (for second-order nonlinearity) and γ (third-order) by increasing the transition matrix elements. This is amply discussed in this volume in the lectures written by C. BUBECK, G. MEREDITH, S. MIYATA and C. L. TANG. Another important class of homogeneous materials is that of the photorefractive crystals discussed in this volume by P. GÜNTER. The materials we will deal with here are heterogeneous, artificial materials. They are made of several (usually two) materials and can show properties which are superior to those of the constituent materials. Exceptional mechanical properties are, for example, exhibited by such composite materials which are widely used. In the field of nonlinear optics, the quantum wells led the way and are discussed in this volume by F. CAPASSO and D. CHEMLA. In the semiconductor wells, the motion of the carriers is free in the plane but confined in the third dimension. We will concentrate here on metal or semiconductor nanocrystals embedded in a dielectric matrix, usually glass. In this case confinement takes place in all three dimensions.

The motivation behind the study of these nanocrystals whose typical size is of the order of a few nanometres is twofold. Firstly, as already indicated in this introduction, we are looking for materials which would allow the development of practical devices. But, secondly, we also aim at a better knowledge of the properties of these nanocrystals which are in an intermediate state between molecules and crystals. Such nanocrystals typically contain a few hundred atoms for the smaller, a few thousand for the larger ones. We recall that semiconductor nanocrystals constitute a very young field[1-3] and that even their linear optical properties are still not fully understood.

The fabrication of these composite materials and their characterization with regard to shape, size, crystallinity, ... of the crystallites are discussed in the lecture by V. DEGIORGIO in this volume and will not be treated here except very occasionally.

2. – General theoretical considerations.

2‘1. *The dielectric constant of a composite.* – Our composites are made of small, approximately spherical, particles of a material of dielectric constant ε, complex and frequency-dependent, in suspension in a dielectric medium of dielectric constant ε_0 which may usually be assumed to be real and dispersionless. The volume fraction occupied by the particles is denoted p and is small compared to 1. The linear optical properties of the composite are described by its effective dielectric constant $\tilde{\varepsilon}$ which is related to ε, ε_0 and p through[4]

$$(1) \qquad \frac{\tilde{\varepsilon} - \varepsilon_0}{\tilde{\varepsilon} + 2\varepsilon_0} = p \, \frac{\varepsilon - \varepsilon_0}{\varepsilon + 2\varepsilon_0} \, .$$

This relationship is easily derived using local-field arguments as follows. The size of the particles being small compared to the wavelength λ, the applied field $\boldsymbol{E}$ in the surrounding medium may be assumed uniform. As indicated in fig. 1, we denote $\boldsymbol{E}_1$ the local field in the vicinity of the sphere and $\boldsymbol{E}_{\text{in}}$ the field inside

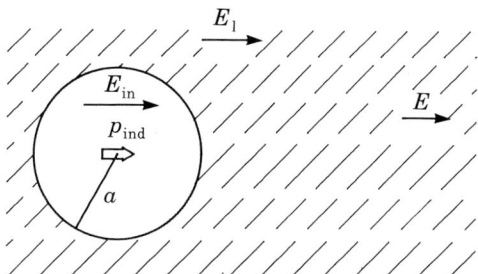

Fig. 1. – A small metal or semiconductor sphere embedded in a dielectric medium. The three fields E, E_1 and E_{in} are defined in the text.

the sphere. As is well known[5],

$$(2) \qquad E_{in} = \frac{3\varepsilon_0}{\varepsilon + 2\varepsilon_0} E_1 .$$

This internal field induces in each sphere of radius a an extra dipole p_{ind}:

$$(3) \qquad p_{ind} = \frac{4\pi}{3} a^3 \frac{\varepsilon - \varepsilon_0}{4\pi} E_{in} ,$$

or

$$(4) \qquad p_{ind} = a^3 \varepsilon_0 \frac{\varepsilon - \varepsilon_0}{\varepsilon + 2\varepsilon_0} E_1 .$$

This leads to an extra polarization P in the composite:

$$(5) \qquad P = N p_{ind} ,$$

where N is the number density of particles ($p = N(4\pi/3) a^3$). This extra polarization is related to $\tilde{\varepsilon}$ through

$$(6) \qquad P = \frac{\tilde{\varepsilon} - \varepsilon_0}{4\pi} E .$$

The local field E_1 is finally given by the usual relationship

$$(7) \qquad E_1 = E + \frac{4\pi P}{3\varepsilon_0} .$$

Using (4), (5), (6) and (7), one then easily obtains the basic relationship (1). Equation (1) is valid up to values of p of the order of $1/3$; for higher values of p, the two materials must be treated on an equal footing and a more symmetrical result obtains[6]. In our case, p being of the order of 10^{-5}–10^{-3}, eq. (1) is perfectly valid and we may even content ourselves with an approximation correct to first order in p:

$$(8) \qquad \widetilde{\varepsilon} = \varepsilon_0 + 3p\varepsilon_0 \frac{\varepsilon - \varepsilon_0}{\varepsilon + 2\varepsilon_0} .$$

We note that $\widetilde{\varepsilon}$ is very close to ε_0.

The field inside the particles is E_{in} and, when we compare it with the applied field, we see that we have a «local-field factor»:

$$(9) \qquad f(\omega) = \frac{E_{\mathrm{in}}}{E} \approx \frac{E_{\mathrm{in}}}{E_1} ,$$

since E is very close to E_1. This «local-field factor» is then approximately given by

$$(10) \qquad f(\omega) = \frac{3\varepsilon_0}{\varepsilon(\omega) + 2\varepsilon_0} .$$

It is a macroscopic concept and the field really acting upon the electrons may differ from the uniform E_{in} by a microscopic local-field factor. Since we will compare the response of nanocrystals with that of the bulk material, we will lump this microscopic local-field factor into the susceptibility χ describing the response of the electrons to E_{in}. We will then drop the quotes and deal only with the local-field factor $f(\omega)$.

When the denominator, $\varepsilon(\omega) + 2\varepsilon_0$, goes through a minimum, $f(\omega)$ may become large. Such a resonance is called the surface plasma resonance and occurs at frequency ω_{s} for which

$$(11) \qquad \varepsilon'(\omega_{\mathrm{s}}) + 2\varepsilon_0 \approx 0 ,$$

where ε' is the real part of ε. Equation (11) is typical of the spherical shape and the surface plasma resonance depends on the particle shape. We may ignore this however, since our particles are approximately spherical. Physically, the surface plasma resonance corresponds to a collective oscillation of the electrons inside the particle.

Such a local-field correction is already important for linear properties. For example, the absorption coefficient of the composite,

$$(12) \qquad \alpha_{\mathrm{abs}}(\omega) = \frac{\omega}{nc} \, \mathrm{Im}\, \widetilde{\varepsilon} ,$$

is easily shown to be given by

$$(13) \qquad \alpha_{\mathrm{abs}}(\omega) = \frac{\omega}{nc} \, p |f(\omega)|^2 \, \varepsilon''(\omega) ,$$

where ε'' is the imaginary part of ε. The index of refraction of the composite n is approximately equal to $\varepsilon_0^{1/2}$. The colour of metal colloids is usually due to the surface plasma resonance [7] through the $|f(\omega)|^2$ term in (13). Figure 2 shows the absorption spectra of a gold colloid and of a silver colloid. The peak, at

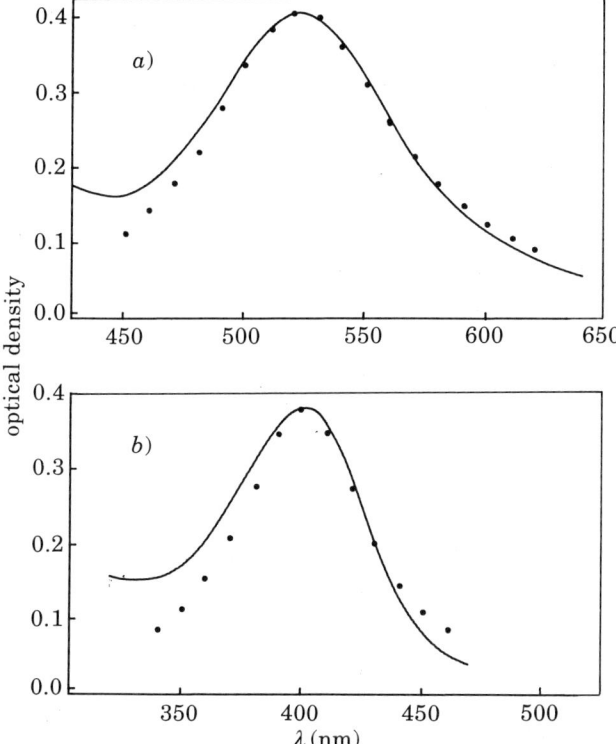

Fig. 2. – Absorption spectra (solid lines) of a gold (a)) and a silver (b)) colloid. The peak corresponds to the surface plasma resonance.

$\sim (520\text{--}530)$ nm for gold and at ~ 400 nm for silver, corresponds to the surface plasma resonance. The beautiful colours of these composites have been used for centuries in the stained-glass art. The absorption coefficient given by eq. (13) is equal to the extinction coefficient calculated using Mie's theory[8] when keeping only the dipolar terms, an approximation justified by the smallness of the particles. We note that scattering losses are negligible in composites containing so small particles.

2'2. *Generalization to the optical Kerr effect.* – Equation (8) governing the linear optical properties of a composite material may easily be generalized to the third-order Kerr susceptibility $\widetilde{\chi}^{(3)}(\omega_1, -\omega_1, \omega_2)$ of the composite when the nonlinearity is due to the crystallites[9]. Responding to the internal field $\boldsymbol{E}_{\text{in}}(\omega_1)$, with the susceptibility $\chi^{(3)}(\omega_1, -\omega_1, \omega_2)$, the electrons of the particles lead to a modification of the dielectric constant ε:

(14) $$\delta\varepsilon(\omega_2) = 4\pi\chi^{(3)}(\omega_1, -\omega_1, \omega_2) E_{\text{in}}(\omega_1) E_{\text{in}}^*(\omega_1),$$

or

(15)
$$\delta\varepsilon(\omega_2) = 4\pi\chi^{(3)} |f(\omega_1)|^2 E(\omega_1) E^*(\omega_1),$$

where simplified notations are used. The small change $\delta\varepsilon$ leads through eq. (8) to a small change $\delta\widetilde{\varepsilon}$ of $\widetilde{\varepsilon}$. Differentiating (8), we get

(16)
$$\delta\widetilde{\varepsilon}(\omega_2) = p[f(\omega_2)]^2 \delta\varepsilon(\omega_2)$$

and, since $\delta\widetilde{\varepsilon}$ and $\widetilde{\chi}^{(3)}$ are related by

(17)
$$\delta\widetilde{\varepsilon}(\omega_2) = 4\pi\widetilde{\chi}^{(3)}(\omega_1, -\omega_1, \omega_2) E(\omega_1) E^*(\omega_1),$$

we finally obtain

(18)
$$\widetilde{\chi}^{(3)}(\omega_1, -\omega_1, \omega_2) = p |f(\omega_1)|^2 [f(\omega_2)]^2 \chi^{(3)}(\omega_1, -\omega_1, \omega_2).$$

This central result has been obtained by generalizing eq. (8). But it could also have been obtained using conventional local-field arguments[10]. For a medium made of molecules possessing the hyperpolarizability $\gamma(\omega_1, \omega_2, \omega_3)$, the susceptibility $\chi^{(3)}(\omega = \omega_1 + \omega_2 + \omega_3)$ is given by

(19) $$\chi^{(3)}(\omega = \omega_1 + \omega_2 + \omega_3) = N f(\omega) f(\omega_1) f(\omega_2) f(\omega_3) \gamma(\omega_1, \omega_2, \omega_3).$$

Usually f is a microscopic local-field factor, but it may as well be a macroscopic one as in our case. N is the number density of microscopic units generally. Equations (18) and (19) are perfectly equivalent, since for our composites $p = NV$ and $\chi^{(3)} = \gamma/V$, where V is the volume of a particle.

The four f factors have also been recovered using a more general, nonlocal approach[11]. Other cases such as a nonlinear response coming from the host matrix or from a nonlinear shell surrounding the particles have also been considered[12]. They will not be discussed here.

2'3. *The figure of merit.* – A nonlinear (Kerr) material is not only characterized by the value of its third-order nonlinear susceptibility $\chi^{(3)}$. Consider, for example, the case of optical phase conjugation. We are in a situation of degenerate four-wave mixing. All the fields oscillate at frequency ω and the Kerr susceptibility is $\chi^{(3)}(\omega, -\omega, \omega)$. Assuming that pump depletion is negligible, the probe and conjugate fields obey, to third order, the equations[13]

(20a) $$\frac{dA_s}{dz} =$$

$$= -\frac{\alpha_{abs}}{2} A_s + i \frac{2\pi\omega}{nc} \chi^{(3)} (2|A_1|^2 + 2|A_2|^2) A_s + i \frac{2\pi\omega}{nc} \chi^{(3)} 2A_1 A_2 A_c^*,$$

(20b) $$\frac{dA_c}{dz} = \frac{\alpha_{abs}}{2} A_c - i \frac{2\pi\omega}{nc} \chi^{(3)} (2|A_1|^2 + 2|A_2|^2) A_c - i \frac{2\pi\omega}{nc} \chi^{(3)} 2A_1 A_2 A_s^*,$$

where the probe and conjugate fields are, respectively,

$$E_s = \frac{1}{2} A_s(z) \exp[i(kz - \omega t)] + \text{c.c.},$$

$$E_c = \frac{1}{2} A_c(z) \exp[-i(kz + \omega t)] + \text{c.c.}$$

with $k = n\omega/c$, A_1 and A_2 are the amplitudes of the forward and backward pump waves. The first term in each of these two equations corresponds to absorption losses, the second one corresponds to two-wave mixing processes between a pump wave and the probe (or the conjugate) wave and is negligible, and the third term corresponds to four-wave mixing. The sample is located between the planes $z = 0$ and $z = L$ and the conjugate wave is such that $A_c(L) = 0$. When it is small, the reflectivity is easily obtained[14]:

$$(21) \qquad r = \frac{A_c(0)}{A_s^*(0)} = iKL2 \sinh\left(\frac{\alpha_{abs}L}{2}\right) \frac{\exp[\alpha_{abs}L]}{\alpha_{abs}L}$$

with $K = (4\pi\omega/nc)\chi^{(3)}A_1A_2$. The maximum reflectivity obtains for a thickness L such that $\alpha_{abs}L \sim 1$ and is thus proportional to $\chi^{(3)}/\alpha_{abs}$.

The previous considerations pertain to the steady-state regime. In such a case, the figure of merit would be $\chi^{(3)}/\alpha_{abs}$. For practical devices, a high repetition rate is also desirable, thus requiring a fast recovery. The material is then best characterized by the figure of merit

$$(22) \qquad F = \frac{\chi^{(3)}}{\alpha_{abs}\tau},$$

where τ is the recovery time. The argument could be made quantitative: for example, with short laser pulses of duration t_p, we are in the transient regime and deal with an effective $\chi^{(3)}$:

$$(23) \qquad \chi_{eff}^{(3)} = \chi^{(3)} \frac{t_p}{\tau},$$

the relevant quantity being $\chi_{eff}^{(3)}/\alpha_{abs}$ or F to within the constant factor t_p.

Introducing the figure of merit F, we thus account for what we already pointed out in the introduction, namely the absorption losses and the finite recovery time. Of course, we may also wish the material to possess other interesting properties such as ease of fabrication, ruggedness, room temperature operating conditions and so forth. These will not really be taken care of in this lecture; we will mainly concentrate on the three basic properties, Kerr susceptibility, absorption losses and recovery time (recombination time in the case of semiconductors).

2'4. *Dielectric confinement and electronic confinement.* – If we consider again the Kerr susceptibility, $\tilde{\chi}^{(3)}$ here, and come back to our central result, eq. (18), we see that it is made of three factors. The first one is the volume fraction p occupied by the particles. Since the absorption coefficient is also proportional to p, the figure of merit F will be insensitive to its value.

The second factor is the local-field correction $|f(\omega_1)|^2 [f(\omega_2)]^2$ and is of relevance since the correction factor is not the same in the expression for $\alpha_{\rm abs}$. This factor comes from the fact that the internal field $\boldsymbol{E}_{\rm in}$ is different from the applied field $\boldsymbol{E}$. It can be larger, near surface plasma resonance, or smaller, off resonance. This effect is sometimes called dielectric confinement. Note that $\boldsymbol{E}_{\rm in}$ depends on the value of ε through eq. (10) and that, because of the optical Kerr effect, ε depends on the magnitude of $\boldsymbol{E}_{\rm in}$. This interdependence may lead to intrinsic optical bistability [15, 16].

The third factor in the expression of $\tilde{\chi}^{(3)}$ is the susceptibility $\chi^{(3)}(\omega_1, -\omega_1, \omega_2)$ which describes the way the electrons in the small particles respond to the internal field $\boldsymbol{E}_{\rm in}$. It may be the same as for the bulk material, but it may also be modified by the three-dimensional confinement. The electrons are confined to the volume $(4\pi/3)\,a^3$ of the spherical crystallite and this may change the eigenfunctions, the eigenenergies and the transition matrix elements [17, 18]. Let us consider the consequences of this electronic-confinement effect in the case of gold or silver colloids. For such noble metals, the conduction band is parabolic to a high degree of accuracy and the effective mass $m_{\rm e}$ is very close to the free-electron mass m_0. Instead of having Bloch wave functions $\exp[i\boldsymbol{k}\cdot\boldsymbol{r}]u_{\rm c}(\boldsymbol{r})$ as in the bulk, the envelope wave function is here a solution of the equation

$$(24) \qquad \left[-\frac{\hbar^2}{2m_0}\,\varDelta + V(\boldsymbol{r})\right]\varphi(\boldsymbol{r}) = E\varphi(\boldsymbol{r})$$

with $V(\boldsymbol{r}) = 0$ if $r < a$ and $V(\boldsymbol{r}) = \infty$ if $r > a$. φ is no longer a plane wave but is proportional to

$$\varphi(\boldsymbol{r}) \propto j_l\!\left(\alpha_{ln}\,\frac{r}{a}\right) Y_l^m(\theta, \varphi)\,,$$

where spherical coordinates have been used, where the Y_l^m's are spherical harmonics and where $j_l(x)$ is the spherical Bessel function of order l, α_{ln} being its n-th zero. $\varphi(\boldsymbol{r})$ vanishes on the surface of the sphere. The eigenenergy is $E_{ln} = \hbar^2\alpha_{ln}^2/2m_0 a^2$. We will come back to the approximations involved in eq. (24), the effective-mass approximation and the infinite potential barrier, but, for the time being, we concentrate on eq. (24) and its solutions. In the bulk case, we have free electrons and the nonlinearity vanishes in the dipolar approximation. In the small-particle case, we have discrete levels and dipolar transitions are allowed between an l level and an $l \pm 1$ one. This leads to a nonvanishing $\chi^{(3)}$ totally due to electronic confinement and, therefore, strongly size dependent [19].

To summarize, the nonlinear Kerr effect in our composites is influenced by mainly two processes: dielectric confinement (or local-field effect) and electronic or quantum confinement. In fact, we will see that dielectric confinement is important only in the case of metal particles but plays a minor role in the case of semiconductor particles. On the other hand, electronic confinement is an important ingredient only in the case of semiconductor particles.

3. – Silver or gold colloids, gold-doped glasses.

3`1. *The linear optical properties.* – We will discuss in this section the properties of composites made of metal nanocrystals in suspension in a liquid, in which case the composite is called a metal colloid, or in glass, what we will call a metal-doped glass. Such a denomination may not be very appropriate since the metal (gold here) is not present as ions but as nanoparticles in the glass matrix. The volume fraction p is very small here, $p \sim 10^{-5}$. The colloids are prepared using a chemical reaction: a silver or gold salt is reduced to the metallic form by a reducing agent, usually sodium citrate[20]. The gold colloids are very stable, but the silver colloids are not and must be studied in the few hours following their preparation.

The linear properties of these silver or gold colloids or of the gold-doped glasses are mainly determined by the local-field effect (dielectric confinement) through the $|f(\omega)|^2$ factor in α_{abs}. We already saw in subsect. 2`1 that the peak at $\lambda \sim$ (520–530) nm for gold colloids and at $\lambda \sim 400$ nm for silver colloids is due to the surface plasma resonance. Notice in fig. 2 that this peak sits on top of a rising edge: α_{abs} increases when one goes to smaller wavelengths for both samples. This rising edge is due to interband transitions whereby d-electrons are promoted to the s-p conduction band. But at the peak position, most of the absorption is due to conduction electrons.

The linear properties may also be influenced by the electronic or quantum confinement effect. Figure 3 shows a series of absorption spectra corresponding to a series of gold-doped glasses having different mean particle sizes. We see that the surface plasma resonance broadens when the particle radius a is reduced. This may be understood on classical grounds[21]. The conduction electrons are almost free electrons and would not absorb in the absence of collisions. Using Drude's classical approach[22], the dielectric constant of the metal is given by

$$(25) \qquad \varepsilon(\omega) = 1 - \frac{\omega_p^2}{\omega^2 + i\,\omega/\tau_0},$$

where ω_p is the plasma frequency and τ_0 is the mean time between two consecutive collisions. When the particle becomes smaller, the electrons also experience collisions with the surface with a mean free path a so that the mean collision

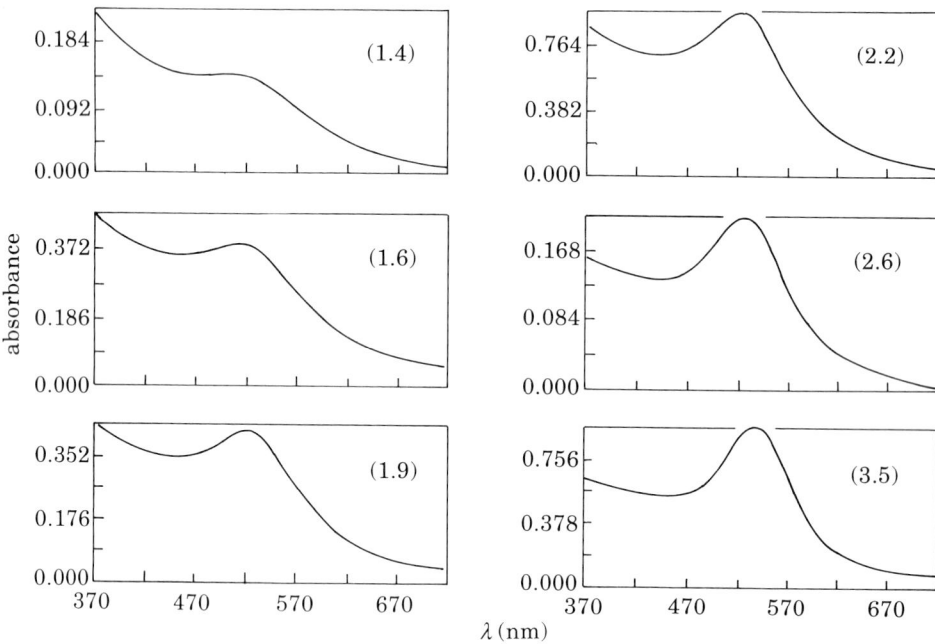

Fig. 3. – Absorption spectra of six gold-doped glass samples. The mean particle radius is indicated in nanometres in the parentheses.

time becomes τ' given by

$$\text{(26)} \qquad \frac{1}{\tau'} = \frac{1}{\tau_0} + \frac{v_F}{a},$$

where v_F is the Fermi velocity. τ_0 should be replaced by τ' in eq. (25), which explains the broadening of the surface plasma resonance. Notice that now absorption is nonzero even for an infinite τ_0.

The broadening may also be understood on quantum-mechanical grounds [19, 23-25]. For a bulk piece of metal, the electron wave functions are the Bloch ones $\exp[i\mathbf{k} \cdot \mathbf{r}] u_c(\mathbf{r})$ with the lattice periodic part u_c and the plane-wave envelope. The dispersion curve $E(k)$ is a parabola. Absorption of a photon leading to a (quasi) vertical transition is thus forbidden and becomes allowed only due to collisions. As was discussed in subsect. 2˙4, in the case of a small sphere the wave functions are now

$$\text{(27)} \qquad \psi_{lmn}(\mathbf{r}) = B_{ln} j_l \left(\alpha_{ln} \frac{r}{a} \right) Y_l^m(\theta, \varphi) u_c(\mathbf{r}),$$

where B_{ln} is a normalization coefficient. The transition matrix element of the momentum operator $\hat{p}$ between ψ_{lmn} and $\psi_{l'm'n'}$ is approximately

$$(28) \qquad \langle l', m', n' \,|\, p \,|\, l, m, n \rangle \approx \int \varphi_{l'm'n'}^{*}(r)\, p \varphi_{lmn}(r)\, \mathrm{d}^3 r \,,$$

and transitions between an l level and an $l \pm 1$ one become possible. Quantitative calculations [19] reproduce, to within a numerical factor of the order of 1, result (26). Here again absorption becomes possible even in the absence of collisions.

3˙2. *Mechanisms for the optical Kerr effect: theory.* – We now want to understand how the optical properties of a small metal particle may be modified by the field E_{in}. We concentrate on the case of gold. The first possibility has already been discussed in subsect. 2˙4. In the bulk, free electrons do not show any dipolar nonlinearity. Because of quantum confinement, dipolar transitions become allowed between the discrete levels giving rise to a nonvanishing dipolar $\chi^{(3)}$ which would be strongly size dependent. Numerical estimates of this contribution show it to be quite small, on the order of 10^{-10} e.s.u. when $a = 50$ Å[26]. This contribution involves only the conduction band.

A second possible mechanism is provided by interband transitions. We are interested in the vicinity of the surface plasma resonance peak $\lambda \sim (520\text{--}530)$ nm. We know that, at this frequency, part of the absorption is due to d-electrons which can be promoted to the s-p conduction band. Most of the d-electrons involved in this transition originate from the vicinity of the X point in the Brillouin zone. As any electronic transition, these transitions between a d-subband and the conduction band saturate at high intensity. They may be considered as being superpositions of two-level transitions. Absorption saturation then leads to first order to a $\chi^{(3)}$ susceptibility which is mainly imaginary, $\mathrm{Im}\,\chi^{(3)}$ being negative. Numerical estimates [26] lead to a value of the order of $(1\text{--}2) \cdot 10^{-8}$ e.s.u. Such a contribution is size independent.

A third possible mechanism is due to the creation of hot electrons. Through absorption, energy is transferred from the laser beams to the conduction electrons. They quickly come to an equilibrium temperature between each other, but this electronic temperature is larger than that of the lattice. It takes on the order of $(1\text{--}2)$ ps for the electrons and the lattice to reach mutual equilibrium [27, 28]. Due to the temperature of the hot electrons, the Fermi-Dirac occupation probabilities are modified near the Fermi energy. This leads to a change in the absorption coefficient for d-electrons, and once again to an imaginary $\chi^{(3)}$, but with a positive imaginary part. The d-electrons involved in this process are located in the vicinity of the L point of the Brillouin zone. Numerical estimates [26] of the hot-electron $\chi^{(3)}$ lead to values of the order of 10^{-7} e.s.u. This third contribution is also size independent.

To summarize, theory shows that the main two mechanisms leading to the Kerr nonlinearity of gold particles are the saturation of the interband transition and the creation of hot conduction electrons. They both involve the interband absorption, but they concern different parts of the Brillouin zone, correspond to different signs for the imaginary part of $\chi^{(3)}$ and show different anisotropies (by which we mean the different components of the $\chi^{(3)}$ tensor).

3'3. *Experimental techniques and results.* – The nonlinear optical properties of silver and gold colloids and of gold-doped glasses have been studied by HACHE *et al.* [26] mainly using the phase conjugation technique. Figure 4 shows the basic experimental set-up. Two counterpropagating pump beams with wave vectors k_1 and k_2 and a probe beam with wave vector k_s meet in the sample. Among the third-order nonlinear polarization terms, the one proportional to $E_1 E_2 E_s^*$ gives rise to a phase-matched conjugate beam propagating with wave vector $-k_s$. The intensity of the conjugate beam is proportional to $|\tilde{\chi}^{(3)}|^2$ (see eq. (21)) and, properly normalizing and taking absorption losses into account, the modulus of $\tilde{\chi}^{(3)}$ may be determined.

The phase conjugation technique is a very powerful one. One may vary the pump beam intensity and study the saturation of the phase-conjugate reflectivity. One may work with copolarized beams and measure the $\tilde{\chi}^{(3)}_{xxxx}$ component of the susceptibility tensor or cross the polarization of any of the three incoming beams to measure the three independent components $\tilde{\chi}^{(3)}_{xyyx}$, $\tilde{\chi}^{(3)}_{xxyy}$ and $\tilde{\chi}^{(3)}_{xyxy}$. One may then check whether the relationship [10]

$$(29) \qquad \tilde{\chi}^{(3)}_{xxxx} = \tilde{\chi}^{(3)}_{xyyx} + \tilde{\chi}^{(3)}_{xxyy} + \tilde{\chi}^{(3)}_{xyxy} \, ,$$

valid for an isotropic medium, is satisfied. Working with short laser pulses and delaying the backward pump pulse, one may study the temporal behaviour of the nonlinear response. Contacting the sample with CS_2 for which $\chi^{(3)}$ is real and positive [29], one may access the phase of $\tilde{\chi}^{(3)}$ for the composite. And one may change the sample or its temperature.

The time response of the Kerr nonlinearity was studied using 5 ps pulses at $\lambda = 527$ nm, obtained by frequency doubling the output of a mode-locked Nd:

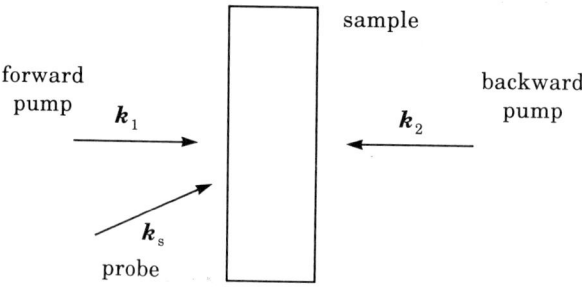

Fig. 4. – The standard geometry for optical phase conjugation. The Kerr nonlinearity gives rise to a conjugate beam which propagates with wave vector $-k_s$.

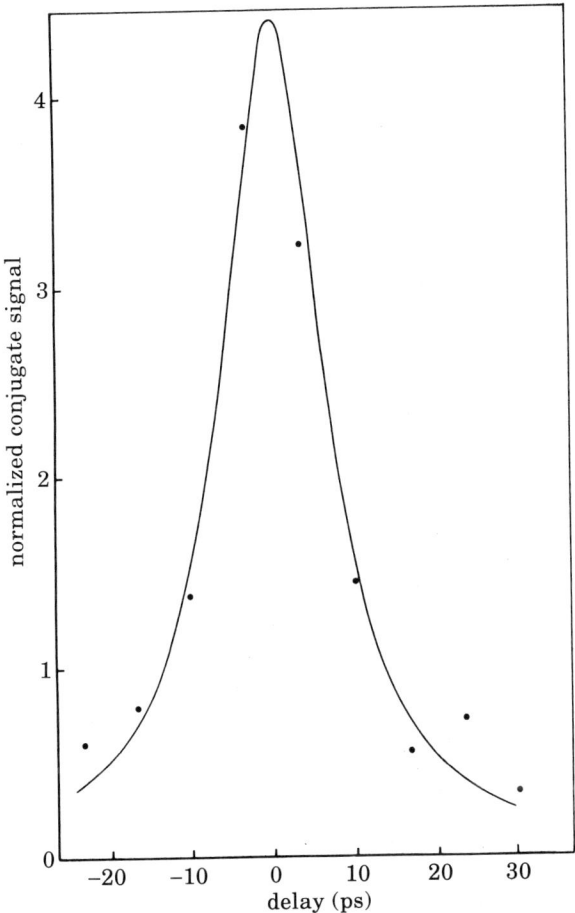

Fig. 5. – The phase-conjugate signal as a function of the backward pump pulse delay for a gold colloid.

glass laser. As shown in fig. 5, the response is very fast, faster than the ~ 5 ps resolution of the experimental set-up. This allows us to assign the nonlinear response to the electrons of the gold nanospheres. The frequency dependence of the phase-conjugate reflectivity was studied in the neighbourhood of the surface plasma resonance. As shown in fig. 6, the role of the local-field correction factor $|f(\omega)|^2 [f(\omega)]^2$ is clearly observed. Using eq. (18), one may then calculate the value of $\chi^{(3)}$ from that of $\widetilde{\chi}^{(3)}$.

Samples containing gold particles of various mean particle sizes were studied and $\chi^{(3)}$ was observed to be size independent. This is in agreement with the predicted weakness of the pure intraband contribution discussed at the beginning of subsect. 3·2. The phase of $\chi^{(3)}_{xxxx}$ was measured to be close to 90°. Using polarization-dependent measurements, the anisotropy of $\chi^{(3)}$ was determined.

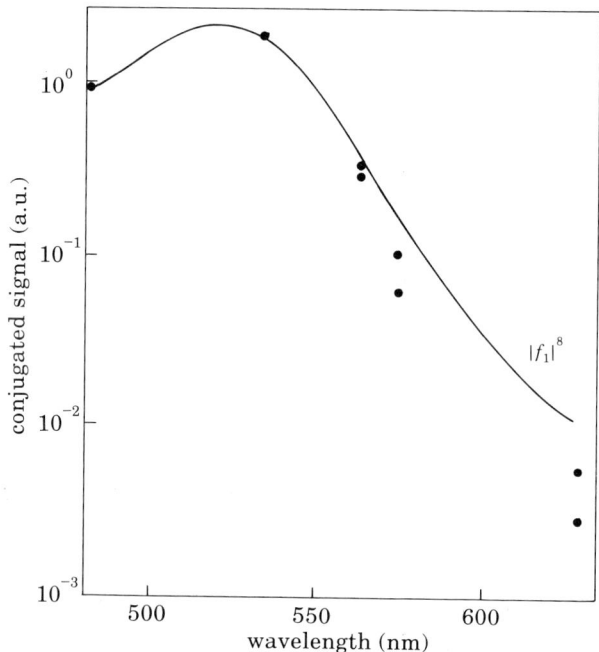

Fig. 6. – The phase-conjugate signal as a function of laser wavelength for a gold colloid. The experimental points are the dots. The solid line gives $|f(\omega)|^8$.

Saturation of phase conjugation was studied for various samples as well as the temperature dependence of $\chi^{(3)}$.

Absorption saturation was also studied using a simple, one-beam set-up: the transmission of the sample was measured as a function of the incident laser beam intensity. Most of these measurements were performed at $\lambda = 532$ nm using 25 ps pulses obtained by frequency doubling a mode-locked Nd:YAG laser.

3˙4. *Discussion.* – These experimental results lead to a complete understanding of the Kerr nonlinearity of gold-doped glasses. $\chi^{(3)}$ is size independent, $\chi^{(3)}_{xxxx}$ is imaginary with a positive imaginary part, whereas $\chi^{(3)}_{xyxy}$ is imaginary with a negative imaginary part. Relationship (29) is obeyed. This is interpreted as follows: the most important mechanism is the hot-electron one, the second contribution, saturation of the interband transitions, is about three times weaker and is responsible for phase conjugation when the probe beam is cross-polarized. In agreement with theory, these are the main two mechanisms. $\chi^{(3)}_{xxxx}$ is of the order of $5 \cdot 10^{-8}$ e.s.u.

For these metal colloids or gold-doped glasses, we observe the role of the local-field correction factor although $|f(\omega)|$ is not very large for gold because of the fairly large value of $\varepsilon''(\omega)$. Near surface plasma resonance, $|f(\omega)| \sim 2$. Nevertheless, away from resonance, $|\widetilde{\chi}^{(3)}|$ decreases very rapidly (see fig. 6). On the other hand, electronic or quantum confinement does not modify $\chi^{(3)}$, since the mechanisms are the same as in the bulk. Dielectric confinement is important, electronic confinement is not.

In terms of applications, the figure of merit $F = |\widetilde{\chi}^{(3)}|/\alpha_{abs}\tau$ is too small to make these materials attractive. Although the recovery time τ is very small, this does not compensate for the weakness of $\chi^{(3)}$ and the large absorption losses. This situation could be improved by using ellipsoidal particles instead of spherical ones. Aligning the ellipsoidal axes[30] would increase the magnitude of the local-field factor $|f(\omega)|$. A more promising way though would be to coat the silver or gold particles with a highly nonlinear shell which would benefit from the local-field enhancement[31]. Such improvements have not yet been experimentally implemented at present.

4. – Semiconductor-doped glasses.

We now consider the case of semiconductor nanoparticles in suspension in glass. Such composites can be made only for a limited number of semiconductors[1,32], some I-VII's such as CuCl or CuBr, some II-VI's such as CdS, CdSe and more seldom CdTe[33]. The technique could not be applied as yet to III-V's such as GaAs. GaAs nanoparticles have been made by depositing a thin layer and then micromachining it using lithographic techniques, but these particles are not as small as those we will discuss. CdS colloids have also been studied[2,3,34]. Here the CdS nanocrystals are made by chemical reaction[35,36] as is also the case for CdS crystallites embedded in a polymer matrix[37]. We will mainly discuss here the case of a glass host matrix. The diameter of the nanocrystals may range from 30 Å to 500 Å.

The Kerr susceptibility $\widetilde{\chi}^{(3)}$ of semiconductor-doped glasses (SDG) is also given by eq. (18). Here the volume fraction p is of the order of 10^{-3}. The local-field factor $f(\omega)$ could show resonances if narrow absorption peaks were present. This is usually not the case, the absorption features being rather broad (we will come back to this broadening problem). The local-field factor may then usually be assumed to be frequency independent. It is furthermore slightly smaller than unity. SDG, therefore, do not show any surface plasma resonance as was the case for silver or gold. We will, therefore, speak indifferently of $\widetilde{\chi}^{(3)}$ or $\chi^{(3)}$ since they are simply proportional although what is measured is, of course, $\widetilde{\chi}^{(3)}$.

4˙1. *The Hamiltonian in effective-mass approximation.* – We now discuss the approximations leading to the electronic wave functions (27). For a particle in a box or a free electron in a piece of metal, the Hamiltonian is simply the sum of the kinetic energy $-\hbar^2/2m_0\Delta$ and of the confining potential $V(r)$. For the bulk case, periodic boundary conditions may be used, leading to the wave functions

(30) $\Psi_k(r) = V^{-1/2} \exp[i\boldsymbol{k} \cdot \boldsymbol{r}]$.

Here the periodic part u is constant. For the spherical confining potential of eq. (24), the wave functions are

(31) $\psi_{lmn}(r) = B_{ln} j_l\left(\alpha_{ln} \dfrac{r}{a}\right) Y_l^m(\theta, \varphi)$.

For a true crystal, one must add to the Hamiltonian a periodic potential accounting for ion-electron and electron-electron interactions[22]. The wave function is now the product of an envelope function $\varphi(r)$ and of a periodic part $\boldsymbol{u}(r)$. In the bulk case, φ is usually taken of the form (30). Near the bottom of the conduction band, i.e. near the centre Γ of the Brillouin zone for direct-gap semiconductors to which we well limit ourselves, the dispersion relation is usually of the parabolic form

$$E(k) = \frac{\hbar^2 k^2}{2m_e} \, ,$$

where the effective mass m_e accounts for the periodic potential (we neglect any possible anisotropy of the effective-mass tensor). Now, when we have for the electron a spatial extension fairly larger than the lattice parameter, such as is usually the case for an impurity state or for a not too small nanocrystal, then we may replace the kinetic energy plus the periodic potential by the operator $-\hbar^2/2m_e \Delta$ acting only on the envelope function $\varphi(r)$, the periodic $\boldsymbol{u}(r)$ being unchanged. This is the effective-mass approximation (EMA) which was used in (24).

The EMA has been shown to be a very good approximation for nanocrystals containing more than a few hundred atoms[38]. For very small nanocrystals which would better be termed aggregates, tight-binding calculations using LCAO have been performed[39,40]. For our purposes, the EMA will be sufficient, but we must keep in mind that such a solid-state-like technique cannot account for the surface states. The surface states do exist, but their study is a very difficult problem since the semiconductor-glass interface is very poorly known.

The second approximation deals with the height of the potential barrier between the inside of the particle and the matrix. For quantum wells such as, for example, GaAs-Al$_x$Ga$_{1-x}$As quantum wells, the potential barrier is finite. It is about half the difference between the barrier energy gap and the well one[41]. In our case, the glass matrix gap is much larger than the semiconductor gap and when the confinement energy ($\propto 1/a^2$) is not too large and when we deal with the lowest levels, we may safely assume the barrier height to be infinite. This is what we do here. A further justification behind this approximation is the large magnitude of the effective mass m_e in glass (assuming such a concept to be meaningful). Finite barriers have nevertheless been considered by several groups [42, 43].

We will, therefore, assume the same confining potential as in eq. (24): $V(r) = 0$ when $r < a$ and $V(r) = \infty$ when $r > a$. We will also use the EMA. What was said about the bottom of the conduction band also applies to the top of the valence band. Things are, however, a bit more complex in semiconductors, since, when an electron is excited from the top of the valence band to the bottom of the conduction band, it leaves behind it a hole, and electron and hole are coupled through Coulomb interaction.

4'2. *Weak confinement and strong confinement.* – Beside the «kinetic energy» terms $-(\hbar^2/2m_e)\Delta_e$ and $-(\hbar^2/2m_h)\Delta_h$ for electron and hole and the confining potentials $V(r_e)$ and $V(r_h)$, the EMA Hamiltonian acting on the envelope function $\varphi(r_e, r_h)$ contains a Coulomb interaction term which assumes the simple form $-e^2/\varepsilon|r_e - r_h|$ only when the particle and the matrix have the same dielectric constant (when $\varepsilon = \varepsilon_0$). Since usually $\varepsilon \neq \varepsilon_0$, this potential is the sum of $-e^2/\varepsilon|r_e - r_h|$ and polarization or solvation terms [44-46]. In simple words, the electron, for example, interacts with the hole but also with its own image and with the hole image. Skipping again this difficulty, we will write for the Schrödinger equation

$$(32) \qquad \left(-\frac{\hbar^2}{2m_e}\Delta_e - \frac{\hbar^2}{2m_h}\Delta_h + V(r_e) + V(r_h) - \frac{e^2}{\varepsilon|r_e - r_h|} \right) \varphi(r_e, r_h) =$$

$$= E\varphi(r_e, r_h).$$

Even with the approximations it involves, eq. (32) cannot be solved analytically. It leads to tractable calculations only in two limiting cases [47].

In the bulk, electron and hole can form a Wannier exciton with the binding energy

$$(33) \qquad E_I = -\frac{\mu e^4}{2\varepsilon^2\hbar^2}$$

and the Bohr radius

$$(34) \qquad a_{exc} = \varepsilon\hbar^2/\mu e^2 \,,$$

where $\mu = m_e m_h/(m_e + m_h)$ is the reduced mass. In our problem, we have two competing processes: confinement which increases the energy by $\sim \hbar^2\pi^2/2\mu a^2$ and Coulomb attraction which lowers the energy by $\sim e^2/\varepsilon a$ or $\sim e^2/\varepsilon a_{exc}$ whichever is larger. When $a \gg a_{exc}$, Coulomb interaction dominates, we are in the weak-confinement regime, the Wannier exciton still exists and is confined as a whole with the total mass $M = m_e + m_h$. When $a \ll a_{exc}$, confinement dominates, we are in the strong-confinement regime, the electron and the hole are

confined independently and the Coulomb term may be treated as a perturbation. Finally, when $a \sim a_{exc}$, we are in the intermediate-confinement
regime.

Experiment shows that, when $a > 3a_{exc}$, we are in the weak-confinement
regime and, when $a < a_{exc}$, we are in the strong-confinement one[48]. In the
first case, taking the zero of energy at the top of the valence band, the energy of
the first excited state is

$$(35) \qquad E = E_g + \frac{\hbar^2 \pi^2}{2M a_{eff}^2} + E_I ,$$

where a_{eff} is slightly smaller than a to account for the finite size of the Wannier
exciton and E_g is the energy gap. π in the second term of (35) is α_{01}, the first
zero of $j_0(x)$. The excitonic line is observable and may lead to large nonlinearities, as will be discussed in the next subsection.

We briefly discuss the intermediate case. Due to their different effective
masses, the electron and hole motions may be adiabatically separated[47]. The
electron is confined according to

$$(36) \qquad \left(-\frac{\hbar^2}{2m_e} \Delta_e + V(r_e) \right) \varphi(r_e) = E_e \varphi(r_e),$$

leading to

$$(37) \qquad \varphi(r_e) = B_{ln} j_l \left(\alpha_{ln} \frac{r_e}{a} \right) Y_l^m (\theta_e, \varphi_e),$$

and the hole moves in the Coulomb potential created by the electron. When the
electron is in an s state ($l = 0$), this potential may be considered as parabolic as a
first approximation and to each electronic state corresponds a series of equally
spaced hole levels. This intermediate-confinement regime may be obtained with
CdS (for which $a_{exc} \approx 32$ Å) or CuBr particles and the hole sublevels have been
observed in the case of CdS particles[49].

4˙3. *Weak confinement and excitonic nonlinearities.* – We now consider
in more detail the weak-confinement case. CuCl particles for which $a_{exc} \approx 7$ Å
are almost always in the weak-confinement regime since the smallest value
of the radius a is usually $\sim (15\text{–}20)$ Å. The exciton binding energy (or
Rydberg) is fairly large and exciton lines are easily observed. The confinement
energy, $\hbar^2 \pi^2 / 2M a_{eff}^2$ or second term of (35), leads to a blue shift of these
lines as the particle size is reduced. This blue shift is small, however,
due to the large value of the total mass M. This blue shift has been observed
for CuCl-doped glasses[1] and also for CuBr-doped ones[32]. We now consider
the nonlinear optical response of such materials when the photon energy

$\hbar\omega$ is close to E given in (35). Such a nonlinearity is termed excitonic nonlinearity since it is mainly due to excitation of Wannier excitons.

We may create one exciton per particle by absorbing a photon or two excitons by absorbing two photons. The electronic system we have to consider is fairly simple with three states: the ground state $|0\rangle$ of energy E_0, the one-exciton state $|1\rangle$ of energy E_1 and the two-exciton state $|2\rangle$ of energy E_2. The hyperpolarizability $\gamma(\omega, -\omega, \omega)$ may be calculated by perturbation theory (diagrams facilitate the calculation) and is given by[50]

$$(38) \quad \gamma = \frac{1}{4\hbar^3}\left\{4A^2\,\frac{T_1/T_2}{[(\omega - \omega_{10})^2 + 1/T_2^2](\omega - \omega_{10} + i/T_2)} - \right.$$

$$-B^2\left[\frac{2T_1/T_2}{[(\omega - \omega_{10})^2 + 1/T_2^2](\omega - \omega_{21} + i/T_2)} - \right.$$

$$\left.\left.-\frac{1}{(\omega - \omega_{10} + i/T_2)(2\omega - \omega_{20} + i/T_2)}\left(\frac{1}{\omega - \omega_{21} + i/T_2} - \frac{1}{\omega - \omega_{10} + i/T_2}\right)\right]\right\},$$

where $A = |\langle 1|d|0\rangle|^2$, $B = |\langle 2|d|1\rangle\langle 1|d|0\rangle|$, d is the electric-dipole moment operator, T_2 is the dephasing time assumed to be the same for the $|0\rangle \to |1\rangle$, the $|1\rangle \to |2\rangle$ and the $|0\rangle \to |2\rangle$ transitions, T_1 is the lifetime of the $|1\rangle$ state, $\omega_{10} = (E_1 - E_0)/\hbar$ and similarly for ω_{21} and ω_{20}. The first or A^2 term in (38) corresponds to saturation of the $|0\rangle \to |1\rangle$ transition. The B^2 terms involve the two-exciton state, the first term in the square bracket corresponding to induced absorption.

In the dipolar approximation that is valid here, we need the matrix elements of d, the electric-dipole moment operator. It is well known that such a matrix element is related to the momentum matrix element through

$$\langle a|d|b\rangle = i\,\frac{q}{m\omega_{ab}}\,\langle a|p|b\rangle,$$

where q and m are the relevant charge and mass. We will, therefore, use the momentum matrix elements in the following. By p, we mean the component of $\boldsymbol{p}$ along the laser field polarization vector.

Because of the Coulomb interaction, the transition frequencies ω_{21} and ω_{10} are different. We denote[51]

$$(39) \qquad\qquad \omega_{\text{int}} = \omega_{21} - \omega_{10}.$$

The matrix elements $\langle 1|p|0\rangle$ and $\langle 2|p|1\rangle$ are approximately equal and proportional to $V^{1/2}$, where V is the volume of the particle. If ω_{int} were zero, *i.e.* if we had noninteracting excitons, then the last term in (38) would vanish. Furthermore, due to the possibility of interchanging the intermediate state in B^2, we would have $B^2 = 2A^2$ and γ would be identically zero. Such a result is due to the boson character of the excitons. To get a nonzero γ, we, therefore, need a nonzero ω_{int}.

When the size of the particles is reduced, two effects tend to counterbalance each other: the numerators (A^2, $B^2 \propto V^2$) decrease, whereas ω_{int} increases so that the A^2 and B^2 terms in (38) do not cancel each other. The size dependence of γ has been discussed quantitatively by HANAMURA et al. [51-53] assuming a size-independent broadening $1/T_2$. Two limiting cases are discussed:

i) $\omega_{int} \gg |\omega - \omega_{10}| > 1/T_2$,

then the dominant mechanism is saturation of the one-exciton line, the first or A^2 term dominates and then $\gamma \propto V^2$.

ii) $|\omega - \omega_{10}| \gg \omega_{int} > 1/T_2$,

then the two mechanisms contribute and the hyperpolarizability is proportional to the product $\omega_{int} A^2$, so that γ is simply proportional to V.

We note that in case i) the susceptibility $\chi^{(3)} = \gamma/V$ increases when the particle size increases. Such a size dependence has been observed [54-56] for CuCl-doped samples although the contribution of trapped carriers makes things more complicated than the simple considerations given here [57]. When the particle size keeps increasing, one gets to the bulk case. In that case, the delocalization of the exciton is confined to a certain coherence zone of volume V_c. What we believe is that, when the particle size increases, $\chi^{(3)}$ may increase (first proportionately to V) until V reaches the coherent volume V_c.

4`4. SDG *in strong confinement: linear properties.* – The remaining of this lecture will be mainly devoted to the strong-confinement regime. CdSe has been the most widely studied case. For CdSe, the exciton Bohr radius is ≈ 56 Å, so that CdSe particles in the strong-confinement regime may easily be made. Ternary alloys $CdS_x Se_{1-x}$ have also been extensively studied, since they are commercially available as colour filters manufactured by Schott, Corning, Hoya or Toshiba. Such commercial filters were shown to have nonlinear properties as early as 1964 [58], but, at that time, their composite nature was not known. They started to be studied as phase-conjugating media by JAIN and LIND [59]. In the commercial filters, the mean radius a is ≈ 40 Å [60]. They may not all correspond to the strong-confinement regime, since, when x varies from 0 to 1, the exciton Bohr radius decreases from the CdSe value ($a_{exc} \approx 56$ Å) to the CdS one ($a_{exc} \approx 32$ Å). Experimental samples with smaller particles may, however, easily be made that correspond to strong confinement.

Varying the chemical composition, the x parameter, one varies the energy gap E_g, thus the cut-off wavelength of the colour filter. But from a physics point of view, it is preferable to work with a pure semiconductor such as CdSe. One thus avoids problems due, for example, to stoichiometric fluctuations. Very nice CdSe samples with a clean silicate matrix have been grown at the Vavilov Institute in Saint-Petersburg[61]. CdSe SDG have also been made by Corning[62].

In the strong-confinement regime, the Coulomb interaction may be omitted as a first approximation in the EMA Hamiltonian of eq. (32). The electron and the hole decouple, the electron envelope function obeying eq. (36). This envelope function assumes the simple form (37) that we already encountered for the metal conduction electrons. Choosing the zero of energy at the top of the valence band, the electron energy is then

$$(40) \qquad E_{ln}^{e} = E_{g} + \frac{\hbar^2 \alpha_{ln}^2}{2m_e a^2} .$$

The levels are degenerate with respect to the m quantum number, a $(2l + 1)$-fold degeneracy. The first level is denoted $1s$, for $l = 0$ and $n = 1$. In this case $\alpha_{01} = \pi$. The second level is $1p$, for $l = 1$, $n = 1$ with $\alpha_{11} = 4.49$. The following ones are $1d$, $2s$ and so on. Notice that since $j_0(x) = (\sin x)/x$, its zeroes are simply given by $\alpha_{0n} = n\pi$. We also recall that the full electron wave function is

$$(41) \qquad \psi_{lmn}(\mathbf{r}_e) = \varphi_{lmn}(\mathbf{r}_e) u_c(\mathbf{r}_e) .$$

In the same way, assuming the valence band to be nondegenerate, the envelope wave function for the hole would obey an equation similar to (36):

$$(42) \qquad \left(-\frac{\hbar^2}{2m_h} \Delta_h + V(\mathbf{r}_h) \right) \varphi(\mathbf{r}_h) = E \varphi(\mathbf{r}_h)$$

and would be of the form

$$(43) \qquad \varphi_{lmn}(\mathbf{r}_h) = B_{ln} j_l \left(\alpha_{ln} \frac{r_h}{a} \right) Y_l^m(\theta_h, \varphi_h) .$$

The confinement energy of the hole would be

$$(44) \qquad E_{ln}^{h} = \frac{\hbar^2 \alpha_{ln}^2}{2m_h a^2} .$$

The selection rules would be determined by the transition matrix elements $\langle \psi_{l'm'n'} | p | \psi_{lmn} \rangle$ between the (n, l, m) hole state and the (n', l', m') electron state. Since the envelope function φ and the periodic part u vary on very different length scales, this matrix element is approximately given by

$$(45) \qquad \langle \psi_{l'm'n'} | p | \psi_{lmn} \rangle = \langle u_c | p | u_v \rangle \int \varphi_{l'm'n'}^*(\mathbf{r}) \varphi_{lmn}(\mathbf{r}) \, d^3r .$$

It is the product of what is traditionally denoted p_{vc} and of the overlap integral of the envelope wave functions. Since the φ_{lmn}'s form an orthonormal basis set, this integral is simply $\delta_{ll'}\delta_{mm'}\delta_{nn'}$ and

(46)
$$\langle \psi_{l'm'n'} | p | \psi_{lmn} \rangle = p_{vc}\delta_{ll'}\delta_{mm'}\delta_{nn'} \;.$$

This leads to very simple selection rules[47]: the only allowed transitions are $1s$-$1s$, $1p$-$1p$ and so on. They occur at frequency ω_{ln} such that

(47)
$$\hbar\omega_{ln} = E_{g} + \frac{\hbar^2\alpha_{ln}^2}{2\mu a^2} - \beta\frac{e^2}{\varepsilon a} \;;$$

μ is the electron-hole reduced mass and the last term is the correction due to Coulomb interaction and obtained by perturbation theory, β is a numerical constant which assumes the value 1.8 for ns-ns transitions[44, 63].

This simple approach allows us to understand, through eqs. (46) and (47), why for such SDG the absorption spectrum becomes structured and the absorption edge is blue-shifted when the size of the crystallites is reduced. This is exemplified by fig. 7, which shows room temperature absorption spectra for three samples containing $CdS_{0.5}Se_{0.5}$ particles of different mean radii. This is also beautifully shown by a bar-shaped sample where the CdSe particles are grown by heat treatment in a temperature gradient. The lower the temperature, the smaller the particles and one gets a pale yellow colour. The higher the temperature, the larger the particles and one gets a deep red colour.

But things are not so simple because the valence band originates from p atomic orbitals which mean a 3-fold orbital degeneracy or a 6-fold degeneracy when spin is taken into account. As in any atom, spin-orbit coupling partly lifts the degeneracy leading to a 4-fold degenerate $P_{3/2}$ level and a twofold degener-

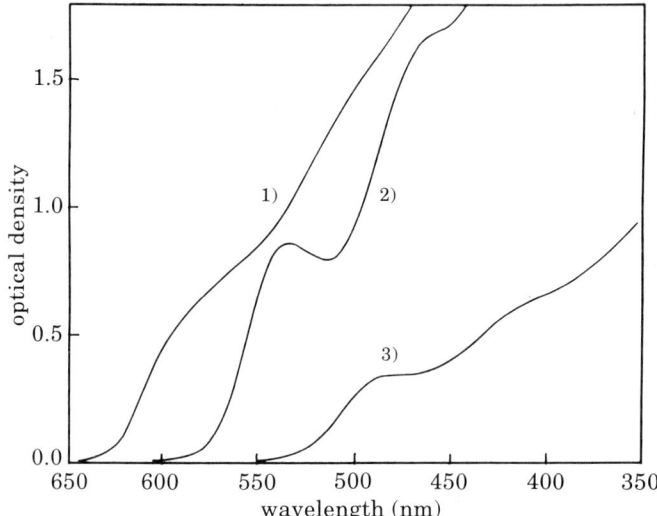

Fig. 7. – Absorption spectra of three $CdS_{0.5}Se_{0.5}$-doped glass samples. The mean particle radius is 1) $60\,\text{Å}$, 2) $23\,\text{Å}$ and 3) $\sim 15\,\text{Å}$.

ate $P_{1/2}$ one. The indices $3/2$ and $1/2$ are the values assumed by the angular momentum J. The energy splitting is denoted Δ. In the bulk, the Hamiltonian was given by LUTTINGER *et al.* [64] and, in the $u_{3/2, 3/2}$, $u_{3/2, 1/2}$, $u_{3/2, -1/2}$, $u_{3/2, -3/2}$, $u_{1/2, 1/2}$ and $u_{1/2, -1/2}$ basis, it assumes the form

(48) $\hat{H} = \dfrac{1}{m_0} \times$

$$
\times
\begin{vmatrix}
P+Q & L & M & 0 & i\sqrt{1/2}\,L & -i\sqrt{2}\,M \\
L^* & P-Q & 0 & M & -i\sqrt{2}\,Q & i\sqrt{3/2}\,L \\
M^* & 0 & P-Q & -L & -i\sqrt{3/2}\,L^* & -i\sqrt{2}\,Q \\
0 & M^* & -L^* & P+Q & -i\sqrt{2}\,M^* & -i\sqrt{1/2}\,L^* \\
-i\sqrt{1/2}\,L^* & i\sqrt{2}\,Q & i\sqrt{3/2}\,L & i\sqrt{2}\,M & P-\Delta & 0 \\
i\sqrt{2}\,M^* & -i\sqrt{3/2}\,L^* & i\sqrt{2}\,Q & i\sqrt{1/2}\,L & 0 & P-\Delta
\end{vmatrix},
$$

where $P = (\gamma_1/2)\,\hat{p}^2$, $Q = \gamma(\hat{p}_\perp^2 - 2\hat{p}_z^2)/2$, $L = -i\sqrt{3}\,\gamma\hat{p}_-\hat{p}_z$, $M = \sqrt{3}\,\gamma\hat{p}_-^2/2$, $\hat{p}_\perp^2 = \hat{p}_x^2 + \hat{p}_y^2$, $\hat{p}_\pm = \hat{p}_x \pm i\hat{p}_y$ and $\hat{p}$ is the momentum operator.

Each operator in this 6×6 matrix operates on the envelope part of the wave function. In writing down eq. (48), we have assumed an isotropic mass tensor. In fact, the CdSe particles have a wurtzite (hexagonal) structure as well as bulk CdSe, but this crystal is not too far from the zincblende (cubic) structure. Assuming a zincblende structure, LUTTINGER [64] introduced three parameters γ_1, γ_2 and γ_3 (plus Δ). Neglecting the small warping of the bands due to the small difference between γ_2 and γ_3, we have set $\gamma = \gamma_2 = \gamma_3$. In the bulk, the Hamiltonian (48) leads to three valence subbands, all parabolic in the vicinity of the Γ point: the heavy-hole band with effective mass $m_{hh} = m_0/(\gamma_1 - 2\gamma)$, the light-hole band with effective mass $m_{lh} = m_0/(\gamma_1 + 2\gamma)$ and the spin-orbit split-off band with effective mass $m_{so} = m_0/\gamma_1$. At the Γ point, the heavy-hole and light-hole bands are degenerate, whereas the split-off band is an energy Δ below.

In two limiting cases, the Luttinger Hamiltonian may be simplified [64, 65]. When $\Delta = 0$, no spin-orbit coupling, spin is irrelevant and $\hat{H}$ reduces to a 3×3 matrix. CdS with $\Delta = 65$ meV is not too far from this case. When Δ is large, one may assume $\Delta = \infty$ and work within the 4-dimensional $J = 3/2$ space with a 4×4 matrix Hamiltonian obtained by keeping only the upper left part of (48). CdSe with $\Delta = 420$ meV is not too far from this case. The whole 6×6 matrix problem has been treated [66], but, in order to simplify the presentation, we will limit ourselves to the $\Delta = \infty$ case [65, 67]. In this case, the Hamiltonian may be given a simpler expression in terms of second-rank irreducible tensor operators [68]. The first time the valence band degeneracy was taken into account was in the case of CuCl [69].

When dealing with a spherical particle, the confining potential $V(r_h)$ must be added to the kinetic-energy operator (the 4×4 Luttinger Hamiltonian). Spherical confinement leads to coupling between the $\boldsymbol{J}$ angular momentum and the $\boldsymbol{L}$ angular momentum of the envelope functions. The hole states are characterized by the total angular momentum $\boldsymbol{F} = \boldsymbol{J} + \boldsymbol{L}$ and the hole wave functions are of the form

(49) $\quad \Psi_{FM}(\boldsymbol{r}_{\mathrm{h}}) =$

$$= \sum_{\substack{l = F + 1/2, \, F - 3/2 \\ L = F + 3/2, \, F - 1/2}} R_l(r_{\mathrm{h}}) \sum_m \sum_\mu \langle l, m, 3/2, \mu | F, M \rangle Y_l^m(\theta_{\mathrm{h}}, \varphi_{\mathrm{h}}) u_{3/2, \mu}(\boldsymbol{r}_{\mathrm{h}}),$$

where $\langle l, m, 3/2, \mu | F, M \rangle$ is a Clebsch-Gordan coefficient. The states with a given value of F are degenerate with respect to the momentum projection M.

Considering the lowest energy levels, when $F = 3/2$, one may have even states which mix $l = 0$ and $l = 2$ and which will be denoted $S_{3/2}$ or odd states which mix $l = 1$ and $l = 3$ and which will be denoted $P_{3/2}$. The hole confinement energy is

$$(50) \qquad\qquad\qquad E_{\mathrm{h}} = \frac{\hbar^2 k^2}{2 m_{\mathrm{hh}}},$$

k being determined by the vanishing of the R_l's at $r = a$. For example, for the $S_{3/2}$ states, k must obey [70]

$$(51) \qquad\qquad j_0(ka) j_2(\sqrt{\beta} ka) + j_0(\sqrt{\beta} ka) j_2(ka) = 0,$$

where $\beta = m_{\mathrm{lh}}/m_{\mathrm{hh}}$ and the radial functions are [70]

$$(52) \quad \begin{cases} R_{0n}(r) = \dfrac{C_n}{a^{3/2}} \left[j_0(k_n r) - \dfrac{j_0(k_n a)}{j_0(\sqrt{\beta} k_n a)} j_0(\sqrt{\beta} k_n r) \right], \\[4mm] R_{2n}(r) = \dfrac{C_n}{a^{3/2}} \left[j_2(k_n r) + \dfrac{j_0(k_n a)}{j_0(\sqrt{\beta} k_n a)} j_2(\sqrt{\beta} k_n r) \right], \end{cases}$$

where n labels the roots of eq. (51) and C_n is a normalization constant. The corresponding levels will be denoted $nS_{3/2}$ [61].

For highly excited electron levels, the nonparabolicity of the conduction band must be taken into account, for example following KANE [71]. This modifies the electron energy E_{e} [61], but the wave function is still given by (37) and (41). To summarize, in semiconductors, one must take into account the degeneracy of the valence band and the nonparabolicity of the conduction band. Figure 8 shows the confinement energy of the hole and of the electron thus calculated with the full 6-fold degeneracy of the valence band.

The selection rules are also modified [61] and, for example, transitions are allowed from all the $nS_{3/2}$ hole levels to the $1S_{\mathrm{e}}(1s)$ electron level although the $1S_{3/2}$-$1S_{\mathrm{e}}$ one is the most intense. Figure 9 shows low-temperature absorption

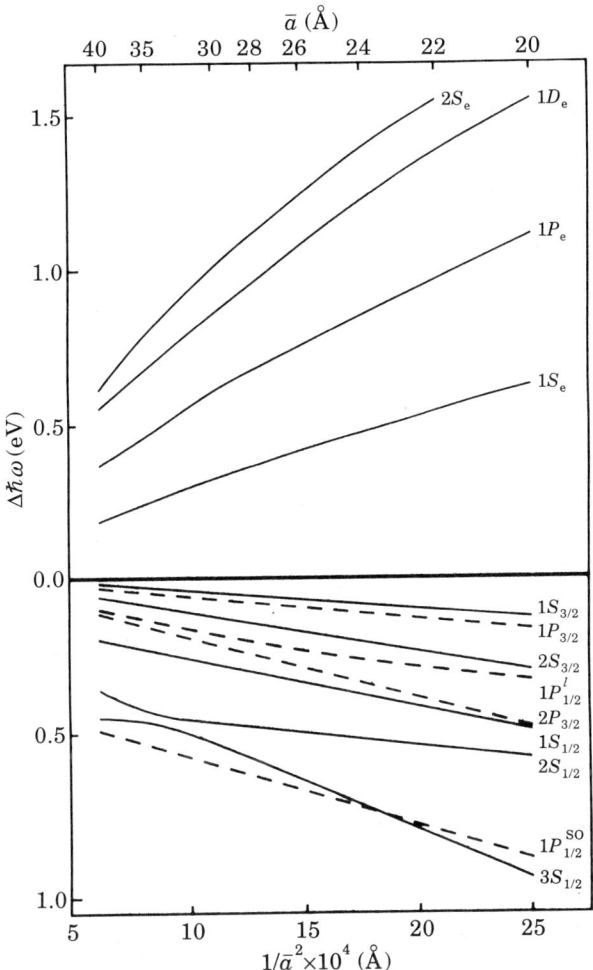

Fig. 8. – Calculated confinement energy of the electron and of the hole for CdSe particles plotted as a function of $1/a^2$. The degeneracy of the valence band and the nonparabolicity of the conduction band have been taken into account.

spectra for three CdSe-doped glasses having mean particle radii of 38 Å, 26 Å and 21 Å. The second derivative of $\alpha_{abs}(\omega)$ clearly shows substructures which have been interpreted within the model just discussed [61]. The first two structures correspond to the $1S_{3/2}$-$1S_e$ and $2S_{3/2}$-$1S_e$ transitions.

The eigenenergy of the electron-hole pair must also take the Coulomb interaction into account. This may be done using perturbation theory [72] or variational calculations [44, 65].

In order to better understand the linear optical properties of these SDG and the levels of relevance in the vicinity of the absorption edge, luminescence

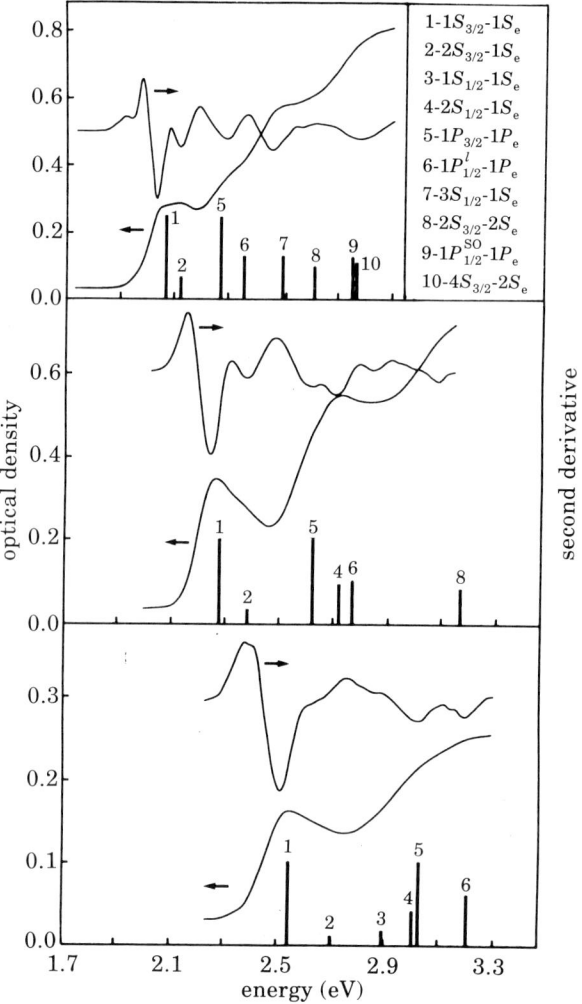

Fig. 9. – Absorption spectra and their second derivative for three CdSe-doped glass samples. The mean particle radius is (top panel) 38 Å, (middle panel) 26 Å and (lower panel) 21 Å. The assignment of the transitions is given in terms of the levels of fig. 8.

measurements are also important. When they are excited with a c.w. laser, the luminescence spectrum is comprised of a narrow peak slightly Stokes-shifted from the $1S_{3/2}$-$1S_e$ absorption peak and a more intense, broader luminescence band having a much larger Stokes shift[61,73]. When they are excited with a pulsed laser, one observes that the first narrow peak is a rather fast component with a lifetime of the order of 1 ns, whereas the broad band is a slow component with a lifetime of the order of microseconds (the larger the Stokes shift, the slower the luminescence). The narrow peak was, therefore, interpreted as direct recombination of an electron in the $1S_e$ level with a hole in the $1S_{3/2}$ one.

The broad luminescence band is thought to be due to trapped-carrier recombination [73, 74].

For commercial Schott filters, this broad band is much more intense than for experimental glasses grown from the same melt. This means that commercial filters contain many more traps than experimental samples. And for experimental samples grown from a same melt, the smaller the particles are, the larger the number of traps is [75]. The traps are, therefore, thought to be mainly located at the semiconductor-glass interface.

When detecting this luminescence with an optical multichannel analyser (OMA) and when using a 4 ns gate in front of the OMA, one gets rid of the broad band and can study the near-edge luminescence. Such luminescence spectra obtained for three CdSe-doped samples at liquid-nitrogen temperature [61] are shown in fig. 10. When the excitation pulse energy is not too large, one ob-

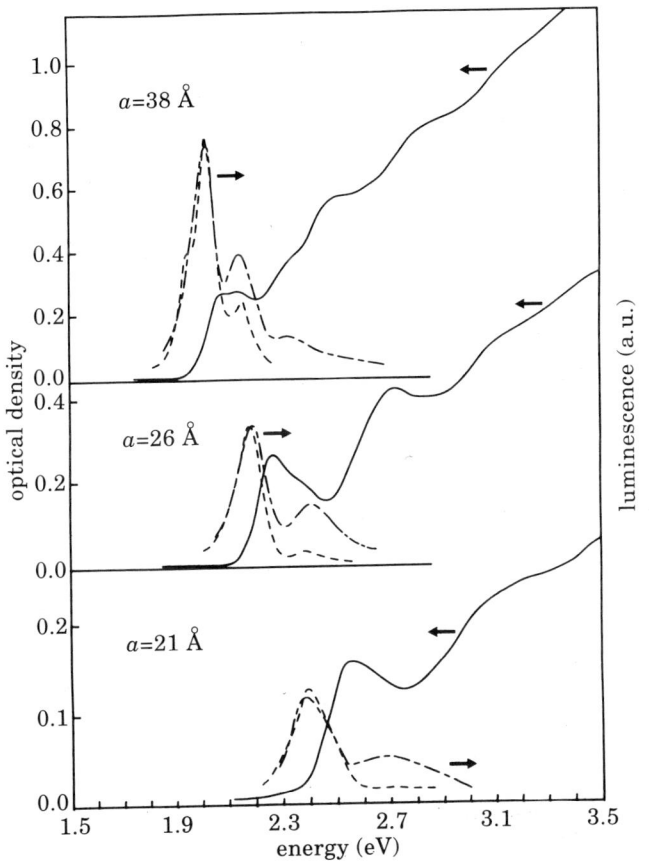

Fig. 10. – Absorption spectra and luminescence spectra for the three samples of fig. 9. Luminescence was excited with medium-intensity laser pulses (dashed lines) and high-intensity laser pulses (long-dashed lines).

serves the narrow feature just discussed above. Increasing the excitation pulse energy, one observes one or two hot-luminescence peaks whose positions exactly coincide with those of higher absorption peaks ($2S_{3/2}$-$1S_e$ and the following one). Hot luminescence corresponds to direct recombination from these higher levels when the population of the traps and of the $1S_{3/2}$-$1S_e$ level saturates.

4˙5. *Broadening mechanisms.* – For SDG in the strong-confinement regime, the electronic transitions are fairly broad. We already discussed the substructure of the first structure due to the $1S_{3/2}$-$1S_e$ and $2S_{3/2}$-$1S_e$ transitions. In fig. 9, this substructure can already be seen in the absorption spectrum itself but is clearer in the second derivative spectrum. For Schott experimental samples, it is barely visible, even in the second derivative spectrum.

The broadening of the transitions may be due to several processes. Firstly, it may be intrinsic as, for example, lifetime broadening or broadening due to electron-phonon coupling. For example, coupling to optical phonons may lead to a vibronic substructure of the electronic transitions. Secondly, it may be extrinsic; this is inhomogeneous broadening which could be due to shape fluctuations, compositional fluctuations in the case of alloys, but much more probable is inhomogenous broadening due to size fluctuations. In the process of nanocrystal growth, particles with different radii a are formed. This growth process which has been studied theoretically by LIFSHITZ and SLEZOV[76] and experimentally [77] gives rise to a size distribution $P(a)$ centred around a mean value $\bar{a}$ and normalized according to

$$(53) \qquad\qquad \int P(a)\,\mathrm{d}a = 1\,.$$

Since the frequency of a given transition is a-dependent, the size distribution leads to an important broadening. For some samples, the size distribution has been shown[78] to be made of several peaks around several «mean» radii, which leads to an even more important broadening.

When the sample temperature is lowered from room temperature to liquid-nitrogen temperature, the transitions get narrower[79] as shown in fig. 11. Further cooling does not lead to significant narrowing. This shows that intrinsic broadening is very important. This result has been confirmed by nonlinear absorption measurements to be discussed below: at room temperature, the dominant broadening mechanism is intrinsic. On the other hand, at low temperature, the dominant mechanism is inhomogeneous broadening. The narrow features exhibited in fig. 9 indicate that the corresponding samples show an unusually small inhomogeneous broadening. Samples grown by Corning also show narrow features[62].

Although this has not yet been done, one could in principle narrow the size distribution. But one would then still be faced with the intrinsic broadening.

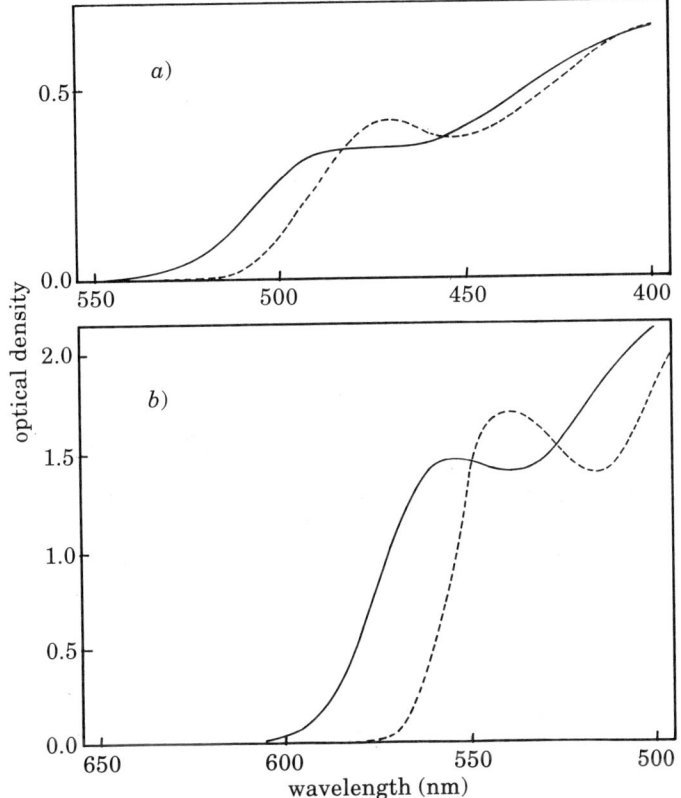

Fig. 11. – Absorption spectra at room temperature (solid lines) and liquid-N_2 temperature (dashed lines) for two $CdS_{0.5}Se_{0.5}$-doped glass samples.

Electron-LO phonon coupling was first put forward as the main cause of intrinsic broadening and as the origin of the Stokes shift of the narrow luminescence peak[79]. To ascertain the role of electron-LO phonon coupling, resonance Raman spectra such as the one shown in fig. 12 have been taken[80-82]. They allow the determination of the coupling strength and lead to the conclusion that it is not very large. Exciting luminescence by photons corresponding to the red tail of the absorption spectrum or looking at luminescence excitation spectra, LO phonon progressions have nevertheless been observed for CdSe[83].

Then coupling to acoustic phonons was claimed to be an important cause of intrinsic broadening[84]. Calculations, however, show their role to be less important than anticipated[85]. We note that, in the calculations dealing with acoustic or optic phonons, the quantum confinement of phonons has to be taken into account[81]. In summary, the origin of the intrinsic or homogeneous broadening of electronic transitions in strongly confined SDG

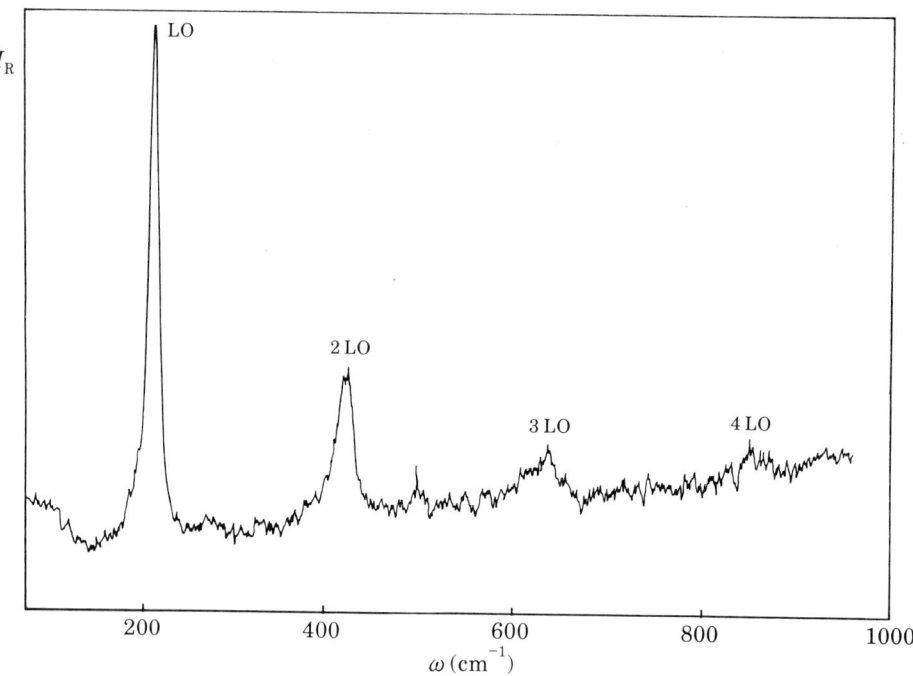

Fig. 12. – Resonant Raman spectrum of a CdSe-doped glass sample. Several orders of scattering by the 210 cm^{-1} LO-phonon are visible.

has not been fully elucidated yet. Furthermore, inhomogeneous broadening is still presently dominant at low temperature.

In the case of weak confinement, the effect of size dispersion is less dramatic for the excitonic lines, since the size dependence is via $1/Ma^2$ instead of $1/\mu a^2$ and since the total mass M is (much) larger than the reduced mass μ. Nevertheless, when the size is reduced, the excitonic lines have been observed to broaden[1, 32].

4˙6. SDG *in strong confinement: the time response*. – In the resonant Kerr effect in SDG, we create photo-excited carriers which modify the optical properties of the material. The response time of the nonlinearity is then determined by the recombination time of the carriers. This recombination time can be strongly reduced by two effects: photodarkening and Auger recombination. With a short laser pulse, we create at time $t = 0$ a number density N_0 of carrier pairs. In semiconductors, the number density N relaxes back to zero according to the equation

$$(54) \qquad \frac{\mathrm{d}N}{\mathrm{d}t} = -AN - BN^2 - CN^3 .$$

The first term is the Shockley-Read term corresponding to nonradiative recombination. When alone, it leads to exponential decay with the time constant $\tau = 1/A$. The second term corresponds to luminescence or pair recombination. The third one corresponds to Auger recombination: for example, an electron and a hole may recombine, the released energy being given to a second electron which is promoted to a higher level. It becomes relevant only when N is large.

As far as luminescence is concerned, things may be slightly different in nanocrystals. When no more than a pair is created per particle, we can only have geminate recombination and the luminescence intensity I_L is proportional to N. On the other hand, when several pairs are created per particle, then the situation becomes bulklike and I_L is proportional to N^2. This N^2 dependence has been observed experimentally under high excitation conditions [86].

Recombination has been studied experimentally using optical phase conjugation and delaying the backward pump pulse, using saturated absorption and delaying the probe pulse or by time-resolving the luminescence [86]. In the first case, $|\chi^{(3)}| \propto N$ is measured. In the second case, the change in absorption coefficient $\Delta \alpha_{abs} \propto N$ is measured. Finally, in the third case, the luminescence intensity $I_L \propto N^2$ is measured. The samples which were studied had a poor luminescence yield, so that the second term in eq. (54) was always negligible. We point out finally that N is the number density of free carriers (trapped carriers recombine differently).

At low excitation, the first term dominates and N decays according to

$$N(t) = N_0 \exp[-At].$$

Low excitation is only affordable when using coherent techniques such as phase conjugation or saturated absorption. In this case, for «fresh» samples, a time constant $\tau = 1/A$ of a few ns is measured. But when the sample has been exposed to a laser beam for some time (the shorter the larger the laser fluence is), it experiences photodarkening [73]. The main consequences of photodarkening are the disappearance of the broad-band luminescence feature and the shortening of the nonradiative decay time $1/A$. Starting from a few ns for a fresh sample, one may get a few tens of picoseconds for a darkened one. Photodarkening is a quasi-permanent photochemical effect. It has been assigned to photoionization of particles, the ejected electrons being trapped in the glass matrix [87, 88]. The sample recovers its original properties after a heat treatment at $\sim 370\,°C$.

We now consider Auger recombination. For a fresh sample, at short delay times t and high N_0, the third term in eq. (54) is dominant, which reduces (54) to

$$(55) \qquad \frac{dN}{dt} = -CN^3,$$

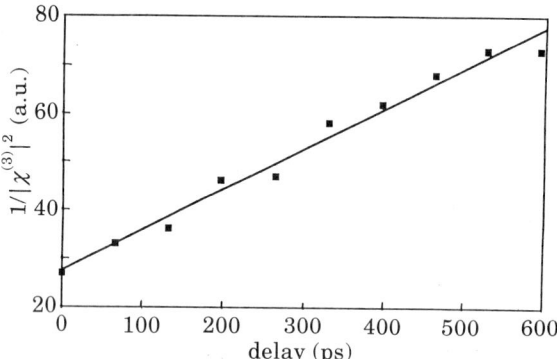

Fig. 13. – Plot of $1/|\chi^{(3)}|^2$ as a function of the backward pump pulse delay in a phase conjugation measurement. The sample is fresh.

whose solution is

$$(56) \qquad \frac{1}{N^2} = \frac{1}{N_0^2} + 2Ct \,.$$

This regime has been observed [86] as is shown in fig. 13, where $1/|\chi^{(3)}|^2$ is plotted *vs.* t (phase conjugation), in fig. 14, where $1/\Delta\alpha_{\mathrm{abs}}^2$ is plotted *vs.* t (saturated absorption), and in fig. 15, where $1/I_L$ is plotted *vs.* t (time-resolved luminescence). When studying luminescence, high excitation is required so that the Auger term is often (as in fig. 15) the dominant one.

For a darkened sample, the nonradiative-decay rate is larger and the more general recombination equation

$$(57) \qquad \frac{\mathrm{d}N}{\mathrm{d}t} = -AN - CN^3$$

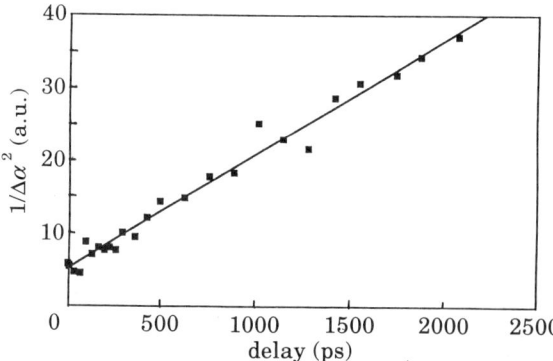

Fig. 14. – Plot of $1/\Delta\alpha^2$ as a function of the probe pulse delay in a nonlinear absorption measurement. The sample is fresh.

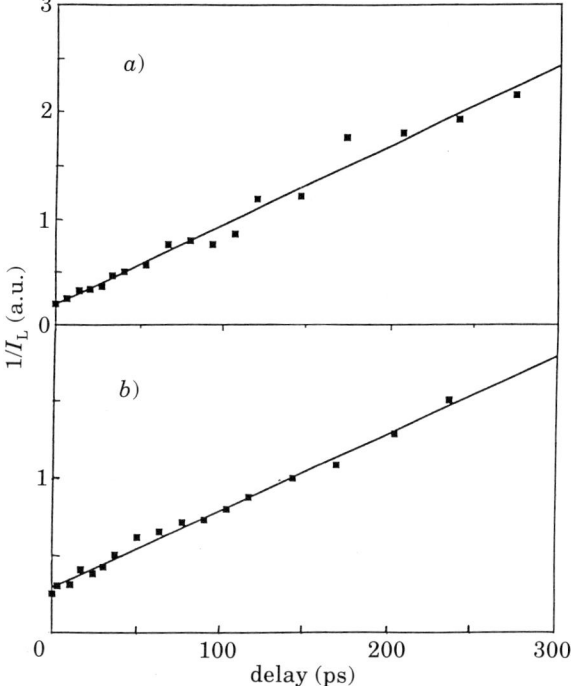

Fig. 15. – Plot of $1/I_L$ as a function of time in a time-resolved luminescence measurement. The sample is fresh.

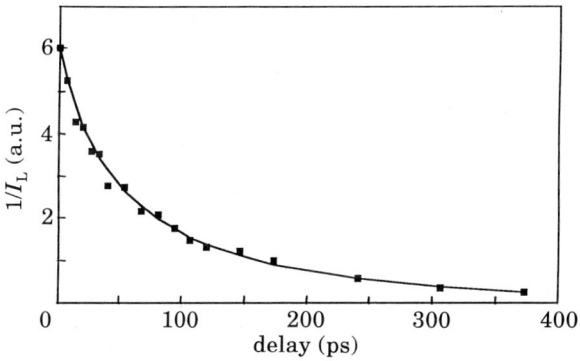

Fig. 16. – Plot of I_L as a function of time in a time-resolved luminescence measurement. The sample is photodarkened. The solid line is a fit using eq. (58).

must be used. Its solution

$$N^2 = \frac{AK \exp[-2At]}{1 - CK \exp[-2At]}$$

(58)

with $K = N_0^2/(CN_0^2 + A)$ provides an excellent fit to the data shown in fig. 16

and obtained by time-resolved luminescence measurements[86]. In this last case, the nonradiative-decay time is ≈ 400 ps. All these data consistently lead to a value of $C \sim (2\text{-}5) \cdot 10^{-30}$ cm^6 s^{-1} for the Auger constant[86,89-91].

To summarize, two processes may drastically reduce the recombination time in SDG: photodarkening and Auger recombination which has been clearly observed. These two processes may lead to very fast recovery, which allowed a very fast bistable device with a response time of 25 ps to be built[92].

4'7. SDG *in strong confinement, electroabsorption.* – A static (or very-low-frequency) electric field $\boldsymbol{E}$ leads to the static Kerr effect which also modifies the optical properties. The change in the absorption spectrum $\delta\alpha_{abs}(\omega)$ is usually measured; it is proportional to the imaginary part of $\chi^{(3)}(0, 0, \omega)$. The real part of $\chi^{(3)}$ could be obtained using the Kramers-Kronig relationships. Such an effect may be used in hybrid devices such as the SEED's made from quantum wells[93]. But the static Kerr effect is also interesting per se. The change in the optical properties is due to the shift of the energy levels and to admixture of neighbouring wave functions. When dealing with atoms for which the spacing between levels is large, perturbation theory is used and the effect is termed

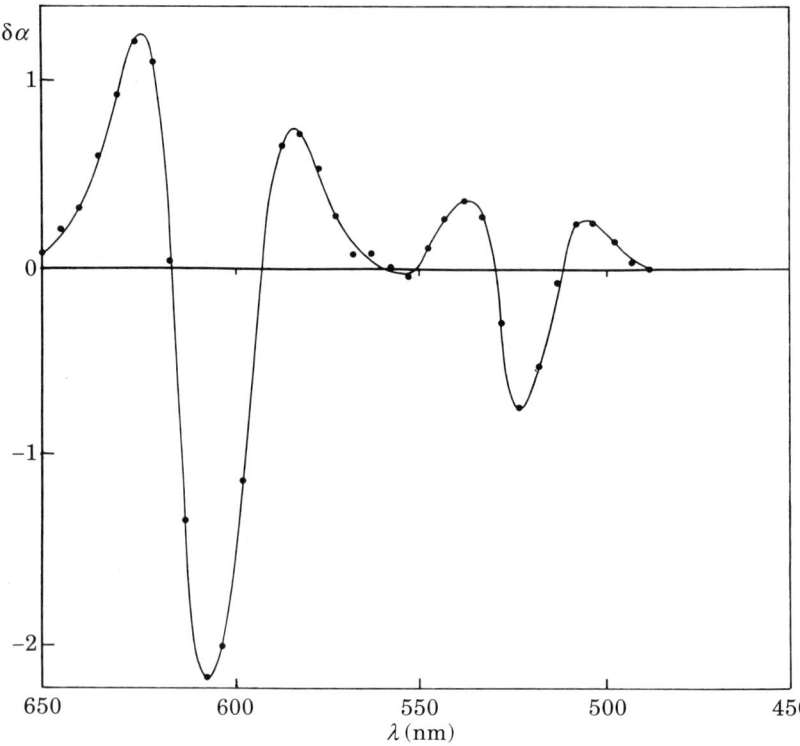

Fig. 17. – Differential absorption spectrum $\delta\alpha(\omega)$ for a CdS$_{0.5}$Se$_{0.5}$-doped glass sample. $\delta\alpha$ is induced by a static electric field of $2 \cdot 10^4$ V/cm.

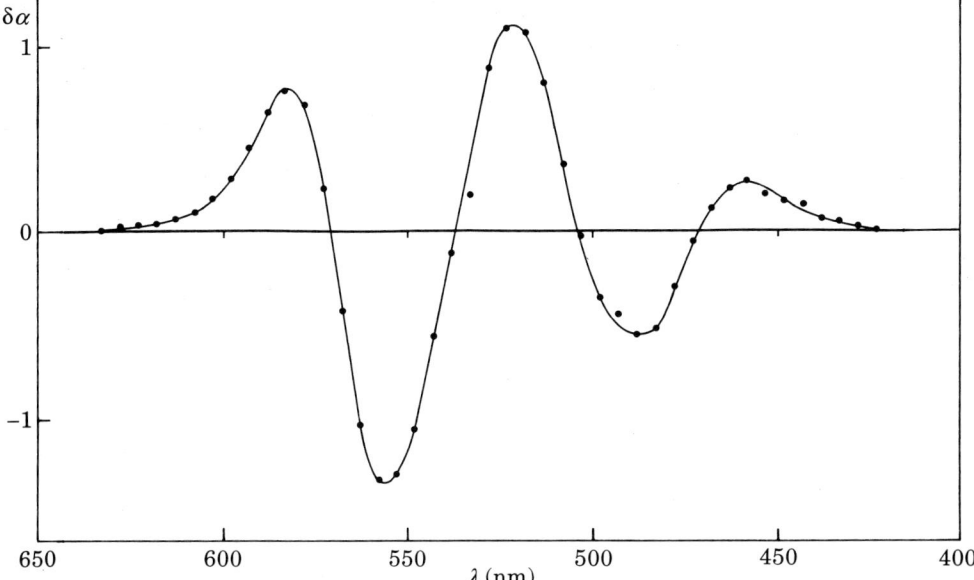

Fig. 18. – Same as fig. 17 but for a sample containing smaller particles. The structure is blue-shifted and of smaller amplitude as compared with fig. 17.

Stark effect. When dealing with a bulk semiconductor for which the levels are very closely spaced compared with eEa, the Hamiltonian

(59) $$H = H_0 + e\boldsymbol{E} \cdot \boldsymbol{r},$$

where H_0 is the zero-field Hamiltonian, must be solved. The effect is then termed Franz-Keldysh effect[94]. This is one and the same effect, only the mathematical approach differs. For SDG, the question is then: how do the confined levels behave?

Electroabsorption measurements were performed on $CdS_{0.5}Se_{0.5}$ samples containing particles of various mean sizes[95]. The electric field with $E_{in} = 2 \cdot 10^4$ V cm^{-1} is applied via indium tin oxide transparent electrodes deposited on the two sides of a thin SDG slab. The applied field oscillates sinusoidally at a frequency of 1 kHz and the change in absorption $\delta\alpha_{abs}$ is measured using lock-in detection at 2 kHz. Typical results are shown in fig. 17 and 18 for two different particle sizes. Oscillations are observed in the vicinity of the first ($1S_{3/2}$-$1S_e$ or loosely speaking $1s$-$1s$) peak with a replica in the vicinity of the split-off band edge.

The position of the oscillations is independent of the magnitude E of the applied field and their amplitude is proportional to E^2. This is typical of a quantum-confined Stark effect: confinement has led to well-resolved discrete levels. Similar results have been obtained by other groups[96,97]. When the particle size is larger, then Franz-Keldysh behaviour is observed[98]. The amplitude of the oscillations shown in fig. 17 and 18 decreases when the particle size is reduced. This is understandable since smaller particles are less polarizable. Using Rayleigh-Schrödinger perturbation theory, these experimental results could be fitted with good accuracy[95]. Variational calculations have also been performed[99]. We note that $\chi^{(3)}(0, 0, \omega) \sim 10^{-12}$ e.s.u., a fairly large value, and that a static field leads to a decrease of the absorption coefficient at the $1S_{3/2}$-$1S_e$ peak and to an increase on each side of it.

4'8. *SDG in strong confinement: the optical Kerr effect.* – We now consider the optical Kerr effect in these SDG in the strong-confinement regime. It has been studied experimentally mainly using two techniques. The first one is optical phase conjugation which allows measurement of $|\chi^{(3)}(\omega, -\omega, \omega)|$ and which has already been discussed above. The second one is nonlinear absorption: an excitation laser pulse at frequency ω_1 creates photocarriers which modify the absorption spectrum of the sample. This change may be monitored by sending a probe pulse at the (tunable) frequency ω_2. We thus measure $\mathrm{Im}\,\chi^{(3)}(\omega_1, -\omega_1, \omega_2)$. A very elegant technique is to use a white-spectrum probe pulse generated by self-phase modulation and to measure the whole transmission spectrum using an OMA[79]. The nonlinear absorption technique may be slightly more powerful since $\chi^{(3)}(\omega_1, -\omega_1, \omega_2)$ contains more information than the degenerate $\chi^{(3)}(\omega, -\omega, \omega)$, but optical phase conjugation is slightly simpler to implement and the degenerate case corresponds to most applications.

The expression for $\gamma(\omega, -\omega, \omega)$ is still of the form (38) except for a few differences. Instead of being a one-exciton state, $|1\rangle$ is now a one-pair state. By this we mean that we have an electron and a hole, for example in the $1S_e$ and $1S_{3/2}$ states, respectively. They are very close to each other, but this is due to quantum confinement and no longer to the Coulomb interaction. Such an entity is sometimes called an exciton, but we prefer to use the name pair to emphasize that it has nothing to do with a Wannier exciton. $|2\rangle$ is then a two-pair state. Another important difference is that here we may have different one-pair states. $|1\rangle$ may correspond to the $1S_{3/2}$-$1S_e$ level, to the $2S_{3/2}$-$1S_e$ one, to the $1P_{3/2}$-$1P_e$ one and so on. So summations over one-pair and two-pair states should be added to eq. (38), but only the resonant terms are finally contributing significantly.

The expression for $\gamma(\omega_1, -\omega_1, \omega_2)$ is slightly more complicated than eq. (38) because of nondegeneracy. Equation (38) was obtained from 24 diagrams. 48 diagrams would be needed for $\gamma(\omega_1, -\omega_1, \omega_2)$[100]. We do not give this expression however, since the physical processes are still basically the same: saturation of a one-photon resonant transition or two-step resonant excitation of a two-pair state. In the case of nonlinear absorption, for example, one may observe saturated (*i.e.* decreasing) absorption towards the $1S_{3/2}$-$1S_e$ or $2S_{3/2}$-$1S_e$ levels or induced absorption from a one-pair level to a two-pair one.

Nonlinear absorption measurements have been performed in the nanosecond [101], picosecond [79] and femtosecond [83, 102] time scales. Figure 19 shows the $-\delta\alpha_{abs}(\omega)$ spectrum obtained for a CdSe-doped glass at low temperature with femtosecond pulses [102]. One can see saturated absorption towards the $1S_{3/2}$-$1S_e$ and $2S_{3/2}$-$1S_e$ levels as well as, on the blue side of these two features, induced absorption towards a two-pair state. In fact, another interpretation is possible. We only discussed until now the contribution of free carriers. Another possible mechanism would be the following: carriers are photoexcited, they rapidly get trapped and, via, for example, the static electric field they create, they modify the absorption spectrum in a way similar to electroabsorption [103]. This interpretation has been put forward to explain the similarity of the $-\delta\alpha_{abs}(\omega)$ spectrum on the femtosecond and nanosecond time scales [83]. We may, therefore, have two mechanisms, one due to free carriers and described by the hyperpolarizability γ, a second one due to trapped carriers and much more difficult to analyse quantitatively.

The result shown in fig. 19 has been confirmed using a tunable dye laser as the probe pulse [104]. We note that, at the positions of the first two transitions, absorption saturates. Delaying the probe pulse, one can monitor absorption recovery as was described in subsect. 4'6.

If we now consider phase conjugation, *i.e.* the degenerate case $\omega_1 = \omega_2 = \omega$, we see that both mechanisms predict absorption saturation. Saturation of a two-level system had been proposed as early as 1987 as the dominant mechan-

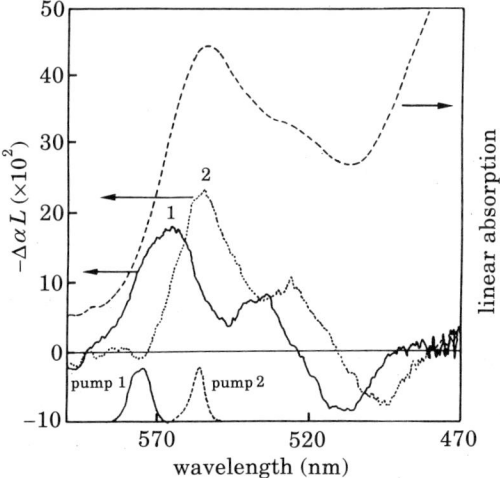

Fig. 19. – Absorption spectrum (dashed line) and two differential absorption spectra (solid line and dotted line) for a CdSe-doped glass sample. $\Delta\alpha$ is induced by a pump pulse whose spectrum is shown in the lower part (from ref. [102]).

ism [105]. For the free-carrier mechanism, in this case the A^2 term in (38) is dominant since $\omega_{21} \neq \omega_{10}$ [72]. For the trapped-carrier one, assuming electroabsorption to be involved, we saw above that, on resonance, absorption decreases.

Undoubtedly, in certain cases, for example for fresh commercial Schott glasses, the trapped-carrier contribution may dominate [106-108]. Figure 20 obtained using phase conjugation with a tunable dye laser shows that, for such a commercial glass, the modulus of $\chi^{(3)}(\omega, -\omega, \omega)$ when plotted as a function of ω is roughly proportional to the absorption coefficient $\alpha_{abs}(\omega)$ [60, 73, 109]. Such a proportionality would be consistent with a trapped-carrier mechanism. It cannot be explained by absorption saturation. However, we note that first, for a darkened sample, trapping centres are in a way inhibited, so that the free-carrier contribution may dominate, and that secondly, by having a closer look at fig. 20, a faint resonance in the vicinity of the first absorption structure is visible.

The main issue with regard to applications is the size dependence of $\widetilde{\chi}^{(3)}$ or more importantly of the figure of merit $F = \widetilde{\chi}^{(3)}/\alpha_{abs}\,\tau$. Is it interesting to go to small sizes? If we think in terms of degenerate four-wave mixing and of free carriers, we are left with the A^2 term of (38). The figure of merit is then easily

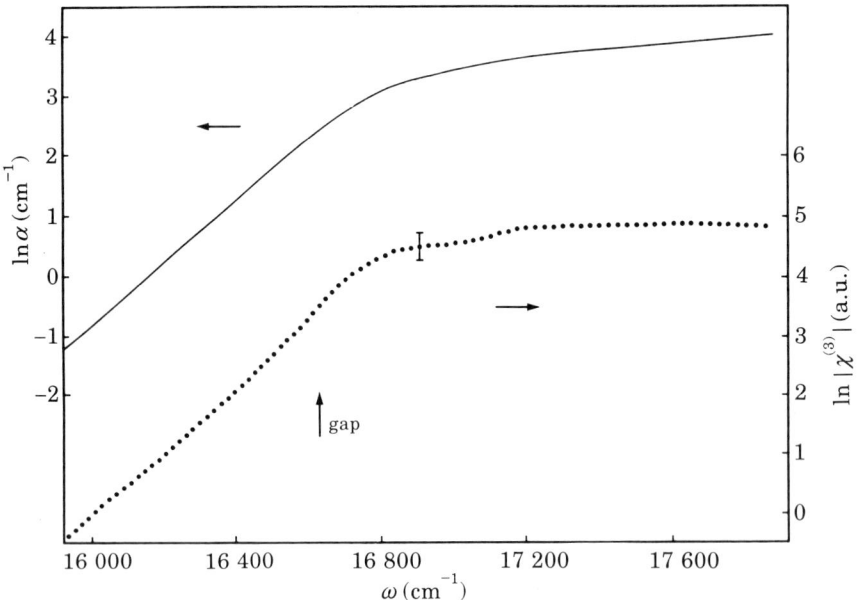

Fig. 20. – Absorption and $\chi^{(3)}(\omega, -\omega, \omega)$ spectra on a logarithmic scale for a $CdS_{0.5}Se_{0.5}$-doped glass sample. $\chi^{(3)}$ was obtained by optical phase conjugation using a nanosecond dye laser.

seen to be proportional to

$$(60) \qquad F \propto \frac{|\langle 1|p|0\rangle|^2 \, T_2}{(1+\delta^2)^{1/2}},$$

where $\delta = (\omega - \omega_{10})\, T_2$ is the normalized detuning. We, therefore, have to consider the size dependence of the transition matrix element $\langle 1|p|0\rangle$ and of the dephasing time T_2 (the full homogeneous linewidth is $\Gamma = 2/T_2$). If the valence band were nondegenerate, the electron and hole envelope wave functions would be given by eqs. (37) and (43) and, as shown by eq. (46), the transition matrix element would be p_{vc}, size independent[110]. If we consider the degeneracy of the valence band in the $\Delta = \infty$ approximation and if we concentrate on the $nS_{3/2}$-$1S_e$ transitions, the hole envelope wave functions are given by eq. (52) and the transition matrix element is the product of p_{vc} and of the overlap integral between $B_{01}j_0(\pi r_e/a)$ and $R_{0n}(r_h)$. Setting $x = r/a$, it is easily shown to be also size independent. In such a case, assuming T_2 to be also size independent, the figure of merit F would be size independent. Optical-phase-conjugation measurements performed on a limited range of sizes show this to be the case experimentally[106, 111] although contradictory results had been reported earlier[110, 112]. There should, therefore, be no point in going to smaller sizes.

Although experiments seem to show that the figure of merit is size independent, a few points remain to be definitely clarified. Firstly, when taking the valence band degeneracy into account in the general case (finite Δ), one then obtains weakly-size-dependent transition matrix elements[61]. Secondly, one must take the size distribution into account. The third-order hyperpolarizability γ is a-dependent, mainly through the energy denominators. The quantity to be averaged is the polarization or the polarizability (α for linear properties and γ for the optical Kerr effect). The Kerr susceptibility of the composite is, for example, given by

$$(61) \qquad \tilde{\chi}^{(3)}(\omega, -\omega, \omega) = N \, |f(\omega)|^2 \, [f(\omega)]^2 \int \gamma_a(\omega, -\omega, \omega) \, P(a) \, \mathrm{d}a \,,$$

where N is the number density of particles,

$$(62) \qquad N = \frac{p}{\overline{V}} \,,$$

where $\overline{V}$ is the mean particle volume. And, in the same way, the absorption coefficient is

$$(63) \qquad \alpha_{\mathrm{abs}}(\omega) = \frac{\omega}{nc} \, 4\pi N \, |f(\omega)|^2 \int \alpha_a''(\omega) \, P(a) \, \mathrm{d}a \,.$$

γ_a and α_a are the hyperpolarizability and the linear polarizability for a sphere of radius a. From (61) and (63), it is clear that $\tilde{\chi}^{(3)}$ is roughly proportional to $1/\overline{V}$. The nonlinearity increases when the size is reduced, what is qualitatively un-

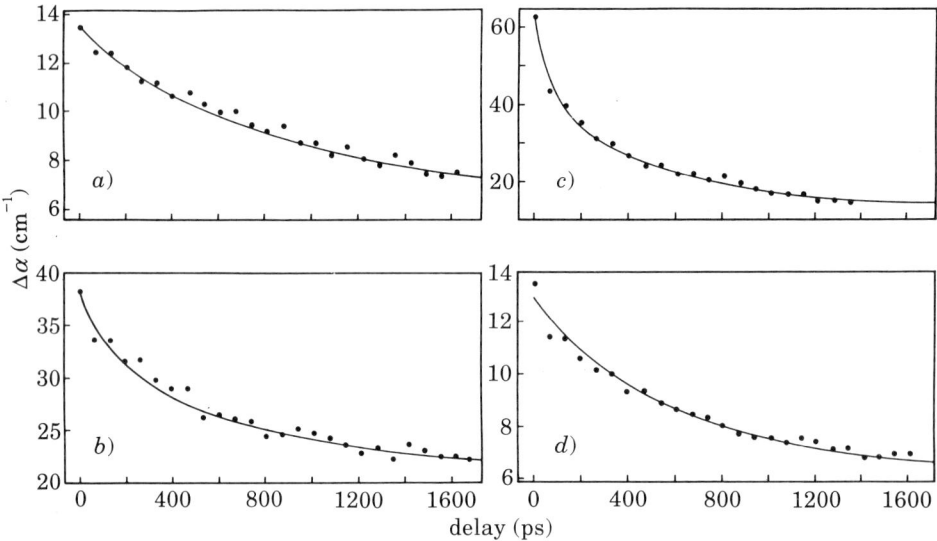

Fig. 21. – Plot of $\Delta\alpha$ or $\chi^{(3)}$ obtained by phase conjugation as a function of the backward pump pulse delay for a Schott commercial filter for several values of the pump fluence. Curves a), b), c), d) were obtained in that order: a) fluence $70\,\mu\mathrm{J/cm^2}$, b) fluence $250\,\mu\mathrm{J/cm^2}$, c) fluence $790\,\mu\mathrm{J/cm^2}$, d) fluence $70\,\mu\mathrm{J/cm^2}$. Note the presence of both a fast (free-carrier) and a slow (trapped-carrier) component.

derstood by saying that quantum confinement leads to a condensation of the oscillator strength into discrete lines[110], but this is exactly compensated for by the increase of α_{abs}. Thirdly, the origin of the nonlinear response remains to be clarified: in some cases, trapped carriers give the dominant contribution, whereas in other cases the contribution of free carriers is clearly evidenced as shown in fig. 21 obtained by delaying the backward pump pulse in optical phase conjugation[90].

The magnitude of the figure of merit F is rather small for the time being. It is about two orders of magnitude weaker than that for excitonic nonlinearities of quantum wells[104]. This is due to the large width of the electronic transitions. Remember that the broadening is mainly due to the size distribution.

5. – Conclusion.

The linear properties of gold colloids, gold-doped glasses, CuCl-, CdS- or CdSe-doped glasses are now very well understood and clearly show the roles of dielectric confinement and electronic or quantum confinement. Such small crystallites, intermediate in size between molecular aggregates and bulk metals or semiconductors, lead to an interesting class of optical composite materials. The optical Kerr effect of these composites is also reasonably well understood.

The optical Kerr effect is fully understood in the case of gold-doped glasses. The local-field correction is important, whereas the confinement-induced $\chi^{(3)}$ is dominated by bulklike mechanisms, interband absorption saturation and hot electrons. The figure of merit is small because of the small value of $\chi^{(3)}$.

Although a few points such as the exact relative magnitude of the free-carrier and trapped-carrier contributions remain to be clarified, we also now have a good understanding of the optical Kerr effect in SDG. The figure of merit is weaker than that of excitonic nonlinearities in quantum wells, but this is only due to the width of the electronic transitions. One may hope that improved fabrication techniques will lead to an improvement in the size dispersion or in the interface-related trapping centres. Growth of particles in a zeolite[113] or of colloids in inverse micelles[114] or electrophoretic size separation[115] may prove interesting techniques. Our basic knowledge has anyway made huge progress, for example, in the understanding of luminescence, broadening mechanisms, photodarkening and Auger recombination.

On the application side, a bistable device has already been developed[92], waveguide devices have been built[116]. On a much slower time scale, advantage has also been taken of the thermal nonlinearity which was not discussed here[117,118]. On the basic-research side, the below-gap nonlinearity of these SDG has also been studied[119] and may show interesting properties. People are now concentrating on the growth mechanisms[120], on the very-low-frequency acoustic modes[121] or on the limits of the effective-mass approximation when very small sizes are reached[122].

* * *

The author gratefully thanks C. FLYTZANIS, M. GHANASSI, F. HACHE, P. ROUSSIGNOL and M. C. SCHANNE-KLEIN for fruitful collaboration.

REFERENCES

[1]   A. I. EKIMOV and A. A. ONUSHCHENKO: *JETP Lett.*, **34**, 345 (1981); **40**, 1137 (1984).

[2]   R. ROSSETTI, S. NAKAHARA and L. E. BRUS: *J. Chem. Phys.*, **79**, 1086 (1983).

[3]   H. WELLER, U. KOCH, M. GUTIERREZ and A. HENGLEIN: *Ber. Bunsenges. Phys. Chem.*, **88**, 649 (1984).

[4]   J. C. MAXWELL-GARNETT: *Philos. Trans. R. Soc. London*, **203**, 385 (1904); **205**, 237 (1906).

[5]   See, for instance, C. J. BÖTTCHER: *Theory of Electric Polarization* (Elsevier, Amsterdam, 1973) or J. D. JACKSON: *Classical Electrodynamics* (John Wiley, New York, N.Y., 1980).

[6]   D. A. G. BRUGGEMAN: *Ann. Phys. (Leipzig)*, **24**, 636 (1935).

[7]   U. KREIBIG and P. ZACHARIAS: *Z. Phys.*, **231**, 128 (1970).

[8] G. MIE: *Ann. Phys. (Leipzig)*, **25**, 377 (1908).

[9] D. RICARD, P. ROUSSIGNOL and C. FLYTZANIS: *Opt. Lett.*, **10**, 511 (1985).

[10] See, for instance, Y. R. SHEN: *Principles of Nonlinear Optics* (John Wiley, New York, N.Y., 1984).

[11] F. HACHE and D. RICARD: *J. Phys. Condensed Matter*, **1**, 8305 (1989).

[12] J. E. SIPE and R. W. BOYD: *Phys. Rev. A*, **46**, 1614 (1992).

[13] See, for instance, *Optical Phase Conjugation*, edited by R. A. FISHER (Academic Press, New York, N.Y., 1983).

[14] B. Y. ZEL'DOVICH, N. T. PILIPETSKY and V. V. SHKUNOV: *Principles of Phase Conjugation* (Springer-Verlag, Heidelberg, 1985).

[15] K. M. LEUNG: *Phys. Rev. A*, **33**, 2461 (1986).

[16] D. S. CHEMLA and D. A. B. MILLER: *Opt. Lett.*, **11**, 522 (1986).

[17] H. FRÖHLICH: *Physica*, **6**, 406 (1937).

[18] R. KUBO: *J. Phys. Soc. Jpn.*, **17**, 975 (1962).

[19] F. HACHE, D. RICARD and C. FLYTZANIS: *J. Opt. Soc. Am. B*, **3**, 1647 (1986).

[20] See, for instance, J. A. A. J. PERENBOOM, J. WIDER and F. MEIER: *Phys. Rep.*, **78**, 173 (1981).

[21] R. H. DOREMUS: *J. Chem. Phys.*, **40**, 2389 (1964); **42**, 414 (1965).

[22] See, for instance, N. W. ASHCROFT and N. D. MERMIN: *Solid State Physics* (Holt-Saunders, Philadelphia, Penn., 1976).

[23] A. KAWABATA and R. KUBO: *J. Phys. Soc. Jpn.*, **21**, 1765 (1966).

[24] L. GENZEL, T. P. MARTIN and U. KREIBIG: *Z. Phys. B*, **21**, 339 (1975).

[25] R. RUPPIN and H. YATOM: *Phys. Status Solidi B*, **74**, 647 (1976).

[26] F. HACHE, D. RICARD, C. FLYTZANIS and U. KREIBIG: *Appl. Phys. A*, **47**, 347 (1988).

[27] G. L. EESLEY: *Phys. Rev. B*, **33**, 2144 (1986).

[28] R. W. SCHOENLEIN, W. Z. LIN, J. G. FUJIMOTO and G. L. EESLEY: *Phys. Rev. Lett.*, **58**, 1680 (1987).

[29] A. OWYOUNG: *Opt. Commun.*, **16**, 266 (1976).

[30] J. W. HAUS, N. KALYANIWALLA, R. INGUVA, M. BLOEMER and C. M. BOWDEN: *J. Opt. Soc. Am. B*, **6**, 797 (1989).

[31] A. E. NEEVES and M. H. BIRNBOIM: *J. Opt. Soc. Am. B*, **6**, 787 (1989).

[32] U. WOGGON and F. HENNEBERGER: *J. Phys. (Paris)*, **49**, C2, 225 (1988).

[33] V. ESCH, B. FLUEGEL, G. KHITROVA, H. M. GIBBS, X. JIAJIN, K. KANG, S. W. KOCH, L. C. LIU, S. H. RISBUD and N. PEYGHAMBARIAN: *Phys. Rev. B*, **42**, 7450 (1990).

[34] H. WELLER, H. M. SCHMIDT, U. KOCH, A. FOJTIK, S. BARAL, A. HENGLEIN, W. KUNATH, K. WEISS and E. DIEMAN: *Chem. Phys. Lett.*, **124**, 557 (1986).

[35] A. FOJTIK, H. WELLER, U. KOCH and A. HENGLEIN: *Ber. Bunsenges. Phys. Chem.*, **88**, 969 (1984).

[36] R. ROSSETTI, R. HULL, J. M. GIBSON and L. E. BRUS: *J. Chem. Phys.*, **82**, 552 (1985).

[37] Y. WANG and W. MAHLER: *Opt. Commun.*, **61**, 233 (1987).

[38] M. G. BAWENDI, M. L. STEIGERWALD and L. E. BRUS: *Annu. Rev. Phys. Chem.*, **41**, 477 (1990).

[39] P. E. LIPPENS and M. LANNOO: *Phys. Rev. B*, **39**, 10935 (1989); **41**, 6079 (1990).

[40] S. V. NAIR, L. M. RAMANIAH and K. C. RUSTAGI: *Phys. Rev. B*, **45**, 5969 (1992).

[41] R. C. MILLER, D. A. KLEINMAN and A. C. GOSSARD: *Phys. Rev. B*, **29**, 7085 (1984).

[42] Y. KAYANUMA and H. MOMIJI: *Phys. Rev. B*, **41**, 10261 (1990).

[43] D. B. TRAN-THOAI, Y. Z. HU and S. W. KOCH: *Phys. Rev. B*, **42**, 11261 (1990).

[44] L. E. BRUS: *J. Chem. Phys.*, **80**, 4403 (1984).

[45] L. BANYAI, P. GILLIOT, Y. Z. HU and S. W. KOCH: *Phys. Rev. B*, **45**, 14136 (1992).

[46] D. B. TRAN-THOAI: *Solid State Commun.*, **85**, 39 (1993).

[47] AL. L. EFROS and A. L. EFROS: *Sov. Phys. Semicond.*, **16**, 772 (1982).

[48] A. I. EKIMOV and AL. L. EFROS: *Phys. Status Solidi B*, **150**, 627 (1988).

[49] A. I. EKIMOV, A. A. ONUSHCHENKO and AL. L. EFROS: *JETP Lett.*, **43**, 376 (1986).

[50] L. BANYAI, Y. Z. HU, M. LINDBERG and S. W. KOCH: *Phys. Rev. B*, **38**, 8142 (1988).

[51] E. HANAMURA, M. KUWATA-GONOKAMI and H. EZAKI: *Solid State Commun.*, **73**, 551 (1990).

[52] E. HANAMURA: *Phys. Rev. B*, **37**, 1273 (1988).

[53] E. HANAMURA: *Opt. Quantum Electron.*, **21**, 441 (1989).

[54] Y. MASUMOTO, M. YAMAZAKI and H. SUGAWARA: *Appl. Phys. Lett.*, **53**, 1527 (1988).

[55] T. ITOH, M. FURUMIYA, T. IKEHARA and C. GOURDON: *Solid State Commun.*, **73**, 271 (1990).

[56] B. L. JUSTUS, M. E. SEAVER, J. A. RULLER and A. J. CAMPILLO: *Appl. Phys. Lett.*, **57**, 1381 (1990).

[57] B. KIPPELEN, R. LEVY, P. FALLER, P. GILLIOT and L. BELLEGUIE: *Appl. Phys. Lett.*, **59**, 3378 (1991).

[58] G. BRET and F. GIRES: *Appl. Phys. Lett.*, **4**, 175 (1964).

[59] R. K. JAIN and R. C. LIND: *J. Opt. Soc. Am.*, **73**, 647 (1983).

[60] P. ROUSSIGNOL, D. RICARD and C. FLYTZANIS: *Appl. Phys. A*, **44**, 285 (1987).

[61] A. I. EKIMOV, F. HACHE, M. C. SCHANNE-KLEIN, D. RICARD, C. FLYTZANIS, I. A. KUDRYAVTSEV, T. V. YAZEVA, A. V. RODINA and AL. L. EFROS: *J. Opt. Soc. Am. B*, **10**, 100 (1993).

[62] D. W. HALL and N. F. BORRELLI: *J. Opt. Soc. Am. B*, **5**, 1650 (1988).

[63] Y. KAYANUMA: *Solid State Commun.*, **59**, 405 (1986).

[64] J. M. LUTTINGER and W. KOHN: *Phys. Rev.*, **97**, 869 (1955); J. M. LUTTINGER: *Phys. Rev.*, **102**, 1030 (1956).

[65] AL. L. EFROS and A. V. RODINA: *Solid State Commun.*, **72**, 645 (1989).

[66] G. B. GRIGORYAN, E. M. KAZARYAN, AL. L. EFROS and T. V. YAZEVA: *Sov. Phys. Solid State*, **32**, 1031 (1990).

[67] J. B. XIA: *Phys. Rev. B*, **40**, 8500 (1989).

[68] N. O. LIPARI and A. BALDERESCHI: *Phys. Rev. Lett.*, **25**, 1660 (1971).

[69] A. I. EKIMOV, A. A. ONUSHCHENKO, A. G. PLYUKHIN and AL. L. EFROS: *Sov. Phys. JETP*, **88**, 891 (1985).

[70] AL. L. EFROS: *Phys. Rev. B*, **46**, 7448 (1992).

[71] E. O. KANE: *J. Phys. Chem. Solids*, **1**, 249 (1957).

[72] Y. Z. HU, M. LINDBERG and S. W. KOCH: *Phys. Rev. B*, **42**, 1713 (1990).

[73] P. ROUSSIGNOL, D. RICARD, J. LUKASIK and C. FLYTZANIS: *J. Opt. Soc. Am. B*, **4**, 5 (1987).

[74] N. CHESTNOY, T. D. HARRIS, R. HULL and L. E. BRUS: *J. Phys. Chem.*, **90**, 3393 (1986).

[75] F. HACHE, M. C. SCHANNE-KLEIN, D. RICARD and C. FLYTZANIS: *J. Opt. Soc. Am. B*, **8**, 1802 (1991).

[76] I. M. LIFSHITZ and V. V. SLEZOV: *Sov. Phys. JETP*, **8**, 331 (1959).

[77] V. V. GOLUBKOV, A. I. EKIMOV, A. A. ONUSHCHENKO and V. A. TSEKHOMSKII: *Fiz. Khim. Stekla*, **7**, 397 (1981).

[78] R. CINGOLANI, C. MORO, D. MANNO, M. STRICCOLI, C. DE BLASI, G. C. RIGHINI and M. FERRARA: *J. Appl. Phys.*, **70**, 6898 (1991).

[79] P. ROUSSIGNOL, D. RICARD, C. FLYTZANIS and N. NEUROTH: *Phys. Rev. Lett.*, **62**, 312 (1989).

[80] A. P. ALIVISATOS, T. D. HARRIS, P. J. CARROLL, M. L. STEIGERWALD and L. E. BRUS: *J. Chem. Phys.*, **90**, 3463 (1989).

[81] M. C. KLEIN, F. HACHE, D. RICARD and C. FLYTZANIS: *Phys. Rev. B*, **42**, 11123 (1990).

[82] J. J. SHIANG, S. H. RISBUD and A. P. ALIVISATOS: *J. Chem. Phys.*, **98**, 8432 (1993).

[83] M. G. BAWENDI, W. L. WILSON, L. ROTHBERG, P. J. CARROLL, T. M. JEDIU, M. L. STEIGERWALD and L. E. BRUS: *Phys. Rev. Lett.*, **65**, 1623 (1990).

[84] K. MISAWA, H. YAO, T. HAYASHI and T. KOBAYASHI: *J. Chem. Phys.*, **94**, 4131 (1991).

[85] S. NOMURA and T. KOBAYASHI: *Solid State Commun.*, **82**, 335 (1992).

[86] M. GHANASSI, M. C. SCHANNE-KLEIN, F. HACHE, A. I. EKIMOV, D. RICARD and C. FLYTZANIS: *Appl. Phys. Lett.*, **62**, 78 (1993).

[87] V. Y. GRABOVSKIS, Y. Y. DZENIS, A. I. EKIMOV, I. A. KUDRYAVTSEV, M. N. TOLSTOI and U. T. ROGULIS: *Sov. Phys. Solid State*, **31**, 149 (1989).

[88] J. MALHOTRA, D. J. HAGAN and B. G. POTTER: *J. Opt. Soc. Am. B*, **8**, 1531 (1991).

[89] K. NATTERMANN, B. DANIELZIK and D. VON DER LINDE: *Appl. Phys. A*, **44**, 111 (1987).

[90] P. ROUSSIGNOL, M. KULL, D. RICARD, F. DE ROUGEMONT, R. FREY and C. FLYTZANIS: *Appl. Phys. Lett.*, **51**, 1882 (1987), and erratum, **54**, 1705 (1989).

[91] J. P. ZHENG and H. S. KWOK: *Appl. Phys. Lett.*, **54**, 1 (1989).

[92] J. YUMOTO, S. FUKUSHIMA and K. KUBODERA: *Opt. Lett.*, **12**, 832 (1987).

[93] D. A. B. MILLER, D. S. CHEMLA, T. C. DAMEN, A. C. GOSSARD, W. WIEGMANN, T. H. WOOD and C. A. BURRUS: *Appl. Phys. Lett.*, **45**, 13 (1984).

[94] W. FRANZ: *Z. Naturforsch. Teil A*, **13**, 484 (1958); L. V. KELDYSH: *Sov. Phys. JETP*, **7**, 788 (1958).

[95] F. HACHE, D. RICARD and C. FLYTZANIS: *Appl. Phys. Lett.*, **55**, 1504 (1989).

[96] A. I. EKIMOV, A. P. SKVORTSOV, T. V. SHUBINA, S. K. SHUMILOV and AL. L. EFROS: *Sov. Phys. Tech. Phys.*, **34**, 371 (1989).

[97] S. NOMURA and T. KOBAYASHI: *Solid State Commun.*, **73**, 425 (1990).

[98] D. COTTER, H. P. GIRDLESTONE and K. MOULDING: *Appl. Phys. Lett.*, **58**, 1455 (1991).

[99] S. NOMURA and T. KOBAYASHI: *Solid State Commun.*, **74**, 1153 (1990).

[100] See, for instance, C. FLYTZANIS: in *Quantum Electronics: A Treatise*, edited by H. RABIN and C. L. TANG (Academic Press, New York, N. Y., 1975), Vol. 1, Part A, p. 9.

[101] A. P. ALIVISATOS, A. L. HARRIS, N. J. LEVINOS, M. L. STEIGERWALD and L. E. BRUS: *J. Chem. Phys.*, **89**, 4001 (1988).

[102] N. PEYGHAMBARIAN, B. FLUEGEL, D. HULIN, A. MIGUS, M. JOFFRE, A. ANTONETTI, S. W. KOCH and M. LINDBERG: *IEEE J. Quantum Electron.*, **25**, 2516 (1989).

[103] A. HENGLEIN, A. KUMAR, E. JANATA and H. WELLER: *Chem. Phys. Lett.*, **132**, 133 (1986).

[104] M. C. SCHANNE-KLEIN: PhD Thesis, University of Paris, unpublished.

[105] S. SCHMITT-RINK, D. A. B. MILLER and D. S. CHEMLA: *Phys. Rev. B*, **35**, 8113 (1987).

[106] M. C. Schanne-Klein, F. Hache, D. Ricard and C. Flytzanis: *J. Opt. Soc. Am. B*, **9**, 2234 (1992).
[107] E. F. Hilinski, P. A. Lucas and Y. Wang: *J. Chem. Phys.*, **89**, 3435 (1988).
[108] B. van Wonterghem, S. M. Saltiel and P. M. Rentzepis: *J. Opt. Soc. Am. B*, **6**, 1823 (1989).
[109] P. Horan and H. Blau: *Semicond. Sci. Technol.*, **2**, 382 (1987).
[110] P. Roussignol, D. Ricard and C. Flytzanis: *Appl. Phys. B*, **51**, 437 (1990).
[111] H. Shinojima, J. Yumoto and N. Uesugi: *Appl. Phys. Lett.*, **60**, 298 (1992).
[112] S. H. Park, R. A. Morgan, Y. Z. Hu, M. Lindberg, S. W. Koch and N. Peyghambarian: *J. Opt. Soc. Am. B*, **7**, 2097 (1990).
[113] Y. Wang and N. Herron: *J. Phys. Chem.*, **91**, 257 (1987).
[114] M. G. Bawendi, A. R. Kortan, M. L. Steigerwald and L. E. Brus: *J. Chem. Phys.*, **91**, 7282 (1989).
[115] L. Katsikas, A. Eychmüller, M. Giersig and H. Weller: *Chem. Phys. Lett.*, **172**, 201 (1990).
[116] C. N. Ironside, T. J. Cullen, B. S. Bhumbra, J. Bell, W. C. Banyai, N. Finlayson, C. T. Seaton and G. I. Stegeman: *J. Opt. Soc. Am. B*, **5**, 492 (1988), and erratum, **6**, 378 (1989).
[117] N. I. Zheludev, I. S. Ruddock and R. Illingworth: *Opt. Quantum Electron.*, **20**, 119 (1988).
[118] G. Mei, S. Carpenter and P. D. Persans: *Solid State Commun.*, **80**, 557 (1991).
[119] D. Cotter, M. G. Burt and R. J. Manning: *Phys. Rev. Lett.*, **68**, 1200 (1992).
[120] B. G. Potter and J. H. Simmons: *Phys. Rev. B*, **43**, 2234 (1991).
[121] B. Champagnon, B. Andrianasolo and E. Duval: *J. Chem. Phys.*, **94**, 5237 (1991).
[122] L. M. Ramaniah and S. V. Nair: *Phys. Rev. B*, **47**, 7132 (1993).

Heterostructure Engineering
of Nonlinear Optical Properties
and Continuum States

F. Capasso and C. Sirtori

AT&T Bell Laboratories - Murray Hill, NJ 07974

1. – Introduction.

During the last decade a powerful new approach for designing semiconductor structures with tailored electronic and optical properties, bandgap engineering, has spawned a new generation of electronic and photonic devices[1]. Central to bandgap engineering is the notion that, by spatially varying the composition and the doping of a semiconductor over distances ranging from a few micrometers down to $\sim 2.5 \, \text{Å}$ (~ 1 monolayer), one can tailor the band structure of a material in a nearly arbitrary and continuous way. Thus semiconductor structures with new electronic and optical properties can be custom-designed for specific applications.

The enabling technology which has made bandgap engineering an exciting reality with far-reaching implications for science and technology is molecular-beam epitaxy (MBE), pioneered by Cho and Arthur in the late 1960s[2].

Large second-order nonlinearities based on intersubband resonant enhancement were first predicted by Gurnick and De Temple in 1983[3]. More recently, several groups have demonstrated that quantum wells made asymmetric by application of an electric field[4] and asymmetric step quantum-well structures[5-10] exhibit large second-order nonlinear susceptibilities for pump wavelengths $\lambda \simeq 10 \, \mu\text{m}$.

In this lecture we discuss our recent work on the nonlinear optical properties of coupled quantum wells[11-17]. Coupled quantum wells represent an excellent model system to investigate optical nonlinearities. The wave functions, the dipole matrix elements and the energy level separation can be tailored over a wide range by varying the well thicknesses and the thickness of the coupling barrier. In sect. 2 we discuss the design of these structures and their optimization for nonlinear optical applications. Section 3 describes intersubband absorption measurements with and without an electric field. Sections 4 and 5 deal with second- and third-harmonic generation (SHG and THG) experiments, respectively. Section 6 deals with multiphoton emission from coupled-well structures. The important role played by continuum resonances in enhancing the emission rate is emphasized. In the last section our recent work on Fabry-Perot electron filters and bound states in the continuum is discussed.

2. – Nonlinear optical properties of coupled quantum wells.

This section describes the design of coupled-well structures with extremely large $\chi^{(2)}(2\omega)$ and $\chi^{(3)}(3\omega)$; these coefficients are the nonlinear susceptibilities for SHG and THG, respectively. Our strategy to optimize $|\chi^{(2)}(2\omega)|$ and $|\chi^{(3)}(3\omega)|$ is, of course, applicable to other nonlinear susceptibilities as well.

Nonlinear susceptibilities of order n, $\chi^{(n)}$, are, in general, the sum of several terms each containing at the numerator the product of $n+1$ dipole matrix elements and at the denominator products of linear combinations involving the photon energies participating in the nonlinear interaction, the energy differences between the excited states and the initial state of the system (usually the ground state), and the linewidths of the transitions[18]. To optimize $|\chi^{(2)}(2\omega)|$ and $|\chi^{(3)}(3\omega)|$ in our quantum wells we need, therefore, to maximize the product of the dipole matrix elements of the transitions and minimize the energy denominators using resonant effects. The latter is achieved by structures that either by design or by application of an electric field have equally spaced energy levels (E_1 through E_3 for SHG and E_1 through E_4 for THG; the subscript 1 denotes the ground state). For the resonant case $\chi^{(2)}(2\omega)$ and $\chi^{(3)}(3\omega)$ are well approximated by the expressions[18]

$$(1) \qquad \chi^{(2)}(2\omega) = \frac{e^3}{\varepsilon_0} N \frac{\langle z_{12}\rangle\langle z_{23}\rangle\langle z_{31}\rangle}{(\hbar\omega - \Delta E_{12} - i\Gamma_{12})(2\hbar\omega - \Delta E_{13} - i\Gamma_{13})} ,$$

$$(2) \qquad \chi^{(3)}(3\omega) =$$

$$= \frac{e^4}{\varepsilon_0} N \frac{\langle z_{12}\rangle\langle z_{23}\rangle\langle z_{34}\rangle\langle z_{41}\rangle}{(\hbar\omega - \Delta E_{12} - i\Gamma_{12})(2\hbar\omega - \Delta E_{13} - i\Gamma_{13})(3\hbar\omega - \Delta E_{14} - i\Gamma_{14})} ,$$

where N is the electron density in the wells, ε_0 the permittivity of the vacuum, e the electron charge, $\Delta E_{ij} = E_j - E_i$ is the separation between subbands j and i of the conduction band quantum well, Γ_{ij} and $\langle z_{ij}\rangle$ are the half-width at half-maximum and the matrix element of the $i \to j$ intersubband transition, respectively. Equations (1) and (2) imply a Lorentzian broadening, $i.e.$ homogeneous, of the transitions. From eq. (1) and (2) it is clear that $\chi^{(2)}$ and $\chi^{(3)}$ exhibit a strong resonant enhancement when the energy levels are equally spaced and the pump photon energy $\hbar\omega$ equals the spacing. Note also that the intersubband matrix elements appearing in eq. (1) and (2) are typically, by design, in the 10 to 20 Å range leading to high oscillator strengths. These matrix elements are about three orders of magnitude greater than the ones found in molecules (a few picometers). These large matrix elements and the multiple-resonance effects are responsible for the extremely large nonlinearities in the infrared. The above matrix elements are somewhat larger than or comparable to (depending on the layer thicknesses, the bandgap of the bulk quantum-well material, etc.) those of interband transitions in the bulk constituents of the quantum wells, such as GaAs and $Ga_{0.47}In_{0.53}As$ [19]. However, in the latter case one cannot exploit resonance effects since the relevant single-photon ($\hbar\omega$), two-photon ($2\hbar\omega$) and three-photon ($3\hbar\omega$) transitions at $\hbar\omega \simeq 120$ meV are strongly detuned from the energy gaps between the various conduction band edges and the valence band edge [19].

Figures 1 a) and b) represent the conduction band diagram of one period of the coupled-well structures designed for SHG and THG at the pump photon energies of the CO_2 laser ((116–134) meV). The states are calculated by solving Schrödinger's equation using Bastard's envelope function approximation [20]. For the $Al_{0.48}In_{0.52}As/Ga_{0.47}In_{0.53}As$ parameters we used $\Delta E_c = 0.51$ eV for the barrier height, $m_e^*(\text{GaInAs}) = 0.043m_0$, $m_e^*(\text{AlInAs}) = 0.07m_0$ for the effective masses and $\gamma = 1.03 \cdot 10^{-18}$ m^2 for the nonparabolicity coefficients. Nonparabolicities were taken into account using the method of ref. [21]. Only the thicker well is assumed to be doped (uniformly) so that space charge effects on the band diagrams of fig. 1 can be neglected. Indicated are also the modulus squared of the wave function for each state. The asymmetry of the two-well structure is essential for SHG since in a symmetric well $\langle z_{31} \rangle = 0$ and, therefore, $\chi^{(2)} = 0$. The energy level separations $E_3 - E_2$ and $E_2 - E_1$ in fig. 1a) are made intention

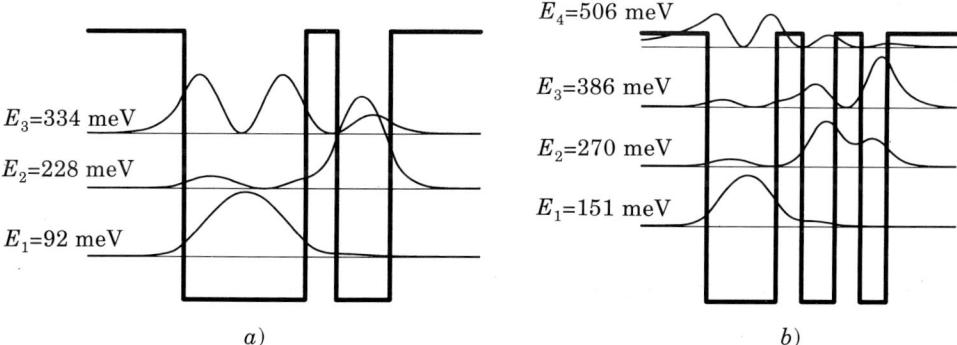

$E_4 = 506$ meV

$E_3 = 334$ meV $E_3 = 386$ meV

$E_2 = 228$ meV $E_2 = 270$ meV

$E_1 = 92$ meV $E_1 = 151$ meV

a)   b)

Fig. 1. – Conduction band energy diagrams of a single period of the AlInAs/GaInAs coupled-quantum-well nonlinear optical structures. Shown are the positions of the calculated energy subbands and the corresponding modulus squared of the wave functions. a) The GaInAs wells have thicknesses of 64 Å and 28 Å and are separated by a 16 Å AlInAs barrier; b) the GaInAs wells have thicknesses of 42, 20 and 18 Å, respectively, and are separated by 16 Å AlInAs barriers.

ally different so that an electric field can be used to achieve equal level spacing for maximum $|\chi_{(2\omega)}^{(2)}|$. For the two-well structure $\langle z_{12} \rangle = 15.4 \,\text{Å}$, $\langle z_{23} \rangle = 22.3 \,\text{Å}$ and $\langle z_{13} \rangle = 12.1 \,\text{Å}$. While $\langle z_{12} \rangle$ and $\langle z_{23} \rangle$ are sensitive to the thickness of the tunneling barrier due to the evanescent wave coupling between the two wells, $\langle z_{13} \rangle$ is not since states 1 and 3 reside essentially in the same well.

To design a structure with a triply resonant $|\chi_{(3\omega)}^{(3)}|$, a harmonic-oscillator potential cannot be used since $\chi^{(3)} = 0$, due to the linearity of electron oscillations. For compositionally graded parabolic or semi-parabolic wells of finite depth the deviation from the ideal harmonic-oscillator potential caused by the finite well depth and band nonparabolicities gives rise to $\langle z_{14} \rangle \neq 0$. Nevertheless, $\langle z_{14} \rangle$ is always much less than $\langle z_{12} \rangle, \langle z_{23} \rangle, \langle z_{34} \rangle$. For the AlInAs/GaInAs structure of fig. 1b) instead, all the relevant matrix elements are large: $\langle z_{41} \rangle = 8.6 \,\text{Å}$, $\langle z_{12} \rangle = 13 \,\text{Å}$, $\langle z_{23} \rangle = 22.7 \,\text{Å}$, $\langle z_{34} \rangle = 22.6 \,\text{Å}$. The calculated energy levels are $E_1 = 151 \,\text{meV}$, $E_2 = 270 \,\text{meV}$, $E_3 = 386 \,\text{meV}$, $E_4 = 506 \,\text{meV}$.

Our AlInAs/GaInAs structures, grown by MBE lattice matched to a semi-insulating (100) InP substrate, consists of 40 coupled-well periods (as in fig. 1a), b)) separated from each other by 150 Å undoped AlInAs barriers. Only the thickest wells are doped n-type with silicon in both structures ($3 \cdot 10^{17} \,\text{cm}^{-3}$ and $1 \cdot 10^{18} \,\text{cm}^{-3}$ for the two-well and three-well structures respectively). Undoped 100 Å GaInAs spacer layers separate the multiquantum well structure from n^+ 4000 Å thick GaInAs contact layers. The layer thicknesses were verified by transmission electron microscopy.

3. – Large linear Stark effect in coupled quantum wells.

In this section the absorption properties of intersubband transitions in coupled wells and the effect of a static electric field normal to the layers are examined.

The sample transmission was measured at room temperature using an infrared Fourier transform interferometer. Only the component normal to the layers of the electric field of the incident wave contributes to intersubband absorption, because the energy levels in fig. 1 correspond to electron motion normal to the layers. Thus, in order to increase the net absorption, we fabricated a multipass (six) waveguide by cleaving a bar and polishing both cleaved ends at 45° angle. One of these edges was then illuminated at normal incidence. The measured absorbance ($= -\log$ transmission) spectrum of the two-well and three-well structures are shown in fig. 2 and 3, respectively. The peaks at 137.2 meV and 238.2 meV in fig. 2 are due to the $1 \to 2$ and $1 \to 3$ transitions, respectively; their position is in excellent agreement with our calculations for the two-well structure (see fig. 1a)). For the peak related to the $1 \to 2$ transition (137.2 meV) one obtains from the data of fig. 2 $I_A = 9.5 \,\text{meV}$. From this value one finds $\langle z_{12} \rangle = 15.5 \,\text{Å}$ in good agreement with the calculated value [11]. From the

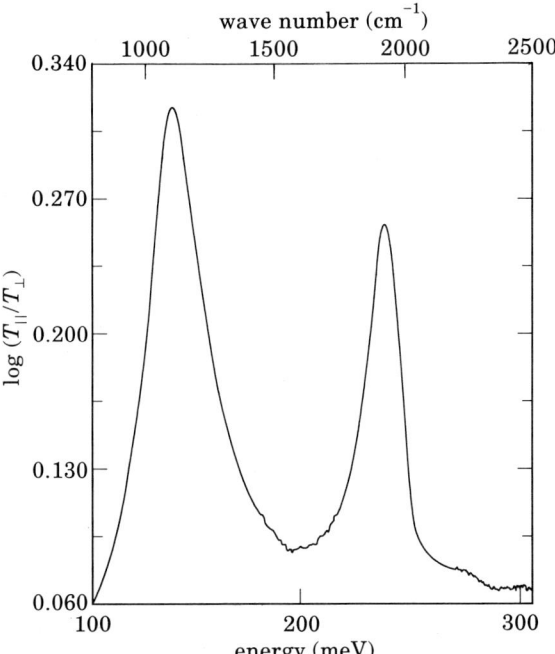

Fig. 2. – Measured absorbance of the two-coupled-well structure at room temperature.

$1 \rightarrow 3$ peak one finds from fig. 2 $I_A = 5$ meV. From the ratio of the integrated
absorption strength (I_A) for the two transitions one obtains $\langle z_{12} \rangle / \langle z_{13} \rangle = 1.38$.
Thus $\langle z_{13} \rangle = 11.2$ Å, in good agreement with the the poretical value. Note that
the $1 \rightarrow 3$ transition is forbidden in a symmetric quantum well. Thus by properly engineering the asymmetry we have been able to create a large dipole matrix
element.

In the absorbance spectrum of the three-coupled-well structure (fig. 3)
the positions of the peaks at 102 meV, 232 meV and 344 meV are related
to $1 \rightarrow 2$, $1 \rightarrow 3$ and $1 \rightarrow 4$ intersubband transitions, respectively, and are
in excellent agreement with the calculated values [16]. From the areas under
the absorbance peaks, fitted with a Lorentzian, and knowing the electron
sheet density in the well, one finds matrix elements in good agreement
with the calculated ones. The «shoulder» at 130 meV in fig. 3 is due to
the $2 \rightarrow 3$ transition. At 300 K the second level is populated by the thermal tail
of the Fermi distribution; in addition the matrix element of the $2 \rightarrow 3$ transition
is large ($\langle z_{23} \rangle = 23$ Å). At 30 K the «shoulder» disappears. At this temperature
the peaks shift to higher energy (by a few percent) and become significantly
narrower (by a factor $\simeq 1.6$).

The coupled-well structures of fig. 1 exhibit large linear Stark effects. Figure 4 shows the calculated intersubband separation energies as a function of

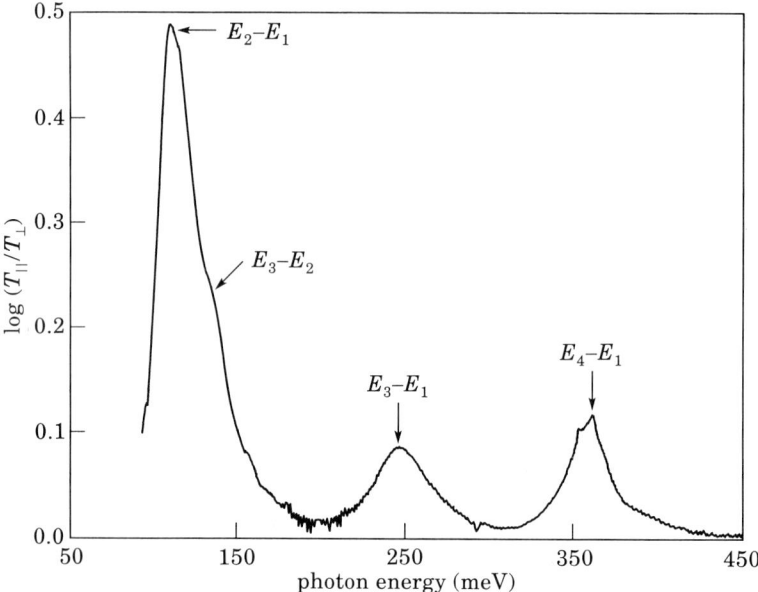

Fig. 3. – Measured absorbance of the three-coupled-well structure at room temperature, $T = 300$ K.

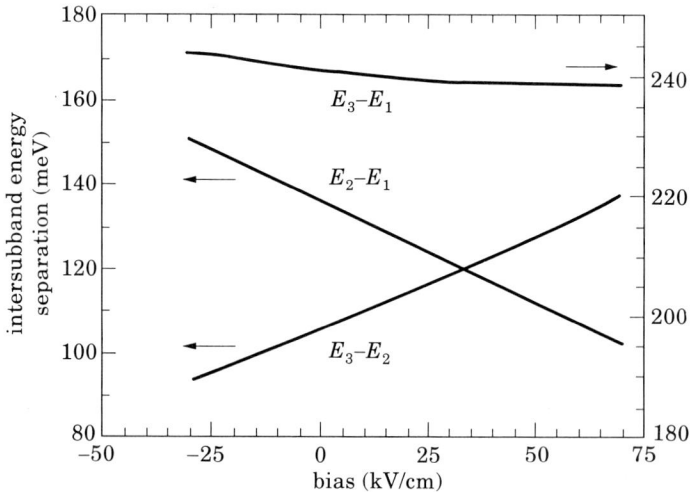

Fig. 4. – Calculated intersubband separation energies as a function of the electric field for the two-coupled-well structure. Positive-bias polarity corresponds to the thin well being lowered in energy with respect to the thick well.

electric field for the two-well structure. Positive field polarity corresponds to the thin well being lowered in energy with respect to the thick well. The inter-subband energy separation ΔE_{13} is found to be weakly dependent on the electric field, while the other two transitions (ΔE_{12} and ΔE_{23}) are strongly affected. Their shift with field is given, with excellent approximation, by the potential drop between the centers of the two wells. This behavior is physically under-stood by noting that, approximately speaking, the first and third states are con-fined by the thick well, while the second state is confined by the thin well (see fig. 1). It is well known that within a first-order approximation the energy of a confined state of a quantum well with respect to the well center is independent of the electric field as long as the potential drop across the well is small com-pared to the confined-state energy, $i.e.$ the linear Stark effect is zero and only a small quadratic effect exists. Thus, as the electric field is increased, the first and third energy levels track the center of the thick well, while the second en-ergy level tracks the center of the thin well. The net effect is that the energy of the $1 \rightarrow 3$ transition is weakly dependent on the electric field, while the en-ergies of the $1 \rightarrow 2$ and $2 \rightarrow 3$ transitions are shifted by an amount equal to the potential drop between the centers of the wells. Our calculations also show that the dipole matrix elements vary weakly (by a few percent) with the electric field in the range used in our experiments.

The Stark shifts were measured with a stabilized CO_2 laser as the source using the 45° wedge geometry of fig. 5. The linearly polarized laser beam, after entering the sample at normal incidence (on one of the polished edges) and traversing the multiquantum well region, is reflected off the top surface of the mesa and passes a second time through the multilayers. The absorption coeffi-cient was measured as a function of the applied electric field, at various laser wavelengths, in the temperature range (10-300) K. Some represenative data are shown in fig. 6a) at 30 K for the two-well and three-well structures. The Stark shifts are virtually independent of temperature since they depend primarily on the wave functions and energy levels, which are weakly dependent on tempera-ture in these structures. In fig. 6a) the peak of the absorption coefficient at a given incident photon energy $\hbar\omega$ corresponds to an electric field F such that $E_2 - E_1 = \hbar\omega$. Thus a plot of $\hbar\omega$ vs. F gives directly the Stark shift (fig. 6b)). We find 4.7 meV/10 kV/cm in very good agreement with the calculated Stark shift

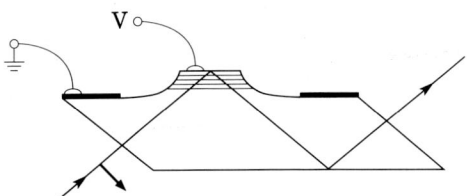

Fig. 5. – Cross-section of multipass sample geometry used in the experiments. Indicated are also the pump beam trajectory and its polarization.

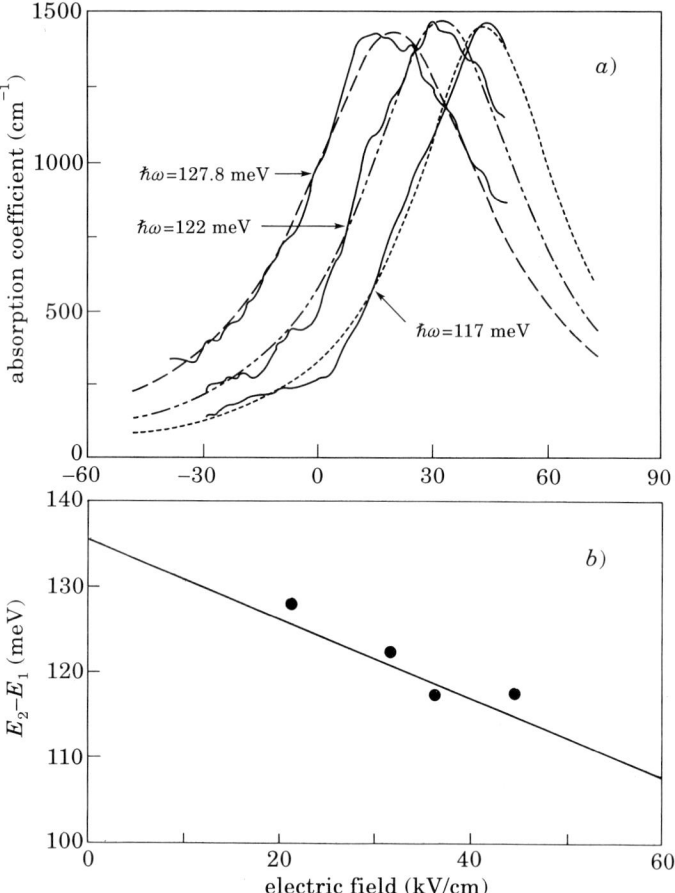

Fig. 6. – a) Measured absorption coefficient as a function of the applied electric field at various pump photon energies for the two-coupled-well structure. The dashed curves are Lorentzian fits to the data. b) Position of the measured absorption peak vs. electric field. Note the large linear Stark effect. The solid line is the calculated $E_2 - E_1$ transition energy taken from fig. 4. $T = 30$ K.

of the $1 \rightarrow 2$ transition for the two-well structure (solid line in fig. 6b)). Similar values of the Stark shift were obtained for the $1 \rightarrow 2$ transition of the three-well structure.

4. – Resonant Stark tuning of second-order susceptibility $\chi^{(2)}(2\omega)$.

For these experiments[12] we used the sample geometry of fig. 5. We adjusted the polarization of the pump beam so as to maximize its component normal to the layers, as shown in fig. 5. The measured second-harmonic power accurately follows the expected square law dependence on the pump power,

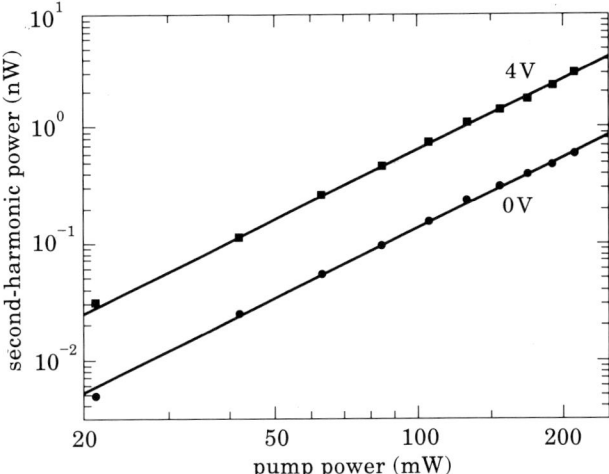

Fig. 7. – Second-harmonic power as a function of pump power at room temperature for different bias conditions for the two-coupled-well structure. The pump wavelength is $\lambda \simeq$ $\simeq 10.15 \, \mu m$. The solid lines represent least-square fits with a quadratic law. The bias polarity is the same as in fig. 6b).

shown in fig. 7 for two bias conditions. For this measurement the laser photon energy $\hbar\omega$ was adjusted to 122 meV. At this energy $2\hbar\omega$ is very close to the measured separation between the third and first energy levels ($E_3 - E_1 =$ $= 238.2$ meV, fig. 2).

The distance traveled by the pump and second-harmonic beams in each pass through the superlattice is much smaller than the coherence length l_c ($\simeq 120 \, \mu m$). The equation for the second-harmonic power $P(2\omega)$ can then well approximated by [11]

$$(3) \qquad P(2\omega) = 2\mu_0^{3/2}\,\varepsilon_0^{1/2}\,(\omega^2/n^3 S)\,|\chi^{(2)}(2\omega)|^2\sin^4\theta\,\mathrm{tg}^2\,\theta L^2\,T(\omega)[T(2\omega)]^{1/2}\,P^2(\omega),$$

where θ is the angle of incidence with respect to the normal to the plane of the layers, L is the thickness of the active region of the superlattice (i.e. the wells and thin coupling barrier) times the number of passes (two), S is the area of the laser spot on the sample (diameter $= 150 \, \mu m$), n is the refractive index ($n =$ $= n(\omega) \simeq n(2\omega)$), $T(\omega)$ and $T(2\omega)$ are the transmission of the CO_2 laser and the second-harmonic signal in traversing twice the coupled-well region, respectively, and $P(\omega)$ is the pump power. The second-harmonic power generated by the bulk and the interference term with intersubband second-harmonic generation are a negligible fraction of the measured second-harmonic power [11]. Note also that the detected fluorescence at 2ω, i.e. the intersubband spontaneous emission occurring when $2\hbar\omega = \Delta E_{13}$, can be shown to be completely negligible also in the most favorable conditions ($\Delta E_{12} = \Delta E_{23}$), i.e. for a bias $= +4$ V. This is because the radiative efficiency of the $3 \to 1$ transition is very small ($\lesssim 10^{-5}$)

since the spontaneous-emission lifetime is many orders of magnitude longer than the intersubband relaxation time and the emission is nondirectional. To verify the selection rules for SHG we use rotate the polarization of the pump beam. The signal is zero when the pump beam has no component of the electric field normal to the layers ($\phi = 0$). As the angle ϕ is increased, the second-harmonic power exhibits the $\sin^4 \phi$ dependence expected for intersubband transitions [11].

From the analysis of the data of fig. 7 and eq. (3) we find $|\chi^{(2)}_{2\omega}| = 3 \cdot 10^{-8}$ m/V at zero bias and $\chi^{(2)}_{2\omega} = 7.5 \cdot 10^{-8}$ m/V at 4.0 V, corresponding to a field $F \approx 3.8 \cdot 10^4$ V/cm. Figure 8 summarizes the results of our experiments at room temperature for various pump photon energies. The second-order susceptibility shows a pronounced peak for positive polarity at fields in the range $(3.5\text{--}4.5) \cdot 10^4$ V/cm.

The enhancement of $\chi^{(2)}_{2\omega}$ by an electric field of the appropriate polarity can be understood in terms of the Stark shifts (fig. 4). For the photon energies shown, $\hbar\omega < \Delta E_{12}$, so that a positive bias must be applied to achieve the resonance condition $\hbar\omega = \Delta E_{12}$ and a peak in $|\chi^{(2)}_{2\omega}|$ (see eq. (1)). The higher the $\hbar\omega$, the lower the field required to achieve this resonance condition and thus the peak of $|\chi^{(2)}_{2\omega}|$ will shift to lower bias; this trend is clearly manifested in the data of fig. 8. The maximum value of $|\chi^{(2)}_{2\omega}|$ will occur when $\Delta E_{12} = \Delta E_{23} = \hbar\omega$. Our calculations predict that this should occur at $\hbar\omega \approx 120$ meV corresponding to an electric field $\approx 3.2 \cdot 10^4$ V/cm (fig. 4), in good agreement with the experimental values (122.2 meV and $3.7 \cdot 10^4$ V/cm). From the curves at 122.2 and 120.4 meV, we obtain a shift of 4 meV for an electric-field increase of 10 kV/cm, in good

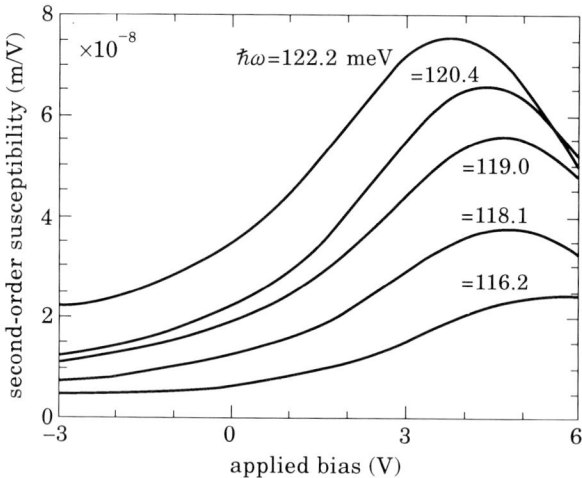

Fig. 8. – Measured second-order susceptibility $|\chi^{(2)}_{2\omega}|$ as a function of bias at different pump photon energies at room temperature. Positive-bias polarity corresponds to the thin well being lowered in energy with respect to the thick well.

agreement with the calculations (fig. 4). Using for Γ the experimental value of 15 meV obtained from the FTIR measurements (fig. 2) and the calculated values of the transition matrix elements ($\langle z_{12} \rangle = 16.1$ Å, ($\langle z_{23} \rangle = 20.4$ Å, ($\langle z_{31} \rangle = 12.4$ Å, at $F = 3.7 \cdot 10^4$ V/cm), one then finds from eq. (1) a maximum value $|\chi^{(2)}_{2\omega}| = 1 \cdot 10^{-7}$ m/V, in good agreement with the experimental data $(0.75 \cdot 10^{-7}$ m/V). The experimental error in the measured $|\chi^{(2)}(2\omega)|$ is estimated to be $\sim \pm 20\%$. Note that typical values for bulk GaAs, InAs and InP are $\sim 5 \cdot 10^{-10}$ m/V at $\lambda \approx 10$ μm pump [18].

5. – Triply resonant nonlinear susceptibility.

The experimental arrangement and sample geometry used to observe THG in the structure of fig. 1b) are identical to those used for the second-harmonic experiments, except that in the present case the CO_2 beam was more tightly focused on one of the 45° edges of the sample to increase the third-harmonic signal.

The third-harmonic power is expected to increase with the third power of the pump beam [18]. In addition, only the component normal to the layers of the electric field of the incident pump wave contributes to intersubband THG. Thus, if we denote with $\phi = 90°$ the polarization direction of the incident pump beam (fig. 5), which maximizes the component of the electric field normal to the layers (as shown in fig. 5), rotating the polarization of the pump will reduce the third-harmonic power according to $\sin^6 \phi$. The measured third-harmonic power (fig. 9 and 10) verifies the above dependence on pump power and polarization

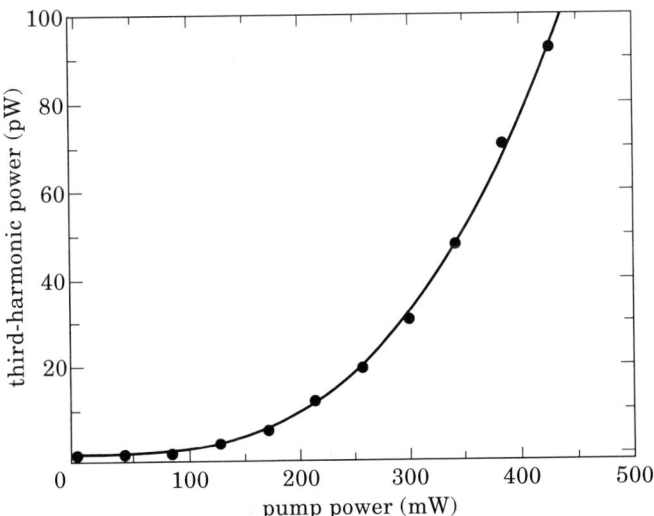

Fig. 9. – Third-harmonic power as a function of pump power $P(\omega)$ in the three-coupled-well structure; the solid line represents a cubic; $\hbar\omega = 118.3$ meV, $T = 300$ K.

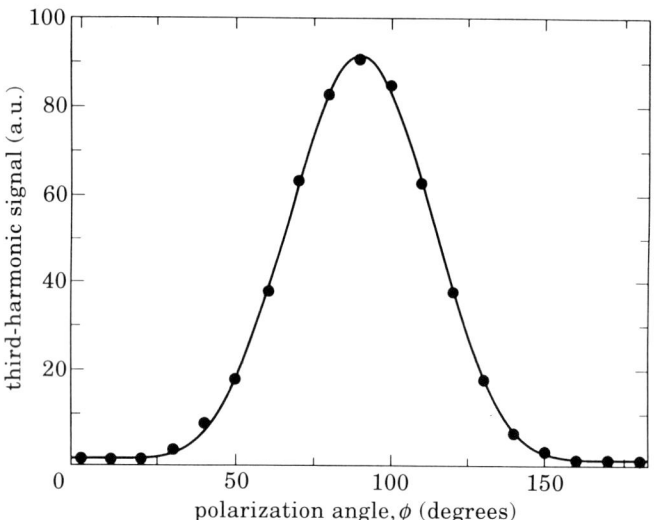

Fig. 10. – Third-harmonic power as a function of pump polarization. The data follow the expected $\sin^6 \phi$ dependence (solid line).

angle. Spontaneous emission at the frequency 3ω is completely negligible due to the extremely small radiative efficiency of the $4 \to 1$ transition.

To derive the expression for the third-harmonic power $P(3\omega)$ generated in our structure, we observe that the distance traveled by the pump and third-harmonic beam in each of the two passes through the superlattice is much smaller than the coherence length ($l_c = 20\ \mu m$ at $\lambda = 10.6\ \mu m$). The expression for $P(3\omega)$ can then be simply derived from Maxwell's equations including the absorption losses at ω and 3ω, for the case of phase matching and no pump depletion [16]

$$(4)\quad P(3\omega) = \frac{3}{4} \frac{\mu_0^2 \omega^2}{n_\omega^3 n_{3\omega}} \frac{\sin^8 \theta}{S^2} \exp\left[-\left(\frac{3}{2}\alpha_\omega + \frac{1}{2}\alpha_{3\omega}\right)L\right] \times$$

$$\times \left[\frac{\sinh(\alpha L/2)}{\alpha/2}\right]^2 |\chi_{3\omega}^{(3)}|^2 P^3(\omega),$$

where $P(\omega)$ is the pump power (i.e. the incident power minus the reflected power), S is the area of the laser spot (diameter $= 42\ \mu m$), n_ω and $n_{3\omega}$ are the refractive indices at ω and 3ω, α_ω and $\alpha_{3\omega}$ are the absorption coefficients for the pump and third-harmonic signal, respectively (obtained from the FTIR data), $\alpha = (3/2)\alpha_\omega - (1/2)\alpha_{3\omega}$, θ is the angle of incidence with respect to the normal to the plane of the layers ($45°$ in our case), L is the interaction length, which for our double-pass structure reduces to $2 \times$(total quantum well thickness plus that of the thin tunneling barriers betweehn the wells)$/\cos \theta$, and μ_0 is the vacuum permittivity. In deriving eq. (4) we have considered the case of a pump linearly polarized so that the component of the electric field normal to the layers is maximized. Since this component is proportional to $\sin\theta$, the contribution of the pump power to THG comes in as $(P(\omega)\sin^2\theta)^3$ in eq. (4). The additional $\sin^2\theta$ factor in eq. (4) is a result of the transformation from the crystal coordinate system to the laboratory frame used to describe the field propagation.

By best-fitting the data of fig. 9 and using eq. (4), one obtains $|\chi_{3\omega}^{(3)}| = 0.6 \cdot 10^{14} \, (\text{m/V})^2$.

The dependence of THG on pump wavelength was also investigated. The data are shown in fig. 11a). The power increases rapidly as the photon energy approaches 115 meV, corresponding to the resonance condition $3\hbar\omega = E_4 - E_1$. The range of photon energies is limited by the CO_2 laser. Note that peaks corresponding to $\hbar\omega = \Delta E_{21}$ and $2\hbar\omega = \Delta E_{31}$ cannot be resolved since the energy differences between ΔE_{12}, ΔE_{23} and ΔE_{34} are less than the broadening, as shown by fig. 3 and 5. From the data of fig. 11a) and substituting in eq. (4) the experimental values of all the quantities involved, one obtains $|\chi_{3\omega}^{(3)}|$ as a function of $\hbar\omega$ (fig. 11b)). Using the calculated $\langle z_{ij}\rangle$ in eq. (2) and the half-widths at half-maximum of the FTIR absorbance peaks ($\Gamma_{12} = 20$, $\Gamma_{13} = 16$ and $\Gamma_{14} = 18$ meV) as an estimate of the broadenings Γ_{ij} in eq. (2), one obtains $|\chi_{3\omega}^{(3)}| \simeq 1.3 \cdot 10^{-14} \, (\text{m/V})^2$ at $\hbar\omega = 115$ meV, which compares favorably with the experimental value, *i.e.* $0.9 \cdot 10^{-14} \, (\text{m/V})^2$ (fig. 11b)).

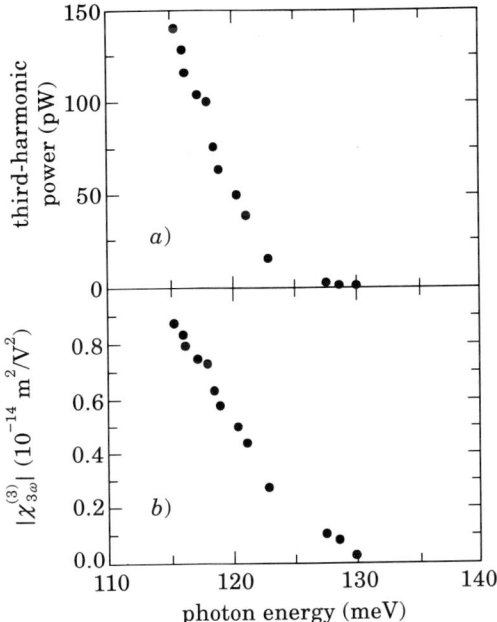

Fig. 11. – a) Third-harmonic power and b) third-order susceptibility $|\chi_{3\omega}^{(3)}|$ as a function of pump photon energy at room temperature. The pump power is 300 mW.

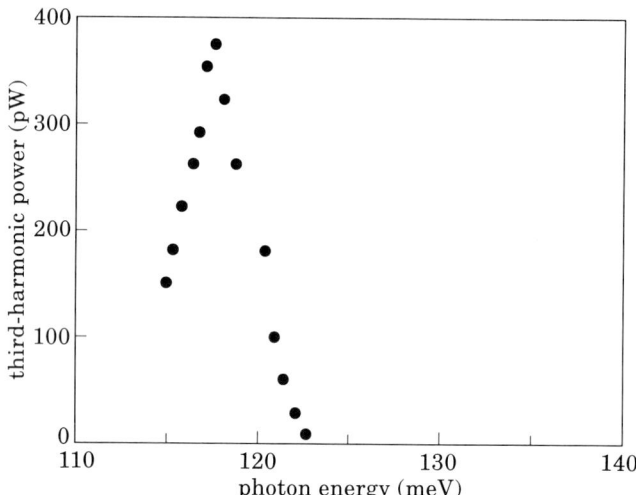

Fig. 12. – Third-harmonic power as a function of pump photon energy at cryogenic tem-
perature (30 K). The pump power is 300 mW; $|\chi^{(3)}_{3\omega}|_{max} = 4 \cdot 10^{-14}$ (m/V)^2.

The above experiment was repeated at 30 K. The measured third-harmonic
power as a function of pump photon energy is shown in fig. 12. Because of the
slight shift of the intersubband transitions to higher energies the peak of the
third-harmonic signal is now observable. It corresponds to a $|\chi^{(3)}_{3\omega}| =$
$= 4 \cdot 10^{-14}$ (m/V)^2. The enhancement is approximately a factor of four compared to
the room temperature and is due to the narrowing of the intersubband transi-
tions, since the peak of $|\chi^{(3)}_{3\omega}|$ scales as $1/(\Gamma)^3$. To our knowledge this is the
largest third-order nonlinear susceptibility reported in any material. It is 5 to 6
orders of magnitude greater than $|\chi^{(3)}_{3\omega}|$ associated with bound electrons in
InAs and GaAs[18], at comparable wavelengths.

WALROD et al.[22] have recently demonstrated intersubband nondegenerate
four-wave mixing in AlGaAs/GaAs quantum wells and GRAVÉ et al.[23] phase
conjugation associated with intersubband nondegenerate four-wave mixing.
These results provide additional evidence of the large enhancement of $\chi^{(3)}$ compared to bulk nonlinear susceptibilities.

6. – Multiphoton electron emission from quantum wells.

In the three-well structure previously discussed (two periods are shown in
fig. 13a)) electrons are promoted to a resonant state above the barriers via a
three-photon transition, giving rise to a photocurrent (fig. 13b)). This process is
conceptually similar to multiphoton ionization of an atom. Depending on the polarity of the applied bias, this effect can be enhanced by the presence of intermediate energy levels (fig. 13b)).

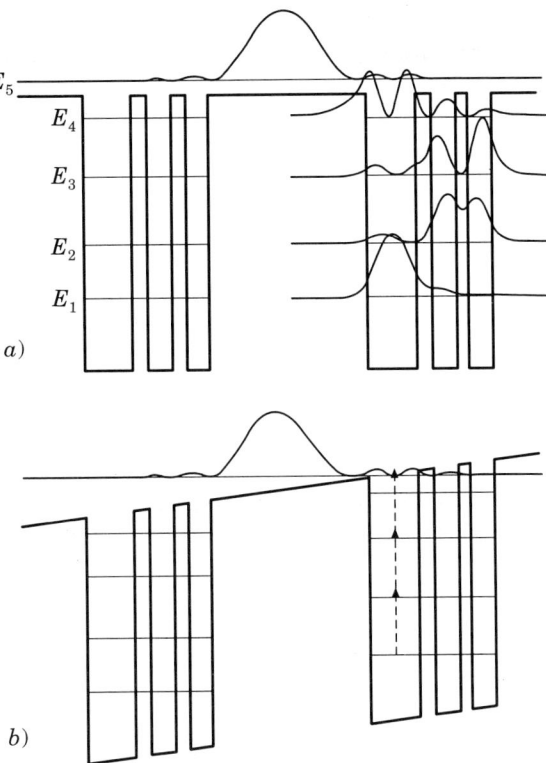

Fig. 13. – a) Energy band diagram at zero bias of two periods of the three-coupled-well structure. Shown are the positions of the calculated energy subbands and the corresponding modules squared of the wave functions. b) Energy band diagram with positive-bias polarity ($+3$ V). Note the Stark shift of the energy levels and the three-photon transition ($3\hbar\omega$ with $\hbar\omega = 130$ meV) into a resonant state localized in the barriers.

The experimental set-up is identical to that used in the resonant Stark tuning experiment (fig. 4), except that in this case the photocurrent is measured. The samples were cooled down to cryogenic temperatures in a Helitran flow dewar to minimize competing effects of electron thermionic emission out of the wells. Since our measurements are performed at relatively high fields ($> 10^4$ V/cm), the Stark effect plays an important role.

To observe the photocurrent generated by the three-photon bound-to-continuum transition in positive-bias polarity, the incident photon energy must be tuned so that $3\hbar\omega \gtrsim \Delta E_c - E_1$, where ΔE_c is the conduction band discontinuity, implying $\hbar\omega \gtrsim 125$ meV. The three-photon transition rate can be strongly increased by the intermediate energy levels E_2 and E_3. In order to exploit this resonant enhancement, we have made use of the Stark effect. At zero bias $\Delta E_{12} < \Delta E_{23}$; ΔE_{12} strongly increases with increasing positive bias, while our calculations show that ΔE_{23} is weakly dependent on the electric field. At $+3$ V and $\hbar\omega = 130$ meV the detunings from the intermediate energy levels are large enough to minimize the population of the second and third levels by absorption ($\hbar\omega - \Delta E_{12} = 13$ meV, $2\hbar\omega - \Delta E_{13} = 15$ meV), but sufficiently small to produce a resonant enhancement of the three-photon transition $1 \to 5$. Note that the final state of the latter is a resonant state above the barriers (fig. 13b)). This state is centered in the barrier layers and resembles a Fabry-Perot cavity resonance in optics, due to the sizable reflection coefficients between the barrier and the three-coupled-well regions. Note, however, that there is still a substantial dipole matrix element between this state and the third confined state of the well ($\langle z_{35} \rangle = 3.8$ Å at $+3$ V). States localized in the barrier layers have previously been theoretically investigated by JAROS and WONG in AlGaAs/GaAs multi-quantum wells [24].

The photocurrent generated by the three-photon transitions is expected to increase with the third power of the laser beam [18]. In addition, only the component normal to the layers of the electric field of the incident wave contributes to the photocurrent. Thus, if $\phi = 90$ represents the orientation of the polarization of the incident laser beam which maximizes the component of the electric field normal to the layers (corresponding to half of the laser power in our $45°$

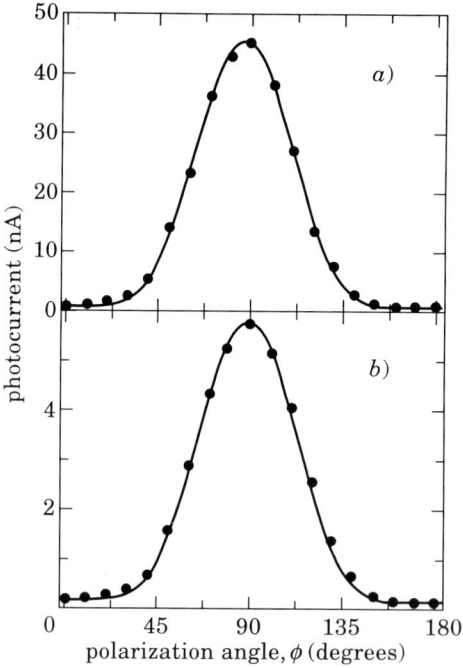

Fig. 14. – Photocurrent as a function of laser polarization for opposite-bias polarities at an incident power of 100 mW. The data follow closely the $\sin^6 \phi$ dependence (solid curve) expected for three-photon intersubband transition. a) Bias $= +3$ V, photon energy $= 130.1$ meV; b) bias $= -3$ V, photon energy $= 135.3$ meV.

waveguide geometry), rotating the polarization of the laser will reduce the photocurrent according to $\sin^6 \phi$, the same dependence observed for the third-harmonic power (fig. 14). The photocurrent measured under the optimum conditions of photon energy and bias described above demonstrates very clearly this angular dependence (fig. 14a)). The same dependence is found with opposite-bias polarity (fig. 14b)), but the signal significantly decreases. This is a result of the asymmetry of the structure; in this polarity the Stark effect increases the energy detunings from levels E_2 and E_3 ($\hbar\omega - \Delta E_{12} = 43$ meV, $2\hbar\omega - \Delta E_{13} = 46$ meV, at -3 V bias), thus decreasing the resonant enhancement of the $1 \rightarrow 4$ three-photon transition.

Figure 15 shows the photocurrent spectral response at $+3$ V bias. The data show that the photocurrent (for photon energies > 124 meV, such that the final state is above the barrier) first rises and reaches a maximum around $\hbar\omega \approx 130$ meV. As previously discussed, at this energy the three-photon transition is strongly enhanced by the intermediate states and is resonant with the state localized in the barrier. At higher energies increased detuning from resonance is responsible for the photocurrent decrease. In reverse polarity (-3 V), we found that the photocurrent increases for $\hbar\omega \gtrsim 119$ meV because at these photon energies the final state is above the barrier. The rise of the photocurrent is gradual and no peak is observed since the multiphoton transition is only slightly enhanced by intermediate states. From an analysis of the photocurrent we were able to deduce a value of the three-photon ionization cross-section $\sigma_3 =$

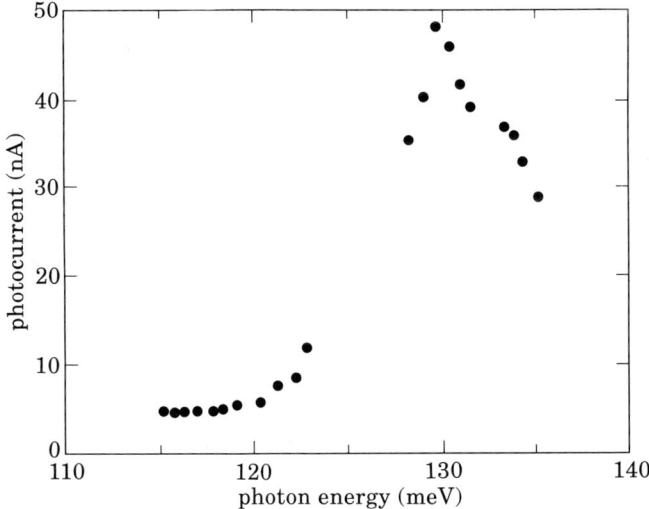

Fig. 15. – Photocurrent spectral response at 15 K for the three-coupled-well structure of fig. 13, obtained with the available transitions of a tunable CO_2 laser, for positive-bias polarity $+3$ V. The incident power (100 mW) is the same at all wavelengths. The peak is a manifestation of the resonant enhancement of the tree-photon electron emission out of the well (see fig. 13b)).

$= 10^{-67} \, \text{cm}^6 \, \text{s}^3$ many orders of magnitude larger than that found in atoms[25].

7. – Fabry-Perot electron filters and bound states in the continuum.

As we have seen in the previous section, the final state of the multiphoton transition in the three-well structure is a resonant state localized primarily in the barrier (fig. 13b)). In this section we discuss an optimum strategy for localization of electron resonant states above a quantum well[26]. We show that it is also possible to create bound states in this energy range (i.e. above the barrier height)[27], a problem dating back to the late twenties[28].

Consider first a conventional rectangular well (fig. 16a)). At energies greater than the barrier height one has a continuum of scattering states. For discrete energies corresponding to a semi-integer number of electron wavelengths across the well one finds transmission resonances. Although at these energies the electron amplitude in the well layer is enhanced, the wave functions do not decay exponentially in the barrier, unlike the confined states of the well, but are plane-wave-like. These states can be localized in the well using as barriers stacks of layers of thickness $\lambda/4$ each, where λ is the de Broglie wavelength in the layer (at the energy of the selected transmission resonance) (fig. 16b), c)). Constructive interference between the waves partially reflected by the heterointerfaces of the $\lambda/4$ stacks leads to the formation of a quasi-bound state above the center well (fig. 16b)). This strongly narrows the transmission resonance in analogy with a Fabry-Perot optical filter where sharp optical resonances are produced using as high-reflectivity mirrors quarter-wave stacks. The degree of localization increases with the number of periods due to the increased reflectivity of the $\lambda/4$ stacks; in the structure with just two-period stacks, the wave function is already highly confined (fig. 16b)). In the superlattice limit and at low temperatures the stacks become Bragg reflectors; a minigap opens up (fig. 16c)) and the localized state becomes a bound state at energies greater than the barrier height. The prediction that certain oscillatory potentials support bound states in the continuum, due to quantum interference, was first put forth by VON NEUMANN and WIGNER in 1929[28]. The formation of such states confined to the barriers of suitably engineered superlattices was proposed in 1977[29].

To design our structure we have used the envelope function approximation and the following conditions:

(5) $$k_\text{W} L_{\text{W, C}} = \pi \,,$$

(6) $$k_\text{W} L_\text{W} = \frac{\pi}{2} \,,$$

(7) $$k_\text{B} L_\text{B} = \frac{\pi}{2} \,,$$

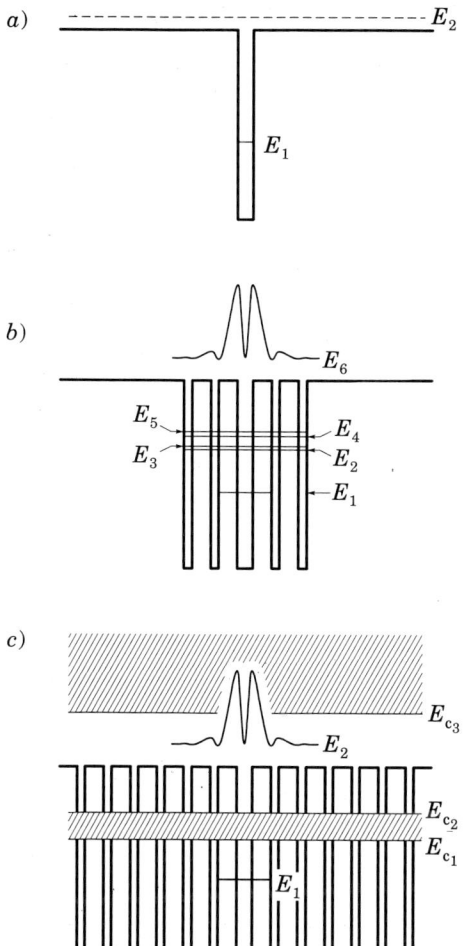

Fig. 16. – Conduction band diagrams of AlInAs/GaInAs heterostructures used in the study of localized continuum states: *a)* Reference sample. Shown are the ground state of the well ($E_1 = 204$ meV) and the position (dashed line) of the first transmission resonance in the continuum ($E_2 = 560$ meV). *b)* Quantum well cladded by two-period quarter-wave stacks (Fabry-Perot electronic filter). Shown is $|\psi|^2$ of the localized quasi-bound state ($E_6 = 560$ meV) formed in correspondence to the transmission resonance and the positions of new states created at lower energies. *c)* In the superlattice limit the $\lambda/4$ stacks behave as Bragg reflectors. The state above the well now becomes a bound state localized by the superlattice minigap ($= 266$ meV).

$L_{W,C}$ is the central-well thickness, L_W and L_B are the thicknesses of the wells and barriers in the cladding regions; k_W and k_B are the wave numbers in the well and barrier materials. Equation (5) represents the transmission resonance condition, while eqs. (6) and (7) are the $\lambda/4$ stack conditions. The starting point is a reference structure (fig. 16*a*)) consisting of a 32 Å Ga$_{0.47}$In$_{0.53}$As QW bound by

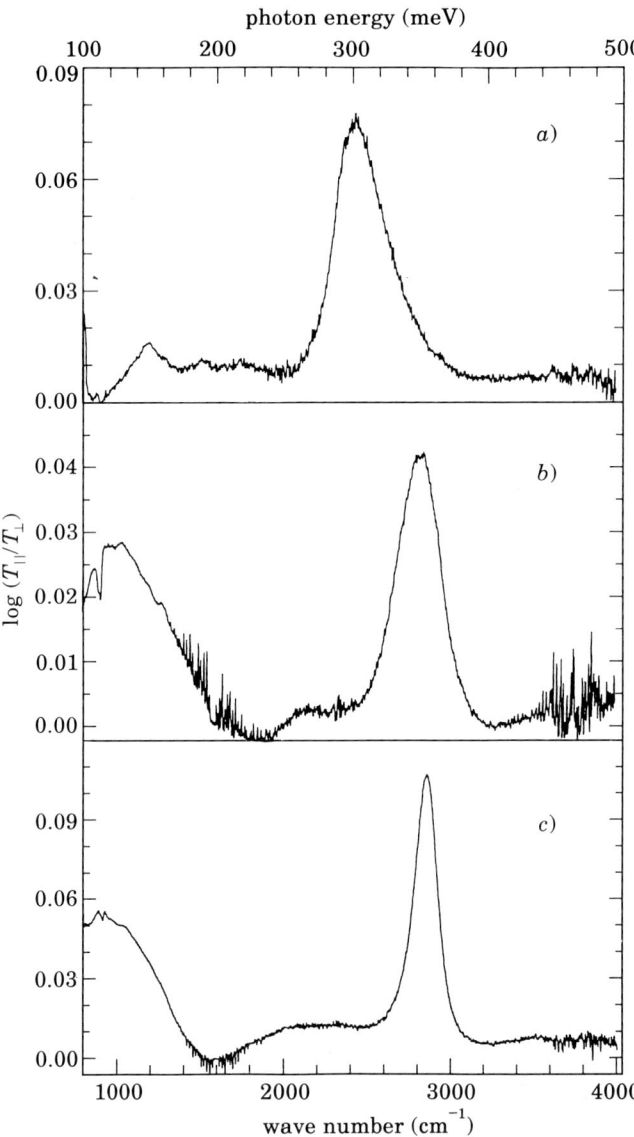

Fig. 17. – Absorption spectra at room temperature for the structures with quarter-wave stacks (fig. 9). The absorbance for polarization parallel to the plane of the layers ($-\log T_{\parallel}$) is subtracted from the absorbance normal to the plane of the layers ($-\log T_{\perp}$), to remove the contribution of free-carrier absorption in the buffer layers. The transition to the confined state above the well corresponds to the peaks at $\simeq 350\,\mathrm{meV}$. The bound-to-continuum transition in the sample without $\lambda/4$ stacks (reference) corresponds to the broad absorption peak at $300\,\mathrm{meV}$. a) Reference sample, b) one-period $\lambda/4$ stack, c) two-period $\lambda/4$ stack.

thick $Al_{0.48}In_{0.52}As$ barriers. The calculations find a single bound state at $E_1 = 204$ meV and continuum resonances. The first of these ($E_2 = 560$ meV) satisfies eq. (5) (fig. 16a)). AlInAs/GaInAs of varying number of periods (1, 2, 6 and infinite) are then introduced in the barrier layers. The thickness of the well and barrier layers in the stacks (16 and 39 Å, respectively) satisfy eqs. (6) and (5), respectively, with k_W and k_B calculated at the energy E_2.

The reference sample had twenty 32 Å InGaAs quantum wells n-type doped to $\simeq 1 \cdot 10^{18}$ cm^{-3}, separated by 150 Å undoped AlInAs barriers. In the other three structures the 32 Å wells, doped to the same level, were cladded, respectively, by 1, 2 and 6 undoped periods, each comprising 39 Å AlInAs barriers and 16 Å GaInAs wells, as described above. The phase coherence length is estimated to be ~ 300 Å at 10 K [30].

The absorption spectra of the reference structure are broad with a long-wavelength cut-off determined by the height of the barrier (fig. 17a)). In the structure with one $\lambda/4$ period the peak is considerably narrower and centered

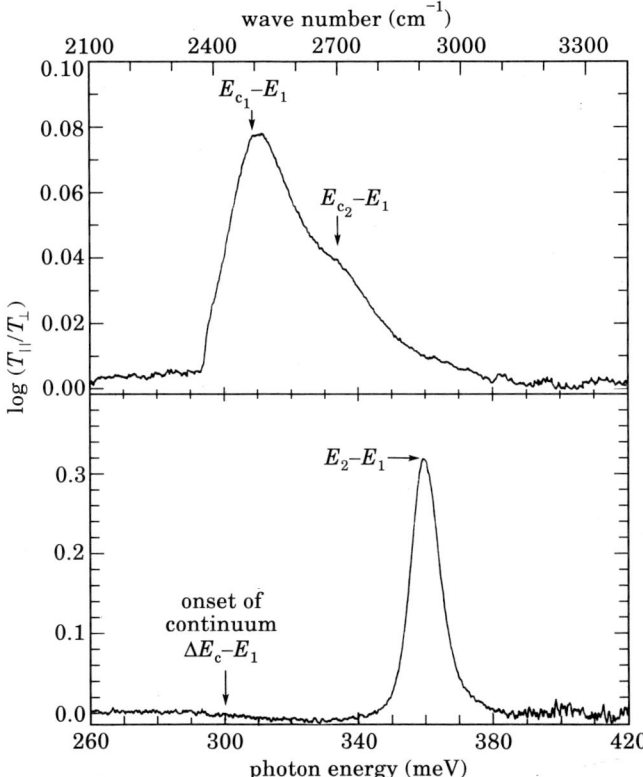

Fig. 18. – Absorption spectra of 10 K for the reference structure (top) and the structure with superlattice Bragg reflefctors of fig. 16c). The transition to the bound state above the well (E_2 in fig. 16c)) corresponds to the peak at 360 meV.

at an energy corresponding to the transition between the ground state of the well and the localized resonant state at the energy E_C (fig. 17b)). As the number of quarter-wave stacks is doubled, the absorption peak does not shift and considerably narrows precisely the behavior expected for a Fabry-Perot (fig. 17c)). In fact the observed narrowing (16 meV) can be quantitatively explained in terms of the reflectivity increase of the $\lambda/4$ stacks. In the structure with six periods at low temperatures, the highly localized state becomes effectively a bound state confined by Bragg reflectors from the superlattice (fig. 16c)) [27]. The absorption spectrum (fig. 18) shows an isolated peak at 360 meV of width ~ 10 meV corresponding to the transition from the state E_1 to the state E_2 in fig. 16b) [27]. It is worth noting that the width of the transition to the confined state above the well in the two- and six-period structure is practically identical to that of the bound-to-bound state transition measured in a conventional 55 Å thick GaInAs well with 300 Å thick barriers.

8. – Conclusions.

To summarize in this lecture we have presented the results of an extensive research program on the i.r. nonlinear optical properties of coupled-quantum-well semiconductors. By careful tailoring of layer thicknesses and compositions the nonlinear susceptibilities can be maximized so that the structures behave effectively as artificial molecules with giant $|\chi^{(2)}|$ and $|\chi^{(3)}|$. One of the most interesting aspects of these new systems is the extremely large linear Stark effect. This phenomenon has been used to electric-field-tune $\chi^{(2)}$ and $\chi^{(3)}$ and is of interest also for the realization of fast i.r. modulators with high on/off ratio. Finally the quantum well structures with quarter-wave stacks as barrier layers have exciting quantum-mechanical properties which might find applications in detectors and other optoelectronic devices.

* * *

We acknoweldge the collaboration of A. Y. CHO and D. L. SIVCO in the growth of the structures and the assistance of A. L. HUTCHINSON in sample preparation. We are grateful to S. N. G. CHU for the TEM measurements and to J. FAIST for collaborating on the experiments on quarter-wave stacks. Partial financial support from Consorzio Technobiochip, Marciana, Italy, is gratefully acknowledged.

REFERENCES

[1]   The concept of bandgap engineering was first introduced in F. CAPASSO: J. Vac. Sci. Technol. B, 1, 457 (1983). For a review, see F. CAPASSO: Science, 235, 172 (1987).
[2]   A. Y. CHO and J. R. ARTHUR: in Progress in Solid State Chemistry, Vol. 10, edited by J. O. McCALDIN and G. SOMORJAI (Pergamon, New York, N.Y., 1975), p. 157.

[3] M. K. GURNICK and T. A. DE TEMPLE: *IEEE J. Quantum Electron.*, QE-19, 719 (1983).
[4] M. M. FEJER, S. J. B. YOO, R. L. BYER, A. HARWIT and J. S. HARRIS: *Phys. Rev. Lett.*, 62, 1041 (1989).
[5] E. ROSENCHER and P. BOIS: *Phys. Rev. B*, 44, 11315 (1991).
[6] P. BOUCAUD, F. H. JULIEN, D. D. YANG, J. M. LOURTIOZ, E. ROSENCHER, P. BOIS and J. NAGLE: *Appl. Phys. Lett.*, 57, 215 (1990).
[7] P. BOUCAUD, F. H. JULIEN, D. D. YANG, J. M. LOURTIOZ, E. ROSENCHER and P. BOIS: *Opt. Lett.*, 16, 199 (1991).
[8] S. J. B. YOO, M. M. FEJER, R. L. BYER and J. S. HARRIS jr: *Appl. Phys. Lett.*, 58, 1724 (1991).
[9] E. ROSENCHER, P. BOIS, J. NAGLE, E. COSTARD and S. DELAITRE: *Appl. Phys. Lett.*, 55, 1597 (1989).
[10] E. ROSENCHER, P. BOIS, B. VINTER, J. NAGLE and D. KAPLAN: *Appl. Phys. Lett.*, 56, 1822 (1990).
[11] C. SIRTORI, F. CAPASSO, D. L. SIVCO, S. N. G. CHU and A. Y. CHO: *Appl. Phys. Lett.*, 59, 2302 (1991).
[12] C. SIRTORI, F. CAPASSO, D. L. SIVCO, A. L. HUTCHINSON and A. Y. CHO: *Appl. Phys. Lett.*, 60, 151 (1992).
[13] F. CAPASSO, C. SIRTORI and A. Y. CHO: *IEEE J. Quantum Electron.*, 30, 133 (1994).
[14] F. CAPASSO and C. SIRTORI: in *Quantum Electronics and Laser Science Conference*, OSA Technical Digest Series, Vol. 13 (Optical Society of America, Washington, D.C., 1992), p. 108.
[15] F. CAPASSO, C. SIRTORI, A. Y. CHO and D. L. SIVCO: in *International Conference on Quantum Electronics Technical Digest Series*, Vol. 9 (Optical Society of America, Washington, D.C., 1992). p. 60.
[16] C. SIRTORI, F. CAPASSO, D. L. SIVCO and A. Y. CHO: *Phys. Rev. Lett.*, 68, 1010 (1992).
[17] C. SIRTORI, F. CAPASSO, D. L. SIVCO and A. Y. CHO: *Appl. Phys. Lett.*, 60, 2678 (1992).
[18] Y. R. SHEN: *The Principles of Nonlinear Optics* (Wiley, New York, N.Y., 1984).
[19] M. G. BURT: *J. Phys. C*, 5, 4091 (1993).
[20] G. BASTARD: *Wave Mechanics Applied to Heterostructures* (Les Éditions de Physique, Paris, 1990).
[21] D. F. NELSON, R. C. MILLER and D. A. KLEINMAN: *Phys. Rev. B*, 35, 7770 (1987).
[22] D. WALROD, S. Y. AUYANG, P. A. WOLFF and M. SUGIMOTO: *Appl. Phys. Lett.*, 59, 2932 (1991).
[23] I. GRAVÉ, M. SEGEV and A. YARIV: *Appl. Phys. Lett.*, 60, 2717 (1992).
[24] M. JAROS and K. B. WONG: *J. Phys. C*, 17, L765 (1984).
[25] S. L. CHU and P. LAMBROPOULOS, Editors: *Multiphoton Ionization of Atoms* (Academic, Orlando, Fla., 1984).
[26] C. SIRTORI, F. CAPASSO, J. FAIST, D. L. SIVCO, S. N. G. CHU and A. Y. CHO: *Appl. Phys. Lett.*, 61, 888 (1992).
[27] F. CAPASSO, C. SIRTORI, J. FAIST, D. L. SIVCO, S. N. G. CHU and A. Y. CHO: *Nature (London)*, 358, 565 (1991).
[28] J. VON NEUMANN and E. WIGNER: *Phys. Z.*, 30, 465 (1929).
[29] F. STILLINGER: *Physica B*, 85, 270 (1977).
[30] F. BELTRAM, F. CAPASSO, D. L. SIVCO, A. L. HUTCHINSON, S. N. G. CHU and A. Y. CHO: *Phys. Rev. Lett.*, 64, 3167 (1990).

Nonlinearities of Conjugated Polymers and Dye Systems.

C. BUBECK

Max-Planck-Institute for Polymer Research
Postfach 3148, D-55021 Mainz, Germany

1. – Introduction.

This lecture shall provide an introduction to nonlinear optical phenomena in polymers and dyes. These materials can have alternating sequences of single and multiple bonds composed of π-electrons. They can be strongly delocalized and form a conjugated system. Such electron systems are easily polarizable in strong electric fields. This can lead to large nonlinearities. The efforts to understand these phenomena are driven by two major reasons. First, it is a question of basic science to study the behaviour of molecular systems at high light intensities in order to understand, for example, multiple photon processes, saturable or induced absorptions, ultrafast molecular excitations, interactions and relaxation processes. Nonlinear optical spectroscopy can give additional information about these processes which is not possible by means of linear spectroscopy. The second reason for this rapidly expanding field is a gold-rush-type hope to find materials which could meet the requirements for potential applications in photonics [1, 2]. A large intensity-dependent refractive index might be used to realize a pure optical information processing, optical switches, phase conjugation [3] or other key techniques.

Advantages of organic materials as compared to inorganic materials are, among others, the considerably larger structural variability, the relative ease to prepare thin films and the fast electronic relaxation processes. Although the initial hope for a quick success is not realized, long-term basic research is necessary—first to understand the physical processes which can yield large nonlinearities in organic solids, second to recognize limitations for certain classes of materials, and third to get ideas how to overcome such limitations with improved materials.

This lecture is written to provide a tutorial introduction to the typical questions and problems concerning nonlinearities of conjugated polymers. To illustrate their outstanding properties, a comparison will be made with some dyes which also have delocalized electron systems, however with well-defined extensions. Because of symmetry reasons, which will be explained in sect. 2, these materials usually do not show second-order nonlinearities. Therefore, this lecture is restricted mostly to third-order nonlinear optical effects.

2. – Basic phenomena and conventions.

2'1. *Nonlinear optical susceptibilities.* – Some basic relations and definitions of nonlinear optics are now shortly introduced. More detailed information can be found in textbooks [1, 3-15]. There are many problems related to an inconsistent use of basic definitions in the nonlinear-optics literature. They are addressed in the excellent book of Butcher and Cotter [11]. Some of them will be pointed out here.

There are two possibilities to write the well-known power series expansion of the macroscopic polarization P in terms of the electric fields E (in SI units, ε_0 is the dielectric permittivity of free space):

$$(1a) \qquad P = \varepsilon_0 (\chi^{(1)} E + \chi^{(2)} E^2 + \chi^{(3)} E^3 + ...),$$

$$(1b) \qquad P = \varepsilon_0 (\chi^{(1)} E + K^{(2)} \chi^{(2)} E^2 + K^{(3)} \chi^{(3)} E^3 + ...).$$

Here $\chi^{(n)}$ are the macroscopic linear and nonlinear optical susceptibilities of order n, respectively. Equations (1a) and (1b) differ by numerical factors $K^{(n)}$. They are related to the kind of the nonlinear optical process and to the number of distinguishable permutations of frequencies [11]. Only for simplicity reasons the frequency dependences of the various quantities and the tensor notations of $\chi^{(n)}$ have been omitted. A survey on some nonlinear optical processes, the related susceptibilities and their K factors is given in table I.

Now we discuss the consequences of the different forms of eqs. (1a) and (1b). If the numerical factors $K^{(n)}$ are not written explicitly in the power series expansion (1a), these factors are included in the definition of $\chi^{(n)}$. Therefore, the susceptibilities measured by means of different processes can differ significantly! Only if the $K^{(n)}$ factors are written explicitly as in (1b), the definition of $\chi^{(n)}$ is *not* dependent on the process.

Mostly the susceptibilities $\chi^{(n)}$ are measured relative to reference materials such as quartz, CS_2 or others. Often it is not quite clear which definition was used in the early absolute measurements of these materials. Unless new improved and generally accepted reference data are available, it is, therefore, absolutely necessary to specify which $\chi^{(n)}$ value was used as reference. This allows later corrections with respect to better data of reference materials. Otherwise the $\chi^{(n)}$ data of different laboratories cannot be compared.

TABLE I. – *Selection of some nonlinear optical processes, common abbreviations, frequency arguments of $\chi^{(n)}$ and their numerical factors K (from [11]).*

Process	Susceptibility	K
second-harmonic generation (SHG)	$\chi^{(2)}(-2\omega; \omega, \omega)$	1/2
electric-field-induced second-harmonic generation (EFISH)	$\chi^{(3)}(-2\omega; \omega, \omega, 0)$	3/2
third-harmonic generation (THG)	$\chi^{(3)}(-3\omega; \omega, \omega, \omega)$	1/4
general four-wave mixing	$\chi^{(3)}(-\omega_4; \omega_1, \omega_2, \omega_3)$	3/2
degenerate four-wave mixing (DFWM)	$\chi^{(3)}(-\omega; \omega, \omega, -\omega)$	3/4

The susceptibilities shown in table I are written in the common notation. In the most general case the incident optical waves have different frequencies ω_1, ω_2, ω_3 and wave vectors $\mathbf{k}_1$, $\mathbf{k}_2$ and $\mathbf{k}_3$. Energy conservation yields the new frequencies ω after the nonlinear interaction: $\omega = \omega_1 + \omega_2$ or $\omega = \omega_1 + \omega_2 + \omega_3$, respectively. The convention to write the frequencies with positive and negative signs is commonly used to indicate the type of electronic interaction and the conservation of momentum of the wave vectors. The frequency before the semi-colon is written with a negative sign. It symbolizes that the generated wave with this frequency can leave the system after the nonlinear optical process.

The conversion between e.s.u. and SI units needs some attention. In the SI system the susceptibilities $\chi^{(n)}$ have the units $(m/V)^{n-1}$. The relations between the susceptibilities are [11]

$$(2) \qquad \chi^{(n)}[\mathrm{SI}]/\chi^{(n)}[\mathrm{e.s.u.}] = 4\pi/(10^{-4}c)^{n-1},$$

with $c = 3 \cdot 10^8$. For example, the relation for $\chi^{(3)}$ data is

$$(3) \qquad \chi^{(3)}[\mathrm{m}^2/\mathrm{V}^2] = 1.4 \cdot 10^{-8}\chi^{(3)}[\mathrm{e.s.u.}].$$

2'2. *Intensity-dependent refractive index.* – Several nonlinear optical effects can be derived easily if we insert the equation $E = E_0 \cos(\omega t - kz)$ for a plane wave that travels in the z-direction and has a frequency ω and wave vector $k = |\mathbf{k}| = 2\pi/\lambda$ (λ = wavelength) into eq. (1b). By use of appropriate trigonometric identities we obtain

$$(4) \qquad \frac{1}{\varepsilon_0}P = \chi^{(1)}E_0\cos(\omega t - kz) + K^{(2)}\chi^{(2)}E_0^2\,\frac{1}{2}[1 + \cos(2\omega t - 2kz)] +$$

$$+ K^{(3)}\chi^{(3)}E_0^3\left[\frac{3}{4}\cos(\omega t - kz) + \frac{1}{4}\cos(3\omega t - 3kz)\right].$$

There are contributions to P with new frequencies 2ω and 3ω, respectively.

They correspond to second- and third-harmonic generation. In addition there is a frequency-independent contribution which is referred to as optical rectification. The contribution $K^{(3)}\chi^{(3)}E_0^3(3/4)\cos(\omega t - kz)$ leads to an intensity-dependent index of refraction $n(I)$. This is a most important consequence of $\chi^{(3)}$ phenomena. It can be derived from eq. (4) if we select only contributions to P which oscillate at the fundamental frequency ω:

$$(5) \qquad \frac{1}{\varepsilon_0} P(\omega) = \left[\chi^{(1)} + \frac{3}{4} K^{(3)}\chi^{(3)}E_0^2\right]E_0 \cos(\omega t - kz).$$

The terms in brackets are treated as an effective susceptibility χ_{eff}. The refractive index n, the relative dielectric constant ε and χ_{eff} are related as

$$(6) \qquad n^2 = \varepsilon = 1 + \chi_{\text{eff}}.$$

If we introduce the linear refractive index n_0 into eq. (5) via $\chi^{(1)} = n_0^2 - 1$, we find

$$(7) \qquad n^2 = n_0^2 + \frac{3}{4} K^{(3)}\chi^{(3)}E_0^2.$$

Approximately the intensity-dependent refractive index n does not differ much from n_0. We use $n^2 - n_0^2 \simeq 2n_0(n - n_0)$ and obtain

$$(8) \qquad n = n_0 + \frac{3}{8n_0} K^{(3)}\chi^{(3)}E_0^2.$$

If the intensity $I = 0.5\ \varepsilon_0 c n_0 E_0^2$ is introduced in eq. (8), we have

$$(9) \qquad n = n_0 + n_2 I$$

in the commonly known form. Since the intensity is usually expressed in units of W/m^2, the relation between n_2 and $\chi^{(3)}$ is[11]

$$(10) \qquad n_2[m^2/W] = 5.26 \cdot 10^{-6} K^{(3)} n_0^{-2} \chi^{(3)}(-\omega; \omega, \omega, -\omega)[\text{e.s.u}].$$

Here $K^{(3)} = 3/4$ has to be inserted, if $\chi^{(3)}$ is defined as in eq. (1b). However, if $\chi^{(3)}$ contains the numerical factor implicitly as in eq. (1a), $K^{(3)} = 1$ must be used. In this case great care is advised if $\chi^{(3)}$ is measured by means of a technique other than DFWM to keep track of the different K factors.

2˙3. *Microscopic polarizabilities and resonances.* – The macroscopic optical susceptibilities $\chi^{(n)}$ are related to the microscopic optical polarizabilities $\alpha^{(1)}, \beta^{(2)}$ and $\gamma^{(3)}$ via

$$(11a) \qquad \chi^{(1)}(-\omega; \omega) = Nf(\omega)\alpha^{(1)}(-\omega; \omega),$$

$$(11b) \qquad \chi^{(2)}(-\omega; \omega_1, \omega_2) = Nf(\omega)f(\omega_1)f(\omega_2)\beta^{(2)}(-\omega; \omega_1, \omega_2),$$

$$(11c) \qquad \chi^{(3)}(-\omega; \omega_1, \omega_2, \omega_3) = Nf(\omega)f(\omega_1)f(\omega_2)f(\omega_3)\gamma^{(3)}(-\omega; \omega_1, \omega_2, \omega_3),$$

where N is the number of molecules per unit volume and $f(\omega_i)$ are dimensionless local-field factors. The indications of order are often omitted or used inconsistently. The microscopic polarizabilities $\beta^{(2)}$ and $\gamma^{(3)}$ are sometimes called «first» and «second» «hyperpolarizabilities», respectively. This leads sometimes to confusion, because $\beta^{(2)}$ and $\gamma^{(3)}$ are related to second- and third-order phenomena as expressed in the generally accepted forms of $\chi^{(2)}$ and $\chi^{(3)}$.

In principle it is possible to calculate the microscopic polarizabilities $\alpha^{(1)}$, $\beta^{(2)}$ and $\gamma^{(3)}$ by means of a summation over states $G, A, B, C, \ldots$ of a molecular system using time-dependent perturbation theory and the density matrix formalism. Following basic work of Bloembergen[16] and others[17,18] the simplified relation for $\gamma^{(3)}$ is

(12) $\gamma^{(3)}(-\omega; \omega_1, \omega_2, \omega_3) =$

$$= P \sum_{A, B, C} \frac{\langle G|\mu|A\rangle\langle A|\mu|B\rangle\langle B|\mu|C\rangle\langle C|\mu|G\rangle}{4\hbar^3(\omega_1 - \Omega_{AG} - i\Gamma_{AG})(\omega_1 + \omega_2 - \Omega_{BG} - i\Gamma_{BG})(\omega - \Omega_{CG} - i\Gamma_{CG})}.$$

Here μ is the operator for the dipole transitions between the quantum states. The symbol P indicates that a summation over all permutations of states and frequencies has to be performed. This would lead to 48 terms. Here only one term is displayed, which dominates in case of resonances. This happens if either one, a combination of two or three incident field frequencies coincide with internal transition frequencies Ω of the system. Therefore, these resonances are called 1-, 2- or 3-photon resonances. At resonance the polarizabilities are complex due to the remaining damping coefficients Γ. To illustrate eq. (12) we consider a three-photon resonance in THG. The resonance denominator of $\chi^{(3)}$ is

(13) $\chi^{(3)}(-3\omega; \omega, \omega, \omega) \sim$

$$\sim [(\omega - \Omega_{AG} - i\Gamma_{AG})(2\omega - \Omega_{BG} - i\Gamma_{BG})(3\omega - \Omega_{CG} - i\Gamma_{CG})]^{-1}.$$

If $3\omega = \Omega_{CG}$, the last term dominates and $\chi^{(3)}$ is imaginary due to the damping term $i\Gamma_{CG}$. Here we see that in resonance $\chi^{(3)}$ is complex. It can be written in the form

(14) $\chi^{(3)} = |\chi^{(3)}| \exp[i\Phi] = |\chi^{(3)}|(\cos\Phi + i\sin\Phi),$

where Φ is called the phase angle. At a three-photon resonance $\Phi = 90°$ follows from relation (13), if the first and second terms are real because they are not in resonance.

2‘4. *Influences of centrosymmetry.* – The structural requirements for even- and odd-order nonlinear optical effects can be easily derived from the power expansion (1). In case of centrosymmetry or a random orientation distribution, the polarization $\boldsymbol{P}$ of a system must be inverted, if the electric field $\boldsymbol{E}$ is reversed

(15) $$\boldsymbol{P}(-\boldsymbol{E}) = -\boldsymbol{P}(\boldsymbol{E}).$$

From this condition it would follow for the second-order polarizability in case of centrosymmetry

$$\boldsymbol{P}^{(2)}(-\boldsymbol{E}) \sim \chi^{(2)}(-\boldsymbol{E})^2 = +\chi^{(2)}\boldsymbol{E}^2 \, ,$$

which contradicts requirement (15) that $\boldsymbol{P}$ should change its sign. This discrepancy can only be solved if $\chi^{(2)}$ vanishes in media with centrosymmetry or random orientations.

Similarly it can be seen that third-order effects are compatible with centrosymmetry or random orientations because

$$\boldsymbol{P}^{(3)}(-\boldsymbol{E}) \sim \chi^{(3)}(-\boldsymbol{E})^3 = -\chi^{(3)}\boldsymbol{E}^3$$

fulfills condition (15). Therefore, it follows for molecules with point symmetry or centrosymmetric crystallographic packing that $\chi^{(2)} \equiv 0$, but nonlinear optical effects based on $\chi^{(3)}$ are possible.

3. – Third-harmonic generation and spectroscopy.

3˙1. *Measurement and evaluation techniques.* – Third-harmonic generation (THG) is a frequently used technique to measure $\chi^{(3)}(-3\omega; \omega, \omega, \omega)$ of molecules or polymers. If performed properly, this technique is rather accurate and sensitive even to small $\chi^{(3)}$ values. Only pure electronic effects have influence on THG. Therefore, thermal or stray-light effects, which can be troublesome in measurements based on the intensity-dependent refractive index, do not contribute to $\chi^{(3)}(-3\omega; \omega, \omega, \omega)$. Furthermore THG can be used as a spectroscopic technique, if the fundamental laser frequency ω is varied. Because of two- and three-photon resonances, THG can give additional information about electronic states not accessible with linear optical spectroscopy.

A detailed introduction to the principles of THG can be found in reviews [19-21] and will not be repeated here. To really understand the propagation of light waves in nonlinear media, the basic works [22-24] should be consulted. In many experiments the harmonic generation in thin films on solid substrates is studied by means of the so-called Maker fringe technique, which goes back to work of Maker *et al.* [25]. An example is shown in fig. 1. The following parameters must be known for the calculation of THG of thin films on substrates: thicknesses of film and substrate as well as their linear absorption coefficients α and refractive indices n at the fundamental and harmonic frequencies, respectively. The dependence of the light intensity $I_{3\omega}$ on the angle of incidence can be calculated

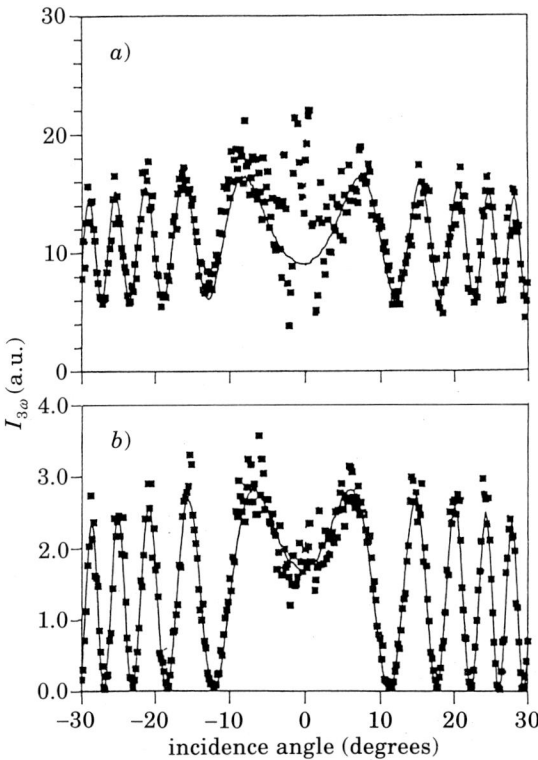

Fig. 1. – Harmonic light intensity as a function of the incidence angle for a) an ultrathin film of a polyphenylacetylene ($d \approx 45$ nm) at the front side of the substrate (fused silica, $d = 1$ mm) and b) for the substrate alone. Both figures are scaled to the same fundamental intensity. The experimental points are shown together with the calculated Maker fringes (from [26]).

only if all fundamental and harmonic waves and their reflections at interfaces are correctly taken into account. Otherwise it cannot be understood that the film placed at the front side or the back side of the substrate with respect to the incident laser beam leads to different Maker fringes [27]. Modulus and phase of $\chi^{(3)}(-3\omega; \omega, \omega, \omega)$ are the only fit parameters used in the calculation of the Maker fringes in fig. 1. The small additional oscillations at incidence angles $< 5°$ are not fitted, because they result from multiple reflections and do not yield additional information.

Although the relative experimental error can be made rather small with this procedure, a systematic error of $\chi^{(3)}$ is still possible, because $\chi^{(3)}$ of the film is evaluated relative to $\chi^{(3)}$ of the fused-silica substrate with a reference value $\chi^{(3)} = 3.11 \cdot 10^{-14}$ e.s.u. at the laser wavelength $\lambda_L = 1064$ nm [24]. This value goes back to earlier reports of Meredith and co-workers at other laser wavelengths [28, 29].

3'2. THG *spectroscopy and energy states in conjugated polymers.* – Considerable resonance enhancements of $\chi^{(3)}(-3\omega; \omega, \omega, \omega)$ can be seen, if the fundamental laser frequency ω is varied. An example with trans-polyacetylene (PA) is shown in fig. 2. Here ω is expressed as the energy E_L which corresponds to λ_L. The major maximum at $E_L \approx 0.6$ eV is attributed to a three-photon resonance with the absorption maximum of PA at $E_{max} \approx 1.8$ eV. The additional maximum in the THG spectrum observed at $E_L \approx 0.9$ eV has no corresponding feature at 2.7 eV in the linear absorption spectrum. This situation seems to be rather typical for conjugated polymers: a strong maximum of $\chi^{(3)}(-3\omega; \omega, \omega, \omega)$ at $3E_L \approx E_{max}$, accompanied by a weaker maximum or shoulder at higher energies with $3E_L > E_{max}$. This is observed similarly in polydiacetylenes (PDA)[30,31], polythiophene (PT)[32] and poly(p-phenylene vinylene) (PPV)[33]. The chemical structures of these polymers are shown in fig. 3.

The understanding of the THG spectra of conjugated polymers is related to the more basic question: what is the physical nature of the ground and excitation states in conjugated polymers? This question is intensively investigated and best understood in PDA single crystals, which serve as a model for this class of materials. Many arguments (see, for example,[34,35]) strongly support a one-dimensional semiconductor model as sketched in fig. 4. It consists of a valence band (VB), a conduction band (CB) and an exciton state (EX) located below the CB. For very general reasons[36], the one-dimensionality leads to a shift of oscillator strength from the VB-CB transition to the VB-EX transition. Therefore, the conduction band edge is usually not visible in the linear absorption

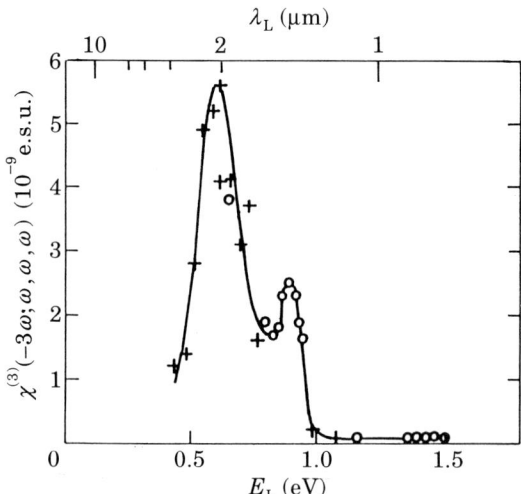

Fig. 2. – Resonance enhancement of $\chi^{(3)}(-3, \omega; \omega, \omega, \omega)$ for trans-polyacetylene, if the fundamental laser wavelength λ_L is varied[37]. The experimental data $(+)$ are from[38] and $(\circ)$ from[39]. The solid line is to guide the eye.

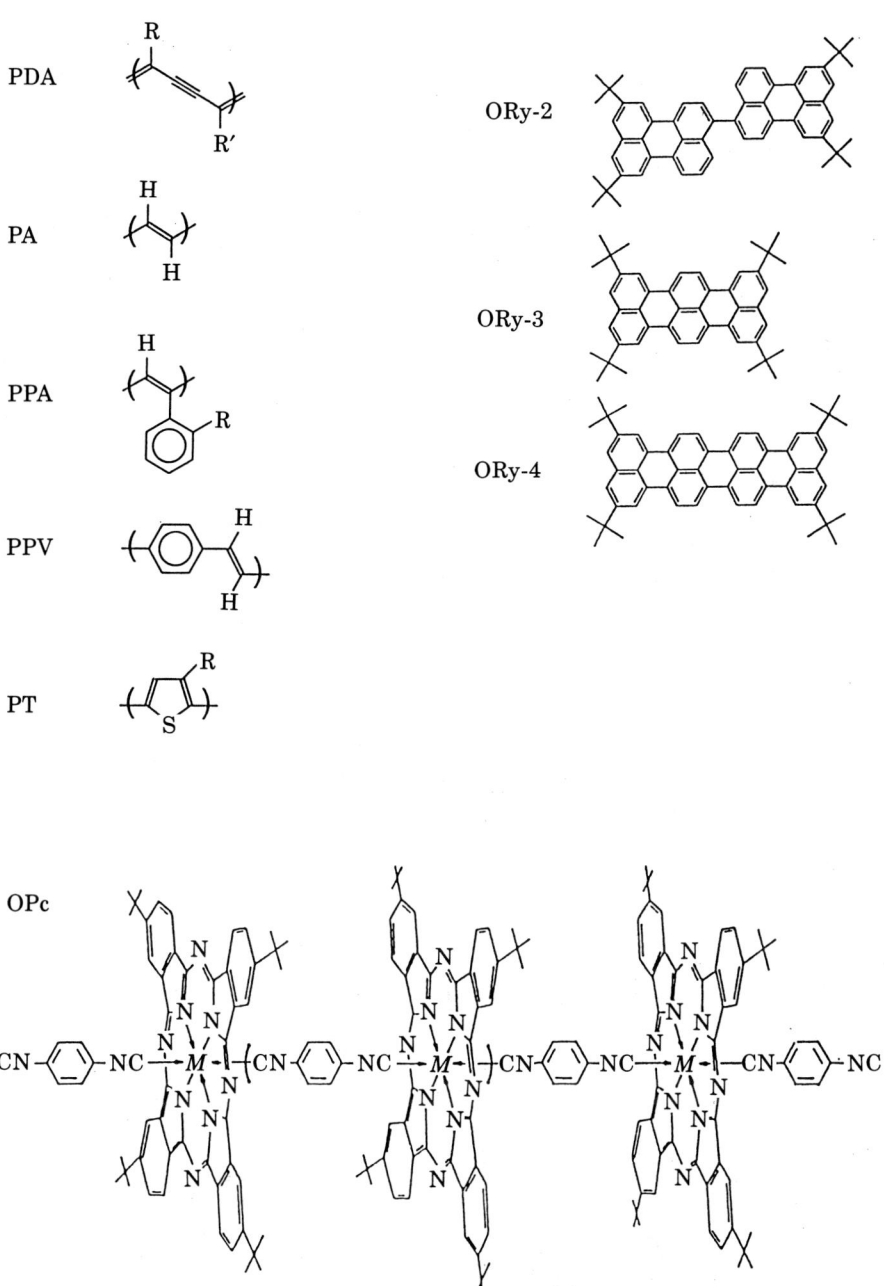

Fig. 3. – Chemical structures and abbreviations of some conjugated polymers and dye systems.

spectrum, which is dominated by the exciton absorption. In PDA the CB edge is located approximately 0.5 eV above the exciton, as can be concluded from photoconductivity and electroreflection experiments [40-42]. This model is widely accepted, at least for PDA. Presumably it can be applied also to other conjugated polymers.

The THG spectra of conjugated polymers are now discussed by means of the energy states and resonances shown in fig. 4. The strong maximum of $\chi^{(3)}(-3\omega; \omega, \omega, \omega)$ is a three-photon resonance with states located at the top of the valence band and the exciton (process 1 in fig. 4). It dominates due to the large oscillator strength of the exciton absorption. The additional resonance at higher energies can have two possible origins as indicated by processes 2 and 3. Process 2 is a three-photon resonance with the continuum of states located at the bottom of the conduction band. Process 3 is a two-photon resonance with a two-photon state located energetically below the EX. In contrast to short-chain polyenes, where the location of two-photon states is well known and understood [43, 44], there is still a debate whether the two-photon states in conjugated polymers are below and/or above the exciton. As sketched in fig. 4, processes 2 and 3 could occur at the same energy E_L of the fundamental laser beam (which corresponds to the short arrows). Therefore, with THG alone, it is not possible to distinguish between processes 2 and 3. Two-photon absorption spectroscopy or other nonlinear optical techniques can help to solve the interesting question which are the essential states in conjugated polymers. Finally it should be mentioned that several detailed theories exist which are able to describe the THG spectra [45-50]. However, in view of the many free fitting pa-

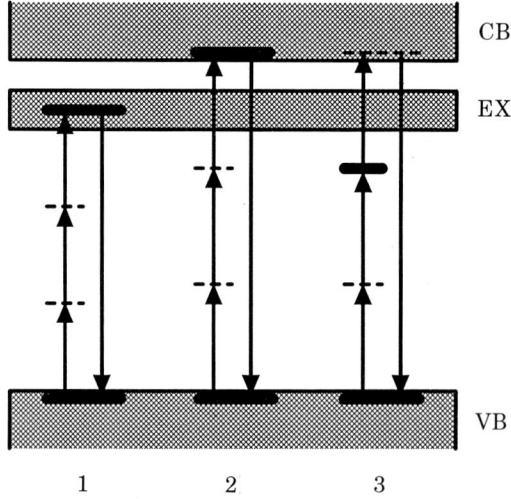

Fig. 4. – Possible multiphoton resonances in third-harmonic generation and energy states of conjugated polymers; VB: valence band; CB: conduction band; EX: exciton. The processes 1, 2 and 3 are explained in the text.

rameters usually involved, the perfect fit to the experimental data is only a necessary but not sufficient condition for the validity of the theoretical model.

4. – Influence of the delocalization of π-electrons.

The basic question is: what is the influence of the delocalization length L_d of the π-electrons on the magnitude of the molecular hyperpolarizability γ and finally on the macroscopic $\chi^{(3)}$ value? It is difficult to exactly define the important length L_d, sometimes also called conjugation length. Following SILBEY[51], L_d may be introduced as a critical length at which the linear polarizability $\alpha^{(1)}$ or the hyperpolarizability $\gamma^{(3)}$ change their functional dependence on the total length L of the system. This will be elucidated below. Strictly speaking, it should be distinguished between a conjugation length L_d^α for the linear polarizability, L_d^γ for the hyperpolarizability γ, or for another physical quantity. In the following we only consider L_d^γ and omit the index γ for simplicity.

Very short chains of length $L < L_d$ with one-dimensional π-electron delocalization are well described by several theories[52-57]. They predict a power law

(16)
$$\gamma \sim L^x$$

with various exponents x in the range between 4.0 and 5.0. These theories describe the experimental results[58-63] of several groups quite well. They are shown in fig. 5. The lengths L of these oligomers are calculated using known bond lengths and angles[44,64]. Although the γ values displayed in fig. 5 are considered to be mostly off-resonant, the scatter of the data is not surprising in view of the difficulties and problems discussed in subsect. 2˙1 and 3˙1. Nevertheless it can be seen that the power law (16) is valid only for L smaller than (2–3) nm. For longer oligomers the γ values seem to «saturate», which means that γ does not increase further according to the power law (16). The critical length L_d at which strong deviations from (16) occur is observed in the range (2–5) nm. The question how L_d relates to the chemical structures of conjugated polymers is an important problem, which is not fully solved yet. Perhaps torsional motions of the polymer backbone[65] or other physical effects limit the delocalization of the π-electrons.

A general theory for extended chains of length $L > L_d$ with one-dimensional π-electron delocalization was developed by FLYTZANIS et al.[53,54]. It relates the linear and nonlinear optical properties to the delocalization length L_d according to

(17)
$$\chi^{(2n'-1)} \sim L_d^{4n'-2},$$

where $n' = 1$ refers to $\chi^{(1)}$ and $n' = 2$ to $\chi^{(3)}$, respectively. The important conse-

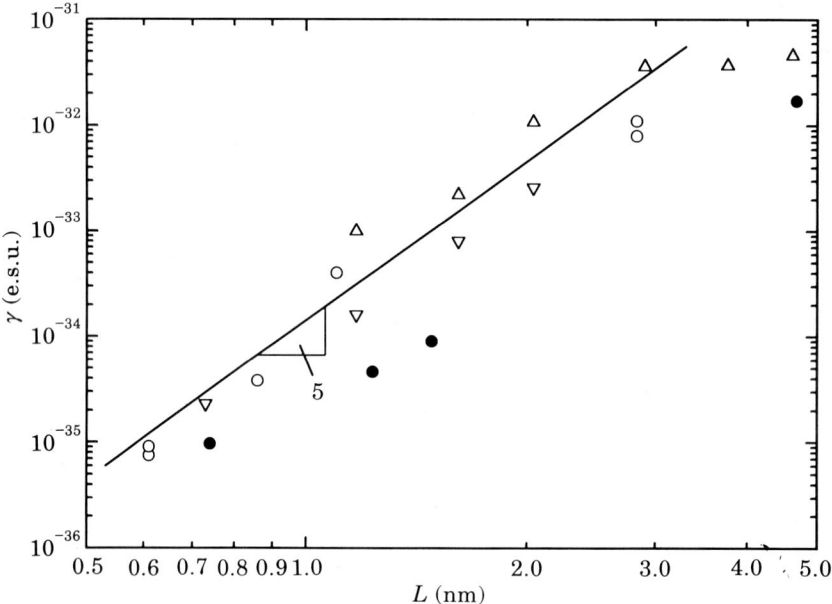

Fig. 5. – Dependence of the molecular hyperpolarizability γ on the length of several short-chain oligomers. Circles refer to polyenes, triangles to thiophene oligomers, respectively. Assignment of the data: open circles[58-60], filled circles[61], triangles facing down[62], triangles facing up[63]. The line with slope 5 corresponds to a power law $\gamma \sim L^5$.

quence $\chi^{(3)} \sim L_d^6$ can be compared with experiments. If L_d is inversely proportional to E_0, which is the optical energy gap[53,54], and $E_0^{-1} \sim \lambda_{max}$, it follows that the off-resonant $\chi^{(3)}$ values of conjugated polymers should scale with λ_{max}^6, irrespectively of their chemical structure.

The master plots of fig. 6 show that such a general relation really exists for conjugated polymers[66,67]. Because of different bulky substituents R which do not contribute much to the nonlinearity (as in the case of alkyl chains), the polymers shown in fig. 3 have a very different number of π-electrons per unit volume. This influences also the absorption coefficient α_{max} at λ_{max} of the thin films. In the normalized plot of $\chi^{(3)}/\alpha_{max}$ against λ_{max} shown in fig. 6b), it can be seen, for example, that PPV has a third-order nonlinearity very similar to that of the other conjugated polymers. This is not as obvious in fig. 6a). Because α_{max} is proportional to $\chi^{(1)}$, the ratio $\chi^{(3)}/\alpha_{max}$ should scale with L_d^4 according to relation (17).

Only on the first view the slopes 6 and 4 of the straight lines in fig. 6a), b) seem to indicate a good agreement between theory and experiment. It should be kept in mind, however, that the $\chi^{(3)}$ values are measured at the fundamental laser wavelength $\lambda_L = 1064$ nm and contain varying resonance enhancements—especially the data with λ_{max} in the range of the three-photon resonance. Recent THG experiments, performed at variable laser wavelengths[33],

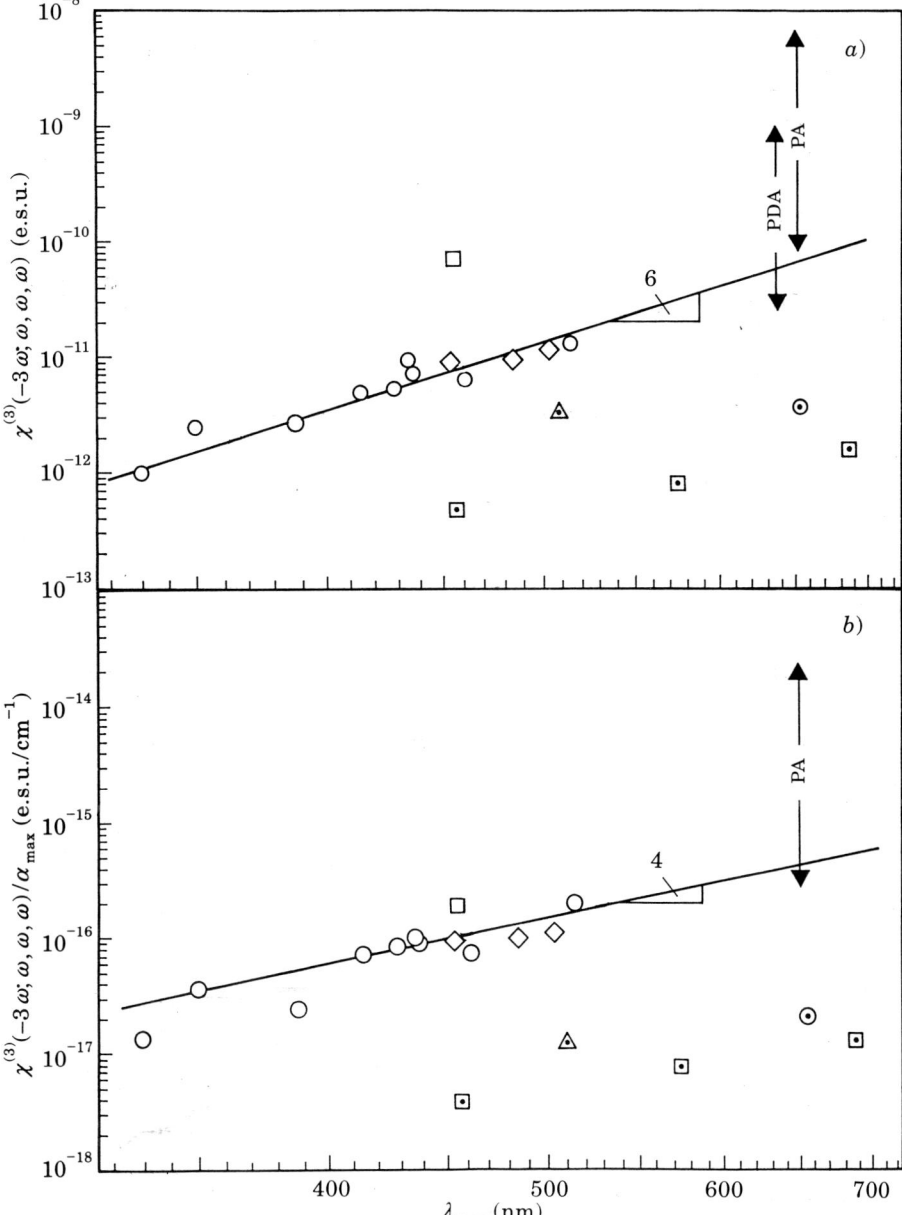

Fig. 6. – Master plots which relate the macroscopic susceptibility $\chi^{(3)}(-3\omega; \omega, \omega, \omega)$, obtained from third-harmonic generation, to linear optical quantities. Several conjugated polymers and dyes with chemical structures given in fig. 3 are displayed according to their wavelengths and absorption coefficients λ_{max} and α_{max} at their absorption maxima, respectively. The individual data points of our own work are determined at the fundamental laser wavelength $\lambda_L = 1064$ nm [26, 67-72] and contain varying resonance contributions to $\chi^{(3)}(-3\omega; \omega, \omega, \omega)$. The range of $\chi^{(3)}$ data for polyacetylene (PA) [31, 32] obtained at variable laser wavelengths is shown for comparison. The lines with slopes 6 and 4 refer to the scaling law of Flytzanis [53, 54]; ○ PPA, ◇ PT, □ PPV, △ R6G, ⊡ ORy, ⊙ OPc.

indicate that the lines which connect the off-resonant $\chi^{(3)}$ data have considerably larger slopes, as those shown in fig. 6. Presumably this discrepancy can be explained with the possibility that the approximation $L_d \sim \lambda_{max}$ is too simple and needs to be refined.

For comparison with the THG data of conjugated polymers [26, 66-68], the results for several dye systems [69-72] are also displayed in fig. 6. The dyes have an order of magnitude smaller $\chi^{(3)}$ values than the conjugated polymers with similar λ_{max}. This is interpreted with the different dimensionality of the π-electron delocalization. The chemical structures of these dyes (see fig. 3) indicate that their π-electron systems are delocalized two-dimensionally, in contrast to the conjugated polymers. The simplest explanation is based on the model of the electron in a one- and two-dimensional potential well [73, 74]. Because of additional degeneracies, the energy levels in the two-dimensional (2D) well are more closely arranged than in the one-dimensional (1D) well. With this argument it can be qualitatively understood that the 2D well of the same maximum lateral length L as the 1D well should have a larger λ_{max} value. The nonlinear optical properties are determined primarily by the maximum length of the potential well and not by λ_{max}. With this argument it can be qualitatively understood that the $\chi^{(3)}$ values of the 2D systems are much smaller as compared to the 1D systems at the same λ_{max}. This underlines that the conjugated polymers with 1D π-electron delocalization are optimally suited to achieve large off-resonant $\chi^{(3)}$ values.

These master plots can be used to recognize severe limitations of this class of materials. In waveguide applications of the intensity-dependent refractive index, the material must not absorb at the laser wavelength. Because the optimum working range is in the red to near-infrared range of light, the usable λ_{max} values of conjugated polymers cannot be much larger than 700 nm. Figure 6 shows that off-resonant $\chi^{(3)}(-3\omega; \omega, \omega, \omega)$ values larger than 10^{-10} e.s.u. cannot be expected for conjugated polymers. By means of eq. (10) this corresponds to an approximate limit for the pure electronic part of the intensity-dependent refractive index $n_2 \lesssim 10^{-15}$ m^2/W.

5. – Degenerate four-wave mixing.

5˙1. *Experiment and evaluation.* – Three incident waves of the same frequency ω can interact in the nonlinear medium and generate a fourth beam at the same frequency in a new direction of space. This process is described by the nonlinear optical susceptibility $\chi^{(3)}(-\omega; \omega, \omega, -\omega)$. Two different geometries are frequently used in DFWM experiments:

In the «phase conjugation» geometry two strong pump beams counterpropagate. The interaction with the third beam from an arbitrary direction can be described by a dual-grating picture [3, 75]. By means of these gratings, a part of the pump beam is diffracted in the reversed direction of third beam. This is called the «conjugate wave». It can be read out by a beam splitter.

In the so-called «folded BOXCARS» geometry[76] three beams are incident on the sample from the same side. By DFWM they generate a fourth beam in the forward direction[77, 78] as shown in fig. 7a). The phase-matching condition of the wave vectors determines the direction of k_4. This configuration can

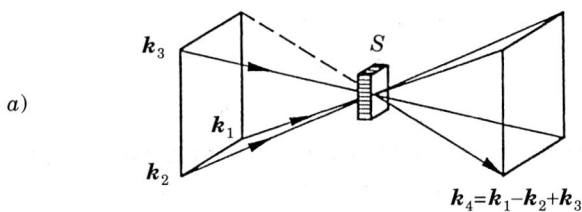

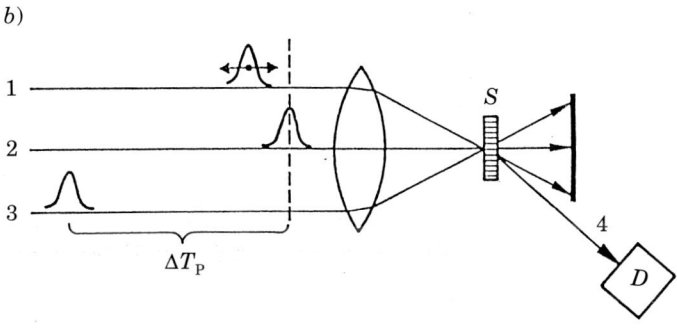

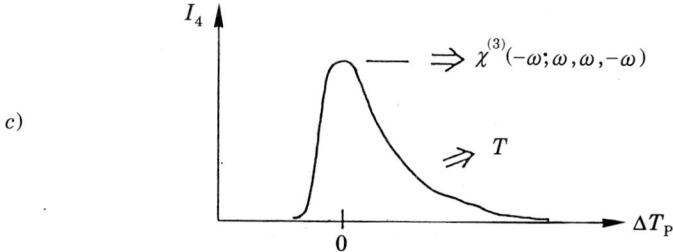

Fig. 7. – Basic principles of degenerate four-wave mixing (DFWM) and transient gratings. a) Folded BOXCARS geometry and phase-matching condition of the wave vectors (S: sample). b) Generation of a transient grating: the pulses 1 and 2 interfere at the sample. The spatial modulation of the light intensity generates a grating, if the sample has an intensity-dependent refractive index. The existence of the grating is read out by the delayed pulse 3, which is diffracted in direction 4 (D: detector). c) Schematic plot of the time dependence of the transient grating. The time-averaged intensity of beam 4 is displayed vs. the delay time of pulse 3.

be adjusted more easily than the phase conjugation setup. The reason why beam 4 propagates in a new direction can be understood by means of the transient-grating picture shown in fig. 7b). It is based on the interference of beams 1 and 2 at the sample and its intensity-dependent refractive index, which can be very large and complex at resonance. This method is sometimes called «transient-grating method» or «real-time holography» [3, 79]. The time dependence of the transient grating can be probed by beam 3 which runs through a variable optical delay. A time modulation method was developed to improve the sensitivity of the technique [80]. A periodic time modulation of beam 1 is used to modulate the grating. Therefore, the intensity of beam 4 oscillates at the same modulation frequency and can be detected by means of a lock-in amplifier. This helps considerably to reduce the influence of stray light and of permanent gratings.

Figure 7c) shows a schematic plot of the time dependence of the transient grating. In the case of an exponential decay, the relaxation time T of the grating can be obtained from the diffracted intensity according to $I_4 \sim \exp[-2\Delta T_P/T]$ [77, 78]. The time resolution of this technique is given by the laser pulse width. The maximum of the intensity I_4 depends on the

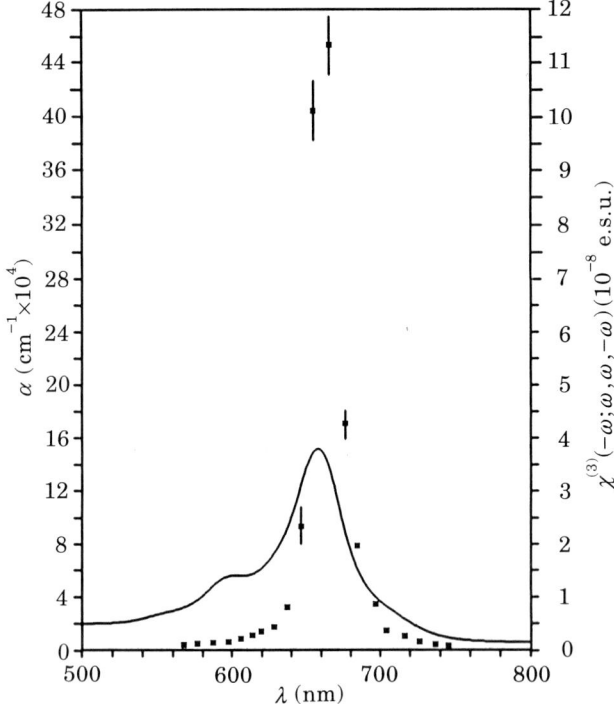

Fig. 8. – Absorption spectrum (full line) and dispersion of $\chi^{(3)}(-\omega; \omega, \omega, -\omega)$ of a 160 nm thick, solution cast, film of the phthalocyanine derivative OPc [72]. The chemical structure of OPc is shown in fig. 3.

$\chi^{(3)}(-\omega;\omega,\omega,-\omega)$ value of the sample. It can be evaluated by comparison with a reference.

The $\chi^{(3)}(-\omega;\omega,\omega,-\omega)$ values of absorbing films can be evaluated by means of

$$(18) \quad \chi^{(3)}(-\omega;\omega,\omega,-\omega) =$$

$$= \left(\frac{I}{I_{\mathrm{ref}}}\right)^{1/2}\frac{d_{\mathrm{ref}}}{d}\left(\frac{n}{n_{\mathrm{ref}}}\right)^{2}\frac{\alpha d\exp[\alpha d/2]}{1-\exp[-\alpha d]}\chi^{(3)}_{\mathrm{ref}}(-\omega;\omega,\omega,-\omega),$$

where α, n and d are the absorption coefficient, refractive indices and effective thicknesses of the sample and reference, respectively. In eq. (18) α refers to the intensity adsorption coefficient, whereas in the corresponding formulae of earlier work [72,80,81] α refers to the attenuation of the field amplitudes. This causes a factor of 2 difference in the formulae. Frequently CS_2 is used as a reference material for DFWM experiments. A reference value $\chi^{(3)} = 6.8\cdot10^{-13}$ e.s.u. is used for CS_2 [82].

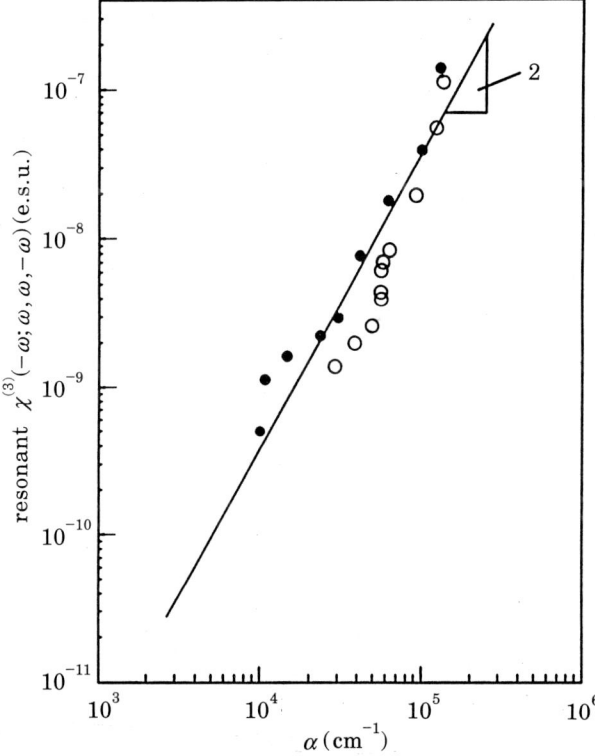

Fig. 9. – Resonant $\chi^{(3)}(-\omega;\omega,\omega,-\omega)$ data of OPc from fig. 8, plotted against the absorption coefficient α of the thin film at the laser wavelength λ, which is varied over the absorption maximum of OPc [72]: ● $\lambda > \lambda_{\max}$, ○ $\lambda < \lambda_{\max}$.

5'2. *Examples of saturable absorption.* – If the laser frequency is tuned to the absorption band of a molecular system, the hyperpolarizability $\gamma(-\omega; \omega, \omega, -\omega)$ increases resonantly because of the first term in the resonance denominator of eq. (12) with $\omega = \omega_1 = \omega_2 = \omega_3$. In resonance we have additionally $\omega = \Omega_{AG}$. This process can be described also in the picture of saturable absorption. Because of the intense laser excitation, a significant population in an excited state of the molecular system can occur, which leads to a reduction of the absorption coefficient for this electronic transition. As long as the incident intensity I_0 is small compared to the saturation intensity I_S [83-85], it is found that the intensity of the DFWM signal I_4 follows $I_4 \sim I_0^3$ [72]. This shows that saturable absorption can be described as a third-order nonlinear optical process.

A characteristic behaviour of saturable absorption in two-level systems has been studied with thin films of rhodamine 6G (R6G), a well-known laser dye, and the oligomeric phthalocyanine derivative OPc shown in fig. 3 [72,81]. The absorption spectrum of OPc and its resonance enhancement of $\chi^{(3)}(-\omega; \omega, \omega, -\omega)$

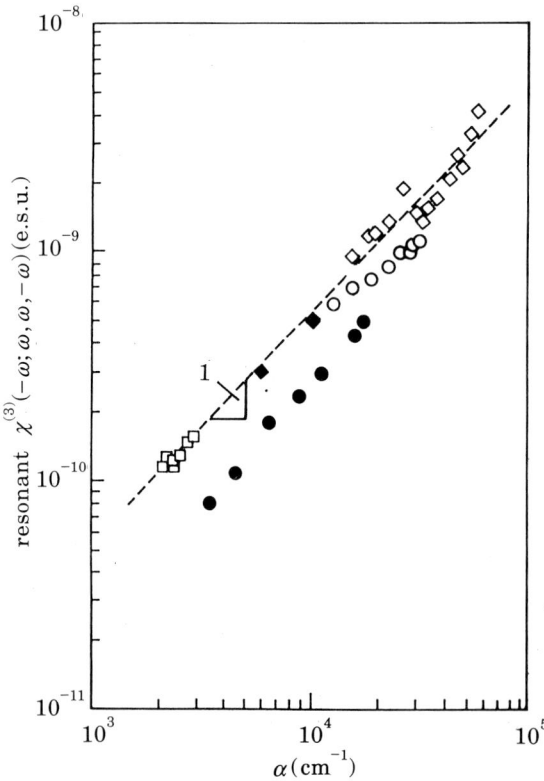

Fig. 10. – Resonant $\chi^{(3)}(-\omega; \omega, \omega, -\omega)$ data of several conjugated polymers, measured and plotted similarly as in fig. 9 [81]. The data of PT(a) are from ref. [86]. □ PPV, ◇ PT, ◆ PT(a), ○ PPA-Si(CH_3)$_3$, ● PPA-C_2H_5.

can be compared in fig. 8. If the resonant $\chi^{(3)}(-\omega; \omega, \omega, -\omega)$ values are plotted as a function of the absorption coefficient $\alpha(\omega)$ at the laser frequency ω, the characteristic dependence $\chi^{(3)}(-\omega; \omega, \omega, -\omega) \sim \alpha^2(\omega)$ can be seen in fig. 9. This quadratic dependence has been found first in thin films of R6G[81] and is a typical result for saturable absorption in two-level systems.

If conjugated polymers are studied with resonant DFWM, a striking difference to the behaviour of small dye systems is found[81]. Conjugated polymers follow a general linear scaling law $\chi^{(3)}(-\omega; \omega, \omega, -\omega) \sim \alpha(\omega)$ shown in fig. 10. It has been explained with phase space filling effects by one-dimensional excitons[81]. This linear scaling has been confirmed recently by other groups[87,88]. Scaling plots like fig. 9 and 10 can be used to distinguish if saturable absorption occurs either in isolated two-level systems or in strongly coupled two-level systems which show excitonic behaviour.

6. – Conclusion.

The scaling laws and master plots described in this lecture indicate systematic relations between structure and third-order nonlinearities. of molecules and polymers. They are quite useful to learn more about fundamental physical processes, which can lead to large nonlinear optical susceptibilities $\chi^{(3)}$ of conjugated polymers and dye systems.

By comparison of the resonant $\chi^{(3)}(-\omega; \omega, \omega, -\omega)$ values with the $\chi^{(3)}(-3\omega; \omega, \omega, \omega)$ data of fig. 6, we see that different measurement techniques and resonances can lead to variations of $\chi^{(3)}$ of the same material over 4-5 orders of magnitude.

* * *

I thank my former and present co-workers Drs. D. NEHER, A. KALBEITZEL, R. SCHWARZ, A. MATHY, A. GRUND, M. BAUMANN and H. MENGES for their valuable contributions and Prof. G. WEGNER for his stimulating interest in this work and helpful discussions. Financial support to this work was given by the BMFT and by Volkswagen-Stiftung.

REFERENCES

[1] B. E. A. SALEH and M. C. TEICH: *Fundamentals of Photonics* (Wiley Interscience, New York, N.Y., 1991).
[2] G. I. STEGEMAN and R. H. STOLEN: *J. Opt. Soc. Am. B*, **6**, 652 (1989).
[3] R. A. FISCHER, Editor: *Optical Phase Conjugation* (Academic Press, New York, N.Y., 1983).
[4] N. BLOEMBERGEN: *Nonlinear Optics* (Benjamin, New York, N.Y., 1965).

[5]   H. RABIN and C. L. TANG, Editors: *Quantum Electronics: A Treatise* (Academic Press, New York, N.Y., 1975).

[6]   A. YARIV: *Quantum Electronics* (Wiley, New York, N.Y., 1975, 1989).

[7]   Y. R. SHEN: *The Principles of Nonlinear Optics* (Wiley Interscience, New York, N.Y., 1984).

[8]   F. A. HOPF and G. I. STEGEMAN: *Applied Classical Electrodynamics*, Vol. II, *Nonlinear Optics* (Wiley, New York, N.Y., 1986).

[9]   M. SCHUBERT and B. WILHELMI: *Nonlinear Optics and Quantum Electronics* (Wiley, New York, N.Y., 1986).

[10] M. D. LEVENSON and S. S. KANO: *Introduction to Nonlinear Laser Spectroscopy* (Academic Press, Boston, Mass., 1988).

[11] P. N. BUTCHER and D. COTTER: *The Elements of Nonlinear Optics* (Cambridge University Press, Cambridge, 1990).

[12] P. N. PRASAD and D. J. WILLIAMS: *Introduction to Nonlinear Optical Effects in Molecules and Polymers* (Wiley, New York, N.Y., 1991).

[13] R. W. BOYD: *Nonlinear Optics* (Academic Press, Boston, Mass., 1992).

[14] D. S. CHEMLA and J. ZYSS, Editors: *Nonlinear Optical Properties of Organic Molecules and Crystals*, Vol. I and II (Academic Press, Orlando, Fla., 1987).

[15] A. C. NEWELL and J. V. MOLONEY: *Nonlinear Optics* (Addison-Wesley Publ. Comp., Redwood City, Cal., 1992).

[16] J. A. ARMSTRONG, N. BLOEMBERGEN, J. DUCUING and P. S. PERSHAN: *Phys. Rev.*, **127**, 1918 (1962).

[17] J. F. WARD: *Rev. Mod. Phys.*, **37**, 1 (1965).

[18] T. K. YEE and T. K. GUSTAFSON: *Phys. Rev. A*, **18**, 1597 (1979).

[19] S. K. KURTZ: in ref.[5], Vol. I, part A, p. 209.

[20] S. SINGH: in *Handbook of Laser Science III*, edited by M. J. WEBER (CRC Press, Boca Raton, Fla., 1986), p. 1.

[21] C. BUBECK: in *Organic Materials for Photonics*, edited by G. ZERBI (North-Holland Elsevier, Amsterdam, 1993), p. 215.

[22] N. BLOEMBERGEN and P. S. PERSHAN: *Phys. Rev.*, **128**, 606 (1962).

[23] J. JERPHAGNON and S. K. KURTZ: *J. Appl. Phys.*, **41**, 1667 (1970).

[24] F. KAJZAR and J. MESSIER: *Phys. Rev. A*, **32**, 2352 (1985).

[25] P. D. MAKER, R. W. TERHUNE, M. NISENOFF and C. M. SAVAGE: *Phys. Rev. Lett.*, **8**, 21 (1962).

[26] D. NEHER, A. KALTBEITZEL, A. WOLF, C. BUBECK and G. WEGNER: *J. Phys. D*, **24**, 1193 (1991).

[27] D. NEHER, A. WOLF, C. BUBECK and G. WEGNER: *Chem. Phys. Lett.*, **163**, 116 (1989).

[28] G. R. MEREDITH, B. BUCHALTER and C. HANZLIK: *J. Chem. Phys.*, **78**, 1533 (1983).

[29] B. BUCHALTER and G. R. MEREDITH: *Appl. Opt.*, **21**, 3221 (1982).

[30] F. KAJZAR and J. MESSIER: *Thin Solid Films*, **132**, 11 (1985).

[31] T. KANETAKE, K. ISHIKAWA, T. HASEGAWA, T. KODA, K. TAKEDA, M. HASEGAWA, K. KUBODERA and H. KOBAYASHI: *Appl. Phys. Lett.*, **54**, 2287 (1989).

[32] W. E. TORRUELLAS, D. NEHER, R. ZANONI, G. I. STEGEMAN, F. KAJZAR and M. LECLERC: *Chem. Phys. Lett.*, **175**, 11 (1990).

[33] A. MATHY, K. UEBERHOFEN, R. SCHENK, H. GREGORIUS, R. GARAY, K. MÜLLEN and C. BUBECK: submitted to *Phys. Rev. B*.

[34] D. BLOOR and R. R. CHANCE, Editors: *Polydiacetylenes*, NATO ASI Ser. E, Vol. **102** (Martinus Nijhoff, Dordrecht, 1985).

[35] M. SCHOTT and G. WEGNER: in ref.[14], Vol. II, p. 3.

[36] H. HAUG and S. W. KOCH: *Quantum Theory of the Optical and Electrical Properties of Semiconductors* (World Scientific, Singapore, 1993), p. 199.

[37] S. ETEMAD and Z. G. SOOS: in *Spectroscopy of Advanced Materials*, Vol. 19, edited by R. J. H. CLARK and R. E. HESTER (Wiley, Chichester, 1991), p. 87.

[38] W. S. FANN, S. BENSON, J. M. J. MADEY, S. ETEMAD, G. L. BAKER and F. KAJZAR: *Phys. Rev. Lett.*, 62, 1492 (1989).

[39] F. KAJZAR, S. ETEMAD, G. L. BAKER and J. MESSIER: *Synth. Met.*, 17, 563 (1987).

[40] K. LOCHNER, H. BÄSSLER, B. TIEKE and G. WEGNER: *Phys. Status Solidi B*, 88, 653 (1978).

[41] L. SEBASTIAN and G. WEISER: *Chem. Phys.*, 62, 447 (1981).

[42] G. WEISER: *Phys. Rev. B*, 45, 14076 (1992).

[43] B. E. KOHLER: *J. Chem. Phys.*, 93, 5838 (1990).

[44] B. E. KOHLER: in *Conjugated Polymers* edited by J. L. BRÉDAS and R. SILBEY (Kluwer Academic Publ., Amsterdam, 1991), p. 405.

[45] S. ABE, M. SCHREIBER and W.-P. SU: *Chem. Phys. Lett.*, 192, 425 (1992).

[46] S. ABE, M. SCHREIBER, W.-P. SU and J. YU: *Phys. Rev. B*, 45, 9432 (1992).

[47] T. HASEGAWA, Y. IWASA, H. SUNAMURA, T. KODA, Y. TOKURA, H. TACHIBANA, M. MATSUMOTO and S. ABE: *Phys. Rev. Lett.*, 69, 668 (1992).

[48] Z. SHUAI and J. L. BRÉDAS: *Phys. Rev. B*, 44, 5962 (1991).

[49] S. N. DIXIT, D. GUO and S. MAZUMDAR: *Phys. Rev. B*, 43, 6781 (1991).

[50] D. GUO, S. MAZUMDAR, G. I. STEGEMAN, M. CHA, D. NEHER, S. ARAMAKI, W. TOR-RUELLAS and R. ZANONI: *Mater. Res. Soc. Symp. Proc.*, 247, 151 (1992).

[51] R. SILBEY: in *Conjugated Materials: Opportunities in Electronics, Optoelectronics, and Molecular Electronics*, edited by J. L. BRÉDAS and R. R. CHANCE (Kluwer Academic Publ., Dordrecht, 1990), p. 1.

[52] K. C. RUSTAGI and J. DUCUING: *Opt. Commun.*, 10, 258 (1974).

[53] G. P. AGRAWAL, C. COJAN and C. FLYTZANIS: *Phys. Rev. B*, 17, 776 (1978).

[54] C. FLYTZANIS: in ref.[14], p. 121.

[55] C. P. DeMELO and R. SILBEY: *J. Chem. Phys.*, 88, 2558 (1988).

[56] J. R. HEFLIN, K. Y. WONG. O. ZAMANI-KAMIRI and A. F. GARITO: *Phys. Rev. B*, 38, 1573 (1988).

[57] Z. G. SOOS and S. RAMASESHA: *J. Chem. Phys.*, 90, 1067 (1989).

[58] J. P. HERMANN, D. RICARD and J. DUCUING: *Appl. Phys. Lett.*, 23, 178 (1973).

[59] J. F. WARD and D. S. ELLIOTT: *J. Chem. Phys.*, 69, 5438 (1978).

[60] S. H. STEVENSON, D. S. DONALD and G. R. MEREDITH: in *Nonlinear Optical Properties of Polymers*, edited by A. J. HEEGER, J. ORENSTEIN and D. R. ULRICH (Materials Research Society, Pittsburg, Penn., 1988), p. 103.

[61] J. P. HERMANN and J. DUCUING: *J. Appl. Phys.*, 45, 5100 (1974).

[62] M. T. ZHAO, B. P. SINGH and P. N. PRASAD: *J. Phys. Chem.*, 93, 7916 (1989).

[63] H. THIENPONT, G. L. J. A. RIKKEN and E. W. MEIJER: *Phys. Rev. Lett.*, 65, 2141 (1990).

[64] H. A. STAAB: *Einführung in die theoretische organische Chemie* (Verlag Chemie, Weinheim, 1975).

[65] G. ROSSI, R. R. CHANCE and R. SILBEY: *J. Chem. Phys.*, 90, 7594 (1989).

[66] D. NEHER, A. KALTBEITZEL, A. WOLF, C. BUBECK and G. WEGNER: in *Conjugated Polymeric Materials: Opportunities in Electronics, Opto-Electronics, and Molecular Electronics*, NATO ASI Series E, Vol. 182, edited by J. L. BRÉDAS and R. R. CHANCE (Kluwer Academic Publ., Dordrecht, 1990), p. 387.

[67] C. BUBECK, A. GRUND, A. KALTBEITZEL, D. NEHER, A. MATHY and G. WEGNER: in *Organic Molecules for Nonlinear Optics and Photonics*, NATO ASI Series E, Vol.

194, edited by J. MESSIER, K. KAJZAR and P. N. PRASAD (Kluwer Academic Publ., Dordrecht, 1991), p. 335.

[68] D. NEHER, A. WOLF, M. LECLERC, A. KALTBEITZEL, C. BUBECK and G. WEGNER: *Synth. Met.*, **37**, 249 (1990).

[69] S. SCHRADER, K. H. KOCH, A. MATHY, C. BUBECK, K. MÜLLEN and G. WEGNER: *Synth. Met.*, **41-43**, 3223 (1991).

[70] S. SCHRADER, K. H. KOCH, A. MATHY, C. BUBECK, K. MÜLLEN and G. WEGNER: *Progr. Colloid Polym. Sci.*, **85**, 143 (1991).

[71] A. GRUND, A. MATHY, A. KALTBEITZEL, D. NEHER, C. BUBECK and G. WEGNER: in *Organic Materials for Nonlinear Optics II*, edited by R. A. HANN and D. BLOOR (Royal Society of Chemistry, Cambridge, 1991), p. 288.

[72] A. GRUND, A. KALTBEITZEL, A. MATHY, R. SCHWARZ, C. BUBECK, P. VERMEHREN and M. HANACK: *J. Phys. Chem.*, **96**, 7450 (1992).

[73] H. KUHN: *J. Chem. Phys.*, **17**, 1198 (1949).

[74] R. S. BERRY, S. A. RICE and J. ROSS: *Physical Chemistry* (Wiley, New York, N.Y., 1980).

[75] R. C. LIND, D. G. STEEL and G. J. DUNNING: *Opt. Eng.*, **21**, 190 (1982).

[76] J. A. SHIRLEY, R. J. HALL and A. C. ECKBRETH: *Opt. Lett.*, **5**, 380 (1980).

[77] G. M. CARTER, J. V. HRYNIEWICZ, M. K. THAKUR, Y. J. CHEN and S. E. MEYER: *Appl. Phys. Lett.*, **49**, 16 (1986).

[78] G. M. CARTER: *J. Opt. Soc. Am. B*, **4**, 1018 (1987).

[79] H. J. EICHLER, P. GÜNTHER and D. W. POHL: *Laser-Induced Dynamics Gratings*, Springer Ser. Opt. Sci., Vol. **50** (Springer, Berlin, 1986).

[80] A. KALTBEITZEL: doctoral thesis (Mainz, 1989).

[81] C. BUBECK, A. KALTBEITZEL, A. GRUND and M. LECLERC: *Chem. Phys.*, **154**, 343 (1991).

[82] N. P. XUAN, J.-L. FERRIER, J. GAZENGEL and G. RIVOIRE: *Opt. Commun.*, **51**, 433 (1984).

[83] R. L. ABRAMS and R. C. LIND: *Opt. Lett.*, **2**, 94 (1978).

[84] R. L. ABRAMS and R. C. LIND: *Opt. Lett.*, **3**, 205 (1978).

[85] R. L. ABRAMS, J. F. LAM, R. C. LIND, D. G. STEEL and P. F. LIAO: in ref.[3], p. 211.

[86] B. P. SINGH, M. SAMOC, H. S. NALWA and P. N. PRASAD: *J. Chem. Phys.*, **92**, 2756 (1990).

[87] N. OOBA, S. TOMARU, T. KURIHARA, Y. MORI, Y. SHUTO and T. KAINO: *Chem. Phys. Lett.*, **207**, 468 (1993).

[88] T. VOGTMANN, W. SCHMID, T. FEHN, F. BAUER, I. BAUER and M. SCHWOERER: *Synth. Met.*, **55-57**, 4018 (1993).

Molecular Design of Third-Order Nonlinear Optically Active π-Conjugated Compounds and Preparation of Thin Films.

T. Wada, M. Hosoda, H. Okawa, A. Terasaki, M. Hara and H. Sasabe

Frontier Research Program, The Institute of Physical and Chemical Research (RIKEN)
2-1 Hirosawa, Wako, Saitama 351-01, Japan

1. – Introduction.

The recent development of optoelectronic devices requires further supplies of novel materials which have large nonlinearities and ultrafast responses. So far inorganic ferroelectric crystals such as $LiNbO_3$, KDP (potassium dihydrogen phosphate) and KTP (potassium titanyl phosphate), or semiconductors such as CdSe and $ZnGeP_2$ have been utilized for those devices, but organic materials essentially fit these requirements. From the materials viewpoint, the intramolecular charge transfer through π-electron conjugation gives large optical nonlinearities on the molecular level, whereas the centrosymmetry of the crystal structure determines the macroscopic second-order nonlinearity $\chi^{(2)}_{ijk}(-\omega_3; \omega_1, \omega_2)$; if the crystal is centrosymmetric, then $\chi^{(2)}$ becomes zero. On the other hand, the third-order optical nonlinearity $\chi^{(3)}_{ijkl}(-\omega_4; \omega_1, \omega_2, \omega_3)$ does not depend on the crystal symmetry but on the microscopic third-order susceptibility $\gamma_{ijkl}(-\omega_4; \omega_1, \omega_2, \omega_3)$ of the constituent molecular unit. Charge transfer complexes and π-conjugated compounds are the most promising materials for nonlinear optical (NLO) applications, but the systematic investigation, such as controls of conjugation length, intermolecular interaction and packing structure, is still open to discussion.

In conjugated linear-chain structures such as polyenes, π-electrons are delocalized in their motion only in one dimension along the chain axis [1]. The major contribution to γ_{ijkl} is the chain axis component γ_{xxxx} with all electric fields aligned along the chain axis (x-axis), and hence the averaged susceptibility $\langle\gamma\rangle$ in isotropic media equals to one-fifth of γ_{xxxx}. There have been various theoretical and experimental studies on linear-chain structures [2]. For example, a power law dependence of γ_{xxxx} for linear polyenes has been found on the number of carbon atom sites with exponents of 5.4 for the *trans* and 4.7 for the *cis* conformer, and it has also been found that γ_{xxxx} is more sensitive to the physical length of the chain than to the conformation[3]. When the dimensionality of the π-electron system is expanded from linear (one dimension) to cyclic chains (two dimensions), theoretical results on cyclic structures such as cyclooctatetraene show a decrease of γ_{ijkl} due to an actual reduction in the effective length available for the π-electron to respond to an optical electric field[4]. In the widely spread two-dimensional π-conjugated systems, other tensor components also contribute to $\langle\gamma\rangle$, as given in

$$(1) \quad \langle\gamma\rangle = \frac{1}{5}[\gamma_{xxxx} + \gamma_{yyyy} + (1/3)(\gamma_{xxyy} + \gamma_{xyxy} + \gamma_{xyyx} + \gamma_{yyxx} + \gamma_{yxyx} + \gamma_{yxxy})].$$

The off-resonant third-order optical susceptibility $\chi^{(3)}_{1111}(-3\omega; \omega, \omega, \omega)$ (hereafter abbreviated as $\chi^{(3)}$) determined by third-harmonic generation (THG) is directly related to $\langle\gamma\rangle$ of the macrocyclic conjugated compound through local-field factors f expressed in terms of the refractive indices n_ω and $n_{3\omega}$ as

$$(2) \qquad \chi^{(3)}_{1111}(-3\omega; \omega, \omega, \omega) = Nf_{3\omega}f_\omega f_\omega f_\omega \langle\gamma\rangle(-3\omega; \omega, \omega, \omega),$$

where N is the number of molecules per unit volume. The $\chi^{(3)}/N$ values show that the isotropically averaged third-order polarizability $\langle\gamma\rangle$ increases with increased size of the macrocyclic conjugated structure with exponent of 4.2 in a manner analogous to the behaviour of conjugated linear chains[5]. We have focused our researches for long on the development of one-dimensional conducting polymer systems such as polythiophene derivatives and polydiacetylenes and of two-dimensional macrocyclic conjugated compounds such as annulenes and metallophthalocyanine derivatives (MPc's). Systematic studies are needed to guide theories concerning the nonlinear optical properties of the two-dimensional π-electron systems. For this purpose, we chose tetra dehydro-methano–annulenes with an 18–28-membered ring size, which were expected to have both a rather rigid perimeter and to exhibit only a small perturbation with respect to π-electron conjugation[6]. On the other hand, MPc's exhibit linear and nonlinear electro-optic responses such as photoconductive, photovoltaic and photocatalytic behaviours, and also have additionally attractive physical properties such as large absorption in the visible region, thermal and chemical stabilities, and thin-film formation. In this lecture, firstly we will summarize our material research approaches[7-11] to these π-conjugated systems with enhanced $\chi^{(3)}$, secondly the techniques for the preparation of ultrathin films, and lastly their application to optical waveguides.

2. – π-conjugated compounds.

2'1. *Polythiophene derivatives.* – Polythiophene was polymerized electrochemically in the solution of thiophene monomer/nitrobenzene with tetra-

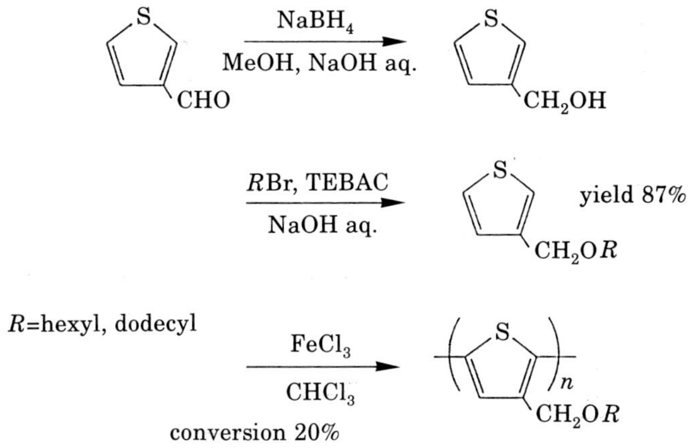

Fig. 1. – Synthetic route of soluble polythiophenes.

methylammonium perchlorate (TMAP) as a supporting electrolyte. The working and counter electrodes were ITO glass and Pt plate, respectively. The synthesized polymer film was washed in nitrobenzene, and then perchlorate ions were extracted from the film in the solvent of nitrobenzene/TMAP. After extraction the undoped film was washed in nitrobenzene again and dried up in a vacuum chamber. Soluble polythiophene derivatives (poly(3-alkyloxymethylthiophene)), on the other hand, were prepared chemically, as shown in fig. 1. Details have appeared elsewhere [12]. In this study $R = 6$:poly(3-hexyloxymethylthiophene) and $R = 12$:poly(3-dodecyloxymethylthiophene) (PDTh) were prepared. Thickness-controlled thin films were made by the spin-coating method from polymer solution. From the optical absorption spectra of polythiophene derivatives, it is clearly observed that the absorption maximum shifts towards shorter wavelength by the introduction of alkyloxymethyl pendants (hyposochromic shift). This indicates the shortening of the conjugation length due to the rotational motion of thiophene rings caused by the longer alkyl chain. The effect of the alkyl chain length between hexyl and dodecyl on hyposochromic shift is almost negligible.

2'2. *Polydiynes.* – Two types of polydiynes were prepared: one was an alkane-bridged polydiyne and the other was an arene-bridged polydiyne. The former is not a π-conjugation system, but by the interchain cross-linking due to thermal treatment it becomes a one-dimensional (1-D) conjugated polymer like polydiacetylene (PDA), as shown in fig. 2. The latter is, on the other hand, a 1-D π-conjugation system and easily converted into the 2-D system by the interchain cross-linking. Poly(1,9-decadiyne) (PDD) and poly(1,4-diethynyl-2,5-dibutoxy-benzene) (PDEDBB) are examples of alkane- and arene-bridged polydiynes, respectively, and prepared as shown in fig. 3 [12]. The solution of

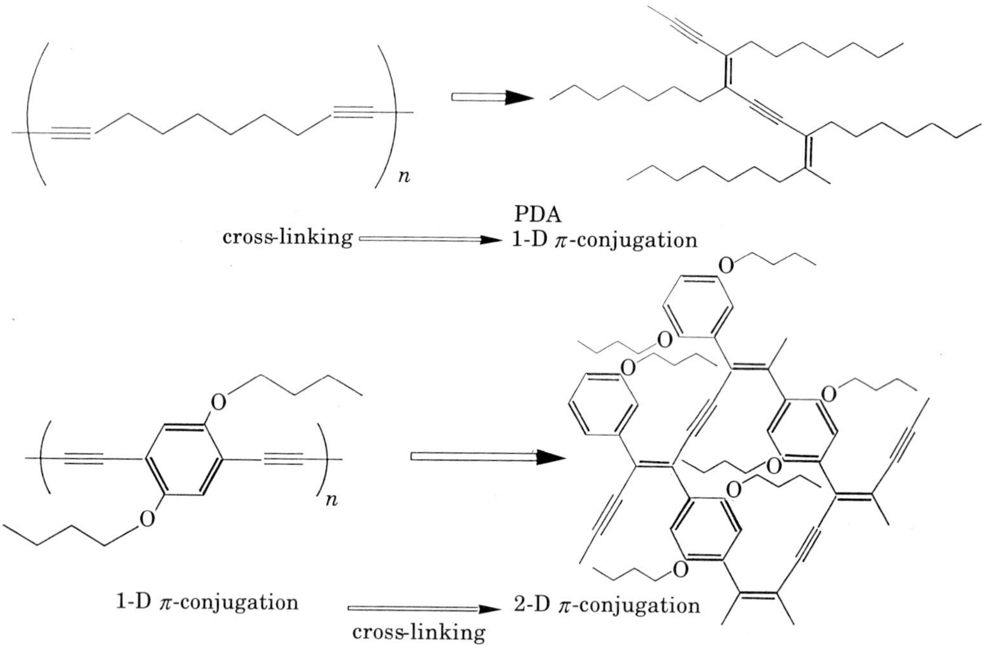

Fig. 2. – Schematic diagram of 1-D and 2-D π-conjugation in polydiynes.

PDD/dichloroethane was spread over the quartz substrate and then a high-quality film could be obtained. Though the as-prepared film was opaque, it turned to be coloured but transparent by annealing at 150 °C for 48 h, which indicates the formation a 1-D π-conjugated system. From the i.r. spectra of this film the formation of C=C bonds can be confirmed. The film was applicable for THG measurement. Under the UV irradiation, however, the PDD film becomes dark purple in colour and less applicable for THG measurement. The film of PDEDBB was also obtained by a spin-coating method. The X-ray diffraction pattern indicates that the PDEDBB film is completely amorphous, though PDEDBB powders show some indication of crystalline diffraction.

2˙3. *Annulenes*. – The annulenes with an 18-28-membered ring size, tetradehydromethano-[18]-(1), -[22]-(2), -[24]-(3), -[28](4)-annulene (fig. 4), were synthesized from the corresponding acyclic compounds obtained by the Wittig reaction between the dicarbaldehydes and the phosphonium salts[13]. The intramolecular oxidative coupling was achieved without employing a high-dilution technique because these molecules possess a favourable configuration for this reaction arising from the 1,6-methano-bridged linkage. The dark coloured annulene crystals exhibit a metallic luster and are thermally stable compared to the corresponding monocyclic annulenes as a result of their rigid molecular skeleton. In spite of Dewar and Gleicher's prediction that the limiting ring size

1,9-decadiyne

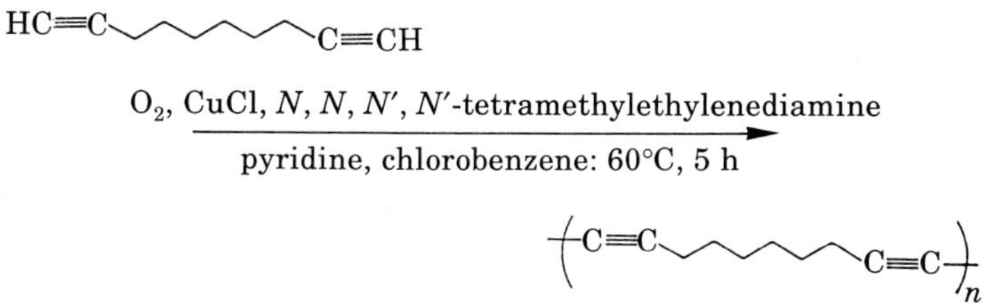

$$HC\equiv C\diagdown\diagup\diagdown\diagup\diagdown C\equiv CH$$

$$\xrightarrow[\text{pyridine, chlorobenzene: }60°C,\ 5\ h]{O_2,\ CuCl,\ N,N,N',N'\text{-tetramethylethylenediamine}}$$

$$\left(\!C\!\equiv\!C\diagdown\diagup\diagdown\diagup\diagdown C\!\equiv\!C\!\right)_n$$

1,4-diethynyl-2,5-dibutoxybenzene

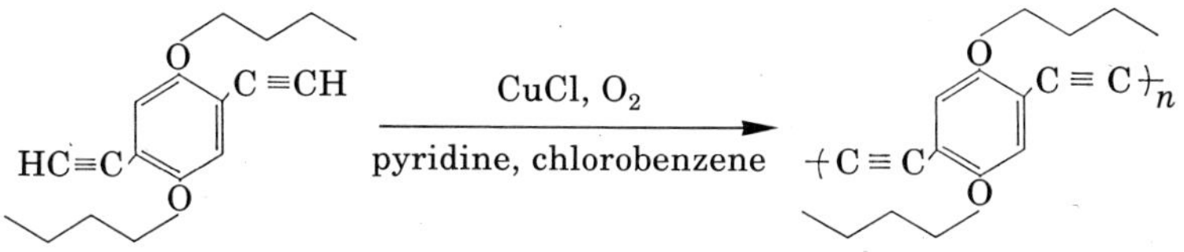

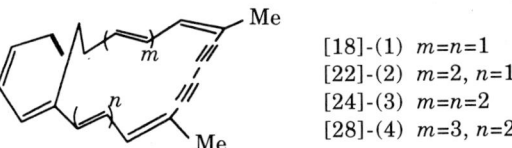

$$\xrightarrow[\text{pyridine, chlorobenzene}]{CuCl,\ O_2}$$

Fig. 3. – Synthetic route of polydiynes.

of Huckel's prediction on aromaticity should occur at a 22-membered ring[14], tetradehydromethano-[28]-annulene (4) shows paratropic properties[13].

The main peaks of the optical absorption spectra for the annulenes (1)-(4) dissolved in poly(methyl methacrylate) (PMMA) are located around 400 nm. The maxima of [4n + 2]-annulenes are located at longer wavelengths than those of [4n]-annulenes. Thus it is evident that in these molecularly doped polymer films the same alternation occurs in the wavelength of the main electronic absorption maxima between [4n + 2]- and [4n]-annulene systems, as has been demonstrated for tetradehydromethanoannulenes and monocyclic annulenes in solution[15].

2'4. *Metallophthalocyanines* (MPc's). – In the electronic spectra of MPc's, two characteristic absorption bands are well assigned, *e.g.* the Soret band ((300–400) nm) and the Q-band (π-π^* transition in the (600–800) nm region). The latter is sensitive to the environment such as orientation and packing of MPc rings. In general, the central metal has little effect on the electronic state of phthalocyanine but a strong influence on the packing arrangement of the

[18]-(1) $m=n=1$
[22]-(2) $m=2,\ n=1$
[24]-(3) $m=n=2$
[28]-(4) $m=3,\ n=2$

Fig. 4. – Molecular structure of annulenes.

phthalocyanine molecules in the condensed state. Therefore, depending on the central metal, features of the Q-band change remarkably in the condensed state. This Q-band has been widely studied as a probe of the phase transitions induced by thermal and/or solvent vapour treatment. MPc's studied in the thin-film preparation by physical vapour depositions were $M = H_2$ (metal free), VO (vanadyl), TiO (titanyl), Sn, Co and Ni. In case of MPc's, polymorphism was observed, hence one can expect that the packing structure of the film grown on the substrate by means of vacuum deposition is easily affected by the interaction with the substrate.

The major problem encountered with unsubstituted MPc systems is poor processability, particularly poor solubility in organic solvents. Taking the waveguide application into account, the materials should be designed to meet several fabrication requirements. To accomplish these, metallization and chemical modification such as introduction of functional groups can be used to improve the physical properties of MPc molecules. For example, by introduction of functional groups, $e.g.$, alkyl, alkoxy, trimethylsilyl and sulfamide groups, into the peripheral position of the benzo rings, an excellent solubility in common organic solvents can be achieved[16]. Tetrakis($tert$-butyl)metallophthalocyanine compounds (MPc(t-bu)$_4$'s : $M = H_2$, VO, Ni) shown in fig. 5 were synthesized from 4-$tert$-butylphthalonitrile according to procedures described in the literature[17-19]. The products were thoroughly purified by column chromatography on silica gel (Merk Kieselgel 60) using chloroform as an eluent, followed by precipitation from chloroform to methanol. The details of the synthetic route have been reported elsewhere[7-10]. MPc(t-bu)$_4$'s show excellent solubility and can be formed into thin films by a spin-coating technique. These can also be dissolved into a polymer solution such as PMMA/chloroform. The

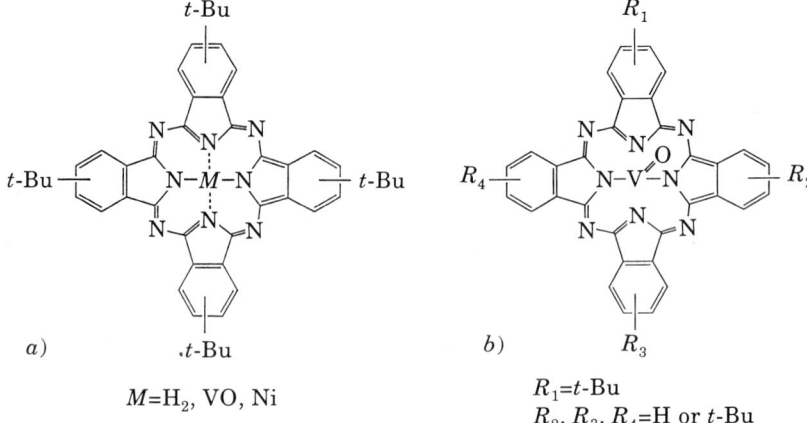

Fig. 5. – Molecular structures of a) tetrakis(t-butyl)phthalocyanines and b) VOPc(t-bu$_n$).

molecularly doped polymer film is formed easily on various kind of substrates by a spin-coating technique and shows excellent optical quality.

We could also synthesize VOPc(t-bu)$_{1.1}$, the mixture of mono- and disubstituted derivatives (fig. 5b)), which was confirmed by field desorption mass spectroscopy (FDMS): m/z 691, 635; UV-VIS (CHCl$_3$) λ_{max} 697, 629 and 346 nm; elemental analysis (%): C 68.23, H 3.94, N 17.32. After VOPc(t-bu)$_{1.1}$ and PMMA at different weight ratios were dissolved in chloroform, the doped PMMA thin films were made by spin-coating onto a 1 mm thick fused-silica substrate. The surface accuracy of the substrate was within 0.5 wavelength at 632.8 nm. Spin-coating was carried out under the condition of about 3000 r.p.m. The obtained sample was divided into two parts, that is, one was as-prepared film, and the other film was treated with dichloroethane vapour in a desiccator at room temperature (see subsect. 4˙2).

3. – Preparation of thin films.

3˙1. *Preparation of organic ultrathin films*. – In general ultrathin-film fabrication processes for organics include techniques from both liquid and gas phases, *i.e.* spin-coating technique, Langmuir-Blodgett method, liquid-phase epitaxy (adsorption), chemical vapour deposition (CVD) and physical vapour deposition (PVD) [20]. The spin-coating technique is a wet process and the simplest way to fabricate a film more than 10 nm in thickness is discussed in sect. 2. CVD and PVD, on the other hand, are dry processes. The CVD process is based on the decomposition and/or radical generation of chemical species (monomers) by stimulating vapour with heat, plasma (discharge) or light (laser), followed by the film formation on a solid substrate. In the case of radical generation, the films form a cross-linked polymer network, whereas, in the case of decomposition of organometallics, the well-organized monolayer of metal and/or semiconductor can be formed on a single-crystal substrate (chemical vapour epitaxy).

The PVD process includes two basic techniques, *i.e.* a vacuum deposition and a sputtering. In the vacuum deposition technique, the substances are heated at higher than melting or sublimation point under a vacuum of 10^{-6} Torr, followed by deposition onto the solid substrate. This technique enables 1) a control of the higher-order structure of the film by varying the evaporation speed and/or substrate temperature, and 2) a creation of hybrid film in which different types of molecules are molecularly mixed by employing multiple evaporation sources and simultaneously depositing. For organic substances the first step of heating/evaporation may give thermal decomposition, and hence particular cares should be taken into account. In the sputtering technique, an inert gas such as Ar and Xe is introduced in the vacuum chamber, ionized to form plasma by applying a high-frequency electric field, accelerated by a d.c. field, then bombards a target material, whose molecules, as a result, fly off and reach the substrate to form a film. When the accelerating field is too large, the decomposition of target substances occurs easily.

The laser ablation technique is an alternative to the vacuum evaporation/deposition technique; instead of heating in a crucible, the target material can be selectively evaporated by a laser beam. This is available not only for the film formation but also for the lithography of a polymer film.

3˙2. *Organic molecular-beam epitaxy.* – The molecular-beam epitaxy (MBE) technique is one of the PVD processes and a most promising technique to control molecular orientation and packing in two dimensions on an atomically flat substrate such as a cleaved single crystal and layered materials (graphite, transition metal dichalcogenides). We have developed an MBE apparatus particularly for organic materials. The primary difficulties of the MBE method for organic molecular systems are their inherent high vapour pressure, thermal instability, multiplicity of the condensed structures (phases) and impurity contents of other substances. A further difficulty in epitaxial growth of organic thin films is the common large lattice mismatch between organic overlayer and inorganic substrate. Compared to the lattice constant of an inorganic substrate, the lattice constant of an organic crystal is quite large and hence the simple concept of lattice mismatch is not sufficient to explain the orientational growth of organic molecules. Various orientational-growth mechanisms for organic molecules have been proposed. In the case of the usual single crystalline substrates which have dangling bonds on the surface, nucleus growth sites are randomly located for large organic molecules, resulting in three-dimensional island structures which are generated by multicrystalline growth. More recently, the van der Waals epitaxy method[21] has shown to provide outstanding inorganic heterostructures, even between materials having a large lattice mismatch. In this heteroepitaxial method, the film growth proceeds *via* van der Waals interactions between layered inorganic materials, such as a transition metal dichalcogenide (MX_2), which has no dangling bonds on its cleaved surface. The first *in situ* observation of van der Waals epitaxy for an organic thin film on a layered inorganic substrate was achieved in combination of MPc molecules with a MoS_2 substrate [22].

Figure 6a) shows the RHEED (reflection high-energy electron diffraction) pattern from the as-prepared KCl surface at the [110] azimuth. The spacing between streaks in the pattern corresponds to the characteristic KCl lattice spacing. The change in the pattern after opening the shutter to the CuPc source is shown in fig. 6b). Although this pattern shows rather spotty features due to diffraction from a three-dimensional structure, the crossed RHEED pattern provides the growth direction, especially the *b*-axis orientation. When the electron beam is parallel to the [100] orientation of KCl, the deduced angle of the *b*-axis from the crossed RHEED pattern is 32° from the normal to the substrate

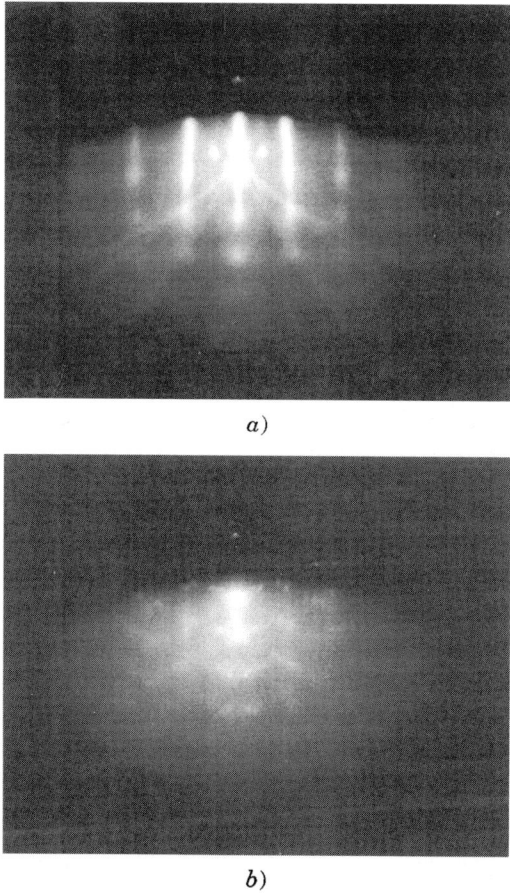

a)

b)

Fig. 6. – a) RHEED pattern of KCl at [110] azimuth and b) crossed RHEED pattern from CuPc layers on KCl at the [110] azimuth.

surface. When the substrate is rotated by 45° in the plane, the angle changes to 24° (fig. 6b)). This indicates that the RHEED pattern can also reveal the growth direction from the initial stage for organic 2-D films during the actual film growth. These data are in very good agreement with earlier electron microscopy results on large crystalline domains. TADA et al. [23] also studied the initial crystal structure of VOPc thin film grown on a KBr substrate and proposed the square lattice with a fourfold symmetry of VOPc and the molecular planes of VOPc parallel to the cleaved (001) face of KBr. In the condensed state, the broadening in the Q-band was observed as a result of exciton splitting in the Q-band due to transition dipole interactions between adjacent VOPc molecules in aggregates.

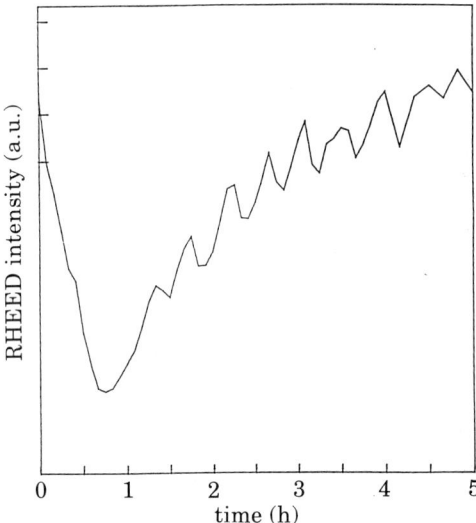

Fig. 7. – RHEED intensity changes of the specular beam during CuPc layer growth.

In order to investigate the possibility of RHEED intensity oscillations during organic-molecular-layer growth, a CCD camera imaging of the RHEED pattern in two dimensions was carried out. Figure 7 shows the intensity change as a function of time of the RHEED pattern from CuPc layers on MoS$_2$. Some periodic structure is apparent in the intensity profile corresponding to the layer growth. While further detailed investigations are required changing the incidence angles of the electron beam and the growth conditions, fig. 7 is a definite indication that the RHEED intensity oscillation occurs even in the organic-molecular-layer growth. If this sort of monitoring is established generally, molecular-layer phase-locked epitaxy (PLE) can be managed by the feedback of results of the intensity analysis even in organic molecular systems. From this point of view, the present result is indeed an encouraging first step toward eventually realizing novel «nanoscopic» material structures such as organic or organic-inorganic molecular superlattices replacing the conventional inorganic ones.

3˙3. *Metallophthalocyanine thin films*. – Figure 8 indicates the various approaches to fabricate a thickness-controlled thin film of MPc's in our laboratory. The vacuum deposition (VD) technique is widely available for unsubstituted MPc's because of their stability. The spin-coating technique, on the other hand, is available for substituted MPc's molecularly dispersed in a polymer matrix such as PMMA. The films prepared by the spin-coating technique are good enough for the optical-waveguide application.

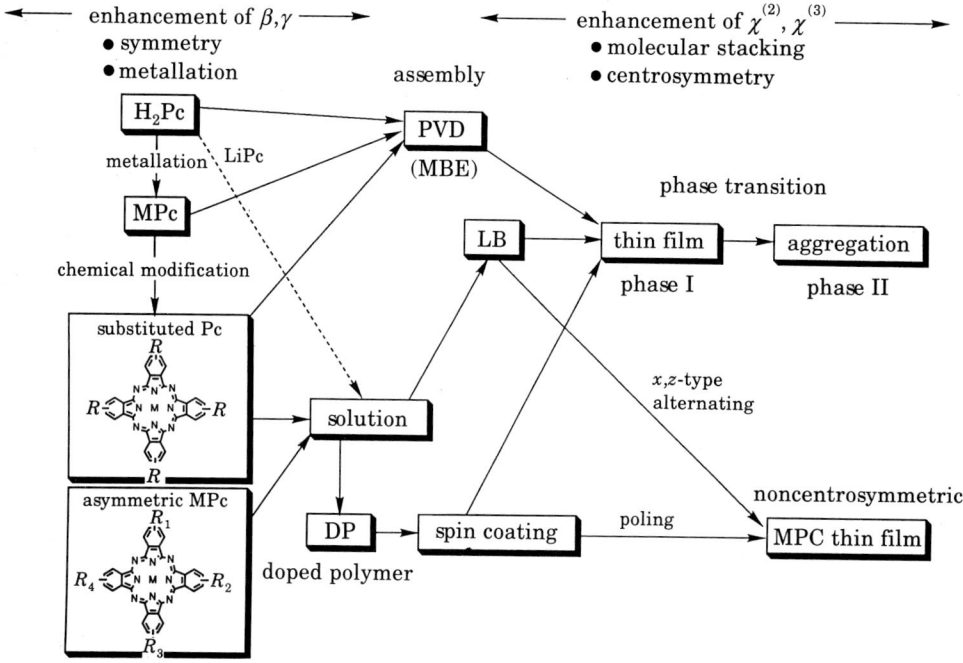

Fig. 8. – Thin-film fabrication of phthalocyanines.

4 – Third-order optical nonlinearity.

4'1. *Third-harmonic generation* (THG) *measurements*. – THG measurements were carried out at fundamental wavelengths of 1907 and 1543 nm in a vacuum of several Torr to eliminate the effect of air. The THG Maker fringe pattern was obtained by rotating the film sample. The optical layout has been described in detail elsewhere[24]. The rotational THG Maker fringes were analysed by standard procedures[25], and the TH intensity $I_{3\omega}$ in an absorbing medium is given as

$$(3) \qquad I_{3\omega} = (64\pi^4/c^2)[A\chi^{(3)}]^2 I_\omega^3 f_a ,$$

where I_ω is the light intensity for the fundamental frequency, c the velocity of light, A the factor arising from transmission and boundary conditions, and f_a is the absorption-dependent factor:

$$(4) \quad f_a = \{[1 - \exp[-\alpha_{3\omega}d/2]]^2 + (\Delta\Psi)^2 \exp[-\alpha_{3\omega}d/2]\}/$$

$$/[(n_{3\omega}^2 - n_\omega^2 - k_{3\omega}^2)^2 + (2n_{3\omega}k_{3\omega})^2],$$

where n and k are, respectively, the real and imaginary parts of the refractive indices, α the linear absorption coefficient, d the sample thickness and $\Delta\Psi$ the phase mismatch between the fundamental and harmonic frequencies. The $\chi^{(3)}$

TABLE I. – $\chi^{(3)}$ *of polythiophenes, polydiynes and annulenes at* 1907 nm (*in* 10^{-13} e.s.u.).

fused silica	0.14
poly(3-dodecyloxymethylthiophene)	25
poly(1,9-decadiyne) (PDD)	0.5
cross-linked PDD	5
poly(1,4-diethynyl-2,5-dibutoxybenzene)	60
annulene (1)/PMMA (< 1.0 wt.%)	1.1
annulene (2)/PMMA (< 1.0 wt.%)	1.9
annulene (3)/PMMA (< 1.0 wt.%)	2.0
annulene (4)/PMMA (< 1.0 wt.%)	1.6

values of the organic films were estimated with reference to the $\chi^{(3)}$ of fused silica ($1.40 \cdot 10^{-14}$ e.s.u. at 1907 nm [26], $1.47 \cdot 10^{-14}$ e.s.u. at 1543 nm estimated using Miller's rule), as summarized in table I.

4‘2. *Enhancement of* $\chi^{(3)}_{1111}(-3\omega; \omega, \omega, \omega)$ *in* MPc's. – The typical result of rotational THG Maker fringes of metal-free H_2Pc(t-bu)$_4$ is shown in fig. 9. An oscillating fringe pattern of the fused-silica substrate as a reference is observed

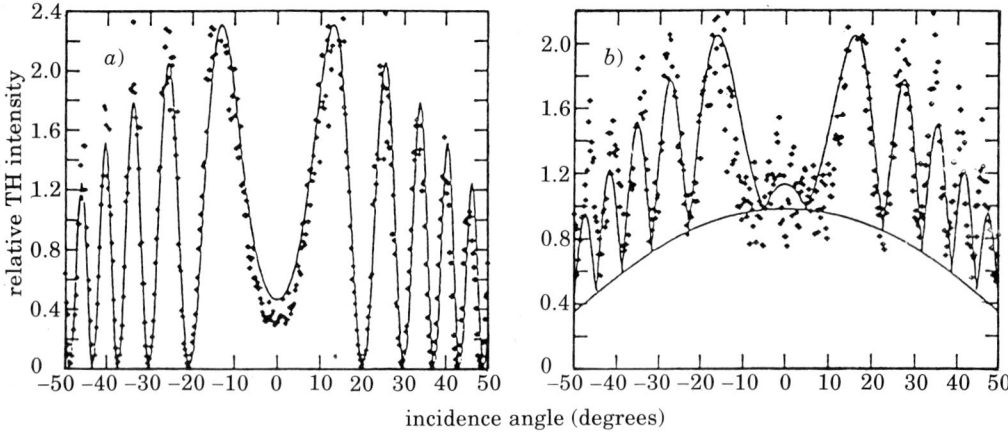

Fig. 9. – THG Maker fringe patterns of *a*) fused-silica substrate (1 mm thick) and *b*) H_2Pc(t-bu)$_4$ (0.15 μm thick) on fused silica.

(fig. 9a)). The solid line indicates the calculated curve of nonabsorbing media [25], i.e.

$$(5) \qquad I_{3\omega} = (256\pi^4/c^2)[A^*\chi^{(3)}/(n_\omega^2 - n_{3\omega}^2)]^2 I_\omega^3 \sin^2(\Delta\Psi/2),$$

where A^* is an overall transmission factor. From the curve-fitting method the $\chi^{(3)}$ value of fused silica can be confirmed. Since the TH intensity of the $H_2 Pc(t\text{-bu})_4$ film is not so large, the fringe pattern of a fused-silica substrate is still observed on the envelope arising from the sample. This envelope curve also fits well eq. (3).

Values of $\chi^{(3)}$ of various phthalocyanine thin films are summarized in table II. The $\chi^{(3)}$ values of unsubstituted MPc's ($M = H_2$ (metal free), Co, Ni, Sn, VO, TiO) are strongly dependent on the central metal species, and a vanadylphthalocyanine (VOPc) film shows the largest value, which is comparable to that of polydiacetylenes. However, MPc(t-bu)$_4$'s show little difference in the $\chi^{(3)}$ values. Investigating the origin of the large $\chi^{(3)}$ value in VOPc is of importance for the understanding of third-order nonlinear responses in macrocyclic compounds. GRIFFITHS et al. reported the polymorphism of VOPc [27]: phase I is attributed to the cofacial packing of VOPc molecules (aligned linearly along the metal-oxo bond: optical absorption peaks at 680 and 740 nm), and phase II to the slipped-stack arrangement (triclinic crystal structure with $P\bar{1}$ space group, optical absorption peak at 820 nm) (fig. 10). The as-prepared VOPc film (VD) has the phase-I arrangement, and is easily converted to the phase II by thermal annealing at 125 °C for more than 15 h. It should be noted this morphological change occurs also in TiOPc. Unlike the VOPc film, the VOPc(t-bu)$_4$ spin-coated film showed no evidence of crystalline structure, even after thermal treatment, presumably due to a steric hindrance of the bulky $tert$-butyl groups. Therefore, in order to reduce the steric hindrance of bulky substituents, an almost monosubstituted VOPc(t-bu)$_{1.1}$ was designed. A spin-coated thin film of

TABLE II. – $\chi^{(3)}$ of various phthalocyanines at 1907 nm (in 10^{-12} e.s.u.).

MPc	H$_2$	Co	Ni	Sn	VO		TiO	
					p–I[*1]	p–II[*2]	p–I[*1]	p–II[*2]
unsubstituted:								
MPc	3.0	7.5	2.3	40	38	81	10	46
substituted:								
MPc(t-bu)$_4$	1.9	—	2.0	—	6.0	—	—	—
MPc(t-bu)$_{1.1}$	—	—	—	—	5.0	10	—	—
MPc(t-bu)$_{1.1}$/PMMA (39.2 wt.%)	—	—	—	—	1.2	6.0	—	—

[*1] phase I, [*2] phase II (thermally annealed at 125 °C for 15 h or treated with dichloroethane vapour for 20 h).

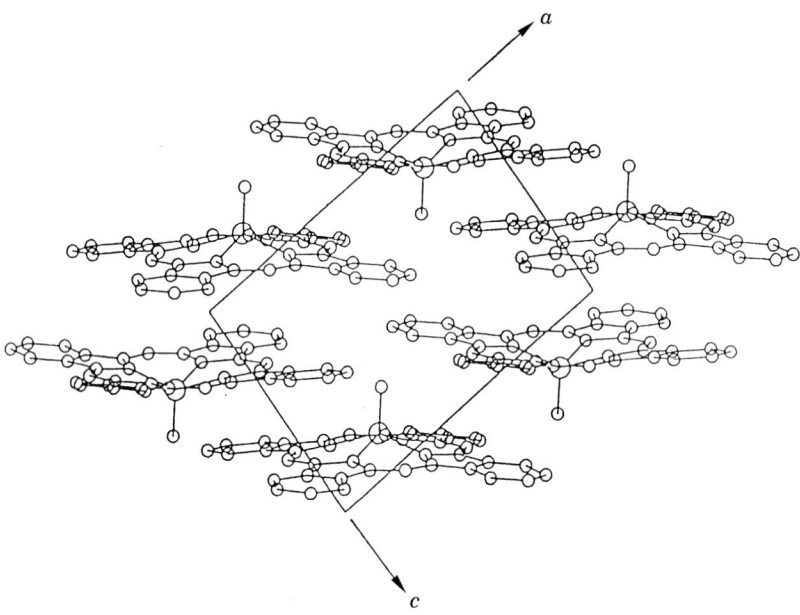

Fig. 10. – Projection of molecular packing of phase II of VOPc along the b-axis, space group $P\bar{1}$, $a = 12.027$ Å, $b = 12.571$ Å, $c = 8.690$ Å, $\alpha = 96.04°$, $\beta = 94.80°$, $\gamma = 68.20°$.

VOPc(t-bu)$_{1.1}$ (9.8 wt.% in PMMA) is optically clear and has absorption maxima at 650 nm and 695 nm (curve a) in fig. 11). These Q-bands are generally influenced by the molecular situation, and hence the aggregation in a condensed state is important in the concentration range studied. A drastic absorption change in the Q-band was observed at various concentrations after the dichloroethane vapour treatment. As illustrated in fig. 11, curve b), a new peak at 810 nm appeared and increased with the decrease of the peak intensity at 650 nm and 695 nm. The spectral change ceased after 20 h under these conditions. These spectral changes are similar to those of a VOPc thin film, which are caused by thermal annealing. The phase transition behaviour of VOPc(t-bu)$_{1.1}$ is shown to be very similar to that of VOPc.

The third-order nonlinear optical properties of them were determined by the rotational THG Maker fringe method at 1543 nm and 1907 nm. TH light at both wavelenghts is generated around the Q-band. Since the absorption change in the Q-band is relatively large before and after treatment, the determination of precise $\chi^{(3)}$ values corrected for absorption factors is considerably important in comparison to different $\chi^{(3)}$ values caused by the packing control. After vapour treatment, the macroscopic $\chi^{(3)}$ values increased by a factor of 2–7 at both wavelengths [28]. These results indicate that the $\chi^{(3)}$'s are larger in the red-shifted phase than in the other phase. It should be added here that the optical quality and mechanical strength of VOPc(t-bu)$_{1.1}$/PMMA are adequate to

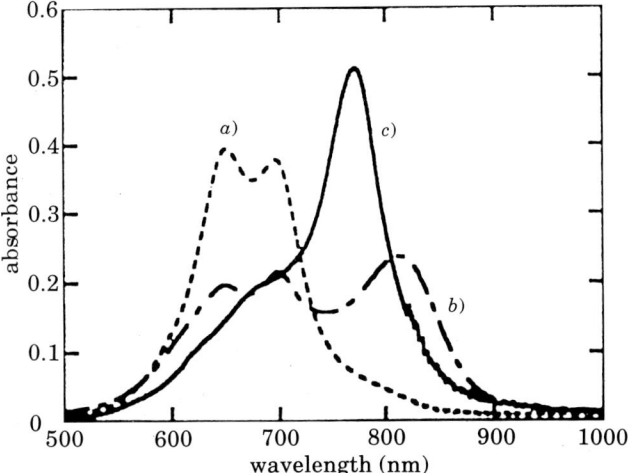

Fig. 11. – Absorption spectra in polystyrene thin film doped with 10 wt. % of VOPc(t-bu)$_{1.1}$: a) phase-I rich, b) phase-II rich and c) VOPc thin film grown by MBE on KBr.

make a thin-film optical waveguide by a conventional polymer processing procedure.

4`3. *Femtosecond response measurement.* – Exciton decay dynamics in VOPc(t-bu)$_n$ thin films were investigated using femtosecond pump-probe spectroscopy. The optical layout and a detailed description of the detection of transmittance changes are described elsewhere [29].

The differential transmission spectra (DTS) of phase-II films of VOPc(t-bu)$_{1.1}$ doped in polystyrene (10%) and of MBE-grown VOPc film on KBr are shown in fig. 12. The key feature of DTS is that a significant bleaching appears at the lowest-energy absorption peak (centred around 810 nm), while the higher-energy absorption peaks bleach only weakly. This indicates either that excitons created by the femtosecond pulses at 620 nm undergo very rapid internal conversion to the lowest excited state, or that the induced excited-state absorption in this energy region cancels the bleaching signals. In the spectral range of (500–600) nm (corresponding to the optical window of the ground-state absorption), excited-state-induced absorption was observed, which was much weaker than the bleaching signal.

The decay curve of the transmittance change consists of three components: a fast (hundreds of femtoseconds), a slower (tens of picoseconds) and a long-lived (longer than hundreds of picoseconds) component. The fast component is interpreted as a bimolecular process, *i.e.* exciton-exciton annihilation, because of its excitation intensity dependence. On the other hand, the slower decay is rather independent of the excitation intensity and can be attributed to a unimolecular

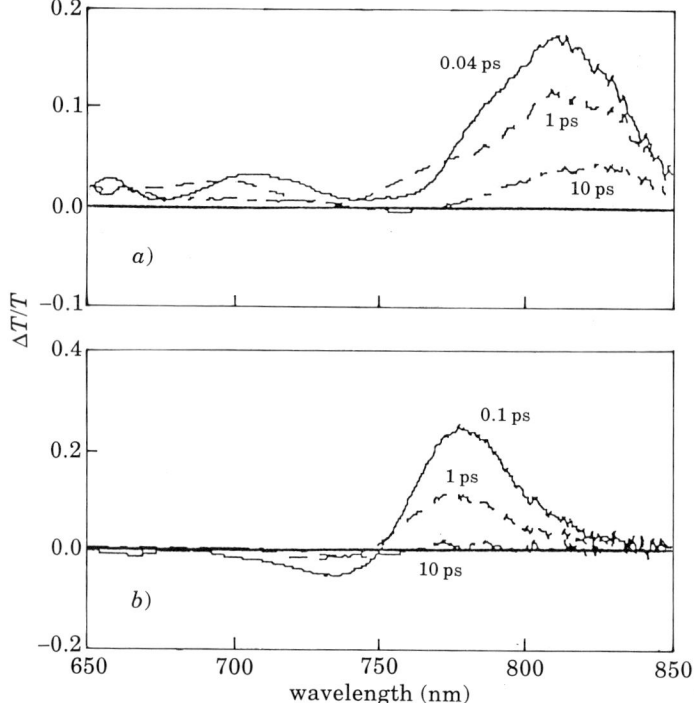

Fig. 12. – Differential transmission spectra of *a*) phase-II film of VOPc(t-bu)$_{1.1}$ doped in polystyrene (10%) and *b*) of MBE-grown VOPc film on KBr.

decay *via* exciton-phonon coupling. The long-lived component is presumably due to triplet-state formation widely observed in MPc's. To fit the initial decay curve we employed the exciton-exciton annihilation model with a time-dependent rate of $t^{-1/2}$ based on the long-range dipole-dipole interaction between excitons, or the motion-limited exciton diffusion. In this model the exciton density (n) was described by the following equation for a delta-function excitation pulse:

$$(6) \qquad\qquad dn/dt = \gamma t^{-1/2} \kappa n \,,$$

where γ and κ are, respectively, a bimolecular and a unimolecular decay constant. Solving eq. (6), the following time dependence of the exciton density is obtained:

$$(7) \qquad n(t)/n_0 = \exp[-\kappa t]/\{1 + (2n_0\gamma/\kappa^{1/2})\,\mathrm{erf}[(\kappa t)^{1/2}]\}\,,$$

where $\mathrm{erf}(x)$ is the error function. We have added a constant term which phenomenologically accounts for the long-lived component in eq. (7). This function can fit the data over a time range up to 100 ps, and yields fitting parameters of $n_0\gamma = 9.1 \cdot 10^5$ s$^{-1/2}$, $\kappa = 2 \cdot 10^{10}$ s^{-1} and a constant term which is ca. 10% of the maximum value of the data. The value of γ of pure VOPc(t-bu)$_{1.1}$ spin-coated film is estimated as $1.6 \cdot 10^{-14}$ cm^3 s$^{-1/2}$. MBE-grown VOPc film, on the other hand, showed a rather fast decay of the bleaching, presumably due to the different packing arrangements. It is, therefore, suggested that the long-range ordering in molecular packing in the MBE-grown VOPc film affects the relaxation excitons.

5. – Implementation of an optical waveguide.

For optically active device application, the intensity-dependent refractive-index change (optical Kerr effect) is essential; the refractive index n of the material depends on the incident light intensity I as

$$(8) \qquad n = n_0 + n_2 I, \qquad n_2 = \chi^{(3)}_{1111}(-\omega; \omega, -\omega, \omega)/n_0^2 \varepsilon_0 c,$$

where n_0 and n_2 are linear and nonlinear refractive indices, respectively. Therefore, large-$\chi^{(3)}$ materials are required for this application in addition to the transparency. For the direct observation of light propagation in the spin-coated doped polymer thin film on the fused-quartz substrate, a c.w. YAG laser beam (1064 nm) was introduced by means of the prism coupling technique and monitored by an i.r.-sensitive CCD camera (fig. 13).

In the case of unsubstituted MPc's dispersed into PMMA with different concentration, the value of $\chi^{(3)}$ increased with increasing MPc concentration, but the aggregation of MPc molecules was observed mostly due to their poor solubility. On the contrary, in the case of substituted soluble MPc's, MPc molecules molecularly dispersed in PMMA and no aggregation was observed. The spin-coated VOPc(t-bu)$_{1.1}$/PMMA film on a fused-quartz substrate has a high optical quality and is available for optical waveguide. As shown in fig. 14, the trace of

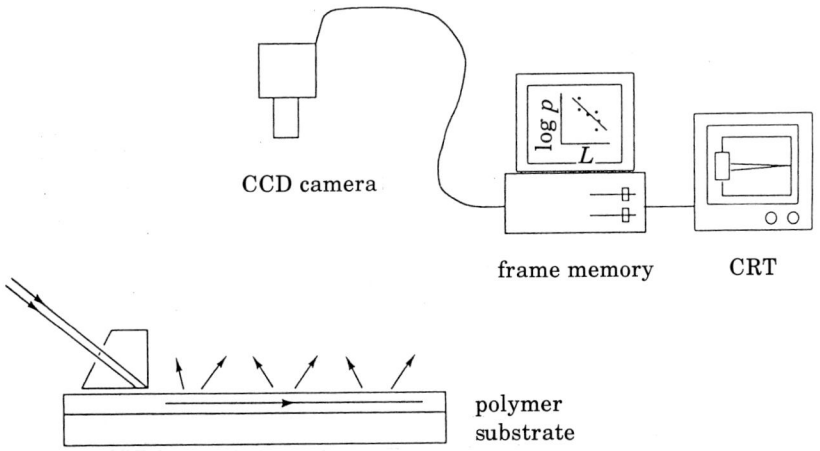

Fig. 13. – Schematic diagram for the measurement of propagating light in the film.

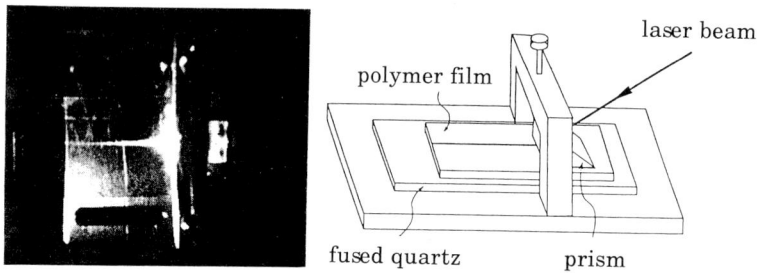

Fig. 14. – The trace of guided light in the VOPc(t-bu)$_{1.1}$/PMMA film introduced by means of the prism coupling technique (left). Schematic representation (right).

the c.w. YAG (1064 nm) laser beam can be detected by an i.r. CCD camera. The propagation loss is estimated as 3.4 dB/cm. As discussed in subsect. 4'2, the $\chi^{(3)}$ value of the VOPc(t-bu)$_{1.1}$/PMMA film can be increased with dichloroethane vapour treatment. Even after the treatment no significant increase in the propagation loss in the guided-wave geometry was observed due to the scattering from the aggregated particles. Here it should be noted that the X-ray diffraction study suggested the domain size of associated VOPc(t-bu)$_{1.1}$ molecules to be on the order of 10 nm. Other types of waveguide configuration such as slab type (four-layered structure) and channel type using substituted MPc/polymer systems are reported elsewhere[30].

6. – Conclusion.

We described our material research on linear and macrocyclic π-conjugated systems, especially metallophthalocyanine (MPc) derivatives, for the third-order nonlinear optics. Both molecular design and control of assembled states of MPc's are important to enhance macroscopic third-order nonlinear optical responses. Enhancement of third-order susceptibility was observed in VOPc vacuum-deposited film and VOPc(t-bu)$_{1.1}$ spin-coated film after a thermally and/or chemically induced phase transition, *i.e.* changes of molecular stacking arrangement. A processable polymeric system based on VOPc(t-bu)$_{1.1}$ was developed for the optical waveguide application.

* * *

This work has been done under the Frontier Research Program, RIKEN (The Institute of Physical & Chemical Research). The authors express their thanks to Prof. A. F. GARITO (University of Pennsylvania) for his critical discussion and to other members of Laboratory for Nano-Photonics Materials of FRP for their help.

REFERENCES

[1] A. F. GARITO, J. R. HEFLIN, K. Y. WONG and O. ZAMANI-KHAMIRI: in *Nonlinear Optical Properties of Polymers*, edited by A. J. HEEGER, J. ORENSTEIN and D. R. ULRICH, *MRS Symposium Proceedings*, Vol. 109 (Materials Research Society, Pittsburgh, Pa., 1988), p. 91.
[2] J. P. HERMANN and J. DUCUING: *J. Appl. Phys.*, 45, 5100 (1974).
[3] J. R. HEFLIN, K. Y. WONG, O. ZAMANI-KHAMIRI and A. F. GARITO: *Phys. Rev. B*, 38, 1573 (1988).
[4] J. W. WU, H. R. HEFLIN, R. A. NORWOOD, K. Y. WONG, O. ZAMANI-KHAMIRI, A. F. GARITO, P. KALYANARAMAN and J. SOUNIK: *J. Opt. Soc. Am. B*, 6, 91, 707 (1988).
[5] T. WADA, A. F. GARITO, H. SASABE, H. HIGUCHI and J. OJIMA: in *Nonlinear Optics*, edited by S. MIYATA (Elsevier Sci. Publ. B. V., Amsterdam, 1992), p. 299.
[6] T. WADA, J. OJIMA, A. YAMADA, A. F. GARITO and H. SASABE: in *Nonlinear Optical Properties of Organic Materials II*, edited by G. KHANARIAN, *SPIE Proc.*, 1147 (SPIE, Bellingham, 1989), p. 286.
[7] T. WADA, S. YAMADA, Y. MATSUOKA, C. H. GROSSMAN, K. SHIGEHARA, H. SASABE, A. YAMADA and A. F. GARITO: in *Nonlinear Optics of Organic and Semiconductors*, edited by T. KOBAYASHI, *Springer Proc. in Physics*, Vol. 36 (Springer-Verlag, Berlin, 1989), p. 292.
[8] H. SASABE, T. WADA, M. HOSODA, H. OHKAWA, M. HARA, A. YAMADA and A. F. GARITO: in *Nonlinear Optical Properties of Organic Materials III*, edited by G. KHANARIAN, *SPIE Proc.*, 1337 (SPIE, Bellingham, 1990), p. 62.
[9] M. HOSODA, T. WADA, A. YAMADA, A. F. GARITO and H. SASABE: *Jpn. J. Appl. Phys.*, 30, 1715 (1991).
[10] M. HOSODA, T. WADA, A. YAMADA, A. F. GARITO and H. SASABE: *Jpn. J. Appl. Phys.*, 30, L1486 (1991).
[11] T. WADA, M. HOSODA, A. F. GARITO, H. SASABE, A. TERASAKI, T. KOBAYASHI, H. TADA and A. KOMA: in *Nonlinear Optical Properties of Organic Materials IV*, edited by K. SINGER, *SPIE Proc.*, 1560 (SPIE, Bellingham, 1991), p. 162.
[12] H. SASABE, T. WADA, M. HOSODA, H. OKAWA, A. YAMADA and A. F. GARITO: *Mol. Cryst. Liq. Cryst.*, 189, 155 (1990).
[13] J. OJIMA, E. EJIRI, T. KATO, M. NAKAMURA, S. KURODA, S. HIROOKA and M. SHUBUTANI: *J. Chem. Soc., Perkin Trans. I*, 831 (1987).
[14] M. J. S. DEWAR and G. J. GLEICHER: *J. Am. Chem. Soc.*, 87, 685 (1965).
[15] J. OJIMA, S. FUJITA, M. MASUMOTO, E. EJIRI, T. KATO, S. KURODA, Y. NOZAWA, S. HIROOKA, Y. YONEYAMA and H. TATEMITSU: *J. Chem. Soc., Perkin Trans. I*, 385 (1988).
[16] E. A. CUELLER and T. J. MARKS: *Inorg. Chem.*, 20, 3766 (1981).
[17] S. K. MIKHALENKO, S. V. BARKANOVA, O. L. LEBEDEV and E. A. LUK'YANETS: *J. Gen. Chem. USSR*, 81, 2770 (1971).
[18] K.-Y. LAW: *Inorg. Chem.*, 24, 1778 (1985).
[19] A. W. SNOW and N. L. JAVIS: *J. Am. Chem. Soc.*, 106, 4706 (1984).
[20] H. SASABE: *Hybrid Materials - Concepts and Case Studies* (ASM International, Ohio, 1988).
[21] A. KOMA and K. YOSHIMURA: *Surf. Sci.*, 174, 556 (1986).
[22] M. HARA, H. SASABE, A. YAMADA and A. F. GARITO: *Jpn. J. Appl. Phys.*, 28, L306 (1989).
[23] H. TADA, K. SAKAI and A. KOMA: *Jpn. J. Appl. Phys.*, 30, L306 (1991).

[24] T. WADA, Y. MATSUOKA, K. SHIGEHARA, A. YAMADA, A. F. GARITO and H. SASABE: in *Photoresponsive Materials*, edited by M. DOYAMA, S. SOMIYA and R. P. H. CHANG, *Proceedings of the MRS International Meeting on Advanced Materials*, Vol. **12** (Materials Research Society, Pittsburgh, Pa., 1989), p. 75.

[25] F. KAZJAR and J. MESSIER: *Thin Solid Films*, **132**, 11 (1985).

[26] J. R. HEFLIN, Y. M. CAI and A. F. GARITO: *J. Opt. Soc. Am. B*, 8, 2132 (1991).

[27] C. H. GRIFFITHS, M. S. WALKER and P. GOLDSTEIN: *Mol. Cryst. Liq. Cryst.*, **33**, 149 (1976).

[28] M. HOSODA, T. WADA, T. YAMAMOTO, A. KANEKO, A. F. GARITO and H. SASABE: *Jpn. J. Appl. Phys.*, **31**, 1071 (1992).

[29] A. TERASAKI, M. HOSODA, T. WADA, H. TADA, A. KOMA, A. YAMADA, H. SASABE, A. F. GARITO and T. KOBAYASHI: *J. Phys. Chem.*, **96**, 10534 (1992).

[30] T. IMAI, H. OKAWA, T. WADA and H. SASABE: in *Nonlinear Optical Properties of Organic Materials VI*, edited by D. WILLIAMS, *SPIE Proc.*, **1775** (SPIE, Bellingham, 1993), p. 402.

SPECIAL TOPICS

Instantaneous Amplitude and Frequency Dynamics of Coherent Wave Mixing in Semiconductor Quantum Wells.

D. S. CHEMLA

Physics Department, University of California at Berkeley
Material Sciences Division, Lawrence Berkeley Laboratory - Berkeley, Cal.

1. – Introduction.

The delocalized electronic excitations of semiconductors are very strongly coupled by the Coulomb interaction. When a semiconductor is excited close to the fundamental band gap, this interaction renormalizes both the band energies and the Rabi frequency, which measures the coupling to the applied electromagnetic field [1, 2]. The renormalizations are, respectively, proportional to the populations excited in the bands and to the interband polarization. They provide a source of optical nonlinearities which are qualitatively different from the nonlinearities of isolated atomic systems. Atomic nonlinearities originate from Pauli exclusion. They are present in all material systems, including semiconductors, and are essentially instantaneous. Conversely, the Coulomb many-body nonlinearities become visible only when the excitation has produced significant population and polarization densities. Their contribution to nonlinear optical response is delayed and dephased with respect to that due to the Fermi statistics. Therefore, many of the specifities of many-body nonlinearities appear in ultrashort-pulse time-resolved nonlinear optical experiments [3-10].

In this lecture we review recent investigations of the specific features of nonlinear optical processes in semiconductors. It is organized as follows. In sect. **2**, we discuss the theory of coherent wave mixing in semiconductors. We emphasize the case where the excitation is resonant with exciton states. In sect. **3**, we review our recent experimental investigations of the *amplitude and phase* of coherent wave mixing resonant with quasi-2*d* excitons (X) in GaAs quantum wells (QW) [8, 10]. In sect. **4**, we discuss these results and conclude.

2. – Resonant excitonic nonlinearities in semiconductors.

In this section we consider the case of a semiconductor resonantly excited close to the lowest exciton state. Within this approximation, we wish to put the equations describing the time evolution of the exciton polarization and population, in a form simple enough to reveal directly the physics of the nonlinear optical response. We start from the usual two-band model of the semiconductor Bloch equation [11]. Neglecting the photon momentum, we only consider vertical transitions. In this case the density matrix $\hat{n}(t)$, breaks into 2×2 blocks:

$$(1.1) \qquad \hat{n}_k(t) = \begin{bmatrix} n_c(t) & \psi(k) \\ \psi^*(t) & n_v(t) \end{bmatrix},$$

where $n_{c,v}(k)$ are the electron populations in the conduction and valence bands, and $\psi(k)$ is the pair amplitude which, as shown below, is proportional to the polarization (when it is possible to neglect the wave vector dependence of the interband dipole moment). The density matrix $\hat{n}(t)$ satisfies the Liouville equation

$$(1.2) \qquad \frac{\partial}{\partial t} \hat{n}_k(t) = -i[\hat{\varepsilon}_k(t), \hat{n}_k(t)] + \frac{\partial}{\partial t} \hat{n}_k(t)\big|_{\text{relax}},$$

where the Hamiltonian matrix

$$(1.3) \qquad \hat{\varepsilon}_k(t) = \begin{bmatrix} \varepsilon_c(k) & 0 \\ 0 & \varepsilon_v(k) \end{bmatrix} + \begin{bmatrix} 0 & -\mu_k E \\ -\mu_k^* E^* & 0 \end{bmatrix} - \sum_{k'} V_{k,k'} \hat{n}_{k'}$$

comprises three contributions. The first one gives the bare-band energies, $\varepsilon_{c,v}(k) = \varepsilon_{c,v}^{(0)} \pm k^2/2m_{c,v}$. The second expresses the coupling of the interband dipole matrix element, μ_k, with the electromagnetic field E. Finally the third one describes the how the Coulomb potential, $V_{k,k'}$, couples states at different wave vector k. Dephasing is accounted for phenomenologically by the term $(\partial/\partial t)\hat{n}_k(t)|_{\text{relax}}$.

From the diagonal and off-diagonal elements of $\hat{\varepsilon}$, one sees clearly how the Coulomb interaction renormalizes the energies and the Rabi frequency. The difference in these quantities when $V_{k,k'}$ is absent or present is

$$(1.4a) \qquad \varepsilon_j(k) \to \varepsilon_j(k) - \sum_{k'} V_{k,k'} n_j(k')$$

and

$$(1.4b) \qquad \mu_k E \to \mu_k E + \sum_k V_{k,k'} \psi(k').$$

For relaxed populations, the Coulomb terms in eq. (1.4a) describe the well-known «band gap renormalization». This equation accounts as well for the dynamic band gap changes induced by coherent or transients populations. Similarly, the second term of eq. (1.4b) describes the dynamic interaction involving polarization waves at different k. To bring out the effects of the nonequilibrium populations, we transform to the electron-hole representation, $n_c(k) \to n_e(k)$ and $n_v(k) \to 1 - n_h(k)$. From eq. (1.1) we obtain the evolution equation of the populations:

$$(1.5) \quad \left(\frac{\partial}{\partial t} + \frac{\partial}{\partial t}\bigg|_{\text{relax}}\right) n_e(k) = -\left(\frac{\partial}{\partial t} + \frac{\partial}{\partial t}\bigg|_{\text{relax}}\right) n_h(k) =$$

$$= 2\,\text{Im}\left[\psi_k\left(\mu_k E + \sum_{k'} V_{k,k'}\psi(k')\right)^*\right],$$

and of the pair amplitude:

$$(1.6) \quad \left[i\left(\frac{\partial}{\partial t} + \frac{\partial}{\partial t}\bigg|_{\text{relax}}\right) - \left(E_g + \frac{k^2}{2m}\right)\right]\psi(k) + \sum_{k'} V_{k,k'}\psi(k') =$$

$$= -(1 - n_e(k) - n_h(k))\mu_k E +$$

$$+ \sum_{k'} V_{k,k'}[\psi(k)(n_e(k') + n_h(k')) - (n_e(k) + n_h(k))\psi(k')].$$

In eq. (1.6) the first driving term in the right-hand side gives the nonlinearity due to the Fermi statistics, *i.e.* it accounts for the weakening of the coupling with the applied field due to Pauli exclusion. The second line expresses the Coulomb nonlinearity which corresponds to the exchange interaction between populations and polarizations at different wave vectors. A further and implicit source of nonlinearity is due to the dependence of the Coulomb potential, $V_{k,k'}$, itself on the electron and hole populations.

These equations are very complex and are usually solved numerically. Often the physical intuition is lost in the computation. In the case where only excitonic resonances are optically excited, the population and polarization of this state dominate over that of the other states and eqs. (1.5) and (1.6) can be greatly simplified [12]. First we Fourier transform these equations to r-space assuming that the interband dipole element is k-independent $\mu_k \to \mu$. Then, noting that $\phi_v(r)$, the solutions of the exciton Wannier equation in r-space, form an complete orthogonal basis set, we develop the polarization, $\psi(r)$, and populations, $n_e(r)$ and $n_h(r)$, on it:

$$(1.7a) \qquad\qquad \psi(r) = \sum_v \psi_v \phi_v(r),$$

$$(1.7b) \qquad\qquad n_{(e,h)}(r) = \sum_v n_{(e,h)v}\phi_v(r),$$

where

(1.8)
$$\left[\frac{-1}{2m}\nabla_r^2 + V(r)\right]\phi_\nu(r) = E_\nu\phi_\nu(r).$$

The «components», ψ_ν and $n_{(e,h)\nu}$, in the exciton representation are not functions of r, they can be, however, functions of time. Their time evolution is found to be given by

(1.9a)
$$\left[i\left(\frac{\partial}{\partial t} + \Gamma_\lambda\right) - \Omega_\lambda\right]\psi_\lambda = -[L^3\phi_\lambda^*(r=0) - n_{e,\lambda} - n_{h,\lambda}]\mu E +$$

$$+ \sum_{\mu\nu} V_{\lambda\mu\nu}(n_{e,\mu}\psi_\nu + n_{h,\mu}\psi_\nu),$$

(1.9b)
$$i\left(\frac{\partial}{\partial t} + \gamma_{e,\lambda}\right)n_{e,\lambda} = i\left(\frac{\partial}{\partial t} + \gamma_{h,\lambda}\right)n_{h,\lambda} = 2\,\mathrm{Im}\,[\mu E\psi_\lambda^*] +$$

$$+ \sum_{\mu\nu} V_{\lambda\mu\nu}(\psi_\mu^*\psi_\nu - \psi_\mu\psi_\nu),$$

where $\Omega_\lambda = E_g - E_\lambda$. We have defined the nonlocal matrix potential, $V_{\lambda\mu\nu}$, by

(1.10)
$$V_{\lambda\mu\nu} = \int d^3r'\,d^3r\,\phi_\lambda^*(r)\phi_\mu(r-r')V(r')\phi_\nu(r'),$$

and we have introduced the phenomenological damping rates of the exciton states: Γ_λ and $\gamma_{(e,h)\lambda}$. These equations show that the exciton states are equivalent to a set of two-level systems obeying Pauli exclusion and coupled by the nonlocal matrix potential $V_{\lambda\mu\nu}$. They are equivalent to the original eq. (1.2).

In this basis the porization is given by

(1.11)
$$P = \mu^* \sum_\nu \psi_\nu \phi_\nu(r=0).$$

In the case where the semiconductor is excited with ultrashort laser pulses close to the exciton resonances, the bound states whose energy falls within the pulse linewidth contribute to the polarization. It is then possible to average eq. (1.9) over these states to obtain the polarization, P, and the population, N. It is found that they satisfy a coupled system of nonlinear wave equations that captures most of the physics and yet is simple enough to be intuitively interpreted [5,8,9]:

(1.12a)
$$\left[\frac{\partial}{\partial t} + \gamma\right]N = -2\,\mathrm{Im}\,[\mu EP^*],$$

(1.12b)
$$\frac{\partial}{\partial t}P = -i(\Omega - i\Gamma)P + i\mu E - i2N\mu E - 2iVNP.$$

Here, γ, Γ, Ω and V are the proper averages of the corresponding quantities in eq. (1.9), and P and N are normalized to incorporate the volume, L^3, and the exciton enhancement factors, $|\phi_v(r = 0)|^2$.

The first equation shows that the exciton states are simply populated by absorption exactly as in the case of a two-level system (2LS). The second shows that the exciton behaves as a driven nonlinear oscillator. The first line describes the linear response of the oscillator driven by the applied field, while the two terms on the second line account for the nonlinear response. The first nonlinear term originates from the Pauli-exclusion reduction of the exciton coupling with the electromagnetic field, $\mu E(t) \rightarrow (1 - 2N)\mu E(t)$. This term is, of course, always present and accounts for the atomic-type nonlinearities. The second nonlinear term is specific to dense-media semiconductors and molecular crystals [4-6, 9, 13]. In semiconductors, it describes the Coulomb mediated exciton-exciton interaction which causes the coupling between populations and polarization waves within the medium. Furthermore, in a semiconductor, the Coulomb potential is sensitive to the nonequilibrium populations through dynamic screening, V itself is a function of N and P. This dependence provides an additional nonlinearity. We call the two nonlinear terms the phase space filling (PSF) and the exciton-exciton interaction (XXI) terms, respectively. For excitation, where $E(t)$ is an ultrashort optical pulse, these two terms have a very different temporal behavior. In particular, when eq. (1.11) is solved in power expansion of $E(t)$, one can clearly see that the XXI contribution appears *delayed and out of phase* with that of the PSF term! Finally let us note that the XXI term has the same form as the term introduced by GYNZBURG and LANDAU to describe the mechanism that, close to the transition temperature, drives metals toward a superconducting state. In the case of small excitation densities where $N \approx P^2$, we can cast eq. (1.11b) in the form

$$(1.13) \qquad \frac{\partial}{\partial t} P = -i(\Omega - i\Gamma)P + i\mu E - i\frac{|P|^2}{P_s^2}\mu E - iV|P|^2 P.$$

This equation was extensively discussed in ref. [5] and used in ref. [9].

3. – Amplitude and phase measurements.

The simplest coherent wave mixing configuration is that of two-beam four-wave mixing (FWM). In such a configuration two ultrashort laser pulses, labeled pulse 2 and pulse 1, separated by a time delay $\Delta t = t_2 - t_1$, and propagating in the directions k_2 and k_1, interfere in a sample to generate a transient grating, which diffracts photons into the background-free direction $k_s = 2k_2 - k_1$. In the case of homogeneously broadened two-level systems the FWM signal is emitted immediately after the second pulse. It originates from the natural decay of the component of the nonlinear polarization $P(t)$, which emits in the direction k_s, and corresponds to free-induction decay (FID)[14]. For inhomogeneously broadened lines, the FWM signal is delayed by Δt after the second pulse and corresponds to a «photon echo». To establish that the FWM signal comprises two phase-shifted contributions, we are faced with the difficult problem of amplitude and the phase recovery of a signal. To achieve this goal we have used a combination of five measurements to determine five quantities: the time-integrated and time-resolved intensities, the power spectrum, the interferometric autocorrelation and cross-correlation with a reference laser pulse. Taken separately, each one is insufficient to retrieve the signal amplitude and phase. All together they give complementary information that allows a good characterization of these two parameters[8-10]. In particular, since the interferometric autocorrelation does not give directly the phase of the FWM signal, we analyze the interferometric autocorrelation data in the following way. For each delay, Δt, we measure the dynamic fringe spacing, *i.e.* the number of interferometric fringes during the interferometer delay τ, FS(τ), for FWM signal *and* FSL(τ) for the laser pulse passing through the same experimental setup and the sample sapphire holder but missing the sample itself. The differential fringe spacing defined as DFS(τ) = FSL(τ) − FS(τ) is then determined numerically. For precise calibration, the laser is operated c.w., thus providing a reference frequency which is used to analyze the laser autocorrelation when operated mode-locked. The DFS(τ) sign is determined by measuring the cross-correlation with the laser. If the phase of the FWM signal is a constant or has a linear time dependence, the DSF(τ) corresponds exactly to the phase difference between the laser and the signal. In the most general case and in the absence of other information, there is no simple mathematical relationship between the DFS(τ) and the phase difference. In our experiments, however, we also know the FWM signal intensity temporal profile and power spectrum, which are smooth and well behaved. We have numerically checked numerous examples which confirm that in this case the DFS(τ) reproduces faithfully the phase difference with the reference.

Excitonic resonances in semiconductors are usually inhomogeneously broadened at low temperatures. In QW structures, however, the quantum confinement in ultrathin layers, narrower than the bulk Bohr radius, stabilizes the quasi-2D excitons up to room temperature[1]. Collisions with the large population of thermal phonons homogenize the resonances and shorten their dephasing time. We have investigated two samples consisting respectively of 47 periods of 98 Å GaAs QWs and 96 Å $Al_{0.3}Ga_{0.7}As$ barrier layers and of 50 periods of 95 Å GaAs QWs and 45 Å $Al_{0.3}Ga_{0.7}As$ barrier layers. The output of a mode-locked Ti:sapphire laser, delivering extremely stable $\approx$ (70–100) fs transform-limited Gaussian pulses at 88 MHz is tuned close to the heavy-hole exciton resonance, $\omega_L \approx \Omega_{hh}$, and split into three beams. Two of these beams were used to generate the FWM signal. This signal could be detected directly as a function of Δt using a slow detector in the conventional way. Alternatively, for a fixed Δt, it could be directed in a Michelson interferometer for the autocorrelation measurements. For the power spectra measurements the signal was directed onto a spectrometer and detected by an optical multichannel analyzer. In order to time resolve the amplitude of FWM signal, for every Δt, the light emitted in the direction k_s was cross-correlated with the third laser beam by sum frequency generation in a highly transparent nonlinear crystal. This cross-correlation determined the temporal profile of the FWM signal $vs.$ the absolute time t. Finally by placing the whole FWM setup inside a Mach-Zender interferometer, the interferometric correlation with the laser could be determined.

In fig. 1 we present time-resolved intensity $vs.$ absolute time, t, for a series of time delays Δt, measured with a laser intensity such that the total (generated by both pulses) exciton density is $N_x \approx 10^{12}\,\mathrm{cm}^{-2}$. The laser pulse duration was $(78 \pm 3)\,\mathrm{fs}$. The weaker pulse 1 acts at $t = 0$ and the stronger pulse 2 acts at $t = \Delta t$. Clearly, the time traces are symmetric both in t and Δt. For all our measurements we have verified from the position of the maximum that the FWM signal is emitted immediately after the second pulse. This behavior confirms that the exciton transition is predominantly homogeneously broadened at room

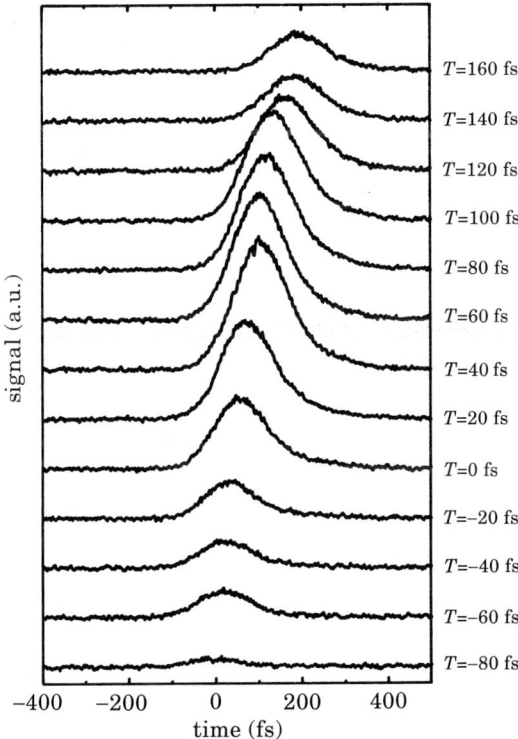

Fig. 1. – Time-resolved four-wave mixing signal $vs.$ absolute time t, for a series of Δt and a total exciton density $N_x \approx 10^{12}\,\mathrm{cm}^{-2}$.

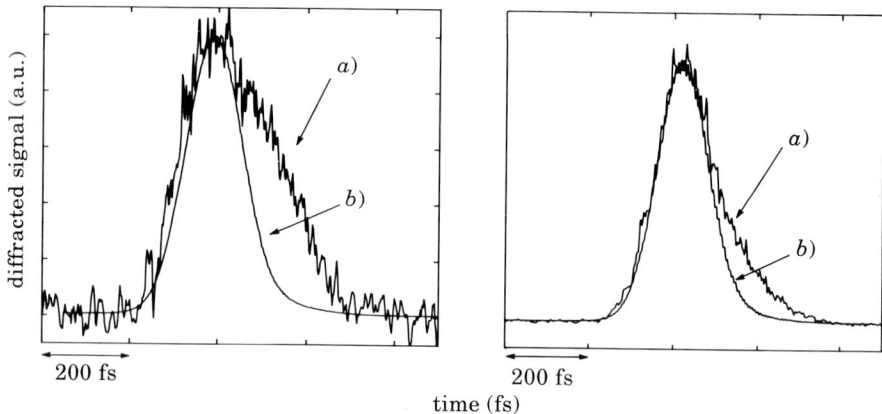

Fig. 2. – Comparison of the temporal profile of *a*) the time-resolved signal at $\Delta t = 0$ and *b*) the time-integrated signal, for two exciton densities, $N_x \approx 10^{11}$ (left) and $N_x \approx 4 \cdot 10^{11}$ cm^{-2} (right).

temperature and, therefore, that the FWM signal corresponds to a free-induction decay. Furthermore, the self-consistency of the data was checked by numerically integrating the time-resolved intensity *vs. t* for each Δt and comparing the result to the time-integrated intensity measured with a slow detector *vs.* Δt. Again, the agreement is excellent.

In fig. 2 we display the temporal profile of *a*) the time-resolved intensity *vs.* absolute time *t* at $\Delta t = 0$ and *b*) the time-integrated intensity *vs.* Δt, for two exciton densities, $N_x \approx 10^{11}$ cm^{-2} and $N_x \approx 4 \cdot 10^{11}$ cm^{-2}. The laser pulse duration was (98 ± 2) fs. Since the relevant information is contained in the line shape, the two curves have been normalized to unity and the unrelated time axes have been shifted to bring the maxima into coincidence. The difference between the two profiles is evident: the former is clearly broader than the latter, with a slower rising edge and a significantly nonexponential trailing edge. This difference is density dependent and shows up noticeably on the trailing side of the profiles. Within a Δt series the total exciton density is constant and, therefore, all the time-resolved traces are expected to have similar line shapes, although their height depends on Δt. This is indeed what is observed at low densities, $N_x \approx 10^{11}$ cm^{-2}. At moderate density, $N_x \approx (2\text{--}4) \cdot 10^{11}$ cm^{-2}, noticeable changes in the temporal profile are seen within a Δt series. In particular the sign of Δt is found to influence the temporal line shape.

In order to be more quantitative, we have solved eq. (1.13) numerically, for the nonlinear polarization, $P^{(3)}(t, k_s)$, radiating in the direction k_s, using Gaussian laser pulses with a duration corresponding to that of our laser and accounting for the effect of upconversion in the time-resolved measurement. This model involves only two fitting parameters: the exciton dephasing time, T_2, and the ratio of the two nonlinearities, $R = VP_s^2$. We impose on the fit the severe constraint that all curves in a Δt series must have the same origin of the absolute time t and a constant calibration. We find that it is impossible to fit the data with such a constraint if we retain only the PSF term. An excellent fit is obtained, however, if both the XXI and PSF contributions are considered. This is shown in fig. 3, where we present the fit of the temporal profiles at low densities, $N_x \approx 10^{11} \text{ cm}^{-2}$. The dash-dotted lines give the PSF contribution, the

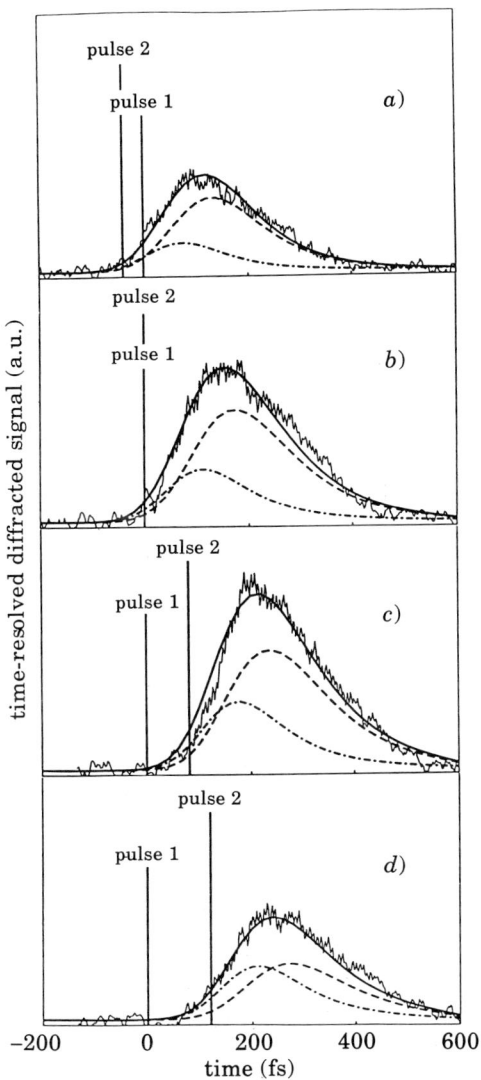

Fig. 3. – Fit of the time-resolved intensity profiles at various Δt for $N_x \approx 10^{11} \text{ cm}^{-2}$, using the model discussed in the text. The dash-dotted lines give the Pauli exclusion contribution, the dashed lines give the exciton-exciton interaction contribution, and the smooth solid lines correspond to their sum.

dashed lines give the XXI contribution, and the smooth solid lines give their sum. The contribution of the XXI to the total energy of the pulse emitted by the sample dominates the emission. It is ≈ 2.2 larger than that due to PSF. As the total exciton density is increased, the XXI contribution is reduced by screening as shown in fig. 4. Interestingly, we find that it dominates as long as the exciton density does not exceed the exciton saturation density in the sample, $N_s \approx$

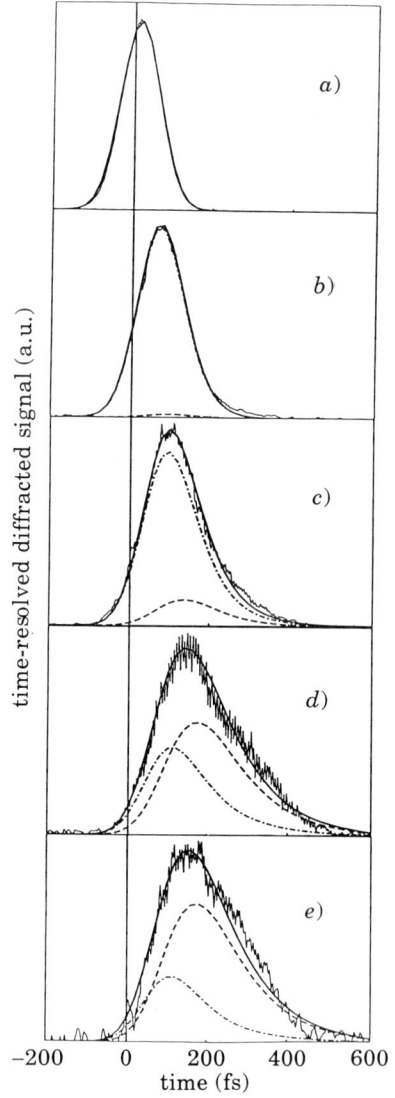

Fig. 4. – Time-resolved intensity profiles and theoretical fit for a) $N_x \approx 10^{13}$ cm^{-2}, b) $N_x \approx$ $\approx 10^{12}$ cm^{-2}, c) $N_x \approx 4 \cdot 10^{11}$ cm^{-2}, d) $N_x \approx 2 \cdot 10^{11}$ cm^{-2}, e) $N_x \approx 10^{11}$ cm^{-2}, showing the effects of screening on the relative strength of the Pauli exclusion and exciton-exciton interaction nonlinearities.

$\approx 3 \cdot 10^{11}$ cm^{-2}[1]. When this critical density is surpassed, the XXI contribution decreases very rapidly and becomes negligible at high densities.

As mentioned above, at low density both T_2 and R remained approximately constant, for all the TRS fits within a single Δt series. At moderate densities, however, this was not found to be the case within a single Δt series (*i.e.* fixed N_x); both T_2 and R had to be varied to fit the data. These observations can be understood in terms of the dynamics of excitons in quantum wells at room temperature[1]. Consider first the case of a total excitation density low enough that photogenerated excitons are the in bound state (binding energy ≈ 10 meV for ≈ 100 Å QW) and are spatially well separated. They interact effectively via the Coulomb potential and their dephasing time is determined by phonon collisions. Their environment is independent of the instantaneous density determined by the order in which the laser pulses arrive in the sample (*i.e.* Δt); therefore, T_2 and R are constant. At moderate densities, however, when the strongest pulse 2 arrives first in the sample, it generates a substantial number of excitons in the bound states. They are ionized by collisions with the energetic thermal phonons (phonon energy ≈ 36 meV for GaAs), in $\approx (100\text{--}200)$ fs, generating e-h pairs in scattering states with a significant excess energy (≈ 25 meV)[15, 16]. The carriers, in turn, both shorten the relaxation time, T_2, owing to their larger effects on the neutral bound states[17, 18] and screen the Coulomb potential. For the reverse time ordering, *i.e.* when the weaker pulse 1 arrives first, less e-h pairs are generated by the first pulse, the effects described above are less pronounced, and the sample remains closer to steady state during the FID emis-

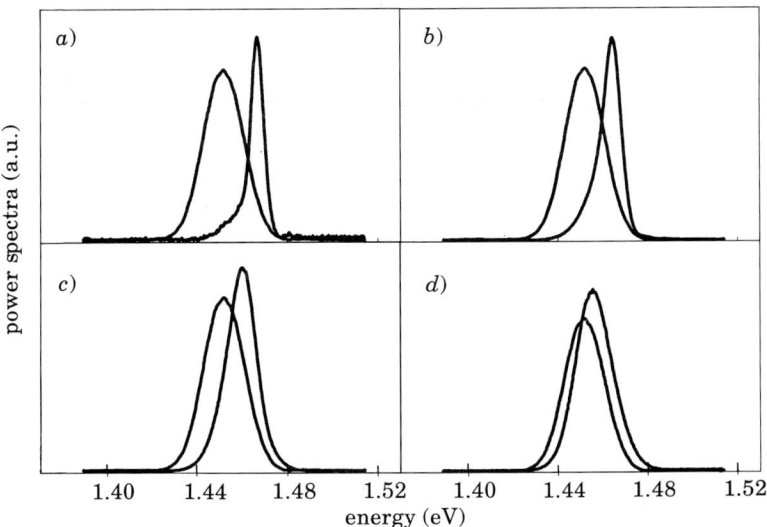

Fig. 5. – Power spectra of the four-wave mixing signal and laser spectra for exciton densities, *a*) $N_x \approx 3 \cdot 10^9$ cm^{-2}, *b*) $N_x \approx 1.2 \cdot 10^{10}$ cm^{-2}, *c*) $N_x \approx 6 \cdot 10^{11}$ cm^{-2}, *d*) $N_x \approx$ $\approx 3 \cdot 10^{11}$ cm^{-2}, when the laser is tuned slightly below the heavy-hole exciton.

sion. Finally, at very high densities, the band gap renormalizes so much during
the laser pulses that the excitons are generated in scattering states, giving free
e-h pairs immediately. They shorten the relaxation time below the experimental resolution. More importantly they screen the Coulomb potential to the point
that the XXI contribution to the emission is eliminated (see fig. 4). The fits also
show that the maximum of the time-resolved intensity is delayed with respect
to the second pulse. We found that the delay is of the order T_2. This indicates
that the time required to *establish* the coherent polarization wave within the
sample (rise time of the emission) is directly related to the dephasing time T_2
which is usually associated only with the negative interferences that produce
the signal decay. As the exciton density increases further, the time-resolved
and time-integrated profiles become more similar. Finally, at very high densi-
ties, $N_x \approx (10^{12}-10^{13})\,\mathrm{cm}^{-2}$, where the band gap renormalization has completely
washed out the exciton resonances, the two profiles become of the order of the
laser pulse and are beyond our resolution. Then, the profiles of all the traces of a
Δt series become similar again.

The power spectra of the FWM signal and of the laser are shown in fig. 5 for
four densities, $N_x \approx 3 \cdot 10^9\,\mathrm{cm}^{-2}$, $1.2 \cdot 10^{10}\,\mathrm{cm}^{-2}$, $6 \cdot 10^{10}\,\mathrm{cm}^{-2}$ and $3 \cdot 10^{11}\,\mathrm{cm}^{-2}$,
when the laser is tuned slightly below the heavy-hole exciton $\omega_\mathrm{L} < \Omega_\mathrm{hh}$. At very
low exciton densities, $N_x \approx 3 \cdot 10^9\,\mathrm{cm}^{-2}$, the FWM power spectra essentially reproduce the line shape of the exciton resonances within the laser spectra. The
line shape is asymmetric, and exhibits only one resonance. As the exciton densi-

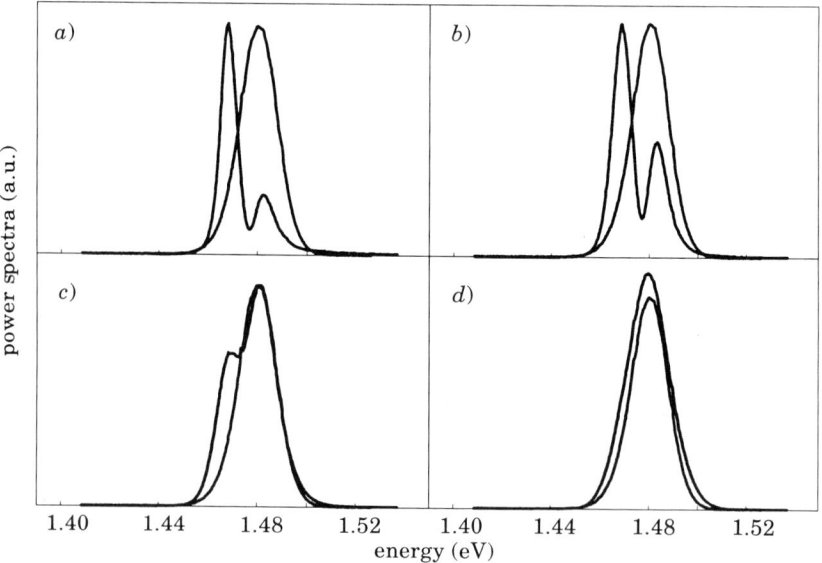

Fig. 6. – Power spectra of the four-wave mixing signal and laser spectra for exciton densi-
ties, *a*) $N_x \approx 4 \cdot 10^9\,\mathrm{cm}^{-2}$, *b*) $N_x \approx 1.2 \cdot 10^{10}\,\mathrm{cm}^{-2}$, *c*) $N_x \approx 6 \cdot 10^{11}\,\mathrm{cm}^{-2}$, *d*) $N_x \approx$
$\approx 3 \cdot 10^{11}\,\mathrm{cm}^{-2}$, when the laser is tuned slightly above the heavy-hole exciton.

ty is increased, the FWM power spectrum evolves toward that of the laser and becomes almost indistinguishable from it at the highest density shown in fig. 5. Figure 6 shows similar data obtained for a laser tuned slightly above the heavy-hole exciton $\omega_L > \Omega_{hh}$, which excites also the light-hole resonance. Two unequal peaks are present at low densities. They evolve toward a single and broader peak, almost in coincidence with the laser, as the density increases. The low-density line shapes in the two figures suggest, by Fourier transform, dynamic nonlinear shifts of the FWM frequency during a single pulse.

Figure 7 shows the $\Delta t = 0$ autocorrelation traces and DFS(τ) for a) the mode-locked laser (calibrated to the reference frequency of the laser operated c.w.), b) the low-density FWM signal and c) the high-density FWM signal for a $\omega_L < \Omega_{hh}$ excitation corresponding to fig. 5. Figure 8 shows the same quantities when $\omega_L > \Omega_{hh}$ as in fig. 6. In both cases the high-density autocorrelation envelopes of the FWM signal are of the order of that of the laser and the DFS(τ) indicates that the FWM instantaneous frequency presents no significant difference with that of the laser. Conversely the low-density envelopes are much longer than that of the laser and, more importantly, the DFS(τ) shows significant nonlinear frequency shifts. For $\omega_L < \Omega_{hh}$, the low-density DFS(τ) starts with a positive linear variation. The slope corresponds exactly to the difference

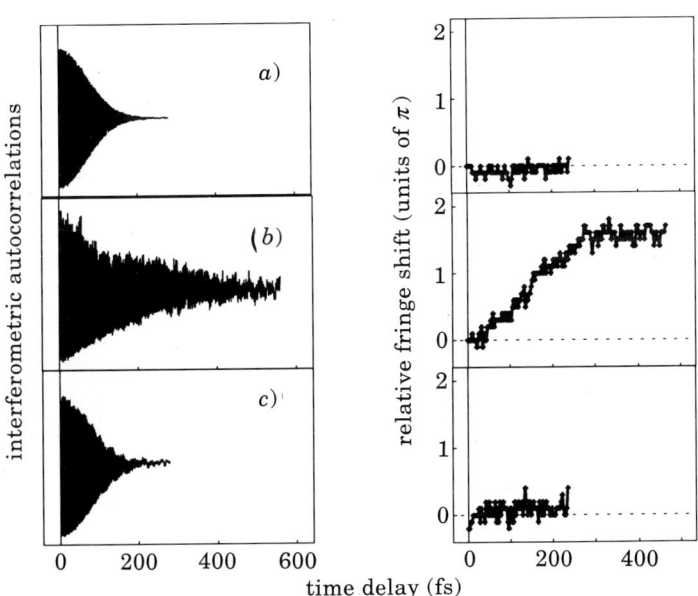

Fig. 7. – Interferometric autocorrelation and differential fringe spacing for a) the laser, b) the low-density and c) the high-density FWM signal in the case where the laser is tuned slightly below the heavy-hole exciton. The conditions of b) and c) are the same as that of the power spectra a) and b) of fig. 5.

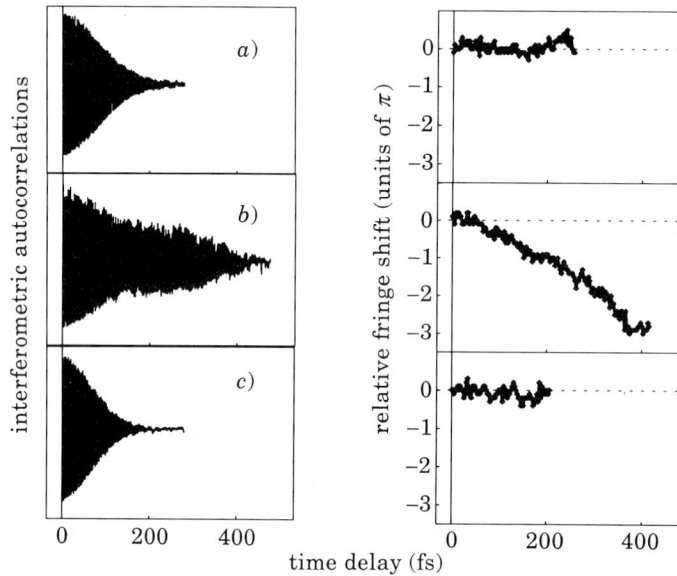

Fig. 8. – Interferometric autocorrelation and differential fringe spacing for a) the laser, b) the low-density and c) the high-density FWM signal in the case where the laser is tuned slightly above the heavy-hole exciton. The conditions of b) and c) are the same as that of the power spectra a) and b) of fig. 6.

in frequencies between the laser and the principal peak seen in the FWM power spectrum of fig. 5a). Then for $300\,\text{fs} < t < 450\,\text{fs}$ the DFS(τ) slope vanishes, indicating that, during the pulse, the instantaneous frequency shifts toward that of the laser. This nonlinear phase dynamic is consistent with the power spectra of fig. 5. In particular, the power spectrum of fig. 5a), besides a main peak at the hh-exciton frequency, exhibits an asymmetric low-frequency tail which extends well into the laser spectra. For $\omega_L > \Omega_{hh}$, again the high-density DFS(τ) has a zero slope showing that the FWM frequency is essentially that of the laser. Conversely, the low-density DFS(τ) starts with a zero slope and then exhibits a negative variation with a curvature. This indicates that, in this case, the FWM frequency starts at the same frequency as the laser but is quickly dominated by a component below the laser central frequency. The dynamics of the instantaneous frequency, however, is complicated and does not correspond to a simple linear variation. Again this is consistent with the power spectrum of fig. 7a), which shows strong but unequal contributions from both excitons. In order to further explore this case, we have adjusted the excitation frequency, $\omega_L \approx \Omega_{lh}$, and intensity to obtain hh-exciton and lh-exciton contributions of roughly the same weight in the FWM power spectrum, as shown in fig. 9a). In this case, the interferometric autocorrelation, fig. 9b), clearly shows several in-

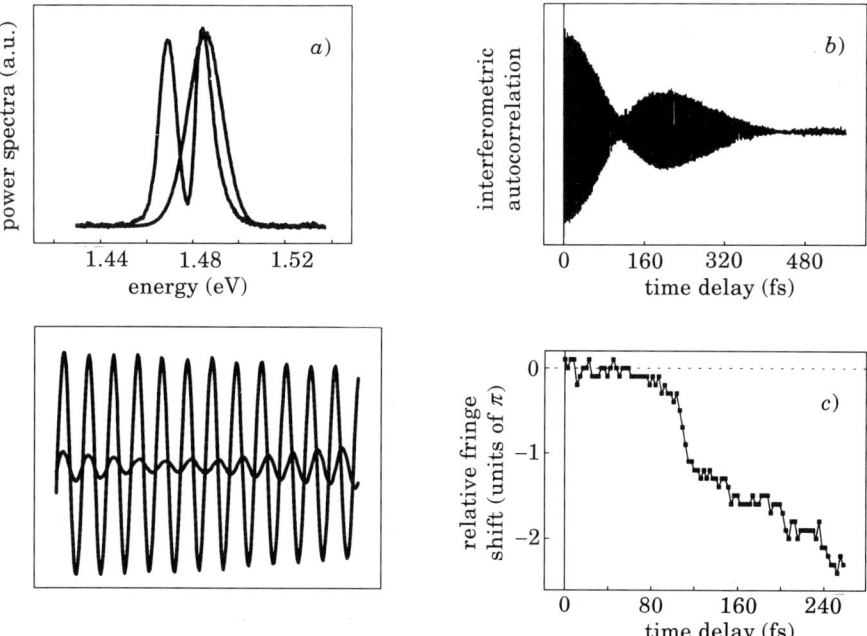

Fig. 9. – *a*) Power spectrum, *b*) interferometric autocorrelation and *c*) differential fringe spacing of the FWM signal in the case where the laser is tuned to give contributions of the hh- and lh-exciton of approximately same weights. The lower left figure is a blow-up of about a dozen of fringes close to the center and the node of the autocorrelation trace showing how the π-shift occurs over eleven fringes only.

terference patterns which are in excellent agreement with the separation of the two peaks of the FWM power spectrum of fig. 9*a*). In order to establish that the observed effect corresponds to the quantum beats of a homogeneous system and not to polarization interferences from independent systems, we have applied the method of ref. [19]. We verified that the asymmetric features seen in the interferometric cross-correlation for various time delays vary as Δt and not as $2\Delta t$. Quantum beats between hh- and lh-excitons have been observed recently as modulations of the decay of FWM signal intensity [20, 21]. The new information provided by the interferometric techniques is shown in fig. 9*c*), where the DFS(τ) is depicted. It starts with a zero slope showing that the FWM frequency is the same as that of the laser, $\omega_{\mathrm{L}} \approx \Omega_{\mathrm{lh}}$. Then it exhibits a negative curvature showing a change toward a lower frequency. At around $\tau \approx 120\,\mathrm{fs}$, the DFS($\tau$) experiences a sudden phase shift of π, before resuming its negative variation. The position of the π-shift corresponds to the middle of the first node in fig. 9*b*). It occurs over a very short interval of about 10 optical fringes. This is shown in the lower left part of the figure where about a dozen of the fringes close to the center and close to the node of the autocorrelation trace have been expanded. The half-fringe shift over eleven fringes is clearly seen. The signal-to-noise ratio is excellent, since one fringe corresponds to 2.8 fs and the measurement is performed with a calibration of 21 stepper motor steps per fringe or an accuracy of ≈ 0.14 fs.

4. – Conclusion.

The time-resolved amplitude measurements are well explained by the model based on a single-resonance excitation approximation of the semiconductor Bloch equations. They show that not only there is a XXI contribution to the nonlinear response, but that, in fact, this contribution dominates over the PSF one whenever the Coulomb interaction is not screened.

The phase measurements provide much more delicate information on the dynamics of the nonlinear polarization through the instantaneous frequency of the FWM signal. The experimental observations obtained in the case of single-resonance contribution to the FWM can be explained qualitatively in terms of the one-resonance approximation of the two-band semiconductor Bloch equations. In general, however, because the system is so nonlinear, it is not enough to use a perturbation expansion, screening has to be accounted for self-consistently. Furthermore because the exciton linewidth and the ultrashort-pulse laser spectrum are both rather broad, the populations of the high-energy states have to be accounted for. Therefore, it is necessary to consider the full three-band (one conduction and two valence) semiconductor Bloch equations with self-consistent screening. In this case one finds two families (heavy holes and light hole) of excitons. Each one possesses «internal» Coulomb and Pauli nonlinearities, but, furthermore, they are coupled via these two mechanisms as well. The coupling of the hh-X and lh-X families by Pauli exclusion originates naturally from the fact they share the same conduction bands. Hence, once an exciton of one family is created, the transition to the conduction band for the other family is affected. The Coulomb coupling between hh-X and lh-X originates from the intervalence band transitions which provide an additional transition channel between the conduction band and either one of the valence bands [13].

* * *

This work was performed in collaboration with M.-A. MYCEK, S. WEISS, J.-Y. BIGOT and R. G. ULBRICH.

It was supported by the Director, Office of Energy Research, Office of Basic Energy Sciences, Division of Materials Sciences of the US Department of Energy, under Contract No. DE-AC03-76SF00098.

REFERENCES

[1] S. SCHMITT-RINK, D. S. CHEMLA and D. A. B. MILLER: *Adv. Phys.*, **38**, 89 (1989).

[2] S. SCHMITT-RINK and D. S. CHEMLA: *Phys. Rev. Lett.*, **57**, 2752 (1986); S. SCHMITT-RINK, D. S. CHEMLA and H. HAUG: *Phys. Rev. Lett. B*, **37**, 941 (1988).

[3] K. LEO, M. WEGENER, J. SHAH, D. S. CHEMLA, E. O. GÖBEL, T. C. DAMEN, S. SCHMITT-RINK and W. SCHÄFER: *Phys. Rev. Lett.*, **65**, 1340 (1990).

[4] M. WEGENER, D. S. CHEMLA, S. SCHMITT-RINK and W. SCHÄFER: *Phys. Rev. A*, **42**, 5675 (1990).

[5] S. SCHMITT-RINK, S. MUKAMEL, K. LEO, J. SHAH and D. S. CHEMLA: *Phys. Rev. A*, **44**, 2124 (1991).

[6] F. JAHNKE and W. SCHÄFER: in *Proceedings of the International Meeting on Optics of Excitons in Confined System, Giardini Naxos, 1991*, Inst. Phys. Conf. Ser. No. 123, p. 261.

[7] A. V. KUZNETSOV: *Phys. Rev. B*, **44**, 8721, 13381 (1991).

[8] M.-A. MYCEK, S. WEISS, J.-Y. BIGOT, S. SCHMITT-RINK, D. S. CHEMLA and W. SCHÄFER: *Appl. Phys. Lett.*, **60**, 2666 (1992).

[9] S. WEISS, M.-A. MYCEK, J.-Y. BIGOT, S. SCHMITT-RINK and D. S. CHEMLA: *Phys. Rev. Lett.*, **69**, 2685 (1992).

[10] J.-Y. BIGOT, M.-A. MYCEK, S. WEISS, R. G. ULBRICH and D. S. CHEMLA: *Phys. Rev. Lett.*, **70**, 3307 (1993).

[11] See, for example, *Optical Nonlinearities and Instabilities in Semiconductors*, edited by H. HAUG (Academic Press, New York, N.Y., 1988).

[12] T. YAHIMA and Y. TAIRA: *J. Phys. Soc. Jpn.*, **47**, 1620 (1979).

[13] D. S. CHEMLA, J.-Y. BIGOT, M.-A. MYCEK, S. WEISS and W. SCHÄFER: *Phys. Rev. B*, **50**, 8439 (1994-II).

[14] See, for example, L. ALLEN and J. H. EBERLY: *Optical Resonances and Two Level Atoms* (Wiley, New York, N.Y., 1975).

[15] W. H. KNOX, C. HIRLIMANN, D. A. B. MILLER, J. SHAH, D. S. CHEMLA and C. V. SHANK: *Phys. Rev. Lett.*, **56**, 1191 (1986).

[16] W. H. KNOX, D. S. CHEMLA, G. LIVESCU, J. E. CUNNINGHAM and J. E. HENRY: *Phys. Rev. Lett.*, **61**, 1290 (1988); W. H. KNOX: in *Hot Carriers in Semiconductor Nanostructures*, edited by J. SHAH (Academic Press, New York, N.Y., 1992), p. 313.

[17] L. SCHULTHEIS, M. D. STURGE and J. HEGARTY: *Appl. Phys. Lett.*, **47**, 995 (1985).

[18] L. SCHULTHEIS, J. KUHL, A. HONOLD and C. W. TU: *Phys. Lett.*, **55**, 1635 (1986).

[19] M. KOCH, J. FEDMAN, G. VON PLESSEN, E. O. GÖBEL, P. THOMAS and K. KÖHLER: *Phys. Rev. Lett.*, **69**, 3631 (1992).

[20] K. LEO, J. SHAH, E. O. GÖBEL, T. C. DAMEN, S. SCHMITT-RINK, W. SCHÄFER, J. MULLER and K. KÖHLER: *Phys. Rev. Lett.*, **66**, 201 (1991); *Mod. Phys. Lett. B*, **5**, 87 (1991).

[21] E. O. GÖBEL, K. LEO, T. C. DAMEN, J. SHAH, S. SCHMITT-RINK, W. SCHÄFER, J. MULLER and K. KÖHLER: *Phys. Rev. Lett.*, **64**, 1802 (1990).

Nonlinear Optical Response of Semiconductor Quantum Wells under High Magnetic Fields.

D. S. CHEMLA

Physics Department, University of California at Berkeley
Material Sciences Division, Lawrence Berkeley Laboratory - Berkeley, Cal.

1. – Introduction.

In semiconductor quantum wells (QW) the electronic states have a reduced, quasi-2D dimensionality. This property alone results in a wealth of new physical properties and interesting applications in the field of electronics and photonics [1]. Motivated by this success, an intense research activity has been directed at further reducing the dimensionality of the electronic states to 1D and 0D. The numerous efforts to make semiconductor structures such as quasi-1D quantum wires and quasi-0D quantum boxes and microcrystallites have, so far, failed to produce samples of quality comparable to that of the available QW's. Two major difficulties are encountered. They are related to i) the atomic control of size fluctuation and ii) the control of surface and interface defects. For these reasons it is interesting to find alternative ways to explore the nonlinear optical properties of quasi-0D systems while avoiding these practical difficulties. As shown below, the dimensionality of the electronic states of semiconductor QW can be tuned continuously from quasi-2D to quasi-0D, in materials with excellent quality and uniformity by application of a magnetic field perpendicular to the QW [2-6]. Further interest in QW's under large magnetic field stems from the prediction of many-body theory, according to which in such systems and at thermodynamic equilibrium an ensemble of spin-polarized electron-hole pairs behave like a gas of noninteracting and nonpolarizable pointlike bosons [7,8]. This two-component quantum system is expected to exhibit new and interesting many-body properties.

In this lecture we review our recent investigations on the nonlinear optical response of semiconductor quantum wells in a strong perpendicular magnetic field, H. The lecture is organized as follows. In sect. 2, we discuss the evolution of the linear optical properties of GaAs QW's as a function of H and examine how the magneto-excitons (MX) extrapolate continuously between quasi-2D QW excitons (X) when $H = 0$ and pairs of Landau levels (LL) when $H \to \infty$. In sect. 3, we present femtosecond time-resolved investigations of their nonlinear optical response. We stress the evolution of MX-MX interactions with increasing H [9,10]. Finally in sect. 4, we study how, as the dimensionality is reduced by application of H, the number of scattering channels is limited and relaxation of electron-hole pairs is affected [11]. We also discuss how nonlinear optical spectroscopy can be exploited to access the relaxation of angular momentum within magneto-excitons [12].

2. – Linear optical response of magneto-excitons.

The theory of the linear optical response of magneto-excitons has been extensively studied [2-6]. When a constant magnetic field is applied perpendicular to a QW, the wave function for the relative e-h coordinate, $r = r_e - r_h$, of pairs with zero center-of-mass momentum satisfies the effective-mass Schrödinger

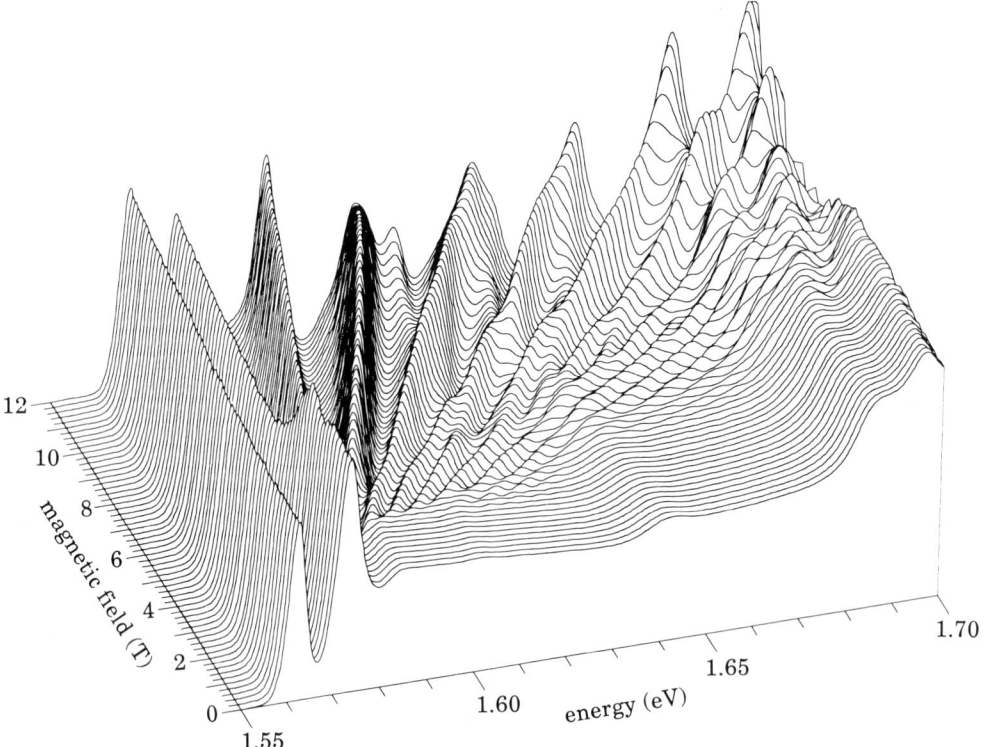

Fig. 1. – Linear absorption of a $L_z = 84$ Å GaAs/AlGaAs quantum well structure for σ_- polarized light for a series of magnetic fields $H = 0$ to 12 T.

equation

(1) $$\left[\frac{1}{2m_e}\left(p+\frac{e}{2c}H\times r\right)^2+\frac{1}{2m_h}\left(p-\frac{e}{2c}H\times r\right)^2-\frac{e^2}{\varepsilon_0 r}\right]\zeta_v(r)=E_v\zeta_u(r).$$

The e-h pairs experience the total potential which consists of the sum of the quadratic potential imposed by H and the $1/r$ Coulomb potential,

(2) $$V(r)=\left[\left(\frac{1}{2}\lambda r\right)^2-\frac{2}{r}\right]R_y,$$

where R_y is the 3D exciton Rydberg. The parameter λ measures the ratio between the magnetic and the Coulomb confinements as measured by the corresponding zero-point energies, $\lambda=\omega_c/2R_y$, where ω_c is the cyclotron frequency. It can also be expressed as $\lambda=(a_0/r_c)^2$, where the Bohr radius, a_0, measures the range of the Coulomb force and the cyclotron radius, r_c, characterizes the magnetic length. For $\lambda=0$ the e-h pairs form usual quasi-2D excitons, $\zeta_v(r)\rightarrow$ $\rightarrow\phi_v(r)$. For $\lambda\rightarrow\infty$ they tend toward e and h in Landau levels, $\zeta_v(r)\rightarrow L_v(r)$. For intermediate values of λ, they form magneto-excitons. It is worth noting that, when $r\rightarrow\infty$, $V(r)\rightarrow\infty$ for any nonzero value of H. Hence, all the MX states originate from the *bound* states of the unperturbed X[2,3], and can be labelled accordingly by the indices $1S$, $2S$, $3S$, etc.

The experimental linear absorption spectra of a high-quality GaAs/AlGaAs QW structure ($L_z=84$ Å) at low temperature, $T=4$ K, and in a perpendicular

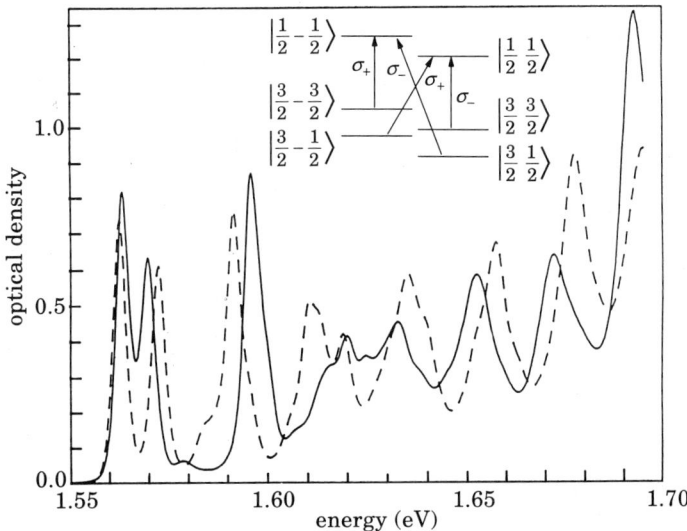

Fig. 2. – Comparison of the $H=12$ T absorption spectra of the $L_z=84$ Å GaAs/AlGaAs quantum well structure for σ_- (dashed line) and σ_+ (solid line) polarization. Inset: schematic of the Γ-point energy levels and one-photon interband transitions.

magnetic field are shown in fig. 1. The spectra were measured with σ_- polarized light as H was tuned from $H = 0$ to 12 T. At the highest field where $r_c \approx$
≈ 71 Å, the magnetic confinement is significantly larger than the Coulomb confinement, $\lambda \approx 4$. The transition from 2D behavior at low fields to 0D behavior at
high fields and the evolution of the MX's from the bound 2D X states are nicely
displayed. On can see the small diamagnetic shift experienced by the lowest-energy $1S$ heavy-hole (hh) and light-hole (lh) excitons. The higher-energy MX's
shift much faster as the H increases. They clearly start close to the hh gap,
which in this sample is approximately located at the lh-X energy, and tend toward the LL's at high field. As the MX's separate away from the gap, the peak
at the lh-X experiences a drop in oscillator strength. This demonstrates that,
when $H = 0$, the hh-X excited states in fact contributed to this peak but were
not resolved. Figure 2 compares the 12 T linear absorption for σ_- (dashed line)
and σ_+ (solid line). They are Zeeman split in agreement with the QW selection
rules at the Γ-point of the Brillouin zone which are shown in the inset of fig. 2.
At such a high field the confinement is strong enough that the absorption
strength is almost zero between the $1s$ and $2s$ MX, indicating that the MX's are
almost diagonal in the LL basis. These spectra are in good agreement with the
theoretical calculations of ref. [6]. It is clearly seen how the oscillator strength
is now concentrated in the sharp MX peaks which directly reflect the quasi-0D
density of states of the eh pairs.

3. – Nonlinear optical response of quasi-0D magneto-excitons.

Pump/probe experiments provide a very powerful technique to measure the
nonlinear response of a medium. In time-resolved experiments one uses a short,
intense and relatively narrow pump to excite the sample. Its transmission is
measured by a short but weak and broad-band probe. The time resolution is obtained by varying the delay, Δt, between the pump and probe pulses. The experimental data are usually collected by measuring the difference between the probe
transmission when the pump is applied and when it is not. Examples of the differential absorption spectra, or DAS, obtained by this method are presented below. In the case of semiconductors the most intuitive way of interpreting pump
/probe experiments is to consider that the pump creates populations of excitons
(or magneto-excitons in the presence of H) whose energy and angular momentum
are determined by the pump photon energy and polarization. The broad-band
probe then measures the change of absorption induced by the presence of these
populations. For excitation in the transparency region, below the gap, the populations are virtual and last only as long as the pump is present in the sample. For
excitation above the gap, the populations are real and can relax after being generated. Exciton populations affect the absorption through several mechanisms [1]. Charge density effects (collisional broadening and Debye screening)

are independent of the angular momentum, but other processes such as phase space filling (PSF), exchange (EXCH) and exciton-exciton interaction (XXI) depend critically on the spin[1]. As the time delay is changed, the dynamics of these effects can be followed on the variations of the DAS *vs.* Δt.

In several theoretical articles[7,8,13-15] it was shown that in the extreme magnetic limit and for a symmetric e-h system ($V_{ee} = V_{hh} = -V_{eh}$) the ground-state energy of an MX gas is just the sum of the energies of the individual MX's. This result implies that at high H the MX-MX interaction disappears. It can be explained intuitively, in a way which captures the essence of the exact many-body theory. For pure parabolic bands (neglecting the Coulomb interaction), the electron and hole wave functions depend only on r_c and are, therefore, mass independent. As H is increased, the magnetic confinement dominates the Coulomb interaction, and the e and h are forced into almost identical and overlapping wave functions. Hence the quasi-0D MX's occupy a much smaller volume at high field and, for the same density, show much less PSF. Furthermore, as the magnetic potential, $(\lambda r/2)^2$, increases, it restricts more and more e's and h's on top of one another making the MX's more rigid and locally neutral and, therefore, much less polarizable. For an exactly symmetric e-h system, $V_{ee} = V_{hh} = -V_{eh}$, this results in a perfect cancellation of the MX-MX interaction. XXI manifests itself directly as a blue shift of the 1s hh-X peak induced by a population of 1s hh-X. This shift can be interpreted as a hard-core repulsion which measures the extra energy cost necessary to create an 1s hh-X in the presence of other 1s hh-Xs[16]. It has been clearly resolved at $H = 0$ during resonant pumping of 1s hh-X or subsequent to the formation of 1s hh-X after excitation of e-h pairs in the continuum[17,18]. At low and moderate X densities the blue shift is proportional to the XXI repulsive potential and to the X density[16-18].

High-density optical nonlinear effects are shown in fig. 3a) and b), where we present the DAS seen by a σ_- probe for excitation resonant with the lowest 1S hh MX ($\omega \approx 1.56\,\text{eV}$) with, respectively, a σ_- pump and a σ_+ pump of the same intensity. Very strong responses are observed at photon energies up to $\omega = 1.67\,\text{eV}$. As previously explained, two types of nonlinearities, spin-dependent and spin-independent, contribute to the DAS. The pump/probe technique is actually able to separate spin-dependent nonlinearities from the spin-independent ones. This is shown in fig. 3c), where the spectra of fig. 3a) are subtracted from those of fig. 3b). One clearly sees that in this difference all the nonlinear response above $\omega \approx 1.59\,\text{eV}$ disappears, demonstrating that the high-energy MX's (2s and above) are only sensitive to the charge density effects (collisional broadening and dielectric screening) induced by the resonantly excited 1s hh σ_- MX's or 1s hh σ_+ MX's. The same figure clearly shows that the 1s MX's ($\omega < 1.59\,\text{eV}$), on the contrary, are very sensitive to the spin of the MX's created by the pump. This behavior is detailed in fig. 4, where we compare an expanded part of the lower portion of the absorption spectra seen by a σ_- probe at

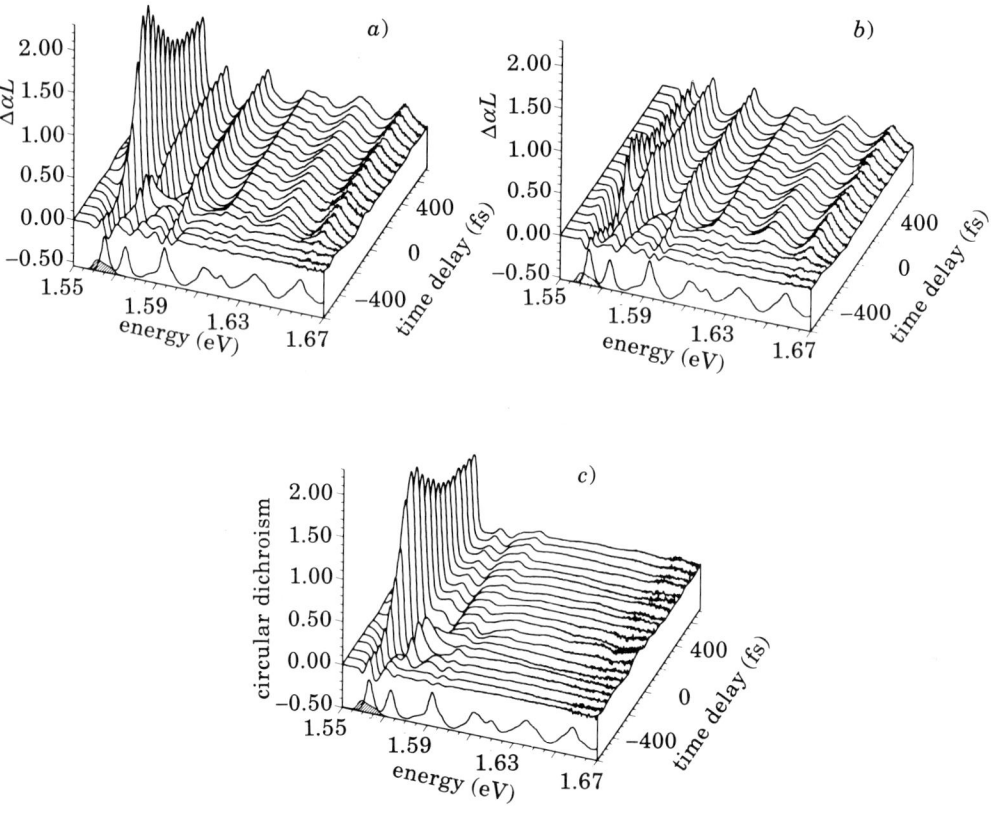

Fig. 3. – Differential absorption spectra *vs.* photon energy and time delay, Δt, for σ_- probe and *a)* σ_- pump, *b)* σ_+ pump. The nonlinear differential circular-dichroism spectra (*c*)) are the differences between the spectra *a)* and the spectra *b)*. They reveal the spin-dependent interactions between magneto-excitons.

$H = 0$ T and $H = 12$ T, for $\Delta t = -660$ fs, 0 fs and $+660$ fs after excitation by a σ_- pump and a σ_+ pump. At $\Delta t = -660$ fs the spectra are essentially those of the unexcited sample. For pump and probe both σ_- polarized, the 1s h-MX response is an instantaneous gain and the blue shift at $\Delta t = 0$, which evolves toward a strong saturation and a smaller blue shift at $\Delta t = 660$ fs. This is due to phase space filling (PSF) and exchange (EXCH) by the coherent ($\Delta t = 0$) and then the relaxed ($\Delta t = 660$ fs) 1s hh σ_- MX's created by the pump. The difference between the $H = 0$ T and the $H = 12$ T cases is only qualitative. The high-field effects are similar to the low-field ones, but significantly attenuated. The instantaneous blue shift changes from 1.9 meV to 0.3 meV. Since the 1s hh and 1s lh originate from distinct e and h states, the 1s lh response is not due to PSF and EXCH produced by the real 1s hh MX's. It comes from the pump-induced virtual populations of 1s lh, *i.e.* the a.c. Stark effect [1]. When the pump polar-

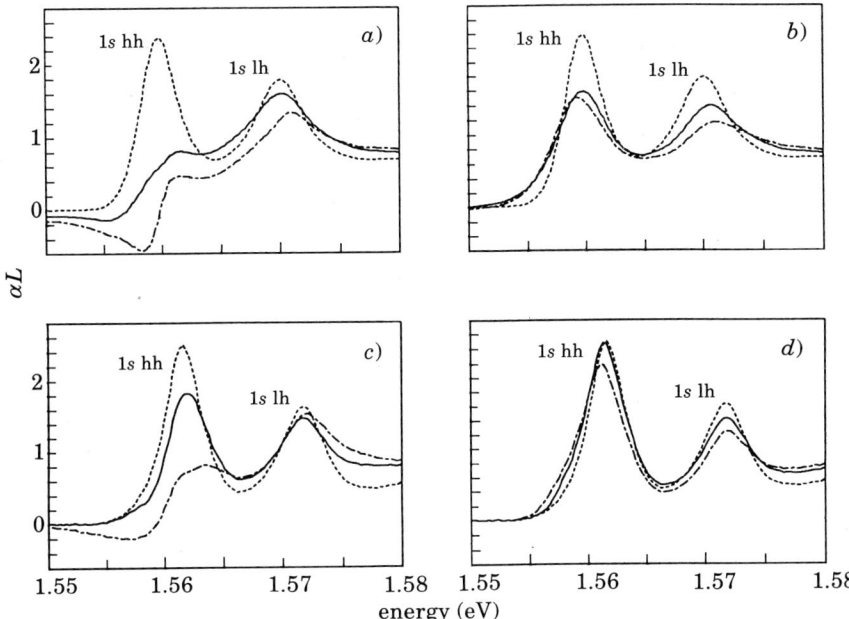

αL

Fig. 4. – Detail of the absorption spectra seen by a σ_- probe near the 1s exciton resonances for $\Delta t = -660$ fs (dashed line), 0 fs (dash-dotted line) and $+660$ fs (solid line); for $H = 0$ T: a) σ_- pump and b) σ_+ pump; for $H = 12$ T: c) σ_- pump and d) σ_+ pump.

ization is reversed to σ_+, the 1s hh σ_- MX exhibits only a small red shift and almost no saturation at $H = 12$ T, whereas at $H = 0$ it still saturates, although much less than for σ_- pump. The 1s lh σ_- MX experiences a small but distinct saturation and blue shift with, however, the $H = 12$ T response attenuated as compared to that at $H = 0$ T.

Two sets of effects have to be distinguished, i) the effects of the polarization of the MX's created by the pump (i.e. the difference between the DAS spectra for σ_- and σ_+ excitations) and ii) the effects of magnetic confinement (i.e. the difference between the DAS spectra for $H = 0$ T and $H = 12$ T). The polarization effects can be understood intuitively by considering [9-12] i) Pauli exclusion and the symmetry of the e and h states out of which the MX's are built and ii) the «molecular» exciton-exciton interaction potential which is attractive in the singlet state and repulsive in the triplet state. PSF and EXCH are strongly active when the pump and probe MX's involve the same e and/or h states. The sign of the shift experienced by the probe MX's derives from the attractive or repulsive character of the MX-MX interaction. The evolution of the nonlinear response vs. H was briefly discussed above. The combination of these two effects explains that i) the MX-MX interaction is reduced at high field, ii) the shifts of the resonances, while keeping the same sign as for $H = 0$ T, decrease as H increases, iii) the inter-MX dielectric screening itself is strongly attenuated. Screening is often associated with charged e-h plasmas which have a very strong effect because there is no gap in their excitation spectrum. A gas of neutral particles can screen as well. In this case the effect is due to the polarization of the particles, as for molecular gases or dielectric media[19]. The gas of MX's generated by the pump reacts exactly in the same way. Because the MX's excitation spectrum has gaps, this dielectric screening is weaker than that of a charged plasma. Nevertheless it is present and significant at high densities. It explains the important reduction of oscillator strength of the $1s$ hh σ_- MX seen at $H = 0$ T for the σ_+ pump. As the magnetic field is increased, the MX's wave function is compressed and they become much less polarizable. At very large field, $H = 12$ T, the MX's are almost rigid and their dielectric screening almost disappears. Hence the $1s$ hh σ_- MX's are no more affected by the σ_+ MX's generated by the pump. To more qualitatively characterize the evolution of the MX-MX interaction we have measured the changes of the blue shift of the $1S$ MX $vs.$ H with linear polarization on another sample and at moderate densities. The variation of the blue shift normalized to the MX density directly measures the interaction potential. The experimental results for $H = 0$ T, 6 T and 12 T, are shown in fig. 5. They clearly demonstrate that the shift and hence the XXI tend to zero as H increases. To the best of our knowledge our experiments represent the first confirmation of the remarkable, exact many-body theory results [7, 8, 13-15].

To go beyond these qualitative arguments, a time-dependent many-body theory of MX nonlinearities was developed[10, 20]. It follows the unrestricted Hartree-Fock theory, introduced by SCHMITT-RINK $et\ al.$ [21], which has been very successful in describing the near-band-gap nonlinear optical effects in semiconductors[22-26]. In the MX case rather than using the Bloch states as a

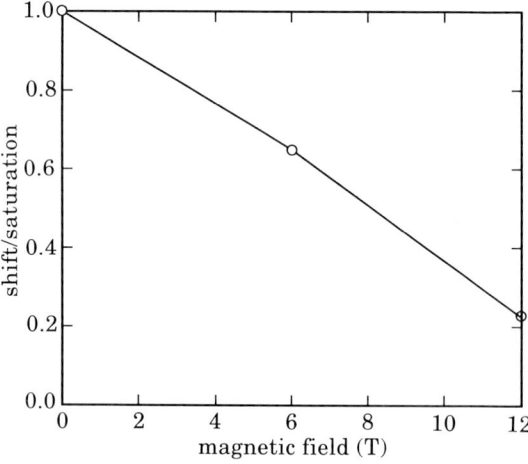

Fig. 5. – Evolution of the blue shift of the $1s$ hh exciton normalized to the exciton density $vs.$ magnetic field H, for moderate exciton densities.

basis, the MX wave functions, $\zeta_\nu(r)$, are expanded on the LL orbitals, L_ν. The Hartree-Fock theories are more appropriate in this case since they become 'exact in the extreme magnetic limit, $\lambda \to \infty$ [7,8]. The formalism follows closely that described in our other lecture. Expressed in terms of the optically connected conduction and valence band LL's, which form a set of two-level systems, the density matrix breaks into 2×2 blocks:

$$
(3) \qquad \hat{n}_\nu(t) = \begin{bmatrix} n_{c,\,\nu}(t) & \psi_\nu(t) \\ \psi_\nu^*(t) & n_{v,\,\nu}(t) \end{bmatrix},
$$

where $n_{(c,\,v),\,\nu}(t)$ are the components of the populations of the conduction and the valence band LL's and $\psi_\nu(t)$ are the components of the pair amplitude. The density matrix obeys the Liouville equation. The difference between this model and a collection of independent two LL systems arises from the coupling between the LL's by the Coulomb interaction, $V_{\nu,\,\nu'}$. This coupling modifies the physics in two ways. Firstly, the conduction and valence band energies are renormalized by the excited populations in direct analogy with the mechanisms responsible for band gap renormalization,

$$
(4a) \qquad \varepsilon_{j,\,\nu}(t) \to \varepsilon_{j,\,\nu} - \sum_{\nu'} V_{\nu,\,\nu'} n_{j,\,\nu}(t)
$$

(j = e or h). Secondly, the coupling with the electromagnetic field, which is expressed by the Rabi frequency, is modified according to

$$
(4b) \qquad \mu E(t) \to \Delta_\nu(t) = \mu E(t) + \sum_{\nu'} V_{\nu,\,\nu'} \psi_{\nu'}(t).
$$

This expresses the fact that the optically connected LL's at ν do not experience the applied field, $\mu E(t)$. Rather they see the self-consistent «local field», $\Delta_\nu(t)$, which is the sum of the applied field and the «molecular» field, due to all the other e-h LL's [10,20]. This is the LL representation of the renormalization which we have expressed in k-space in our other lecture. The interaction between MX's appears in the Liouville equation as an «exchange» term:

$$
(5) \qquad \sum_{\nu'} V_{\nu,\,\nu'} [\psi_{\nu'} n_\nu - \psi_\nu n_{\nu'}].
$$

It is important to note that this term vanishes identically on the diagonal, $\nu = \nu'$, giving the clue which explains the observed difference in the interaction between MX's in the same state and in different states at low and high magnetic fields. In the Hilbert space, the effect of H can be viewed as a rotation that aligns the MX's along the LL's basis vectors, as we mentioned when discussing the spectra of fig. 1 and 2. For small fields, the MX's have components on many LL's and at large field they have components only on a small subset of LL's. Eventually for $H \to \infty$ the MX's are exactly aligned with the LL's. Therefore, for a small H there is a strong interaction between all the MX's, through their components on different LL's, L_ν and $L_{\nu'}$, whenever one MX state is photoexcited. At high field, the interaction within one MX state vanishes and yet persists between different MX states. The residual interaction is mostly due to the self-energy corrections, which explain the high-field disappearance of the $1s$ MX blue shift found experimentally. The attractive inter-MX self-energy correction explains the experimental red shift of the $2s$ MX induced by a population of photoexcited $1s$ MX. At high field the MX's behave like two LL systems and their nonlinear optical response becomes dominated by Pauli exclusion. Numerical solutions of the Liouville equation reproduce the experimental results fairly well [10, 20]. In particular they show that indeed the Coulomb correlation is completely quenched in the extreme magnetic limit.

4. – Relaxation of electron-hole pairs.

An important aspect of a quasi-0D DOS is that the states are lumped together in narrow energy bands as compared to the 1D, 2D and 3D DOS which contain continua. Therefore, for any transition in quasi-0D, the number of available initial and final states is limited. This is expected to strongly influence relaxation and can have important consequences in the dynamics of devices. In this section we discuss how a reduced quasi-0D dimensionality affects the thermalization carrier populations by carrier-carrier scattering. We also discuss nonlinear optical spectroscopy investigations of the relaxation of angular momentum within MX's via transition to one-photon forbidden states.

Using ≈ 100 fs circularly polarized pump pulses, nonthermal populations were generated at about 25 meV above the lowest $1s$ hh resonance of the QW sample at a lattice temperature of 4 K. The DOS was controlled by varying the applied magnetic field from $H = 0$ T to $H = 12$ T [11, 12]. The spin-dependent PSF, which determines the energy location of the carriers, was separated from the spin-independent charge density effects as discussed in sect. 2. The $\sigma_{\pm}$ probe DAS measured for a σ_{+} pump is subtracted from the $\sigma_{\pm}$ probe DAS measured for σ_{-} pump [11]. Figure 6 compares these nonlinear differential circular-dichroism spectra, for two pump/probe delays, $\Delta t = 0$ fs and 200 fs, and for the two cases, $H = 0$, fig. 6a), and $H = 12$ T, fig. 6b). In agreement with previous observations [27, 28], in the absence of a magnetic field one observes at $\Delta t = 0$ an instantaneous spectral hole burning in the continuum at an energy slightly lower than the pump photon central energy and no response at the $1s$ hh exciton. Hole burning in the continuum is the signature of the PSF induced by the nonthermal populations generated by the pump. At $\Delta t = 200$ fs the carriers have thermalized by collision among each other and occupy the states at the bottom of the band, out of which the $1s$ hh X's and $1s$ lh X's are made. They thus block the transitions to the X's. Hence the DAS reproduces the profile of these two resonances and the spectral hole in the continuum has disappeared. Because the excess energy is smaller than the energy of optical phonons, the ther-

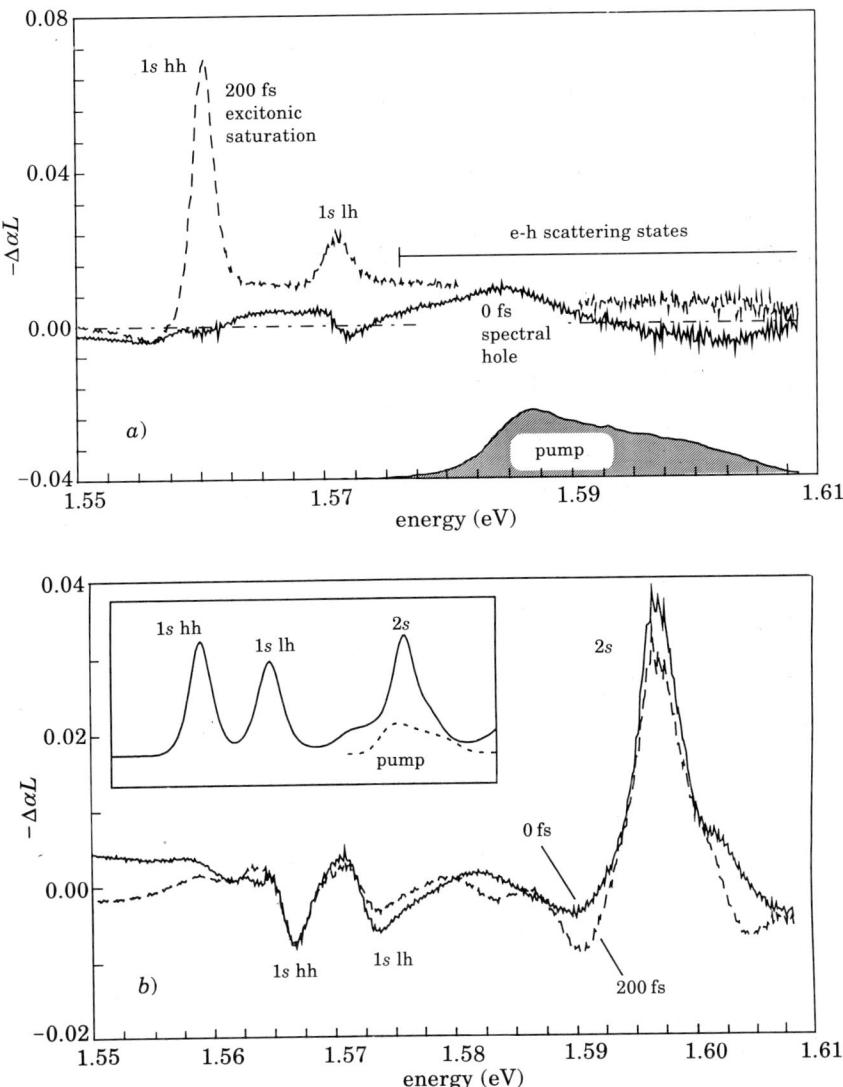

Fig. 6. – Circular-dichroism spectra, for excitation 25 meV above the 1s hh resonance and at $\Delta t = 0$ fs (solid line) and $\Delta t = 200$ fs (dashed line) for a) $H = 0$ and b) $H = 12$ T.

malization is internal to the plasma, *i.e.* the average energy remains constant as the e-h populations quickly establish a thermal distribution among each other at a temperature different from that of the lattice[28]. When the magnetic field is applied, the same pump excites the 2s MX as shown in the inset of fig. 6b). There is immediately a strong PSF signal at the 2s MX and a small signal due to dielectric screening at the 1s hh MX and 1s lh MX (note the different scales of the fig. 6a) and 6b)). The important result is that the spectra at $\Delta t = 0$ fs and

$\Delta t = 200\,\mathrm{fs}$ are almost identical. The nonthermal populations are blocked at $25\,\mathrm{meV}$ above the lowest states. The inelastic carrier-carrier scattering is quenched because of the limited number of final states available in the quasi-0D DOS. In addition the carriers cannot emit optical phonons because their energy is too small. The confinement in quasi-0D causes a dramatic reduction of the thermalization rates which has, of course, important implications for device applications [11, 12].

The polarization-resolved pump/probe techniques also allow a study of relaxation processes associated with spin. The level diagram of the inset of fig. 2 shows that under magnetic field the lowest-energy exciton state, $\mathrm{MX_{min}}$, is built up of a $|3/2, -3/2\rangle$ hole and a $|1/2, 1/2\rangle$ electron. This transition, with $\Delta m = 2$, is forbidden by one-photon absorption and, therefore, it is usually not observed. At $H = 12\,\mathrm{T}$ the $\mathrm{MX_{min}}$ is about $1.4\,\mathrm{meV}$ below the lowest one-photon

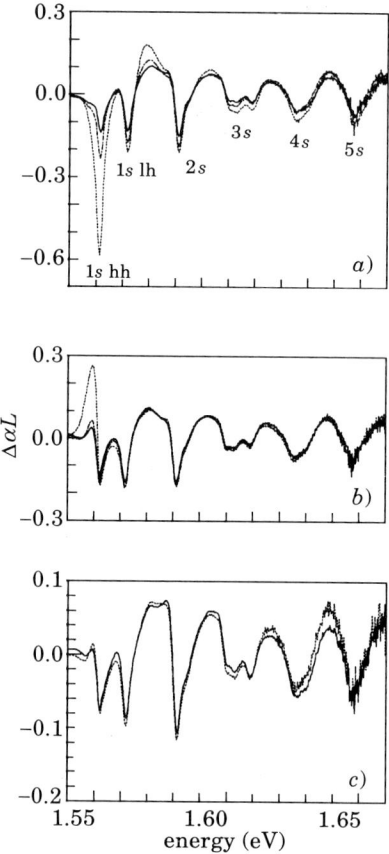

Fig. 7. – Differential transmission spectra of a σ_- probe for excitation resonant with the $1S$ exciton at $12\,\mathrm{T}$ by a) a σ_- pump and b) a σ_+ pump. In a) and b) the spectra correspond to $\Delta t = 0.66\,\mathrm{ps}$ (solid line), $40.66\,\mathrm{ps}$ (dotted line) and $66.66\,\mathrm{ps}$ (dash-dotted line), $40\,\mathrm{ps}$ per curve. After $\Delta t = 200\,\mathrm{ps}$ the spectra become identical (c)).

active MX. It can be generated indirectly, however, from one-photon active MX undergoing a spin-flip of the electron or the hole. This can be the case for the electron spin-flip of the $1S$ MX_{hh}^+, made of a $|3/2, -3/2\rangle$ hole and a $|1/2, -1/2\rangle$ electron, or the hole spin-flip of $1S$ MX_{hh}^-, made of a $|3/2, 3/2\rangle$ hole and a $|1/2, 1/2\rangle$ electron. We have observed such spin relaxation by following the evolution of the DAS measured with a σ_- probe after resonant excitation of the $1S$ MX_{hh}^- and $1S$ MX_{hh}^+ by σ_- pump and σ_+ pump, respectively [12]. The spin relaxation is found to occur on a very long time scale as compared to all the other processes discussed so far. As shown in fig. 7c), it is only after $\Delta t = 200$ ps that the two DAS become identical. For shorter delays, fig. 7a) and b), the two DAS show very distinct features at the $1S$ transition and almost identical profiles at higher energy. This response is in agreement with the discussion of sect. 2, where we stressed that for $1S$ resonant excitation the high-energy MX's are affected by charge density effects only. Furthermore, it is seen that the DAS profiles of these MX's do not change after $\Delta t \approx 1$ ps. This behavior shows that, on the time scale of the experiment, the charge density in the sample is constant after it has been generated. From the exponential decay of the $1S$ MX_{hh}^+'s DAS we infer a time of (65 ± 5) ps for flipping angular momentum of the $1S\,MX_{hh}^-$ hole and a time of (105 ± 10) ps for flipping the $1S\,MX_{hh}^+$ electron. Such lomg times are consistent with the spin relaxation times which have been reported recently [29-32]. A direct comparison, however, is difficult since we measure the flipping time of an electron or a hole within a MX, whereas the other measurements are relative to the spin-flip of free carriers.

5. – Conclusion.

We have explored the femtosecond dynamics of the nonlinear optical response of magneto-excitons, as the quasi-2D quantum well electronic states are further confined in quasi-0D by a strong magnetic field. We have observed, in agreement with the exact many-body theory result, that at high field the Coulomb correlation between magneto-excitons in the same state is quenched. It persists, however, between magneto-excitons in different states giving strong nonlinear responses. These results show that a gas of magneto-excitons is a unique two-component many-body system that behaves very differently from the one-component systems such as the fractional quantum Hall effect condensate or the Wigner crystal. We have also found that the magnetic field induces a restriction on the number of states available for transition, which almost completely quenches the relaxation of nonthermal populations. This produces qualitatively different carrier dynamics that must be accounted for in any quasi-0D system or device. Finally we have observed relaxation of angular momentum of the electron and hole within magneto-excitons, which results in a transition toward a one-photon forbidden state.

* * *

This work was performed in collaboration with J. B. STARK, W. H. KNOX and S. SCHMITT-RINK.

The work of DSC is supported by the Director, Office of Energy Research, Office of Basic Energy Sciences, Division of Materials Sciences of the US Department of Energy, under contract No. DE-AC03-76SF00098.

REFERENCES

[1] S. SCHMITT-RINK, D. S. CHEMLA and D. A. B. MILLER: *Adv. Phys.*, **38**, 89 (1989).

[2] O. AKIMOTO and H. HASEGAWA: *J. Phys. Soc. Jpn.*, **22**, 181 (1967).

[3] M. SHINADA and K. TANAKA: *J. Phys. Soc. Jpn.*, **29**, 1258 (1970).

[4] S. R. E. YANG and L. J. SHAM: *Phys. Rev. Lett.*, **58**, 2598 (1987).

[5] G. E. W. BAUER and T. ANDO: *Phys. Rev. B*, **38**, 6015 (1988).

[6] H. CHU and Y. C. CHANG: *Phys. Rev. B*, **40**, 5497 (1989).

[7] I. V. LERNER and YU. E. LOZOVIK: *Ž. Eksp. Teor. Fiz.*, **80**, 1488 (1981) (*Sov. Phys. JEPT*, **53**, 763 (1981)).

[8] D. PAQUET, T. M. RICE and K. UEDA: *Phys. Rev. B*, **32**, 5208 (1985).

[9] J. B. STARK, W. H. KNOX, D. S. CHEMLA, W. SCHÄFER, S. SCHMITT-RINK and C. STAFFORD: *Phys. Rev. Lett.*, **65**, 3033 (1990).

[10] S. SCHMITT-RINK, J. B. STARK, W. H. KNOX, D. S. CHEMLA and W. SCHÄFER: *Appl. Phys. A*, **53**, 491 (1991).

[11] J. B. STARK, W. H. KNOX and D. S. CHEMLA: *Phys. Rev. Lett.*, **68**, 3080 (1992).

[12] J. B. STARK, W. H. KNOX and D. S. CHEMLA: *Phys. Rev. B*, **46**, 7919 (1992).

[13] A. B. DZYUBENKO and YU. E. LOZOVIK: *J. Phys. A*, **24**, 415 (1991).

[14] V. M. APALKOV and E. I. RASBHA: *JETP Lett.*, **53**, 442 (1991).

[15] G. E. W. BAUER: *Phys. Rev. Lett.*, **64**, 60 (1990).

[16] S. SCHMITT-RINK, D. S. CHEMLA and D. A. B. MILLER: *Phys. Rev. B*, **32**, 6601 (1985).

[17] N. PEYGHAMBARIAN, H. M. GIBBS, J. L. JEWELL, A. ANTONETTI, A. MIGUS, D. HULIN and A. MYSYROWICZ: *Phys. Rev. Lett.*, **53**, 2433 (1984).

[18] D. HULIN, A. MYSYROWICZ, A. ANTONETTI, A. MIGUS, W. T. MASSELINK, H. MORKOC, H. M. GIBBS and N. PEYGHAMBARIAN: *Phys. Rev. B*, **33**, 4389 (1986).

[19] H. HAUG and S. SCHMITT-RINK: *Prog. Quantum Electron.*, **9**, 3 (1984).

[20] C. STAFFORD, S. SCHMITT-RINK and W. SCHÄFER: *Phys. Rev. B*, **41**, 10000 (1990).

[21] S. SCHMITT-RINK and D. S CHEMLA: *Phys. Rev. Lett.*, **57**, 2752 (1986); S. SCHMITT-RINK, D. S. CHEMLA and H. HAUG: *Phys. Lett. B*, **37**, 941 (1988).

[22] W. SCHÄFER: *Adv. Solid State Phys.*, **28**, 63 (1988); W. SCHÄFER, K. H. SCHULDT and R. BINDER: *Phys. Status Solidi B*, **150**, 407 (1988).

[23] C. ELL, J. F. MULLER, K. EL SAYED, L. BANYAI and H. HAUG: *Phys. Status Solidi B*, **150**, 393 (1988); *Phys. Rev. Lett.*, **62**, 304 (1989).

[24] R. ZIMMERMANN: *Phys. Status Solidi B*, **146**, 545 (1988); R. ZIMMERMANN and M. HARTMANN: *Phys. Status Solidi B*, **150**, 365 (1989).

[25] I. BALSLEV, R. ZIMMERMANN and A. STAHL: *Phys. Rev. B*, **40**, 4095 (1989).

[26] R. BINDER, S. W. KOCH, M. LINDBERG, N. PEYGHAMBARIAN and W. SCHÄFER: *Phys. Rev. Lett.*, **65**, 899 (1990).

[27] W. H. KOX, C. HIRLIMANN, D. A. B. MILLER, J. SHAH, D. S. CHEMLA and C. V. SHANK: *Phys. Rev. Lett.*, **56**, 1191 (1986).

[28] W. H. KNOX, D. S. CHEMLA, G. LIVESCU, J. E. CUNNINGHAM and J. E. HENRY: *Phys. Rev. Lett.*, **61**, 1290 (1988); W. H. KNOX: in *Hot Carriers in Semiconductor Nanostructures*, edited by J. SHAH (Academic Press, New York, N.Y., 1992).

[29] A. TAKEUCHI, S. MUTO, T. INATA and T. FUJII: *Appl. Phys. Lett.*, **56**, 2213 (1990).

[30] M. KOHL, M. R. FREEMAN, D. D. AWSCHALOM and J. M. HONG: *Phys. Rev. B*, **44**, 5923 (1991).

[31] T. C. DAMEN, L. VIÑA, J. E. CUNNINGHAM and J. SHAH: *Phys. Rev. Lett.*, **67**, 3432 (1991).

[32] S. BAR-AD and I. BAR-JOSEPH: *Phys. Rev. Lett.*, **68**, 349 (1992).

Nonlinear Magneto-Optical Materials: Photo-Induced Gyrotropy.

S. Hugonnard-Bruyère, C. Buss, R. Frey and C. Flytzanis

Laboratoire d'Optique Quantique du Centre National de la Recherche Scientifique
Ecole Polytechnique - 91128 Palaiseau Cédex, France

1. – Introduction.

The need to develop nonlinear optical materials that can be used to rapidly and significantly modify the laser beam characteristics such as intensity, temporal and spatial frequency spectra or polarization state is widely recognized and justifies the ongoing effort there [1,2]. Indeed several nonlinear optical effects allow one to modify these characteristics when an intense laser beam is sent through an isotropic transparent nonlinear medium. Thus the probe intensity can be modified through multiphoton processes, the spatial and temporal beam profiles through self-modulation and focusing, respectively, the polarization state, finally, through photo-induced linear or circular birefrigence. For all these photo-induced effects the effort has been directed at delineating the underlying mechanisms and relating them to material features, on the one hand, and, on the other, at improving the performances of the nonlinear materials or synthesizing new ones.

Among these photo-induced effects the ones that permit a modification and control of the polarization state of a beam are quite subtle and require a particular attention. The photo-induced linear birefrigence is simply related to phase changes that occur through the optical Kerr effect, a well-studied third-order optical process [3]; the photo-induced circular birefrigence, on the other hand, is connected to rotations of the polarization state of the probe beam which can be obtained through the photo-induced gyrotropy or the photo-induced Faraday rotation. Their study was initiated only recently [4] and their potential applications are now recognized.

In this lecture we concentrate our attention on the photo-induced gyrotropy obtained in experiments that exploit magneto-optical interactions in isotropic paramagnetic materials in the Faraday configuration: the static magnetic field is applied along the propagation of the laser beam. It arises through the combined effect of optical Kerr nonlinearity and Faraday rotation and as a consequence the choice of the materials must be such that these two processes are simultaneously enhanced. We show that this situation is encountered in semi-magnetic semiconductors, namely binary II-VI semiconductors doped with magnetic impurities where the giant photo-induced Faraday rotation was evidenced when the photon energy is close to the gap[4,5]; in contrast this effect is minuscule in the undoped compound under similar resonance conditions[6].

In sect. 2 we briefly recall the physical background of the Faraday effect. The material features of the semi-magnetic semiconductors and the physical mechanisms that underly the giant photo-induced Faraday rotation observed in these materials are succinctly analysed in sect. 3, while the main observations and behaviour in the single-beam configuration are presented in sect. 4. Finally in sect. 5 we present results obtained in the pump and probe geometry and also introduce an analysis of the figure of merit and the preliminary performances of switching devices based on these materials.

2. – Physical background.

2'1. *The Faraday rotation.* – In optically active and isotropic materials such as most of the useful paramagnetic materials, the susceptibility tensor χ for a magnetization M_0 along with the z-axis is given by

$$(2.1) \qquad \chi = \begin{bmatrix} \chi_1 & i\chi_1' & 0 \\ -i\chi_1' & \chi_1 & 0 \\ 0 & 0 & \chi_1 \end{bmatrix},$$

where the elements χ_1 and χ_1' are real when the laser frequency is off resonance. In the Faraday configuration in which the laser beam propagates along the magnetization direction, the two counterrotating circulary polarized waves σ_+ and σ_- propagate without deformation with refractive indices n_+ and n_-, respectively. As a consequence the polarization direction of a linearly polarized input wave E of frequency ω after a propagation through a length $\mathscr{L}$ in such a material rotates by an angle θ_F which is called the Faraday rotation angle:

$$(2.2) \qquad \theta_F = \frac{\omega \mathscr{L}}{2c}(n_- - n_+),$$

where c is the light velocity in vacuum.

The elements of the susceptibility tensor χ can be calculated using a classical description for the movement of the bound electrons of charge e which are submitted to the electric-field force $e\,E$ and to the Lorentz force $e\,v \wedge H$, where v is the electron velocity and H the applied static magnetic field[7]. These elements can also be determined quantum mechanically noting that the Zeeman effect splits the degenerate spin sublevels and modifies the energy spacings and oscillator strengths of electric-dipole-allowed transitions for left and right circularly polarized light. Adopting this latter approach, the Faraday rotation angle θ_F can be written in first-order approximation as

$$(2.3) \qquad \theta_F = \sum_i a_i \Delta E_i \,,$$

where ΔE_i and a_i are the Zeeman splitting and a constant depending on the laser frequency, the summation being performed over all the states involved in the calculation of $n_\mp$. In the conventional Zeeman effect, ΔE_i is proportional to the applied magnetic field H and to the Landé factor g_i of state i. Generally, these Landé factors are relatively small (one to ten) so that the corresponding Faraday rotation angles are small even when the laser frequency is chosen near the band gap resonance of the semiconductor medium. As shown in the next subsection, this is no longer the case in semi-magnetic semiconductors.

2'2. *Semi-magnetic semiconductors.* – Semi-magnetic semiconductor compounds [8] $A_{1-x}^{II} M_x B^{VI}$ are obtained from a II-VI semiconductor by substituting a proportion x of cations A^{II}(Cd, Zn, Hg, ...) with a same amount of magnetic impurities M(Mn, Co, Fe, Ni, Eu, ...), B^{IV} being generally S, Se or Te. The maximum value of the mixing parameter x depends on matching requirements of the structural lattice parameters and, therefore, strongly varies from one compound to another. Large values can nevertheless be obtained, such as $x = 0.77$ and $x = 0.86$ for $Cd_{1-x} Mn_x Te$ and $Zn_{1-x} Mn_x Te$, respectively, where the magnetic impurity is the element manganese (Mn) with $4s^2\, 3d^5$ outer electronic structure. Manganese is a transition metal that differs from the A^{II} element by the presence of a valence half-filled $3d$-shell but otherwise has a $4s^2$-shell too whose electrons get delocalized and involved in the s-p^3 bonding in the same manner as those of the replaced A^{II} element preserving thus the initial tetrahedral bond configuration and cristalline structure. Further the five electrons of the half-filled $3d$-shell remain unbound and localized on the parent Mn^{++} ion; in the presence on a strong static magnetic field they are responsible of a strong magnetization M_0 because of Hund's rule whereby all five electrons are aligned in the same direction, *i.e.* ($\uparrow \uparrow \uparrow \uparrow \uparrow$). In the paramagnetic regime at the temperature T this magnetization is given by [9]

$$(2.4) \qquad M_0 = \bar{x} N_0 g_{Mn} \mu_B S B_S [g_{Mn} S \mu_B H / k_B (T + T_{AF})]$$

with the spin $S = 5/2$. In eq. (2.4), N_0 is the number of anions per unit volume, $g_{Mn} \approx 2$ the Landé factor for the Mn^{++} ion, μ_B the Bohr magneton, k_B the Boltzmann constant, and $B_S[\eta]$ is the standard Brillouin function for parameter η. Finally, $\bar{x}$ and T_{AF} are coefficients which represent an effective manganese concentration and an antiferromagnetic temperature, respectively. At high temperatures ($T > 80$ K), $\bar{x}$ is equal to the real manganese concentration x, and

T_{AF}, which is calculated from the antiferromagnetic interactions between Mn^{++} ions[10], is given by

$$(2.5) \qquad T_{AF} = -\frac{2}{3} x\, S(S + 1)\, ZJ/k_B \, ,$$

where Z is the number of nearest neighbours ($Z = 12$ for zinc blende and wurtzite semi-magnetic semiconductors), and J is the nearest-neighbour exchange integral. At low temperature ($T < 30$ K), $\bar{x}$ and T_{AF} are only phenomenological parameters which depend on the concentration x of Mn^{++} ions[11] and on the magnetic-field amplitude[12].

The high magnetization existing in these semi-magnetic semiconductors introduces considerable modifications in the interband transitions when a static magnetic field is applied to the material. Without magnetic field, close to the band gap which is a direct one and lies in the centre of the Brillouin zone, the bands are parabolic with conductor states being of s-type, hence spin 1/2, and the valence states being of p-type, which because of a spin-orbit coupling split into two subbands of spins 3/2 and 1/2, respectively. Of these two spin-orbit split-off subbands here we shall be concerned with the states of the former one, which at the top of the valence band are of symmetry Γ_8, while those of the directly above lying bottom of the conduction band possess the Γ_6 symmetry.

When a magnetic field is applied to the material, the spin sublevels in the conduction and valence bands are split off by magnetic interactions: these are the Landau and Zeeman interactions as in conventional semiconductors, and, moreover, in the case of semi-magnetic semiconductors, the spin exchange interaction between the sp bands and the $3d$-impurity electrons. Denoting by z the direction of the magnetic field, and s and S the spin operators for the band and localized electrons and introducing $s_\pm = (s_x \pm is_y)/\sqrt{2}$ and $S_\pm = (S_x \pm iS_y)/\sqrt{2}$, the spin exchange interaction can be written as

$$(2.6) \qquad H_{ex} = \sum_{R_i} J(\boldsymbol{r} - \boldsymbol{R}_i)\, \boldsymbol{S}_i \cdot \boldsymbol{s} = \sum_{R_i} J(\boldsymbol{r} - \boldsymbol{R}_i)(S_{iz}s_z + S_{i+}s_- + S_{i-}s_+),$$

where $\boldsymbol{R}_i$ is the position of the i-th magnetic impurity and $\boldsymbol{r}$ the band electron coordinate. Because of the large extension of the wave functions of the band electrons the latter probe a large number of Mn impurities and one can introduce two approximations: firstly, the mean-field approximation by replacing the summation over S_i in eq. (2.6) by a thermal average $\langle S \rangle = \langle S_z \rangle$, and, secondly, the virtual crystal approximation by substituting $J(\boldsymbol{r} - \boldsymbol{R}_i)$ in eq. (2.6) by $x\, J(\boldsymbol{r} - \boldsymbol{R})$, where $\boldsymbol{R}$ now runs over all sites of the sublattice partially occupied by Mn^{++} and hence carrying the summation over all sites. The compound effect of these two simplifications amounts to replacing eq. (2.6) by

$$(2.7) \qquad H_{ex} = x\, J_0 s_z \langle S_z \rangle,$$

where the exchange operator $J_0 = \sum_{R} J(\boldsymbol{r} - \boldsymbol{R})$ only depends on the structure of

the semi-magnetic semiconductor studied and $\langle S_z \rangle$ is related to the magnetization M_0 given in eq. (2.4) by

(2.8) $$M_0 = -x \, N_0 \, g_{Mn} \mu_B \langle S_z \rangle.$$

In fact, among the three magnetic contributions, at Mn^{++} concentrations
larger than one percent the exchange interaction is the dominant process and
the Landau and Zeemann effects can be safely neglected. Calculations [13] indicate that the spin degeneracies in the Γ_6 conduction band and Γ_8 valence band
are raised as in the case of the pure Zeeman effect with the only difference that
the spin sublevel splittings are now much larger than for the conventional Zeeman effect, and the sign is even changed for the conduction band. Figure 1 represents a schematic picture of the spin splitting of the relevant conduction and
valence states in a magnetic field and also shows the electric-dipole-allowed
transitions for the Faraday configuration.

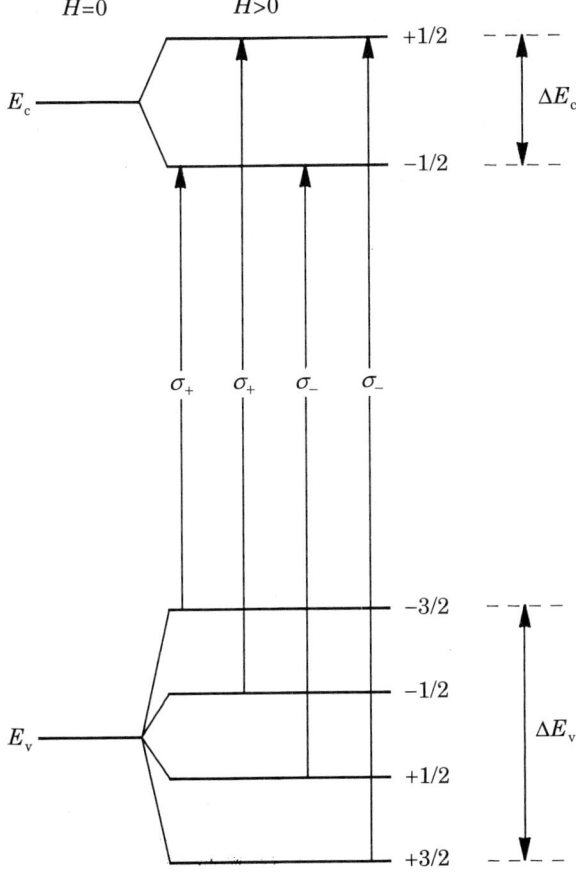

Fig. 1. – The equivalent system used for the calculation of the photo-induced Faraday rotation angle.

The energy splitting indicated in this diagram are

(2.9a)
$$\Delta E_c = \frac{\alpha M_0}{g_{Mn}\mu_B},$$

(2.9b)
$$\Delta E_v = -\frac{\beta M_0}{g_{Mn}\mu_B}$$

for the conduction and valence states, respectively. In eq. (2.9) $\alpha = = \langle s|J_0(r)|s\rangle/\Omega_0$ and $\beta = \langle p_z|J_0(r)|p_z\rangle/\Omega_0$ are the exchange integrals for the conduction and valence bands, respectively, Ω_0 being the volume of an elementary cell. Due to the large values of α and β ($N_0\beta \approx -4N_0\alpha \approx -0.88$ eV for $Cd_{1-x}Mn_xTe$, for instance), the spin sublevel splittings are approximately two orders of magnitude larger in a semi-magnetic semiconductor than in its parent binary compound. Thus the magnetic impurities indeed act as «local amplifiers» of the applied magnetic field and induce giant pseudo-Zeeman splittings of the conduction and valence bands[14]. Since the Faraday rotation angle is directly proportional to the Zeeman splittings (see eq. (2.3)), giant Faraday rotations as large as 5000 degrees cm^{-1} kG^{-1} at liquid-helium temperature can be observed experimentally for optical frequencies near the band gap resonance[15].

3. – Physical origins of the photo-induced gyrotropy in semi-magnetic semiconductors.

As shown in eq. (2.2), the Faraday rotation angle θ_F depends on the refractive indices $n_{\mp}$ corresponding to the two counterrotating circular polarizations $\sigma_{\mp}$. As a consequence, any of the parameters which enters the expression of $n_- - n_+$ and depends on the pump laser intensity I_p can substantially modify θ_F and thus produces photo-induced Faraday rotation angles θ_{NL} defined as

(3.1)
$$\theta_{NL} = \theta_F(I_p) - \theta_F(I_p = 0),$$

where $\theta_F(I_p)$ and $\theta_F(I_p = 0) = \theta_L$ represent the Faraday rotation angles at high and low pump intensities, respectively. These parameters are the energy and population of the levels which are of interest for calculating n_- and n_+ (in our case the conduction and valence bands), and the magnetization which enters in θ_F through the spin sublevel splittings given in eqs. (2.9).

Phenomenologically, the intensity-dependent refractive indices $n_{\mp}$ can be written in the form

(3.2)
$$n_{\mp}^2 = 1 + 4\pi[\chi^{(1)} + \Delta\chi_{\mp}^{(1)} + \chi^{(3)}I_p + \Delta\chi_{\mp}^{(3)}I_p],$$

where $\chi^{(3)}$ is the third-order susceptibility which enters the optical-Kerr-effect coefficient n_2. $\Delta\chi_{\mp}^{(1)}$ and $\Delta\chi_{\mp}^{(3)}$ are the modifications of the above coefficients, $\chi^{(1)}$ and $\chi^{(3)}$, respectively, induced by the magnetic field.

For the calculation of $\Delta\chi_{\mp}^{(1)}$ we use the equivalent system defined in fig. 1

since only the interband transitions are strongly modified by the exchange interaction. The linear susceptibility is calculated through the density matrix technique[16] applied to this six-level system. Detailed calculations are given in ref.[12,17]. In the following, we will concentrate on the physical origin and on the response and lifetimes of the different contributions existing in the photo-induced Faraday rotation.

3'1. *The magneto-optical Kerr contribution.* – At high laser intensities the six spin sublevels represented in fig. 1 are shifted through the dynamic Stark effect by amounts which are proportional to the light intensity. Furthermore, a corresponding change in the position of the levels relative to one another also occurs. As the Faraday rotation angle is proportional to the energy spacing of these spin sublevels in the conduction and valence bands, there is a magneto-optical Kerr contribution θ_K to θ_{NL} which is given by[12]

$$(3.3) \qquad \theta_K = \frac{\pi \omega \mathscr{L}}{n_0 c} \Delta A \, I_p M_0 \, ,$$

where n_0 is the refractive index at low laser intensity with no applied magnetic field, and ΔA is a factor which depends on the laser frequency but is independent of the laser intensity and magnetization.

For this contribution to θ_{NL}, whose origin is the dynamic Stark effect, the response and lifetimes (τ_{RK} and τ_{LK}, respectively) are related to electronic motions and are virtually instantaneous ($\tau_{RK} \approx \tau_{LK} \approx 10^{-14} \mathrm{s}$).

3'2. *The refractive-index saturation contribution.* – If the pump laser frequency is larger than the band gap resonance, or if the pump laser intensity is high enough, free carriers can be generated in the conduction and valence bands through one- or two-photon absorption. As the populations of the spin sublevels of both bands are of importance for determining the value of the refractive index, a high free-carrier concentration (greater than 10^{17} cm^{-3}) leads to a saturation of these values of the refractive index and, consequently, to a decrease of the Faraday rotation angle. This effect provides a refractive-index saturation contribution θ_p to θ_{NL} which is given by[12]

$$(3.4) \qquad \theta_p = \frac{2}{3} \frac{\pi \omega \mathscr{L}}{n_0 c} \Delta n_T(t) \, d^2 \, F[3B_{3/2}(\eta_v) + 2B_{1/2}(\eta_c)],$$

where $\Delta n_T(t)$ is the density of thermalized free carriers at the time t, d is the unperturbed dipolar transition moment, and $F = E_g/[E_g^2 - (\hbar \omega)^2]$ with E_g the band gap energy at zero magnetic field. In eq. (3.4), $B_{x/2}(\eta_y)$ is the Brillouin function for electrons in the conduction band ($x = 1$ and $y = c$) and holes in the valence band ($x = 3$ and $y = v$). The argument of this Brillouin function is given by $\eta_y = \Delta E_y/2k_B T_y$, where T_y is the effective temperature for electrons ($y = c$) or holes ($y = v$) in the conduction and valence bands, respectively. We point out that eq. (3.4) is written under the assumption of quasi-equilibrium for electron and hole populations in the conduction and valence bands, respectively; this assumption is valid for times longer than a few 10^{-13} s since the intraband relaxation is very fast in semiconductor media.

This fast intraband relaxation also implies that the response time τ_{RP} for the refractive-index saturation contribution is almost instantaneous ($\tau_{RP} < $ < few 10^{-13} s). Conversely, the lifetime τ_{LP} which is determined by the interband recombination can vary from less than one nanosecond, for the stimulated recombination occurring at high free-carrier concentrations[12], to more than one microsecond if the recombination is dominated by traps.

3'3. *The magnetic contribution.* – Since the Faraday rotation angle depends on the spin sublevel splittings ΔE_c and ΔE_v, a magnetization change ΔM occurring due to the pump laser pulse can cause a magnetic contribution θ_M to θ_{NL}, which is given by[12]

$$(3.5) \qquad\qquad \theta_M = \theta_L \frac{\Delta M}{M_0}.$$

Two different physical origins of the magnetization change are of importance here: one is due to spin flip-flops occurring during the intraband relaxation, and gives rise to a pure magnetic contribution θ_{PM} to θ_M, and another is due to thermal Mn^{++} spin relaxation caused by the heating of the material, and gives rise to a thermal magnetic contribution θ_{TM} to θ_M.

3'3.1. The pure magnetic contribution. When free carriers are generated through one- or two-photon absorption, spin-flip processes occur for these free carriers during the intraband relaxation. In large-gap semi-magnetic semiconductors such as $Cd_{1-x}Mn_xTe$ these spin-flip processes occur most likely through the transverse part of the exchange interaction (related to the term $s_- S_+ + s_+ S_-$ in eq. (2.6)), and conserve the total spin[18]. As a consequence, the Mn^{++} magnetization M_0 changes by a quantity ΔM_{PM} which is related to the magnetization modifications of electrons and holes. If we consider the case of free carriers generated through two-photon absorption with identical densities for each spin sublevel, ΔM_{PM} can be written as

$$(3.6) \qquad\qquad \Delta M_{PM} = M_0 \frac{\Delta n_T(t)}{xN_0} \frac{B_{1/2}(\eta_c) - 3B_{3/2}(\eta_v)}{5B_{5/2}(\eta_{Mn})},$$

where $\eta_{Mn} = 5\, g_{Mn}\mu_B H/2k_B(T + T_{AF})$. At complete saturation of the $B_{x/2}(\eta_y)$ functions, the maximum magnetization decrease is then of the order of $M_0 \Delta n_T(t)/xN_0$, so that θ_{PM} is negligible at high Mn^{++} concentrations or at low free-carrier densities.

Since the exchange-interaction–induced intraband relaxation is the origin of the pure magnetic contribution, the corresponding response time τ_{RPM} is virtually instantaneous ($\tau_{RPM} \approx$ few 10^{-13} s). Alternatively, the lifetime is related to the thermal relaxation of the Mn^{++} spins and can, therefore, strongly vary from one nanosecond to ten microseconds depending on the Mn^{++} concentration and on the sample temperature [19].

3`3.2. The thermal magnetic contribution. Since optical transitions in solids occur between bands rather than between discrete levels, with optical and acoustical phonons being emitted during the intraband relaxation, these processes, undoubtedly, lead to heating of the material with a temperature increase $\Delta T = E_{ph}/C_p$, where E_{ph} and C_p are the total energy per volume unit deposited in phonons and the volume heat capacity of the material at temperature T, respectively. Since the magnetization M_0 is related to T through eq. (2.4), the magnetization decrease ΔM_{TM} corresponding to a small temperature increase ΔT is given by [17]

$$(3.7) \qquad \Delta M_{TM} = M_0 \frac{\partial}{\partial T}\left[\ln\left(B_{5/2}\left[\frac{5}{2}g_{Mn}\mu_B H/k_B(T+T_{AF})\right]\right)\right]\Delta T .$$

Since the heat capacity is very small at low temperatures ($T < 20$ K) [20], the thermal magnetic contribution θ_{TM} to θ_M may be large in this temperature range.

In this thermal magnetic process, the spin-phonon relaxation of the Mn^{++} ions provides the response time τ_{RTM} of the contribution θ_{TM}, while the thermal diffusion is responsible for the lifetime τ_{LTM} (typically from 1 µs to 1 ms depending on the experimental conditions).

TABLE I. – *Recapitulation of the different physical origins of the photo-induced gyrotropy in semi-magnetic semiconductors and the associated response and lifetimes.*

Parameter	Spin sublevel energies	Spin sublevel populations	Magnetization	
			Pure magnetic	Thermal
physical process	magneto-optical Kerr effect	saturation of refractive indices	exchange interaction induced spin flip-flops	material heating induced magnetization change
response time	instantaneous ($\approx 10^{-14}$ s)	instantaneous ($\approx 10^{-13}$ s)	instantaneous ($\approx 10^{-13}$ s)	1 ns-10 µs ([Mn^{++}], T)
lifetime	instantaneous ($\approx 10^{-14}$ s)	< 1 ns (stimulated) (1–10) ns (spontaneous) up to 10 µs (traps)	1 ns-10 µs ([Mn^{++}], T)	1 µn-1 ms

3'4. *Recapitulation.* – For each of the physical origins of the photo-induced gyrotropy in semi-magnetic semiconductors reviewed in this section, table I summarizes and draws together the physical origin and the corresponding response and lifetimes. It can be readily seen from this table that, except for the thermal magnetic contribution, the rise time is almost instantaneous, a property which can be useful for potential applications in optical signal processing. On the other hand, the lifetime can strongly vary depending on the relaxation mechanism, the Mn^{++} concentration and the sample temperature. All these features are important for the design and operation characteristics of commutation devices based on these materials.

In the light of the above considerations we summarize below the main results and conclusions drawn from recent experimental investigations of the photo-induced Faraday rotation in semi-magnetic semiconductors in the case of a single (pump) beam (sect. 4) followed by that of the pump and probe beam configuration (sect. 5).

4 – Single-beam (pump) configuration.

This is the case in which θ_{NL} is measured for the pump beam that sets up the photo-induced Faraday rotation. In such a case, the polarization direction of a picosecond duration neodymium-YAG laser pulse of intensity I_p is measured with an analyser and a detector placed behind the analyser.

At low I_p, when the photo-induced effect is negligible, the measured rotation angle is θ_L. Under the prevailing experimental conditions, 1 mm thick $Cd_{0.75}Mn_{0.25}Te$ sample at 10 K, θ_L being proportional to the magnetization, M_0 is also proportional to the magnetic-field intensity for $H < 5\,T$ as expressions (2.3), (2.4) and (2.8) easily imply.

At high I_p, the dependence of the photo-induced Faraday rotation angle θ_{NL} *vs.* the applied magnetic-field intensity up to 4 T is shown in fig. 2 for a laser peak intensity of about 2 GW/cm^2. As seen there, for low magnetic fields, less than $H_v = 0.5\,T$ for our sample, the relative photo-induced change of the Faraday rotation amounted to 14%, whereas for magnetic fields larger than $H_c = 2.25\,T$ this amounted to only 6%, implying that saturation sets in at different stages as the magnetic field is increased. The observations displayed an excellent fit to three straight lines of decreasing slope as the magnetic-field intensity is increased.

Due to the short duration of the laser pulse (≈ 30 ps) the thermal process could not grow and only the magneto-optical Kerr and refractive-index saturation terms significantly contributed to θ_{NL}. Indeed, due to the high Mn^{++} concentration used in this experiment ($x = 0.25$), the pure magnetic term cannot give a contribution larger than a few $10^{-5}\,\theta_L$, even for a

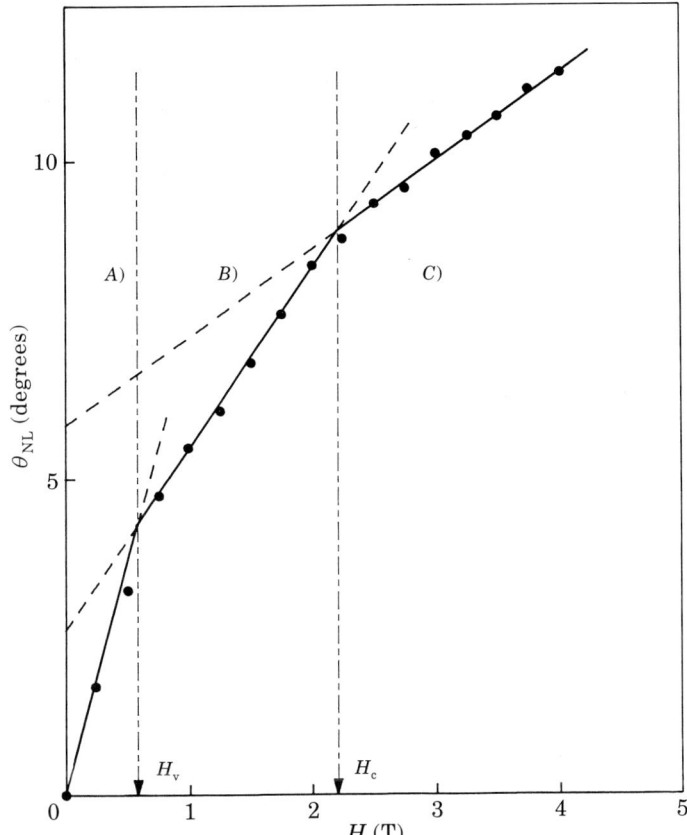

Fig. 2. – Photo-induced Faraday rotation angle as a function of the applied magnetic field. H_v and H_c are the values of H limiting regions A), B) and C) (see ref.[12]); $T = 10$ K, $I = 2$ GW/cm^2, Cd$_{0.75}$Mn$_{0.25}$Te.

high free-carrier density (10^{18} cm^{-3}), a value which is considerably smaller than the measured one ($\approx 10^{-1}\,\theta_L$).

In order to give a simpler physical description of the behaviour, let us reduce the Brillouin function $B_J(\eta_y)$ appearing in eq. (3.4) to two straight lines of equations $B_J(\eta_y) = (J+1)\eta_y/3J$ and $B_J(\eta_y) = 1$ for small and large values of η_y, respectively. In these expressions $J = 3/2$ and $y = $ v, and $J = 1/2$ and $y = $ c for the valence and conduction bands, respectively. Since $|\alpha| < |\beta|$, or $\eta_c < \eta_v$, $B_{3/2}(\eta_v)$ saturates before $B_{1/2}(\eta_c)$ when the magnetization is increased. For small values of H both $B_{3/2}(\eta_v)$ and $B_{1/2}(\eta_c)$ are proportional to M_0 so that θ_{NL} is given by

$$(4.1) \qquad \theta_{\text{NL}} = [\lambda_K I_p + (\lambda_p^{(c)} + \lambda_p^{(v)})\Delta n_T(t)] M_0 .$$

In eq. (4.1), $\lambda_K = \pi\omega\mathscr{L}\Delta A/n_0 c$ is related to the magneto-optical Kerr contri-

bution, while $\lambda_p^{(c)}$ and $\lambda_p^{(v)}$ are related to the refractive-index saturation processes for the conduction electrons and valence holes, respectively. Note that the
constants $\lambda_p^{(c)}$ and $\lambda_p^{(v)}$, which can be easily calculated owing to eqs. (3.4) and
(2.9), are independent of the laser intensity and applied magnetic field. As the
magnetization is proportional to H, θ_{NL} is also proportional to H; this corresponds to zone $A)$ in fig. 2.

As H increases beyond a value H_v such that $\eta_v = 9/5$, $B_{3/2}(\eta_v)$ saturates at
the value one, while $B_{1/2}(\eta_c)$ is still proportional to η_c and θ_{NL} becomes

$$(4.2) \qquad \theta_{NL} = [\lambda_K I_p + \lambda_p^{(c)} \Delta n_T(t)] M_0 - \frac{18 g_{Mn} \mu_B k_B T_v}{5\beta} \lambda_p^{(v)} \Delta n_T(t).$$

As a consequence, the photo-induced Faraday rotation exhibits a linear dependence *vs.* the applied magnetic field with a smaller slope than that given by
eq. (4.1); this is zone $B)$ of fig. 2.

Finally, beyond a value H_c such that $\eta_c = 1$, $B_{1/2}(\eta_c)$ also saturates at the
value one and

$$(4.3) \qquad \theta_{NL} = \lambda_K I_p M_0 + 2 g_{Mn} \mu_B k_B \left[\frac{\lambda_p^{(c)} T_c}{\alpha} - \frac{9 \lambda_p^{(v)} T_v}{5\beta} \right] \Delta n_T(t).$$

The photo-induced Faraday rotation has, therefore, a linear dependence *vs.* H
with the slope being solely dependent on the Kerr contribution; this corresponds to zone $C)$ in fig. 2.

5. – Pump and probe experiments.

Pump and probe experiments were performed in order to verify the theoretical predictions concerning the response and lifetimes. The schematic diagram
of the pump and probe experimental setup is presented in fig. 3. The semi-magnetic semiconductor sample was placed inside the same magnetic cryostat as in
the single-beam experiment. Although some experiments were performed with
a picosecond resolution [21], the results presented in this section were obtained
with a nanosecond response time for the electronics. As a consequence, the
probe beam is a monomode c.w. laser (He-Ne, semiconductor, or Ti-sapphire
laser depending on the experiment). This probe beam was linearly polarized by
a polarizer and its time-dependent intensity behind an analyser ($I_T(t)$) was
measured with a fast detector and a transient digitizer and stored in a personal
computer.

When no pump pulse was applied to the sample, the transmitted intensity is
given by

$$(5.1) \qquad \qquad I_T(t) = I_0 \cos^2 \theta_0,$$

where I_0 is the intensity incident onto the analyser and θ_0 the angle existing be-

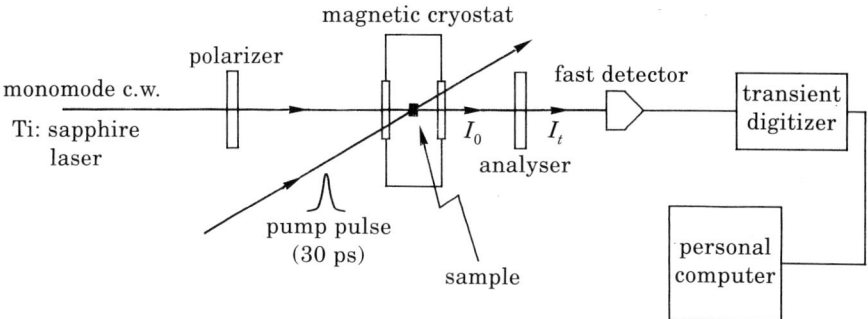

Fig. 3. – Schematic diagram of the pump and probe setup.

tween the analyser direction and that of the probe polarization with no pump pulse. When a pump pulse of picosecond duration was incident on the sample at instant $t = 0$, photo-induced absorption, phase change and polarization rotation could occur such that $I_T(t)$ was given by

$$(5.2) \qquad I_T(t) = \frac{I_0}{2} \exp[-\alpha_{NL} \mathscr{L}][1 + \cos(2\varphi_{NL})\cos 2(\theta_0 + \theta_{NL})],$$

where α_{NL} and φ_{NL} are the photo-induced absorption coefficient and phase change, respectively.

5'1. *Thermal magnetic contribution.* – In order to isolate the thermal magnetic contribution from the others, the experiment was performed at low temperatures ($T < 4$ K) to maximize the temperature increase ΔT. The magneto-optical Kerr contribution which is only present during the excitation process can be safely neglected on a time scale larger than one nanosecond used in this experiment. In order to eliminate the refractive-index saturation contribution, we chose the semi-magnetic semiconductor $Zn_{1-x}Mn_xTe$ and excited the material through single-photon absorption on the intra-Mn^{++} transition at 2.2 eV [22]. Here, as the frequency of the intra-Mn^{++} transition was smaller than the band gap resonance (2.4 eV to 3.2 eV depending on the Mn^{++} concentration), no free carriers were created. Such a situation cannot be achieved in $Cd_{1-x}Mn_xTe$ and accordingly the thermal contribution cannot be isolated because the band gap resonance frequency is smaller than that of the intra-Mn^{++} transition. This point was confirmed by the response signal to a picosecond duration excitation in both cases of $Zn_{1-x}Mn_xTe$ and $Cd_{1-x}Mn_xTe$. In zinc manganese telluride the rise time was slow since it was governed by the spin-phonon relaxation of Mn^{++} ions, while in cadmium-manganese telluride there was, in addition, an instantaneous contribution due to the refractive-index saturation term [21].

The maximum value of the photo-induced Faraday rotation angle (occurring

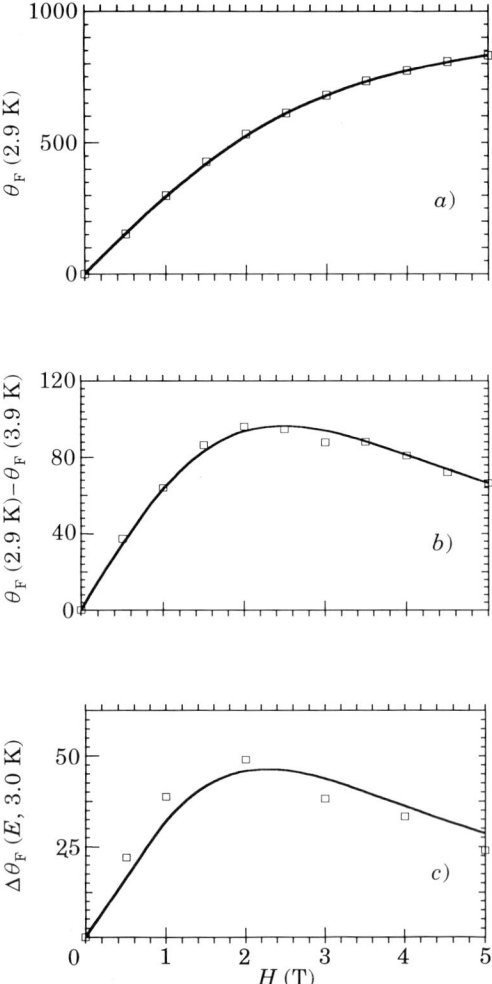

Fig. 4. – Linear Faraday rotation angle (a)), difference between linear Faraday rotation angles at different temperatures (b)), and photo-induced thermal magnetic Faraday rotation angle (c)) vs. the magnetic field. Squares correspond to experimental data and continuous lines are fits obtained with $B_{5/2}$ Brillouin functions using $T_{AK} = 2.0$ K (see ref. [18]).

about 5 µs after the excitation pulse) can be readily calculated by using eq. (5.2) with $\alpha_{NL} = \varphi_{NL} = 0$ since no significant photo-induced absorption and phase change were observed.

Figure 4 shows the experimental results concerning the linear Faraday rotation (fig. 4a)), the difference between the Faraday rotation angles at two different temperatures (fig. 4b)) and the photo-induced Faraday rotation (fig. 4c)). All these quantities which are plotted as a function of the magnetic field were obtained for a 1 mm thick sample of $Zn_{1-x}Mn_x Te$ with $x = 8\%$. The experimental data represented by squares in figures 4a) and b) are very well fitted with the $B_{5/2}$ Brillouin function with $T_{AF} = 2.0$ K. The good quality of the fit is attested by fig. 4a) and even more by fig. 4b), where the difference between Faraday rotation angles measured at 2.9 and 3.9 K is plotted as a function of H. Indeed, in the difference the sensitivity is about ten times higher than in the direct measurement. Obviously, the quality of the fits demonstrates the relevance of the fitting function. The continuous line of fig. 4c) is the best fit obtained when using eqs. (3.5) and (3.7) with the same $B_{5/2}$ Brillouin function and the same antiferromagnetic temperature ($T_{AF} = 2.0$ K) as in the case of fig. 4a) and b). In this case, the only variable is the temperature increase ΔT, with a mean value of $\Delta T = 0.5$ K.

Figure 5 shows oscilloscope traces giving the probe intensity transmitted by the analyser as a function of time. Here, the experiment was performed at a constant magnetic field of 2 T in two different $Zn_{1-x}Mn_x Te$ samples ($x = 8\%$ and 2% in figures 5a) and b), respectively). For a given Mn^{++} concentration, three different sample temperatures were chosen (7 K, 5 K and 3 K). As shown in fig. 5, the initial rise time of the signal is faster at higher temperatures and larger Mn^{++} concentrations, and the last feature appears to be almost independent of these parameters. This last feature could be due to the thermal diffusion process which in our experiment lasted longer than 200 µs, whereas the initial feature is most likely due to the spin relaxation of Mn^{++} ions by phonons which is the relevant process for magnetization changes occurring during thermal

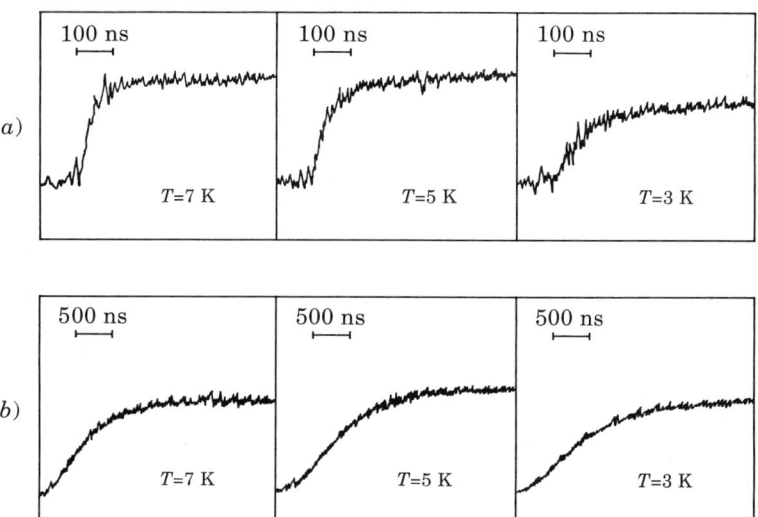

Fig. 5. – Oscilloscope traces showing the rising part of the probe laser transmission through the analyser at different temperatures for two different Mn^{++} concentrations of the $Zn_{1-x}Mn_x Te$ sample (a) $x = 8\%$ and b) $x = 2\%$) (see ref. [18]).

heating. Indeed, the spin relaxation is known to be faster both at higher temperatures, since there are more phonons available for this process, and at higher Mn^{++} concentrations, since Mn^{++} clusters are expected to be much more efficient for this purpose than isolated ions [19].

5.2. *Refractive-index saturation contribution*. – In order to isolate the refractive-index saturation contribution from the others, the experiment was performed at high temperatures ($T > 70$ K) where the thermal magnetic contribution is negligible. Furthermore, the pure magnetic contribution could be neglected since the Mn^{++} concentration was high enough ($x > 0.03$), and the magneto-optical Kerr term could be discounted with the nanosecond resolution of these experiments. Free carriers generated through two-photon absorption of 1.06 µm picosecond duration laser pulses in $Cd_{1-x}Mn_xTe$ samples gave rise to the refractive-index saturation contribution. Its amplitude was measured with a probe beam consisting of a c.w. Ti: sapphire laser the frequency of which could be adjusted near the band gap resonance in order to enhance the effect. Evidently, such an enhancement could not have been obtained in $Zn_{1-x}Mn_xTe$ (band gap energy between 2.4 and 3.2 eV) due to the strong intra-Mn^{++} absorption occurring at frequencies larger than 2.2 eV.

Actually provisions must be made to extract the polarization rotations θ_{NL} from contributions due to photo-induced absorptions and phase changes that are also present at such high temperatures. The experimental procedure was the following: the time evolution of the photo-induced coefficient α_{NL} was first measured at a given pump laser intensity with no analyser and the corresponding values recorded by the computer. $I_T(t)$ was then registered for five different values of θ_0 (typically 30, 60, 90, 120 and 150 degrees). θ_{NL} and α_{NL} were then extracted as a function of time from combinations of pairs of these results thanks to eq. (5.2) with no adjustable parameter.

Experiments were performed for several values of the Mn^{++} concentrations ($x = 0.03$, 0.10, 0.15 and 0.30), sample temperatures (90 K $< T <$ 300 K) and magnetic fields ($0 < H < 5$ T) with pump pulses of energies up to 500 µJ and for probe wavelenghts ranging from near to far away from the band gap resonance. The results strongly depend on the sample and the temperature used, especially with respect to the lifetime of the process (see below). Nevertheless, the main trends were apparent from these numerous experiments. Firstly, the maximum value of θ_{NL} ($\theta_{NL}(t = 0)$ since the rise time is almost instantaneous) was proportional to be absorbed energy, no saturation being observed, thus implicating that higher photo-induced Faraday rotations could be obtained at higher pump intensities. Secondly, $\theta_{NL}(t = 0)$ was observed to increase more so than was the observation for θ_L near the band gap resonance, a situation which is rather current in nonlinear optics. Thirdly, $\theta_{NL}(t = 0)$ was found to be proportional to H, no saturation having been observed up to 5 T. Finally, $\theta_{NL}(t = 0)$ was observed to decrease more so than was the observation for θ_L with in-

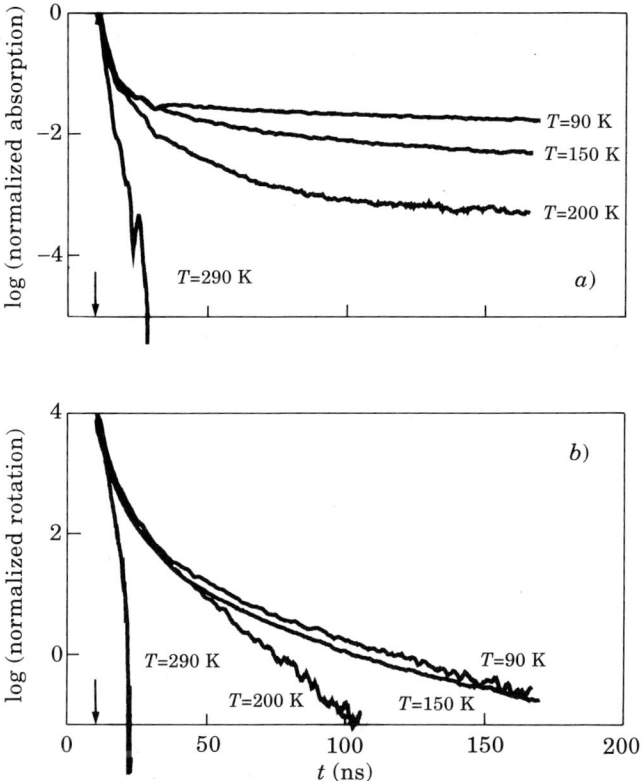

Fig. 6. – Logarithm of the normalized photo-induced absorption coefficient (a)) and Faraday rotation angle (b)) as a function of time at different temperatures. The arrows indicate the instant when the pump pulse was applied.

creasing temperatures. These two last points can be well understood by noting that, at high T, the magnetization is proportional to $H/(T + T_{AF})$ (see. eq. (2.4) for a small value of the argument of the Brillouin function). As the linear Faraday rotation angle is proportional to M, θ_L is also proportional to $H/(T + T_{AF})$. Furthermore, at high temperature the photo-induced Faraday rotation angle due to the refractive-index saturation process is proportional to M/T (see eq. (4.1)) such that $\theta_{NL}(t = 0)$ is proportional to $H/T(T + T_{AF})$. This result demonstrates that $\theta_{NL}(t = 0)$ is proportional to H as observed experimentally and that $\theta_{NL}(t = 0)/\theta_L$ is inversely proportional to the sample temperature. However, as the band gap resonance also changes with changing T, the exact relation between the linear and photo-induced Faraday rotation angle could not be demonstrated and only the global trend described above was observed experimentally.

Figures 6a) and b) show the time evolution of the logarithm of the normalized photo-induced absorption coefficient and polarization rotation, respectively. The experiment was performed in $Cd_{0.85}Mn_{0.15}Te$ at four different temperatures ($T = 90, 150, 200$ and 290 K) in a magnetic field of 5 T. For both series of curves one can observe a fast and a slow component, the contribution of the slower decreasing with increasing temperatures. However, let us note that the time dependences of the slow component of α_{NL} and θ_{NL} are markedly different. These results together with the long time scale (100 ns) observed in this experiment evidently suggest that most likely traps of different origins play an important role in the photo-induced gyrotropy in semi-magnetic semiconductors at high temperatures. This seems to be also verified by experiments performed on other samples which give results on time scales which strongly vary from sample to sample.

5˙3. *First performances for the photo-induced gyrotropy in semi-magnetic semiconductors.* – Measurements were made in order to determine the pump fluence, as a function of the sample temperature, which was needed to obtain a photo-induced Faraday rotation of 45 degrees at $H = 5$ T for a 1 mm thick sample. When the analyser was set perpendicular to the probe polarization in the absence of a pump pulse, a commutation device could have been built with a contrast of 10^2 to 10^3 and a maximum transmission of 50%. In order to minimize the pump fluences, the probe frequency was chosen close to the band gap resonance, so that only $Cd_{1-x}Mn_xTe$ was considered for potential devices. The corresponding results are reported in table II together with the associated response and lifetimes. It can be readily seen from this table that the response time is almost always instantaneous, while the lifetime largely varies from 10 ns to 1 ms depending on the sample and temperature used; such a property could be of importance to tailor the opening duration of the commutation device, if a controlled fabrication of $Cd_{1-x}Mn_xTe$ samples was found. Let us further note that, at low temperatures, the pump fluence needed for a photo-induced 45 degree rotation is extremely low (1 µJ/cm^2). This pump fluence is much

TABLE II. – *First performances obtained for the photo-induced gyrotropy in* $Cd_{1-x}Mn_xTe$ *semi-magnetic semiconductors. The values are taken for a 45°photo-induced polarization rotation.*

Temperature (K)	Pump fluence (µJ/cm^2)	Response time (ns)	Lifetime
3	1	< 1	ms
30	30	< 1	(10–100) ns
77	100	< 1	(10–100) ns
300	10 000	< 1	(10–500) ns

higher (10 mJ/cm^2) at room temperature, but this corresponds only to a pump energy of 1 nJ for a 10 μm^2 laser spot which could be used in a practical device.

6. – Conclusion.

Semi-magnetic semiconductor crystals have been demonstrated to be very efficient for the purpose of the photo-induced gyrotropy. Several physical mechanisms have been identified that allow large photo-induced rotations of the linear-polarization direction of a probe beam when a high-intensity pump pulse also propagates through the material. At low temperatures ($T < 30$ K) the thermal magnetic contribution consecutive to the sample heating dominates; this contribution is long lived with slow response and lifetimes. At high temperatures ($T > 70$ K), when the thermal magnetic contribution is negligible, the refractive-index saturation contribution is preponderant for times longer than the pulse duration; this contribution has an instantaneous rise time and a lifetime which strongly varies (from 0.1 ns to 1 μs) according to the interband relaxation process considered (stimulated to trap-assisted recombination, respectively). During the pump pulse an instantaneous contribution also exists due to the magneto-optical Kerr effect. A pure magnetic contribution has also been predicted which is due to spin flip-flops occurring between free carriers and the $3d$ electrons of the Mn^{++} ions during the intraband relaxation; however, this contribution is probably negligible at high Mn^{++} concentrations.

Single-beam experiments have confirmed these predictions and shown that saturations, due to the free-carrier statistics in the conduction and valence bands, occurred for the photo-induced rotation angles when the magnetic field was increased.

Pump and probe experiments have also confirmed the theoretical predictions. In particular, the thermal magnetic contribution was isolated from the others and investigated in detail in $Zn_{1-x}Mn_xTe$ at very low temperatures. In the same manner, the refractive-index saturation contribution has been studied free of thermal effects at higher temperatures ($T > 70$ K) in $Cd_{1-x}Mn_xTe$. These experiments have demonstrated the dominant role played by traps in this temperature range and proved that, even at the early stage of this study, interesting performances could be obtained for potential commutation devices. However, in order to elucidate further the potential use of such devices in the field of optical information processing, considerable work remains to be carried out both in bulk materials, for which the role played by traps at high temperature must be characterized in much great details, and in quantum confined structures since multiquantum wells [8] and dots embedded in glasses [23] have been manufactured for semi-magnetic semiconductors.

REFERENCES

[1] See, for instance, Y. R. SHEN: *Principles of Nonlinear Optics* (John Wiley, New York, N.Y., 1984).

[2] See, for instance, *Nonlinear Optics: Materials and Devices*, edited by C. FLYTZANIS and J. L. OUDAR (Springer-Verlag, Berlin, 1986).

[3] P. D. MAKER, R. W. TERHUNE and C. M. SAVAGE: *Phys. Rev. Lett.*, **12**, 507 (1964).

[4] J. FREY, R. FREY, C. FLYTZANIS and R. TRIBOULET: *Opt. Commun.*, **84**, 76 (1991).

[5] J. FREY, R. FREY, C. FLYTZANIS and R. TRIBOULET: *J. Opt. Soc. Am. B*, **9**, 132 (1992).

[6] K. KUBOTA: *J. Phys. Soc. Jpn.*, **29**, 998 (1970).

[7] See, for instance, G. FOWLES: *Introduction to Modern Optics* (Holt Rinehart and Winston, New York, N.Y., 1975).

[8] For a review see J. K. FURDYNA: *J. Appl. Phys.*, **64**, R29 (1988), or O. GOEDE and W. HEIMBRODT: *Phys. Status Solidi B*, **146**, 11 (1988), and references therein.

[9] J. A. GAJ, R. PLANEL and G. FISHMAN: *Solid State Commun.*, **29**, 435 (1979).

[10] J. SPALEK, A. LEWICKI, Z. TARNAWSKI, J. K. FURDYNA, R. R. GALAZKA and Z. OBUSZKO: *Phys. Rev. B*, **33**, 3407 (1986).

[11] D. L. PETERSON, D. U. BARTHOLOMEW, U. DEBSKA, A. K. RAMDAS and S. RODRIGUEZ: *Phys. Rev. B*, **32**, 323 (1985).

[12] J. FREY, R. FREY and C. FLYTZANIS: *Phys. Rev. B*, **45**, 4056 (1992).

[13] J. A. GAJ, J. GINTER and R. R. GALAZKA: *Phys. Status Solidi B*, **89**, 655 (1978).

[14] G. REBMANN, C. RIGAUX, G. BASTARD, M. MENANT, R. TRIBOULET and W. GIRIAT: *Physica B*, **117** and **118**, 452 (1983).

[15] D. U. BARTHOLOMEW, J. K. FURDYNA and A. K. RAMDAS: *Phys. Rev. B*, **34**, 6934 (1986).

[16] C. FLYTZANIS: in *Quantum Electronics*, edited by H. RABIN and C. L. TANG (Academic, New York, N.Y., 1975), Vol. **1**, part A.

[17] S. HUGONNARD-BRUYÈRE, R. FREY and C. FLYTZANIS: *Opt. Commun.*, **94**, 357 (1992).

[18] H. KRENN, K. KALTENEGGER, T. DIETL, J. SPALEK and G. BAUER: *Phys. Rev. B*, **39**, 10918 (1989).

[19] D. SCALBERT, J. CERNOGORA and C. BENOÎT À LA GUILLAUME: *Solid State Commun.*, **66**, 571 (1988).

[20] R. R. GALAZKA, S. NAGATA and P. H. KEESOM: *Phys. Rev. B*, **22**, 3344 (1980).

[21] S. HUGONNARD-BRUYÈRE, J. FREY, R. FREY and C. FLYTZANIS: *Mater. Sci. Eng. B*, **16**, 239 (1993).

[22] N. T. KHOI and J. A. GAJ: *Phys. Status Solidi B*, **83**, K133 (1977).

[23] K. YANATA, K. SUZUKI and Y. OKA: *J. Appl. Phys.*, **73**, 4595 (1993).

All-Optical Switching in Integrated Structures.

GAETANO ASSANTO

Dipartimento d'Ingegneria Elettronica della Terza Università
via Eudossiana 18, Roma, Italia

1. – Introduction.

Nonlinear optical effects require high intensities in the source fields. This simple consideration stems from the size of the nonlinear coefficients involved in any higher-than-first-order expansion of the polarization field driving the wave equation. For this reason, materials with larger nonlinear coefficients and geometries apt to maximize the involved intensities have to be sought. Integrated-optics configurations are ideal candidates for the study of nonlinear optical interactions, because waveguides provide the confinement of the fields in transverse dimensions of the order of the wavelength and for propagation lengths limited only by scattering and absorption losses, beating the detrimental effect of diffraction in bulk optics[1-4]. For a given input power, therefore, the guided-wave geometry yields the maximum intensity over the longest interaction distance, and this permits the experimental investigation of a number of nonlinear optical effects. Some of them are mutuated from the corresponding bulk interactions, as it is the case for degenerate four-wave mixing, coherent Stokes and anti-Stokes Raman scattering, second- and third-harmonic generation, sum and difference frequency generation, etc.[1]. Others are peculiar of the guiding geometry, and among them nonlinear distributed grating and prism couplers[5,6], mode conversion[7], contrapropagating second-harmonic generation[8], power-dependent directional coupling[9,10], etc.

In this lecture we will introduce a specific subset of nonlinear guided-wave optics, namely all-optical switching in integrated structures. Due to its relevance in the area of telecommunications and all-optical computing, all-optical switching (AOS) has grown in importance towards the realization of devices able to operate with reduced input powers and fast response times for pipeline signal processing. Excellent reviews are available on nonlinear optics in fibres[11] and on other nonlinear waveguide interactions

457

which do not deal specifically with optical switching [1,4]. Second-harmonic generation will be addresses only with reference to AOS.

After summarizing the main properties of a waveguide, in sect. 3 and 4 we will describe two representative integrated AOS geometries based on the optical Kerr effect, the Mach-Zehnder interferometer and the coherent coupler. Some taxing material problems in the successful realization of third-order devices will be addressed in sect. 5, defining some material figures of merit. Finally, we will discuss nonlinear phenomena due to cascading of second-order effects and the implementation of AOS devices in noncentrosymmetric materials, outlining their peculiarities with respect to the Kerr case.

2. – Summary of waveguide concepts.

Guiding an electromagnetic field in a dielectric structure can be described as the result of a geometric resonance involving plane waves. For the 1-dimensional case, i.e. a planar waveguide, a film of refractive index higher than that of the substrate and the superstrate (cladding) is, above a minimum thickness, able to provide such a resonance due to total internal reflection. For an asymmetric geometry (substrate ≠ cladding), this happens above a cut-off value for either the thickness, or the index increment with respect to the surrounding media, or the frequency of light. Above the cut-off, a finite number of guided modes are eigensolutions of Maxwell equations in the given structure, and propagate with given polarization, wave vector and transverse field distribution [12].

In a planar guide, i.e. a structure invariant along the direction y and with propagation along z, modes polarized with the electric field parallel (TE) or perpendicular (TM) to the interfaces are allowed and form a complete orthogonal set. Their propagation eigenvalues β are linked to the wavelength λ and the material parameters by a dispersion relation $\beta = \beta(\lambda, n_s, n_f, n_c, h)$, with n_i ($i = $ = s, c, f) the various refractive indices and h the film thickness.

Figure 1 shows the sketch of a planar guide, the field distribution of the lowest-order modes with typical dispersion curves.

A channel waveguide provides confinement in both x and y, thereby maximizing the field intensity in the region of highest refractive index. Even in this case orthogonal modes are allowed in a discrete set, although the exact analytical solutions are not always available. Figure 2 is the sketch of a channel with some modal distributions.

A few points need to be stressed when describing the propagation of guided modes:

a) Guided modes exhibit maxima of the field amplitude within the region of highest index (film or core), whereas the transverse distribution extends in the surrounding media with an evanescent profile. The degree of such penetra-

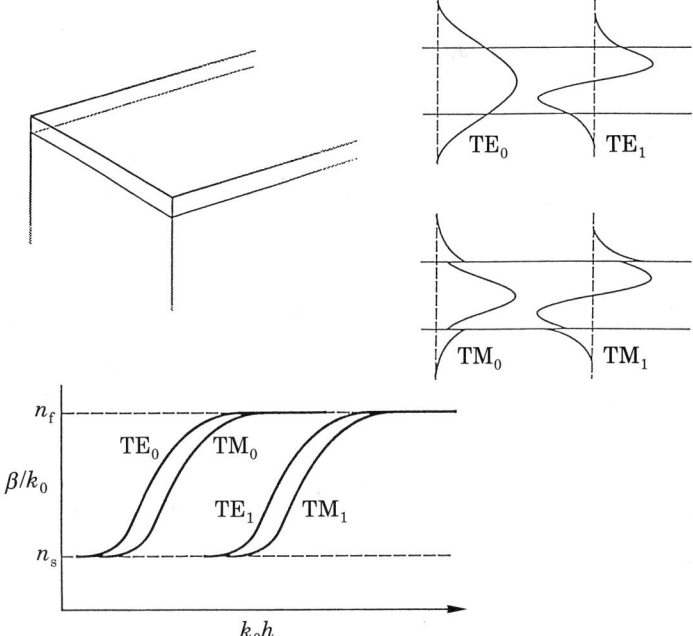

Fig. 1. – Planar waveguide: sketch of a slab guide, transverse E-field profiles of the low-est-order modes and typical dispersion curves. We assume $n_f > n_s \geqslant n_c$, and h is the film thickness.

tion outside the film or core depends on the index increment defining the guiding region with respect to substrate and cladding. For a given guide, a specific set of dimensions minimizes the effective cross-section or area A_{eff}.

b) Every discrete mode is characterized by specific polarization directions of its field and by propagation constants given by the dispersion relation.

c) Every interaction between eigenmodes is forbidden, unless linear or nonlinear perturbations couple them with one another.

d) In a regime of weak perturbation, the modal eigencharacteristics can be taken as invariants, and coupled-mode theory is applicable; the transverse distributions are taken into account via an overlap integral involving two or more fields, and a phase mismatch between the propagation constants is defined.

A given interaction is most efficient when the overlap is maximum and the mismatch is minimum, $i.e.$ momentum is conserved. The existence of distinct modes and waveguide dispersion provide extra degrees of freedom for satisfying wave vector conservation.

e) When the perturbation cannot be regarded as weak, $i.e.$ when the superimposed variations in one or more of the parameters relevant to the guiding

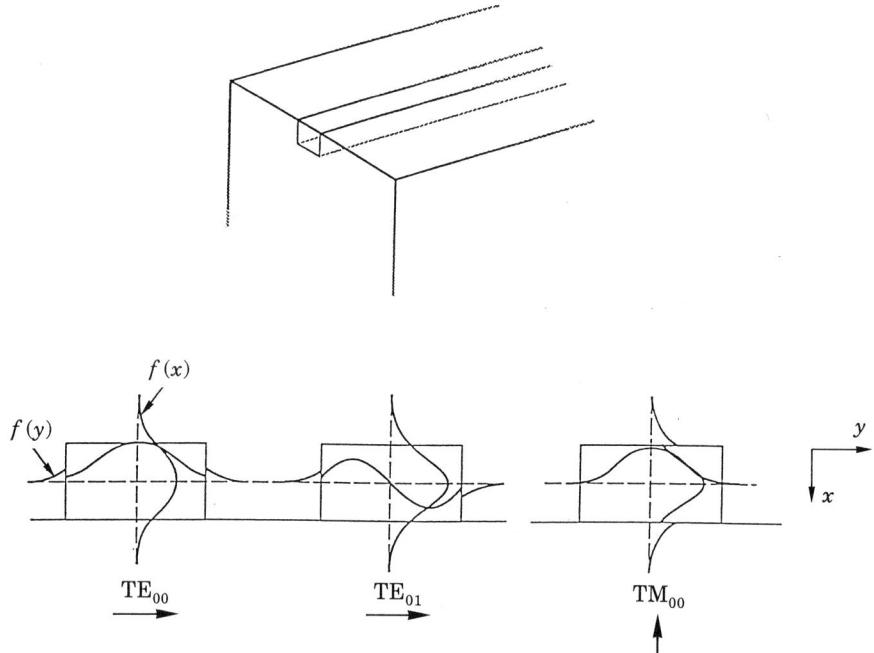

Fig. 2. – Channel waveguide: sketch of a channel and representative E-field profiles of the lowest-order modes. We assumed y parallel to the s-c interface.

properties of the structure are comparable to their linear values, then new high-intensity eigensolutions to Maxwell equations need to be found. This is, for instance, the case of modes degenerate in propagation constants and of spatial solitons [13,14]. We will not deal with such cases, which go beyond the scope of this introductory overview.

2'1. *The intensity-dependent wave vector.* – Indicating by $P(x, y)$ the transverse distribution of a perturbing polarization field $P(\boldsymbol{r}, t) = (1/2) P(x, y) a_p(z) \cdot \exp[i(\omega t - \beta_p z)] + \text{c.c.}$ and by $E(x, y)$ a guided-wave modal profile with $E(\boldsymbol{r}, t) = (1/2) E(x, y) a_{wg}(z) \exp[i(\omega t - \beta_0 z)] + \text{c.c.}$, applying coupled-mode theory, the evolution of the normalized slowly varying amplitude $a_{wg}(z)$ can be described by

$$(1) \qquad \frac{\mathrm{d}}{\mathrm{d}z} a_{wg}(z) = -\frac{i\omega}{4} \int \mathrm{d}x \, \mathrm{d}y \, P(x, y) E^*(x, y) a_p(z) \exp[i(\beta_0 - \beta_p)z].$$

A phase-matched interaction corresponds to $\beta_p = \beta_0$. While this expression is quite general and holds for linear and nonlinear perturbations, it takes a specific form for degenerate third-order interactions, *i.e.* those involving only one frequency and the product of three fields. If the interacting fields are co-polarized and co-propagating, *i.e.* nonlinear effects are induced by a single strong field via the third-order polarization, $P_i^{NL}(x, y) = \varepsilon_0 \chi_{iiii}^{(3)} E_i E_i E_i^*$ with $a_p(z) = a_{wg}^2(z) a_{wg}^*(z)$, and the index of refraction becomes field/intensity dependent:

$$(2) \qquad n(E) = n_0 + \frac{3\chi_{1111}^{(3)}}{8n_0} |E|^2 = n_0 + n_2 I$$

for isotropic media, with I the field intensity.

In a waveguide, the inclusion of such nonlinear dependence in (1) results in a phase change for the propagating field, *i.e.* a modification of the linear propagation constant β. If $a_{wg}(z)$ is normalized in such a way that $a_{wg} a_{wg}^*$ represents the guided-wave power associated to a particular mode, the wave vector becomes in turn intensity dependent [15]:

$$(3) \qquad \beta = \beta_0 + \beta_2 |a_{wg}|^2 = \beta_0 + \frac{k_0 n_2}{\sigma} |a_{wg}|^2$$

with $k_0 = 2\pi/\lambda$ and σ an effective area or height for channel or planar guides, respectively.

It is important to emphasize that, although the index and the wave vector do change due to the nonlinearity via self- (a beam acting upon itself) or cross- (a beam acting upon another beam) phase modulation, the most relevant quantity is the phase change accumulated over a given interaction/propagation length L, $\phi^{NL}(L)$. This notion is fundamental to AOS both via third-order frequency-degenerate interactions and via cascaded $\chi^{(2)}$ effects. In the case of self-phase modulation $\phi^{NL}(L) = \int_0^L dz \beta_2 |a_{wg}|^2$.

In the next sections we will briefly describe a couple of significant examples of AOS devices based on the nonlinear phase shift obtainable via a third-order susceptibility, *i.e.* the optical Kerr effect. The first, an interferometer, is the most direct approach to utilize a phase change due to the intensity. An entire class of AOS structures is, indeed, based upon nonlinear interference. The second is a coupler, where the changes in phase alter the coupling conditions between modes and consequently the energy exchange between them.

3. – The nonlinear Mach-Zehnder interferometer.

A modal field propagating down a guide of length L will acquire a nonlinear phase shift dependent on its intensity. For a lossless waveguide, $\phi^{NL} = \beta_2 |a_{wg}|^2 L = (2\pi/\lambda) n_2 (|a_{wg}|^2/\sigma) L$. Coherent fields of different intensities and/or with different effective nonlinearities and/or travelling different optical-path lengths can, therefore, accumulate a relative phase difference with respect to their linear (*i.e.* at low excitation levels) phase offset. For two guides a and b:

$$(4) \qquad \Delta\phi^{NL} = \phi_a(I_a) - \phi_b(I_b) = k_0 \Delta(n_2 |a_{wg}|^2 L/\sigma).$$

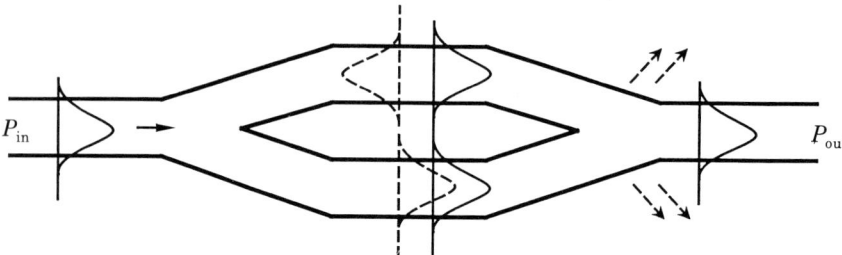

Fig. 3. – Sketch of an integrated nonlinear Mach-Zehnder interferometer formed by single-mode channels. An injected TE_0 mode excites the two arms after the first Y-junction.
If the fields remain in phase (solid line), unitary transmission is obtained. Otherwise, if the
fields enter the second Y-junction in phase opposition (dashed line), they tend to excite the
second-order mode in the output, and result in radiation (dashed arrows) at the output
stem.

Two such fields will then undergo intensity-dependent interference if superimposed in space after a given distance, thereby transducing a phase variation
into an amplitude modulation. The integrated nonlinear Mach-Zehnder interferometer (NLMZ), sketched in fig. 3, operates on this principle. An input
beam of power P_{in} is split at a Y-junction into two modal fields (preferably of the
lowest order) which propagate down two parallel well-separated (*i.e.* decoupled) waveguides. After propagation, the two fields are recombined via another
Y-junction in order to interfere at the output guide. The Y-junctions operate
similarly to plane-wave beamsplitters, and the whole device can be regarded as
a volume interferometer in which one of the outputs is the radiation into the
substrate/cladding at the stem of the output Y.

A differential phase shift can be obtained with unequal power splitting at
the junctions, or with channels of unequal effective areas, or lengths, or with
guides of different nonlinearities (as in the case of a nonlinear overlayer on one
of the arms). A linear or nonlinear asymmetry is, in any event, required.

The transfer characteristic of a c.w. NLMZ is a sinusoidal function of $\Delta\phi$.
For zero phase bias and equal powers in the two arms, $P_{out} = P_{in} \cos^2(\Delta\phi/2)$. Notice, however, that the nonlinear phase required for switching is larger than the
differential π value. For example, if $\sigma_b = 2\sigma_a$, one must be able to achieve
$\Delta\phi_a^{NL} = 2\pi$ with the available power and nonlinearity. For identical guides with
zero phase bias and an unequal power splitting ratio δ at the first junction, if the
second Y is a 3 dB coupler the transmission is $T = (1/2)[1 - 2\sqrt{\delta(1-\delta)} +$
$+ 4\sqrt{\delta(1-\delta)}\cos^2(\Delta\phi/2)]$, with contrast $(T_{max} - T_{min})/(T_{max} + T_{min}) = [\delta(1-\delta)]^{1/2}$.
100% modulation can be retrieved using an output junction with the same split-
ting ratio interchanged between guides[16]. In this case $T = 4\delta(1 -$
$-\delta)\cos^2(\Delta\phi/2)$. Here $\Delta\phi^{NL} = k_0 n_2 P_{in} L(1 - 2\delta)/\sigma$.

The first experiment on an AOS guided-wave NLMZ was reported in lithium niobate, with a limited output modulation[17]. Recently, a successful dem-

onstration has been reported in MBE-grown AlGaAs waveguides [18]. 330 fs pulses at 1.52 μm were optically switched using average input powers of 8.5 mW and working far from one- and two-photon material resonances. The modulation of normalized transmission was as high as 85%, due to the use of pulses with a continuous intensity distribution. A Gaussian (or similar) power distribution in time has a detrimental effect on the overall switching performance when utilizing a quasi-instantaneous nonlinearity, because a continuous distribution in intensity is always mapped onto a nonlinear phase and an amplitude response varying during each pulse. This problem is common to all AOS devices utilizing the Kerr effect, and can be overcome by the use of temporally rectangular pulses or solitons. In the latter case, in fact, the nonlinear phase shift is constant across the field envelope and the whole entity behaves as a unit or a particle [19-21].

3'1. *Other AOS devices based on intensity-dependent interference.* – A power-controlled phase term can modify an interference condition in any con-

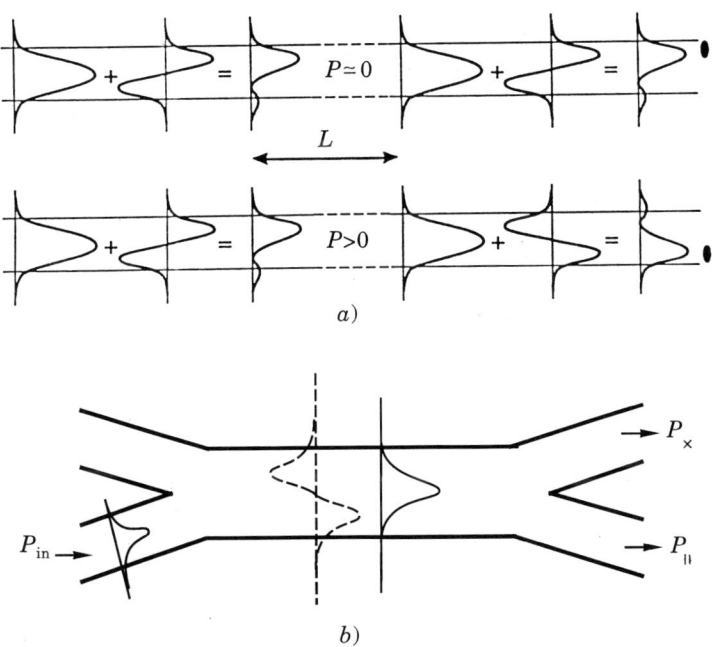

Fig. 4. – Sketch of a nonlinear mode mixer: *a)* the two lowest-order modes of the same waveguide overlap in phase (upper panel, low power) or in opposition of phase (lower panel, high power) after propagation over a distance L. The two cases result in different spatial locations of the field maxima at the output. *b)* Example of a four-port NLMM using a bimodal guide in the central region.

figuration where fields spatially overlap. Other examples of geometries for AOS based on such mechanisms are[2]:

1) The nonlinear mode mixer (NLMM), where the interference condition between two distinct eigenmodes of the same guide varies with input power. The nonlinear differential phase is due to unequal intensity-dependent wave vectors associated with the corresponding transverse profiles. A sketch of a NLMM with the two lowest-order TE modes is shown in fig. 4a), with the indication of the bright spots at the output for low and high powers. NLMMs have been demonstrated in GaAs/AlGaAs multiple-quantum-well waveguides near the material resonance[22].

A NLMM becomes a four-port device by combining the bimodal guide with single-mode Y-junctions, in order to excite one or another output branch depending on the interference condition at the output stem. Figure 4b) shows an example of such a geometry. A careful design may result, indeed, in good switching contrast with the possibility of maximizing the nonlinear effects associated with a specific material. A four-port NLMM has been recently proposed in a composite glass-polymer geometry which lowers the switching threshold by reducing the field amplitude of the odd supermode in the nonlinear region to nearly zero[23].

2) The nonlinear distributed coupler, either of the prism or of the grating type[5,6]. Input coupling relying on the phase-matched excitation of a guided mode can be altered by the presence of a phase shift growing with the guided-wave intensity along the coupling region, where the phase mismatch becomes intensity and z dependent, i.e. $\Delta\beta(I) = \Delta\beta_0 - \beta_2 |a_{wg}(z)|^2$. The initial phase mismatch is $\Delta\beta_0 = k_0 n_p \sin\theta_p - \beta_0$ for a prism (subscript p) and $\Delta\beta_0 = k_0 n_c \sin\theta_c \pm \pm 2\pi/\Lambda - \beta_0$ for a grating of period Λ and a cladding of index n_c. For optimum low-power coupling efficiency, $\Delta\beta_0 = 0$. The initial phase relationship between the radiation (input) field and the guided mode (excited) is, therefore, modified, while the coupling takes place in a distributed fashion in space[24]. This nonlinear effect results in pulse narrowing, limiting, switching and even bistability if diffusion is present, and has been employed to evaluate in magnitude and sign the Kerr coefficient of planar structures with doped glass, liquid crystals, organic polymers and semiconductors[2 and references therein, 25-30].

4. – The nonlinear directional coupler.

The nonlinear directional coupler (NLDC) is, by far, the most-studied AOS device to date, because it represents the basic four-port building block of any AOS circuitry, including all-optical computing. An extensive literature is available on the NLDC, detailing its various models, modes of operations and features[1-3, 9, 10, 31-43]. Here, we will give a rather essential description, trying

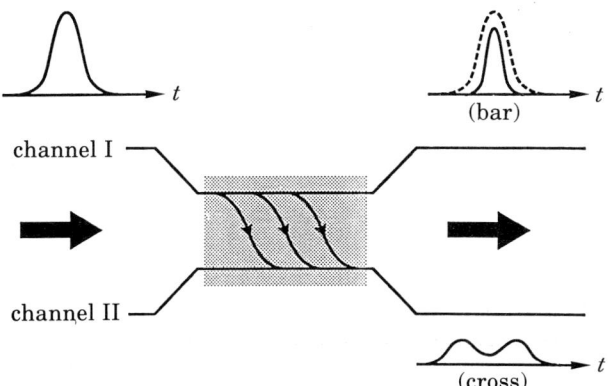

Fig. 5. – Schematic representation of a half-beat-length nonlinear directional coupler, with a pulsed input in the bar channel.

to elucidate those characteristics which are in common to other AOS geometries based on nonlinear coupling.

The standard NLDC consists of two parallel straight waveguides in close proximity, such that the overlap between the transverse field distributions of each guide gives rise to mutual coupling. The two guides, independent of and unperturbed by each other when infinitely distant, become a four-port structure capable of energy transfer between the two arms. The degree of such coupling and the length of the interaction will dictate the power redistribution occurring in propagation towards the output ports. A linear coupler of length $L = L_b/2$ will route the power injected in the input (bar) channel to the other (cross)-channel, whereas for $L = L_b$ all the power is emitted by the bar channel after a coupling cycle. Here L_b is the beat length, related to the strength Γ of the coupling by $L_b = \pi/\Gamma$. The two cases are usually referred to as «half-beat length» and «one-beat length» couplers. The former is schematically represented in fig. 5, with the indication of its nonlinear behaviour: when the intensity is sufficiently high, the coupling dynamics is altered enough to re-couple most of the power to the «bar» channel, while the low-power tails of the input pulse keep travelling as in the linear case.

The evolution of the slowly varying field amplitudes $a(z)$ and $b(z)$ for TE modes in the two guides can be described by the coupled-mode equations [9]

(5)
$$
\begin{cases}
\dfrac{\mathrm{d}}{\mathrm{d}z}\, a(z) = i\Gamma b(z) \exp[i\Delta\beta z] + i[(\beta_2)_a\, |a|^2 + 2(\beta_2)_{ab}\, |b|^2], \\[2mm]
\dfrac{\mathrm{d}}{\mathrm{d}z}\, b(z) = i\Gamma a(z) \exp[-i\Delta\beta z] + i[(\beta_2)_b\, |b|^2 + 2(\beta_2)_{ba}\, |a|^2]
\end{cases}
$$

with $\Delta\beta$ the mismatch between the (linear) wave vectors in a and b, $(\beta_2)_a$ and

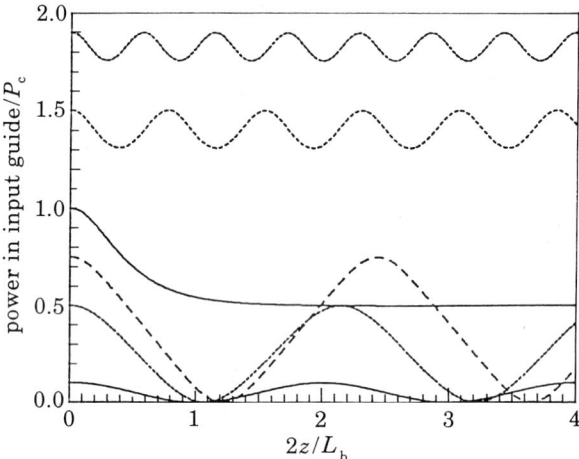

Fig. 6. – Graphs of the normalized power in the bar (input) channel *vs.* propagation distance. At $P = P_c$, the effective beat length becomes infinite (unstable point). For $P < P_c$ the power is periodically exchanged between channels, but for $P > P_c$ the induced nonlinear mismatch is large enough to prevent any significant energy exchange.

$(\beta_2)_b$ the self- and $(\beta_2)_{ab} = (\beta_2)_{ba}$ the cross-phase modulation coefficients, respectively.

When the power (intensity) at the input channel is increased, the phase terms in square brackets in (5) will progressively detune an initially matched NLDC ($\Delta\beta = 0$), making its effective beat length longer and longer until a critical value is reached for which $L_b = \infty$, as shown in fig. 6. This critical power is

(6)
$$P_c = \frac{8\Gamma}{(\beta_2)_a + (\beta_2)_b - 4(\beta_2)_{ab}} \cdot$$

For this input power the NLDC acts as a 3 dB splitter, but this is an unstable point of operation. If P_c is exceeded, the power will predominantly remain in the input channel, due to the large mismatch nonlinearly induced. By varying the level of the excitation in the input guide one can, therefore, all-optically switch the signal from the cross to the bar channel. Notice, however, that while a half-beat-length NLDC tends to operate as a straight-through guide at high powers, a one-beat-length coupler will operate as a cross-switch only over a limited range of excitations, as one can easily see in fig. 7. If $\Delta\beta \neq 0$, the critical power becomes larger but complete switching is still obtainable if the mismatch can be nonlinearly cancelled.

As pointed out before, the key quantity is the nonlinear phase shift which each channel can accumulate. Since the power oscillates with z between the two waveguides, the intensity-driven phase shift in each arm is reduced with respect to the single-guide case. Because of this power exchange, in order to accu-

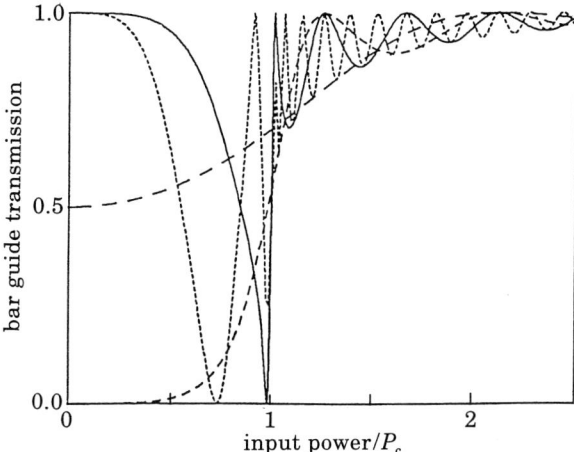

Fig. 7. – Normalized transmission of the bar channel in NLDCs of various lengths: -----
$L = (1/2)L_b$, $- - - L = (1/4)L_b$, —— $L = 1L_b$, ····· $L = 3L_b$. Notice how, the longer
the device, the lower the power required for switching. This is apparent in the 3-beat-
length NLDC as compared to the 1-beat-length device, despite their identical low-power
responses.

mulate a differential shift of π, each guide of a half-beat-length NLDC must
achieve $\phi^{NL} = 4\pi$. The requirement is more stringent than in a Mach-Zehnder.

Recent theoretical contributions include linearly and nonlinearly asymmetric couplers [35, 36, 38, 40, 43], NLDCs with gain [34, 41], multiphoton absorption [37] and with multiple guides [42, 43]. Coherent switching can also be obtained controlling the phase of a small signal in the second input channel, transducing a phase modulation into the routing of a large signal. Finally, as already
pointed out in a previous section, pulse break-up occurs unless square pulses or
solitons are employed.

To the best of our knowledge, the most successful experimental demonstra-
tions of NLDCs have been reported in AlGaAs and GaAs/AlGaAs quantum-
well channel waveguides excited by 450 fs pulses at 1.545 mm, below half the
band gap of the material [44]. Peak powers used to switch 4.8 and 6 mm long
couplers were in the kW range, but values as low as 100 W were employed in
the same material using longer (2 cm) NLDCs in order to accumulate larger
phase shifts [45].

4'1. *Other devices based on nonlinear coupling.* – The equations seen above
are, with minor adjustments, representative of several AOS devices based on
nonlinear coupling [31]. Similar effects are indeed obtained if, instead of two
modes in adjacent waveguides, one considers two distinct modes (or super-
modes) of the same structure, realizing a nonlinear mode converter

(NLMC) [7]. In this case linear coupling is absent and switching powers are relatively high, unless a periodic grating is introduced in order to substantially reduce the modal phase mismatch and re-introduce a coupling mechanism [46].

If the coupling is realized between two counterpropagating modes, as is the
case in a distributed Bragg reflector, feedback is added even for an instantaneous and local nonlinearity [47-50]. This results from a geometrical resonance
in the direction of propagation, giving rise to a gap in the dispersion diagram.
Nonlinear distributed feedback gratings (NLDFBG) on planar or channel
waveguides can, therefore, exhibit switching and/or bistability via the formation of gap solitons depending on power and on the detuning with respect to the
resonance centre wavelength [50]. The latter, in fact, becomes an intensity-dependent variable when using the Kerr effect. The NLDFBG requires a π phase
shift for AOS operation, but having the reflected beam collinear to the input

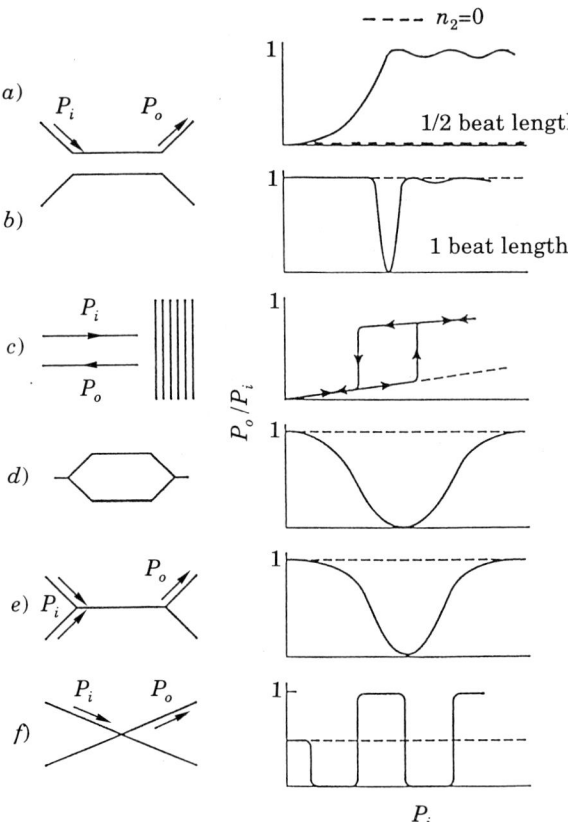

Fig. 8. – Summary of integrated AOS structures and their idealized Kerr law responses. a)
Half- and b) full-beat-length nonlinear directional coupler, c) nonlinear distributed feedback grating, d) nonlinear asymmetric Mach-Zehnder interferometer, e) nonlinear mode
mixer, f) nonlinear X-junction (adapted from [3]).

represents a practical obstacle for implementing switching networks. Its bistable response, however, is important in analogy to the role of an electronic memory gate, a flip-flop. To date a few and only partially successful NLDFBGs have been demonstrated, in slab [26, 51,52] and channel [53] waveguides, using Si on insulator [51], indium antimonide [26] and polymeric [52,53] films with electronic [51] and/or thermo-optic nonlinearities.

A combination of nonlinear interference and coupling describes more complex AOS devices, such as nonlinear Y (NLY) and X (NLX) junctions. These structures are best analysed using numerical propagation algorithms, rather than coupled-mode theory. In these cases channel waveguides couple to each other before merging into a wider stem where various modes interact and interfere with relative phases depending upon their intensities and transverse profiles [2, 54-57]. The junctions behave then as a NLDC and a NLMM in proximity to and at the stem, respectively, with the possibility of nonlinearly addressing one or the other output ports. Symmetric and asymmetric NLXs have been demonstrated in AlGaAs channels forming small 0.1° or 0.2° angles at the stem in order to fully exploit the NLMM effect [58].

In fig. 8 we sketchily summarize geometry and c.w. response of basic integrated AOS devices with a Kerr law nonlinearity. As we will see in the next section, real materials exhibit properties often removed from such a simple model.

5. – Non-Kerr effects and material considerations.

While a Kerr refractive index $n = n_0 + \Delta n(I) = n_0 + n_2 I$ is the kind of third-order nonlinearity which appears desirable for AOS, the anharmonic response of an electron within its potential well will contain higher-order terms leading to an upper limit (saturation) of the refractive-index change Δn_{sat}. Optical damage can be regarded as a limiting case of such a restriction, with unrecoverable effects. In addition, excitations close to an absorption band via one- or many-photon transitions will induce thermal effects leading to noninstantaneous and nonlocal index changes. Nonlocal nonlinear effects are able to modify the response of an AOS device via spatial averaging and feedback. It is clear, therefore, that nonideal features of materials employed in AOS devices must be accounted for. To this effect, some criteria have been developed and some figures of merit defined in order to provide simple guidelines in the evaluation of material candidates for AOS [2,3].

Index saturation. A linear intensity dependence of the refractive index is an unphysical approximation at large excitations. A two-level system provides a simple expression for such an upper limited wave vector change:

$$(7) \qquad \Delta\beta(|a_{\mathrm{wg}}|^2) = \frac{\gamma\Delta\beta_{\mathrm{sat}}|a_{\mathrm{wg}}|^2}{1+\gamma|a_{\mathrm{wg}}|^2}$$

with a similar expression holding, where appropriate, for a change in absorption. An upper limit on $\Delta\beta(|a_{\mathrm{wg}}|^2)$ imposes an upper limit on the phase shift achievable over a given length L. In addition, since linear (one-photon) absorption is detrimental to the operation of an AOS device because it reduces the intensity available for the interaction and the device fan-out, a figure combining saturation and linear absorption is

$$(8) \qquad W = \frac{\phi^{\mathrm{NL}}}{2\pi\alpha_0 L} = \frac{\Delta\beta_{\mathrm{sat}}}{2\pi\alpha_0} \approx \frac{\Delta n_{\mathrm{sat}}}{\alpha_0\lambda} \,.$$

A half-beat-length NLDC requires $W > 2$ for switching, while for a Mach-Zehnder interferometer we need $W > 1$. Larger throughputs can be obtained with reduced linear absorption in materials with a higher W. Notice that this figure of merit is independent of waveguide parameters, and can be employed in a preliminary evaluation of materials for AOS. The inclusion of waveguide scattering losses in α_0 effectively reduces W values estimated via experiments in bulk or in solution.

Table I lists nonlinear phase shifts and W required for various AOS devices, and includes W values for throughputs of 50 and 80%.

TABLE I. – *Third-order devices and main characteristics, including the nonlinear phase required for switching and the W figure of merit for various throughputs.*

Device	L	$\Phi^{\mathrm{NL}}(/\pi)$	No. ports	W	W (50%)	W (80%)
NLDC	$0.5\,L_{\mathrm{b}}$	4	4	2	2.9	9
NLDC	L_{b}	≈ 3.3	4	1.7	2.4	7.4
NLMZ	—	2	2	2	1.44	4.5
NLX	—	7.4	4	3.7	5.34	16.6
NLDFBG	—	1	3	0.5	0.7	2.25

Multiphoton absorption. If the linear absorption α_0 is negligible and the limiting factor is two-photon absorption (TPA), related to the imaginary part of a complex third-order susceptibility, a figure of merit T can be defined as

$$(9) \qquad T = \frac{4\pi}{\phi^{\mathrm{NL}}} \approx \frac{2\lambda\alpha_2}{n_2}$$

with $\alpha \approx \alpha_2 I$. T must be < 1 to permit a phase shift $> 4\pi$. Due to TPA limita-

TABLE II. – *Best nonlinear directional couplers and Mach-Zehnder interferometer realized in AlGaAs. The W parameter was estimated using $I_{sat} = 10\,GW/cm^2$. The throughput lowered at high excitations due to multiphoton absorption.*

Device	Material	Thru %	$\lambda\,(\mu m)$	$L\,(mm)$	$n_2\,(cm^2/W)$
NLDC	AlGaAs	> 30	1.55	$6 \approx 0.4\,L_b$	$1.1 \cdot 10^{-13}$
NLDC	QW-AlGaAs/ /GaAs	> 25	1.55	$4.8 \approx 0.4\,L_b$	$1.3 \cdot 10^{-13}$
NLMZ	AlGaAs	> 12	1.55	8.6	$1 \cdot 10^{-13}$

Device	α_2 (cm/GW)	$\Delta t\,(ps)$	$I_{peak}\,(W)$	W	T	α_3 (cm^3/GW^2)
NLDC	0.08	0.5	≈ 1000	> 9	0.22	0.05
NLDC	0.08	0.5	≈ 1000	> 11	0.18	0.08
NLMZ	0.08	0.4	≈ 1000	> 3.9	0.24	0.25

tions, when working with semiconductors one should operate below the band gap to reduce linear absorption losses, but below half the band gap if two-photon transitions are permitted. The best AOS experiments performed in AlGaAs waveguides and mentioned above were achieved by operating below the two-photon absorption band edge, in order to satisfy both criteria in terms of W and T figures of merit [44, 45, 58].

Criteria on three-photon absorption limits have been recently developed to interpret some of the experimental data from AOS in AlGaAs below TPA limits [37].

In table II we have listed W and T and other relevant parameters evaluated in some AlGaAs AOS devices, namely NLDCs and NLMZ. To calculate W, a saturation intensity of $10\,GW/cm^2$ has been assumed.

Thermal effects. Another detrimental effect of material absorption is the conversion of energy into heat and, consequently, a temperature increase. This induces an index change $\Delta n \sim (\partial n/\partial T)\,\alpha_0 I$ via the thermo-optic effect, which can be approximated by a Kerr law [6]. Such Δn is in some cases opposite in sign to the electronic one, and tends to cancel it. Furthermore, thermal time responses range in the $(0.1-1)\,\mu s$ region, *i.e.* are much slower than electronic nonlinearities and lead to time-integrating phenomena when short pulses are employed [3, 6, 31]. If ultrafast switching is sought and the relaxation time is comparable to the interval between pulses, the device response will depend on the whole excitation history, exhibiting a slowly decaying memory. In addition,

TABLE III. – *List of third-order materials with some measured parameters relevant to AOS. The figure W was calculated using the maximum intensity employed without optical damage to the structure or a conservative value of 1GW/cm². λ is the wavelength of operation, while λ_x refers to the main absorption peak of the material.*

Material	n_2 (cm²/W)	τ (ps)	α_0 (cm⁻¹)	λ (μm)	W	T	F	λ_x (μm)
SiO₂ glass	$3 \cdot 10^{-16}$	< 0.01	10^{-6}	> 1	> 1000	≪ 1	> 10^6	—
Pb-glass	$4 \cdot 10^{-15}$	—	$5 \cdot 10^{-3}$	> 1	> 7.5	< 0.7	> 10^5	—
CdSSe-glass	≈ 10^{-14}	20	3	0.58	0.3	—	≈ 10^5	0.52
RN-glass	$1.3 \cdot 10^{-14}$	—	0.01	> 1	≈ 13	< 0.1	—	—
As₂S₃-glass	$1.4 \cdot 10^{-13}$	< 1	0.03	1.06	—	0.4	—	0.53
AlGaAs	$1.1 \cdot 10^{-13}$	< 0.5	0.8	1.55	≈ 9	0.22	≈ 1000	0.79
QW-AlGaAs	$1.3 \cdot 10^{-13}$	< 0.5	0.8	1.55	≈ 11	0.18	≈ 1000	0.79
poly4BCMU	$5 \cdot 10^{-14}$	—	1.7	1.32	0.3	< 1	—	0.5
poly4BCMU	$1.5 \cdot 10^{-13}$	—	0.88	1.06	0.6	5.6	—	—
polysilane	$1.3 \cdot 10^{-13}$	—	—	0.53	—	—	—	0.34
PDA	$4.7 \cdot 10^{-13}$	—	6.9	0.53	—	—	—	—
PDA/PTS	≈ $3 \cdot 10^{-11}$	< 80	0.5	1.06	4	—	—	0.67
PDA/PTS	≈ $2 \cdot 10^{-12}$	< 80	0.8	1.06	40	4	—	—
PPV	$1.3 \cdot 10^{-12}$	—	1	0.65	≈ 30	—	—	0.49
MP-PPV	$6 \cdot 10^{-13}$	—	1	1.06	3.7	—	—	0.42
DANS	$0.8 \cdot 10^{-13}$	< 30	≈ 0.2	1.06	> 4	1	—	0.43
DANS	$0.8 \cdot 10^{-13}$	< 50	< 0.2	1.32	> 5	≈ 0.2	—	0.43
DAN2	$2 \cdot 10^{-13}$	< 30	< 1	1.06	> 2	1	—	0.44
DAN	$5 \cdot 10^{-12}$	—	3	0.63	26	—	—	—
Pt-PhC	$1.3 \cdot 10^{-11}$	—	—	1.06	—	—	—	0.65

each individual pulse will give rise to a distorted asymmetric temporal response due to the temperature change during its own evolution.

For pulse widths ≪ thermal relaxation time, it is possible to define a figure of merit F as the ratio of the Kerr nonlinear index to the thermal effective nonlinear index [59]. This represents the number of fast switching operations which can be performed before thermal effects become of comparable entity. Clearly, F must be as large as possible for good AOS, typically > 1000.

Table III is a collection of parameters evaluated for various nonlinear materials experimentally tested. Most of these materials have been employed to realize waveguides for AOS.

Nonlocal effects. Nonlocality is another deviation from the ideal Kerr effect, which is assumed to be perfectly local. In materials where carrier diffusion or similar effects take place, the phase shift will result from a spatial integration, in analogy to the time-integrating action of thermal nonlinearities. A nonlocal nonlinearity tends to average the effect of a Kerr index and reduces, therefore, its effectiveness [31, 60]. In addition it provides a source of spatial feedback which, in some cases, allows bistability [61, 62].

6. – Cascading of second-order susceptibilities.

The key issue in AOS is the ability of a guided wave to accumulate a large enough nonlinear phase shift during propagation. This phase shift allows then one to amplitude modulate or route the signal. While a change in the refractive index is certainly a direct way to achieve such a shift, it relies on a third-order process often characterized by small nonlinear coefficients in those materials with a fast response far from absorptive resonances. A phase shift is also the by-product of nonlinear effects based on second-order interactions, first among them second-harmonic generation (SHG) [63, 64]. Although $\chi^{(2)}$ phenomena require noncentrosymmetric media, these materials benefit from more developed chemistry and technology, mainly because of their applications in electro-optics. The use of anisotropic crystals and/or poled polymers does not represent, therefore, a limitation in comparison to $\chi^{(3)}$ materials.

Using the slowly varying approximation and normalized field amplitudes, the coupled-mode equations describing the c.w. propagation of a fundamental wave and its second harmonics in a lossless medium can be written in the form [64, 65]

(10)
$$\begin{cases} \dfrac{\mathrm{d}}{\mathrm{d}z}\, a_{2\omega}(z) = -i\kappa(-2\omega;\,\omega,\,\omega)\, a_\omega^2(z)\exp\left[i\Delta\beta z\right], \\[2mm] \dfrac{\mathrm{d}}{\mathrm{d}z}\, a_\omega(z) = -i\kappa(-\omega;\,2\omega,\,-\omega)\, a_{2\omega}(z) a_\omega^*(z)\exp\left[-i\Delta\beta z\right] \end{cases}$$

with $\Delta\beta = \beta(2\omega) - 2\beta(\omega)$ the phase mismatch and $\kappa(-2\omega;\,\omega,\,\omega) = \kappa(-\omega;$

of the modal profiles $E_\omega(x, y)$ $E_\omega(x, y)$ $E_{2\omega}(x, y)$ (assumed real) as it stems from (1).

Equations (10) are well known and admit exact solutions [66]. The nonlinear system they represent exhibits stable and unstable nonlinear eigenmodes and, for a fixed ratio between the amplitudes of the interacting waves, the evolution depends on their initial relative phase [67]. We want to focus, however, on the case of an initially zero second-harmonic wave with a nonvanishing phase mismatch. The process described corresponds to an effective $\chi^{(3)}$ nonlinearity, with $\{\chi^{(3)}\}_{\text{eff}} = \chi^{(2)}(-\omega; 2\omega, -\omega)\chi^{(2)}(-2\omega; \omega, \omega)$. For negligible pump depletion, this leads to an effective z- and intensity-dependent wave vector coefficient:

$$(11) \qquad (\beta_2)_{\text{eff}} \approx -\frac{\kappa^2(1 - \cos \Delta\beta z)}{\Delta\beta}$$

in $m^{-1}W^{-1}$ for a channel and W^{-1} for a slab, respectively, with a maximum value $(\beta_2)_{\text{eff}} \approx \pm\kappa^2/\Delta\beta$ at $z = |\pi/\Delta\beta|$. These values can be quite high as compared to intrinsic $\chi^{(3)}$. For example, in a poled polymer with $d^{(2)} \approx 50\,\text{pm/V}$, $n \approx 1.6$ and $L/\lambda = 10^4$, $(n_2)_{\text{eff}} = \sigma(\beta_2)_{\text{eff}}/k_0 \approx 2d_{\text{eff}}^{(2)}L/n^3 c\varepsilon_0\lambda \approx 4.6 \cdot 10^{-11}\,\text{cm}^2/\text{W}$.

When depletion is substantial, this Kerr law description is no longer appropriate. The fundamental field creates a 2ω wave which propagates with a different phase velocity, and the product $a_{2\omega}(z)\,a_\omega^*(z)$ in (10) produces a polarization source in quadrature with the fundamental, with a resulting phase delay or lag depending on the sign of $\Delta\beta$. The fundamental wave periodically retrieves the energy up-converted during the propagation, acquiring an additional phase shift due to this velocity mismatched interation. Figures 9a)-b) show the calculated nonlinear phase of the fundamental beam for various values of $\Delta\beta L$ and a given product κL. ϕ^{NL} accumulates in a stepwise fashion both vs. z and vs. the input power (intensity) at ω. The maximum step size is $\pi/2$ for $|\Delta\beta L| \ll 1$, and $\phi^{\text{NL}}(I, z)$ becomes progressively smoother and with smaller steps as $|\Delta\beta L|$ is made larger. The evolution of the fundamental amplitude is shown in fig. 10a)-b), where cyclic depletion is apparent.

A number of features characterize this phase shift due to up- and successive down-conversion of a wave:

a) The stepwise growth of ϕ^{NL} with z and with input excitation, with plateaux in extended ranges of powers.

b) The oscillation in throughput for the ω wave, although unity transmission can always be restored. If not restored, the output at 2ω corresponds to a two-photon loss for the fundamental.

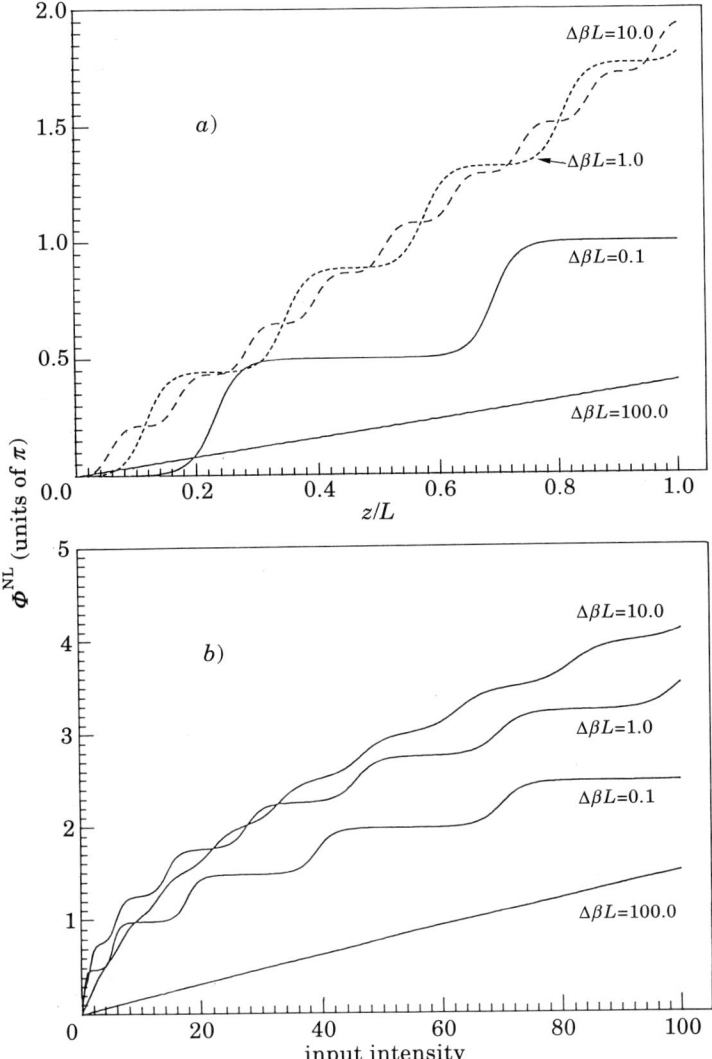

Fig. 9. – Cascaded second-order effects in second-harmonic generation: nonlinear phase of a wave at ω for various degrees of phase mismatch $\Delta\beta L$ in a structure with $\kappa L = 4$: a) vs. propagation distance for $|a_\omega(0)|^2 = 25$, b) vs. input excitation.

c) The sign of $\dot{\phi}^{\mathrm{NL}}$ depends on the sign of $\Delta\beta$. Self-focussing or defocussing nonlinearities can be emulated acting upon the crystal (or field polarization) orientation.

d) Due to the coherent nature of the interaction, both phase and amplitude of the fundamental wave can be modulated by a weak signal (a seed) injected in $z = 0$ at frequency 2ω.

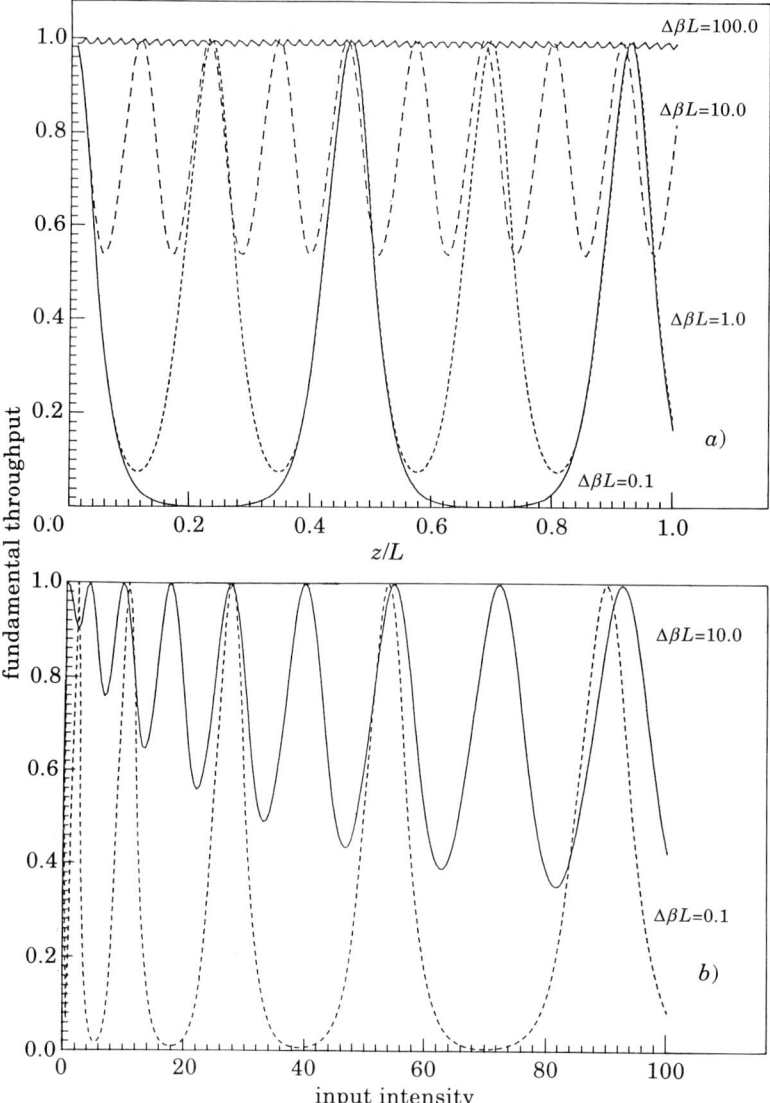

Fig. 10. – Cascaded second-order effects in second-harmonic generation: throughput of the fundamental *a*) *vs.* propagation with $|a_\omega(0)|^2 = 25$ and *b*) *vs.* input power. In these calculations, $\kappa L = 4$.

A straightforward configuration for utilizing this $\chi^{(2)}$ phase shift for AOS is an integrated Mach-Zehnder. Since the maximum phase step is $\pi/2$, we will describe the operation of a NLMZ with two identical arms (same effective area, nonlinearity and length) with opposite phase mismatches. This choice gives us the potential of achieving a π differential phase shift within the first plateau

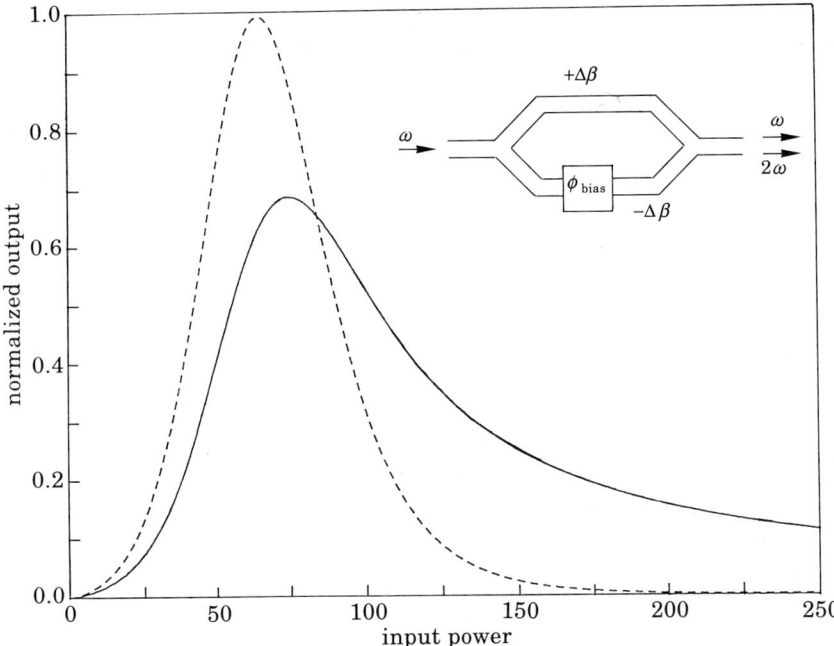

Fig. 11. – Cascaded Mach-Zehnder (insert): calculated response of a device with a phase offset of π, $\kappa L = 1$ and $\Delta\beta L = \pm 0.1$ for c.w. (dashed line) and pulsed (solid line) excitation. In the latter case, the horizontal axis represents peak powers of Gaussian pulses.

(fig. 9), the amount required to completely switch an interferometer. Opposite $\Delta\beta$'s can be obtained by quasi-phase-matching, *i.e.* by introducing gratings of the appropriate periodicities Λ_1 and Λ_2 in the two channels. In this case the two phase mismatches, $\Delta\beta_1 = \beta(2\omega) - 2\beta(\omega) \pm 2\pi/\Lambda_1$ and $\Delta\beta_2 = \beta(2\omega) - 2\beta(\omega) \pm 2\pi/\Lambda_2$, can be made such that $\Delta\beta_1 = -\Delta\beta_2$.

If a phase offset of π is introduced in one channel, the device will be in the off (nontransmitting) state at low powers, and will reach 100% transmission only at high powers, *i.e.* when $\phi^{NL} = \pm\pi/2$ per arm operating on the first phase plateau (see fig. 9*b*) and fig. 12 insert). The resulting c.w. response is shown in fig. 11 (dashed line), together with the pulse response (solid line) obtained by integrating over a Gaussian pulse envelope with peak powers as in the horizontal axis. Notice how, due to the instantaneous (c.w.) transfer characteristic typical of the NLMZ, the maximum throughput is $\approx 30\%$ lower than in the c.w. case. However, in contrast to the Kerr case, no pulse break-up occurs for the chosen set of parameters without resorting to square pulses or solitons. Figure 12 shows the NLMZ time response to a Gaussian envelope with peak power of 65 W (first plateau in the insert), in order to emphasize the temporal narrowing with unity peak transmission.

A similar study can be conducted for other AOS geometries, introducing the

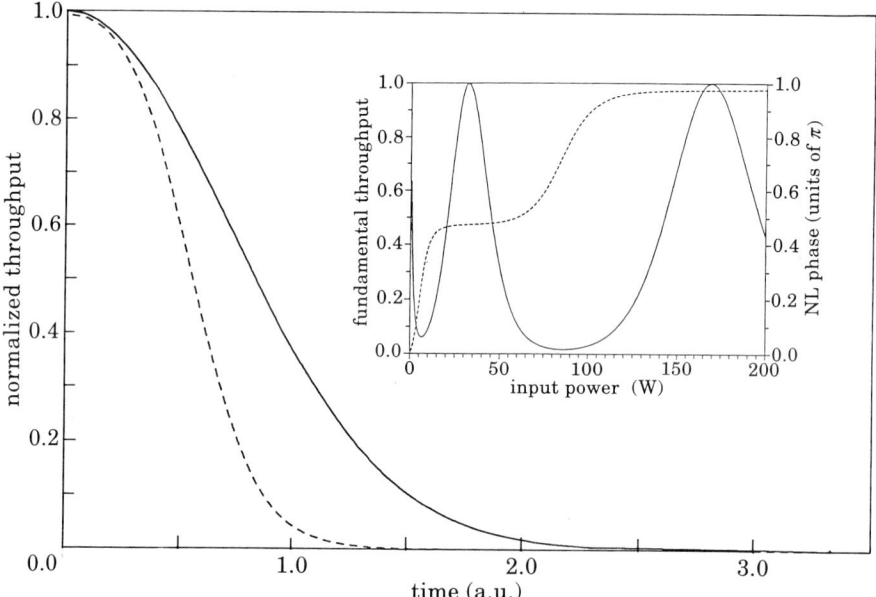

Fig. 12. – Cascaded Mach-Zehnder: normalized time response (dashed line) of the same NLMZ (fig. 11) to a Gaussian pulse (solid line) with a peak power of 65 W. This value corresponds to the first plateau in the phase *vs.* input diagram, shown in the insert (dotted line).

SHG equations in a coupled set describing the device [65]. Figures 13*a*)-*b*) show the main features of a $\chi^{(2)}$ half-beat-length NLDC with channels decoupled at 2ω. This NLDC exhibits good switching characteristics with oscillations due to the periodic up-conversion of the ω beam. The output from the cross-state decreases to zero with increasing input excitation, and the power reverts back to the bar channel. The oscillations are reduced but not eliminated in the pulsed case, due to the averaging. Such a geometry could find applications in the generation of trains of short pulses from a quasi-c.w. input.

Finally, we like to briefly mention the possibilities offered by a 2ω seed at the input of a channel waveguide. In this case, a strong modulation of a fundamental beam can be obtained even in the case of perfect phase matching, simply acting on the phase or the amplitude of the coherent weak 2ω field [68]. Figures 14*a*)-*b*) show throughput and phase of the ω beam at the output of the waveguide *vs.* the phase of the injected seed 1000 times weaker and for $\Delta\beta = 0$. It is apparent the degree of modulation achievable under conditions for which, without field at 2ω, the fundamental would be completely up-converted.

AOS via a $\chi^{(2)}$-$\chi^{(2)}$ process is a relatively novel approach and appears very promising even though some important issues such as spectral limitations, group velocity dispersion and walk-off need to be carefully addressed when employing pulses. Since the refractive index is not actually changed by the interaction,

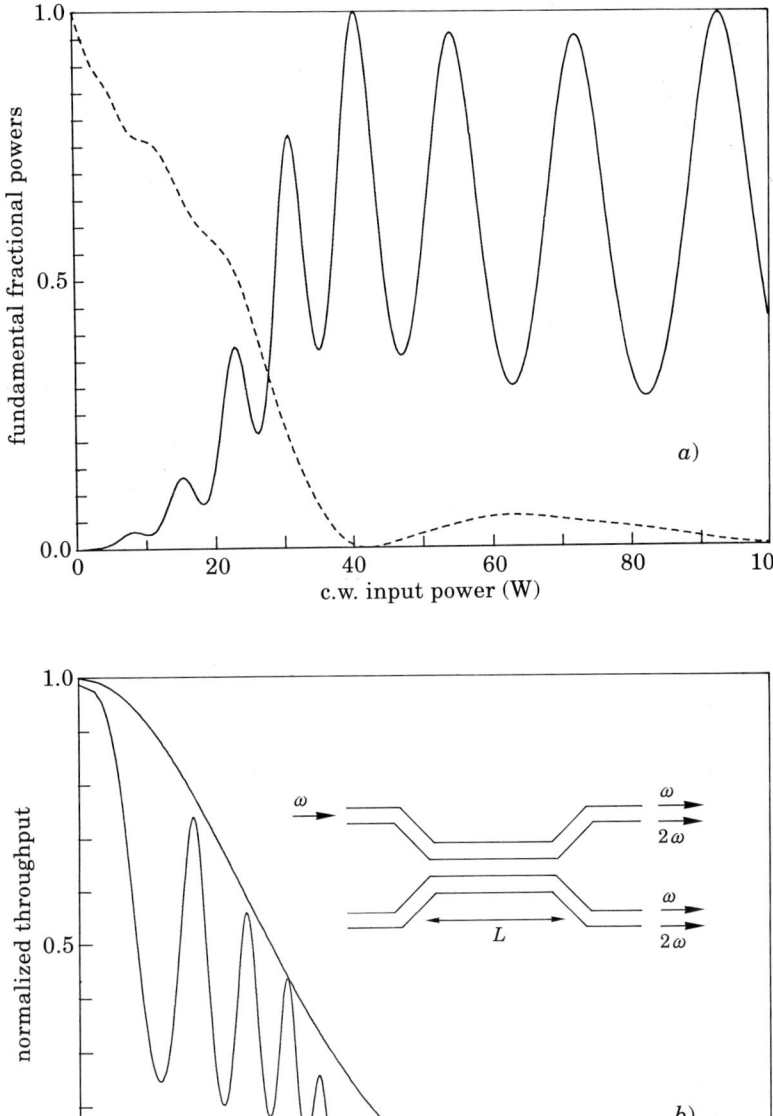

Fig. 13. – Cascaded NLDC (insert): a) cross (dashed line) and bar (solid line) fractional output *vs.* input power, b) temporal profile of input and output pulses in the bar (solid line) and in the cross (dashed line) channels for a peak power of 93 W. The smooth solid line represents the Gaussian input pulse. Here $\kappa L = 4$ and $\Delta\beta L = 8$ in both guides.

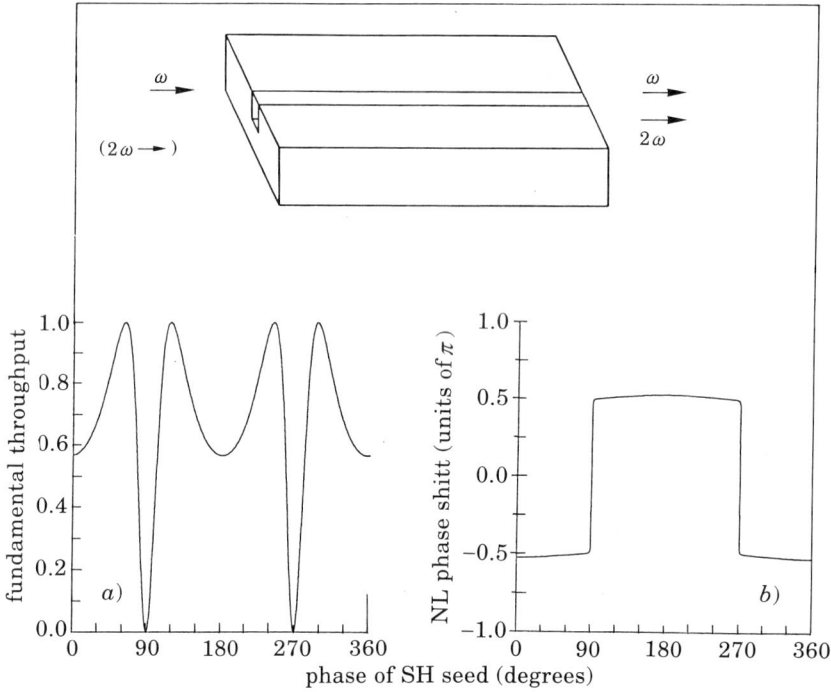

Fig. 14. – Cascaded second-order effects in a seeded channel (insert) in phase matching: *a)* throughput and *b)* phase of the fundamental beam *vs.* the phase of a 2ω seed 1000 times weaker. Here $\kappa L = 2$ and $|a_\omega(0)|^2 = 10$.

most of the problems associated with $\chi^{(3)}$ and mentioned in sect. 5 do not apply here, with the exception of linear and nonlinear absorption at the frequencies involved. This issue has an important bearing on phase matching, whenever the detuning is affected by a temperature change.

Large $\chi^{(2)}$-$\chi^{(2)}$ nonlinear phase shifts have been measured in KTP crystals, both in bulk by z-scan[69] and in a segmented waveguide configuration [70], and in BBO crystals via self-diffraction [71]. In the waveguide case ϕ^{NL} exceeding π were measured by self-phase modulation spectral broadening when using picosecond pulses from a mode-locked Ti:sapphyre laser [70]. Nondegenerate interactions have also been proposed for AOS[72] in a dual-wavelength NLMZ.

Conclusions.

All-optical switching in integrated-optics configurations appears promising and in rapid expansion, a field which has certainly left its infancy. While the theoretical understanding is more advanced than its experimental counterpart, now that picosecond and femtosecond lasers have become widely available in the near infrared at telecommunication wavelengths, the characterization of nonlinear materials is certainly progressing at a fast pace and new material candidates for AOS in waveguides keep entering this arena. The developed figures of merit and switching criteria make a screening possible in terms of device applications, and the novel concept of cascaded second-order effects is going to have a large impact by re-introducing noncentrosymmetric crystals and poled polymers in the game. In this scenario and in conjunction with progress in material science, dedicated AOS configurations operating with excitation levels typical of semiconductor lasers are likely to be demonstrated in the near future.

REFERENCES

[1] G. I. STEGEMAN and C. T. SEATON: *J. Appl. Phys.*, **58**, R57 (1985).
[2] G. ASSANTO: *J. Mod. Opt.*, **37**, 855 (1990).
[3] G. I. STEGEMAN, E. M. WRIGHT, N. FINLAYSON, R. ZANONI and C. T. SEATON: *J. Lightwave Technol.*, **6**, 953 (1988).
[4] D. B. OSTROWSKY and R. REINISCH, Editors: *Guided Wave Nonlinear Optics*, NATO ASI Series E, Vol. 214 (Kluwer Acad. Publ., Dordrecht, 1992).
[5] G. ASSANTO, M. B. MARQUES and G. I. STEGEMAN: *J. Opt. Soc. Am. B*, **8**, 553 (1991).
[6] G. ASSANTO, R. M. FORTENBERRY, C. T. SEATON and G. I. STEGEMAN: *J. Opt. Soc. Am. B*, **5**, 432 (1988).
[7] Y. SILBERBERG and G. I. STEGEMAN: *Appl. Phys. Lett.*, **50**, 801 (1987).
[8] R. NORMANDIN and G. I. STEGEMAN: *Opt. Lett.*, **4**, 58 (1979).
[9] S. M. JENSEN: *IEEE J. Quantum Electron.*, QE-18, 1580 (1982).
[10] A. MAIER: *Sov. J. Quantum Electron.*, **12**, 1490 (1982).
[11] G. P. AGRAWAL: *Nonlinear Fiber Optics* (Academic Press, New York, N.Y., 1989).
[12] H. KOGELNIK: in *Integrated Optics*, edited by T. TAMIR (Springer-Verlag, New York, N.Y., 1979), p. 13.
[13] See, for example, D. MIHALACHE, M. BERTOLOTTI and C. SIBILIA: in E. WOLF: *Progress in Optics XXVII*, Chapt. 4 (Elsevier Science Publ. B.V., Amsterdam, 1989); A. D. BOARDMAN, K. BOOTH and P. EGAN: in ref. [4] above, p. 201.
[14] Y. S. KIVSHAR: *IEEE J. Quantum Electron.*, **29**, 250 (1993).
[15] G. I. STEGEMAN: *IEEE J. Quantum Electron.*, QE-18, 1610 (1982).
[16] R. H. REDIKER and F. J. LEONBERGER: *IEEE J. Quantum Electron.*, QE-18, 1813 (1982).
[17] A. LATTES, H. A. HAUS, F. J. LEONBERGER and E. P. IPPEN: *IEEE J. Quantum Electron.*, **19**, 1718 (1983).
[18] K. AL-HEMYARI, J. S. AITCHISON, C. N. IRONSIDE, G. T. KENNEDY, R. S. GRANT and W. SIBBETT: *Electron. Lett.*, **28**, 1090 (1992).
[19] N. J. DORAN and D. WOOD: *J. Opt. Soc. Am. B*, **11**, 1843 (1987).
[20] S. TRILLO, S. WABNITZ, E. M. WRIGHT and G. I. STEGEMAN: *Opt. Lett.*, **13**, 672 (1988).

[21] A. B. ACEVES, J. V. MOLONEY and A. C. NEWELL: *Phys. Rev. A*, **39**, 1809 (1989).

[22] P. LI KAM WA: *Opt. Quantum Electron.*, **23**, S925 (1991).

[23] J. Y. CHEN, S. I. NAJAFI and S. HONKANEN: *Opt. Commun.*, **98**, 201 (1993).

[24] G. M. CARTER and Y. J. CHEN: *Appl. Phys. Lett.*, **42**, 643 (1983).

[25] G. ASSANTO: in *Nonlinear Optics and Optical Computing* (Plenum Press, New York, N.Y., 1990), p. 229.

[26] J. E. EHRLICH, G. ASSANTO, G. I. STEGEMAN and T. H. CHIU: *IEEE J. Quantum Electron.*, **27**, 809 (1991).

[27] M. B. MARQUES, G. ASSANTO, G. I. STEGEMAN, G. R. MOHLMANN, E. W. P. ERD-HUISEN and W. H. G. HORSTUIS: *Appl. Phys. Lett.*, **58**, 2613 (1991).

[28] U. BARTUCH, A. BRAUER, P. DANNBERG, H. H. HORHOLD and D. RAABE: *Int. J. Optoelectron.*, **7**, 275 (1992).

[29] M. BERTOLOTTI, F. MICHELOTTI, C. SIBILIA, E. FAZIO, L. SCHIRONE, A. FERRARI, F. PALMA and G. C. RIGHINI: *Appl. Opt.*, **31**, 737 (1992).

[30] C. CACCIATORE, D. CAMPI, C. CORIASSO, G. MENEGHINI and C. RIGO: *Electron. Lett.*, **28**, 1624 (1992).

[31] G. ASSANTO: in ref. [4] above, p. 257, and references therein.

[32] F. J. FRAILE-PELAEZ, G. ASSANTO and D. HEATLEY: *Opt. Commun.*, **77**, 402 (1990).

[33] A. ANKIEWICZ and G.-D. PENG: *Int. J. Optoelectron.*, **6**, 15 (1991).

[34] J. WILSON, G. I. STEGEMAN and E. W. WRIGHT: *Opt. Lett.*, **16**, 1653 (1991).

[35] A. W. SNYDER, D. J. MITCHELL, L. POLADIAN, D. R. ROWLAND and Y. CHEN: *J. Opt. Soc. Am. B*, **8**, 2102 (1991).

[36] Y. CHEN, A. W. SNYDER and D. N. PAYNE: *IEEE J. Quantum Electron.*, **28**, 239 (1992).

[37] C. C. YANG, A. VILLENEUVE, G. I. STEGEMAN and J. S. AITCHISON: *Opt. Lett.*, **17**, 710 (1992).

[38] C. C. YANG: *Opt. Lett.*, **16**, 1641 (1991).

[39] A. T. PHAM and L. N. BINH: *J. Opt. Soc. Am. B*, **8**, 1914 (1991).

[40] J. ATAI and Y. CHEN: *J. Appl. Phys.*, **72**, 24 (1992).

[41] A. W. SNYDER, L. POLADIAN and D. J. MITCHELL: *Opt. Lett.*, **17**, 118 (1992).

[42] C. SCHMIDT-HATTENBERGER, U. TRUTSCHEL and F. LEDERER: *Opt. Lett.*, **16**, 294 (1991).

[43] W. D. DEERING, M. I. MOLINA and G. P. TSIRONIS: *Appl. Phys. Lett.*, **62**, 2471 (1993).

[44] J. S. AITCHISON, A. H. KEAN, C. N. IRONSIDE, A. VILLENEUVE and G. I. STEGEMAN: *Electron. Lett.*, **27**, 1709 (1991); A. VILLENEUVE, C. C. YANG, P. G. J. WIGLEY, G. I. STEGEMAN, J. S. AITCHISON and C. N. IRONSIDE: *Appl. Phys. Lett.*, **61**, 147 (1992).

[45] A. VILLENEUVE, K. AL-HEMYARI, J. U. KANG, C. N. IRONSIDE, J. S. AITCHISON and G. I. STEGEMAN: *Electron. Lett.*, **29**, 721 (1993).

[46] S. TRILLO, S. WABNITZ and G. I. STEGEMAN: *J. Lightwave Technol.*, **6**, 971 (1988).

[47] H. WINFUL, J. H. MARBURGER and E. GARMIRE: *Appl. Phys. Lett.*, **35**, 379 (1979).

[48] A. MECOZZI, S. TRILLO and S. WABNITZ: *Opt. Lett.*, **12**, 1008 (1987).

[49] G. VITRANT: in ref. [4] above, p. 285.

[50] C. MARTIJN DE STERKE and J. E. SIPE: *J. Opt. Soc. Am. B*, **6**, 1722 (1989); J. SIPE: in ref. [4] above, p. 305.

[51] N. D. Sankey, D. F. Prelewitz and T. G. Brown: *J. Appl. Phys.*, **73**, 1 (1993).

[52] K. Sasaki, K. Fujii and T. Tomioka: *J. Opt. Soc. Am. B*, **5**, 457 (1988).

[53] S. Aramaki, G. Assanto and G. I. Stegeman: *Opt. Commun.*, **94**, 326 (1993).

[54] Y. Silberberg and B. G. Sfez: *Opt. Lett.*, **13**, 1132 (1988).

[55] J. P. Sabini, N. Finlayson and G. I. Stegeman: *Appl. Phys. Lett.*, **55**, 1176 (1989).

[56] H. Fouckhardt and Y. Silberberg: *J. Opt. Soc. Am. B*, **7**, 803 (1990).

[57] T.-T. Shi and S. Chi: *Opt. Lett.*, **16**, 1077 (1991).

[58] J. S. Aitchison, A. Villeneuve and G. I. Stegeman: *Opt. Lett.*, **18**, 1 (1993).

[59] S. R. Friberg and P. W. Smith: *IEEE J. Quantum Electron.*, QE-23, 2089 (1987).

[60] D. R. Heatley, E. M. Wright, J. Ehrlich and G. I. Stegeman: *Opt. Lett.*, **13**, 419 (1988).

[61] G. Vitrant, R. Reinisch, J. Cl. Paumier, G. Assanto and G. I. Stegeman: *Opt. Lett.*, **14**, 898 (1989).

[62] G. Assanto and G. I. Stegeman: *Appl. Phys. Lett.*, **56**, 2285 (1990).

[63] N. R. Belashenkov, S. V. Gagarskii and M. V. Inochkin: *Opt. Spectrosc. (USSR)*, **66**, 1383 (1989).

[64] G. I. Stegeman, M. Sheik-Bahae, E. Van Stryland and G. Assanto: *Opt. Lett.*, **18**, 13 (1993).

[65] G. Assanto, G. I. Stegeman, M. Sheik-Bahae and E. Van Stryland: *Appl. Phys. Lett.*, **62**, 1323 (1993).

[66] J. A. Armstrong, N. Bloembergen, J. Ducuing and P. S. Pershan: *Phys. Rev.*, **6**, 1918 (1962).

[67] S. Trillo, S. Wabnitz, R. Chisari and G. Cappellini: *Opt. Lett.*, **17**, 637 (1992).

[68] G. Assanto, G. I. Stegeman, M. Sheik-Bahae and E. W. Van Stryland: in *Nonlinear Optics: Materials, Fundamentals, and Applications*, OSA 1992 Tech. Dig. Series, Vol. **18**, pap. PD11, 1 (1992).

[69] R. DeSalvo, D. J. Hagan, M. Sheik-Bahae, G. Stegeman and E. W. Van Stryland: *Opt. Lett.*, **17**, 28 (1992).

[70] M. Sundheimer, Ch. Bosshard, E. W. Van Stryland, G. I. Stegeman and J. D. Bierlein: *Opt. Lett.*, **18**, 1379 (1993).

[71] R. Danielius, P. Di Trapani, A. Dubietis, A. Piskarskas, D. Podenas and G. P. Banfi: *Opt. Lett.*, **18**, 574 (1993).

[72] D. C. Hutchings, J. S. Aitchison and C. N. Ironside: *Opt. Lett.*, **18**, 793 (1993).

INDICE ANALITICO

PROCEEDINGS OF THE INTERNATIONAL SCHOOL
OF PHYSICS «ENRICO FERMI»

Nuova Tipografia Compositori, Bologna